Fibre Reinforced Cementitious Composites

T0187832

Modern Concrete Technology Series
A series of books presenting the state-of-the-art in concrete technology
Series Editors

Arnon Bentur
Faculty of Civil and Environmental Engineering
Technion-Israel Institute of Technology, Israel

Sidney Mindess
Department of Civil Engineering
University of British Columbia, Canada

Fibre Reinforced Cementitious Composites
A. Bentur and S. Mindess

Concrete in the Marine Environment
P.K. Mehta

Concrete in Hot Environments
I. Soroka

Durability of Concrete in Cold Climates
M. Pigeon and R. Pleau

High Performance Concrete
P.C. Aïtcin

Steel Corrosion in Concrete
A. Bentur, S. Diamond and N. Berke

Optimization Methods for Material Design of Cement-based Composites
Edited by A. Brandt

Special Inorganic Cements
I. Odler

Concrete Mixture Proportioning
F. de Larrard

Sulfate Attack on Concrete
J. Skalny, J. Marchand and I. Odler

Fibre Reinforced Cementitious Composites

Second edition

Arnon Bentur and
Sidney Mindess

CRC Press
Taylor & Francis Group
Boca Raton London New York

CRC Press is an imprint of the
Taylor & Francis Group, an **informa** business
A TAYLOR & FRANCIS BOOK

Contents

Preface

Plain concrete is a brittle material, with low tensile strength and strain capacities. To help overcome these problems, there has been a steady increase over the past 40 years in the use of fibre reinforced cements and concretes (FRC). The fibres are not added to improve the strength, though modest increases in strength may occur. Rather, their main role is to control the cracking of FRC, and to alter the behaviour of the material once the matrix has cracked, by bridging across these cracks and so providing some post-cracking ductility.

It has been 16 years since the publication of the first edition, and an enormous amount of progress has taken place in that time. Much work has been done in optimizing the properties of the composite material: new fibre types and geometries have been developed, and surface treatments have been employed to make the fibres more compatible with the cementitious matrix. As a result, such composites can be better 'tailored' for specific applications for which conventional cementitious systems are not suitable. Also, a new generation of FRC materials has been developed, the so-called 'high performance' FRCs, which exhibit multiple cracking and strain hardening beyond the point of first cracking, with a concomitant increase in energy absorption capacity. Finally, there is now an increasing use of FRC in truly structural applications, as described in the final chapter.

The object of this book remains the same: to develop, in some detail, the fundamental scientific principles which govern the performance of FRC and to describe the production processes and properties of specific systems prepared with different types of fibres, such as steel, glass, polypropylene, natural fibres and various types of high performance polymeric fibres. To achieve these aims, the traditional materials science approach is used:

1   characterization of the microstructure;
2   relationships between the microstructure and engineering properties;
3   relationships between microstructural development and the processing techniques; and
4   selection of materials and processing methods to achieve FRC composites with the desired characteristics.

The book is divided into two parts:

1   The first part deals with the theoretical background underlying the behaviour of FRC, including an intensive treatment of the mechanics of fibre reinforced brittle matrices and its implications for cementitious systems, taking into account the special bulk and interfacial microstructure of FRC.
2   The second part describes the principal types of fibre – cement composites, from the point of view of production processes, physical and mechanical properties, durability and applications. These characteristics are discussed in terms of the basic principles governing the behaviour of FRC.

This book is designed not only for scientists and graduate students, but also for practicing engineers. It includes an extensive and up-to-date reference list, and numerous graphs and tables describing the engineering properties of the different FRC systems. The intent is to provide the reader with information important for engineering applications, as well as a proper background for assessing future developments.

Arnon Bentur
Sidney Mindess

Acknowledgements

Many individuals and organizations have generously provided photographs and have given us permission to adapt material from their publications for this book, and we thank them for this courtesy. They are as follows: Aedificatio Publishers, Germany; Akademick Forlag, Denmark; AMEC, Canada (Dr D.R. Morgan); American Ceramic Society; American Concrete Institute; American Shotcrete Association; American Society for Testing and Materials; American Society of Civil Engineers; Austrian Chemical Institute; Bekaert Corporation; Blackwell Publishing; Building Research Establishment, UK; Building Research Station, UK; Cambridge University Press, UK; Clarendon Press; The Concrete Society, UK; The Construction Press, UK; EFNARC, UK; Elsevier Science Publishers Ltd., UK; Fibermesh Co., USA; Grace Construction Products, USA; Institut für Bauforschung der RWTH, Aaachen, Germany; International Association for Housing Science; International Ferrocement Information Centre, Thailand; International Society for Stereology; IOP Publishing, UK; Japan Concrete Institute; Japan Society of Civil Engineers; B.H. Levelton and Associates, Vancouver, Canada (Mr P.T. Seabrook); Materials Research Society, USA; National Building Research Institute, Technion, Israel; National Physical Laboratory, UK; National Research Council of Canada; Oxford University Press, UK; Palladian Publications Ltd., UK; Pilkington Brothers PLC, UK; Plastic and Rubber Institute, UK; Plenum Press, USA; Prestressed Concrete Institute, USA; Mr Y. Rechter, Israel; RILEM, France; The Royal Institute of Chemistry, UK; The Royal Society, UK; SIPI srl, Italy; Springer, The Netherlands; Starrylink-Editrice, Italy; Swedish Cement and Concrete Institute; Taylor and Francis Group, UK; Thomas Telford Publications, UK; University of British Columbia, Canada; van Nostrand Reinhold; John Wiley and Sons, Inc.; J. E. Williden Publ., UK; Dr R.F. Zollo.

Notations

a	acceleration due to energy
Δa	crack extension
A_f	cross-sectional area of fibre
A_m	cross-sectional area of matrix
A_{mp}	amplitude of crimped fibre
a_θ	proportion of fibres oriented at angle θ
ACK	Aveston Cooper Kelly model
AR	alkali resistant (glass fibres)
B	brittleness ratio
b	beam width
	or
b	length of debonded zone
c	crack (flaw) length (or 1/2 of the length)
CH	calcium hydroxide ($Ca(OH)_2$)
CMOD	crack mouth opening displacement
CSH	calcium silicate hydrate
d	fibre diameter
d_F	equivalent fibre diameter
DFRCC	ductile fibre reinforced cement composite
DSP	densified with small particles
ECC	engineered cementitious composite
E_b	modulus of elasticity of the composite in bending
E_c	modulus of elasticity of the composite in tension
E_f	modulus of elasticity of the fibre in tension
E_m	modulus of elasticity of the matrix in tension
E_{mo}	hypotehtical modulus of elasticity of void-free matrix
E_t	modulus of elasticity in tension
dE_s	change of surface energy due to the creation of a new crack surface
E_{1-2}	the energy consumed in opening of the first crack in the multiple cracking zone
f	snubbing friction factor
f_c	uniaxial compressive strength of concrete

f'_t	splitting tensile strength of concrete
$f_{ct,1}$	limit of proportionality of reinforced concrete beam
$f_{R,i}$	residual flexural strength of a beam
$f_{Rm,1}$	average residual flexural strength of SFRC beam at the moment a crack is expected to occur
FIER	fibre intrinsic efficiency factor
F_{brdg}	bridging load of fibre across a crack
FCM	fictitious crack model
FRC	fibre reinforced cement or concrete
G_c	critical strain energy release rate
G_{db}	fracture energy of the fibre–matrix interface
G_F	energy absorbed during the fracture process due to additional deformation of the damage zone
G_m	fracture energy of the matrix
	or
G_m	shear modulus of the matrix
G_{tip}	crack tip critical energy release rate of the composite
G_{II}	second mode fracture energy, that is, in a shearing fracture process of the fibre–matrix interface
GFRC	glass fibre reinforced concrete (USA)
GRC	glass fibre reinforced cement (Europe)
HPFRCC	high performance fibre reinforced composite
h	beam height
h_f	value representing the surface roughness of the reinforcement
h_{sp}	distance between tip of notch and top of specimen
I	pressure gradient of water flowing through a crack
I_f	moment of inertia of the fibre
$I_{5....30}$	toughness indices
ITZ	Interfacial Transition Zone
J_b	contribution of fibre bridging to fracture energy
J_m	contribution of microcracking damage to fracture energy
J_t	total fracture energy
J_{tip}	composite crack tip toughness
k	matrix foundation stiffness
k_m	curvature of non-linear hinge in a beam in bending
k_1	elastic curvature of the uncracked part of a hinge in a beam in bending
k_2	curvature of the cracked part of the hinge in a beam in bending
K_c	critical stress intensity factor
K_f	plain strain bulk modulus of the fibre
K_{Ic}^m	effective fracture toughness of the FRC matrix
K_{Ic}^i	fracture toughness of the fibre–matrix interface
$K_{Ic,p}$	fracture toughness of the paste

K_m	plain strain bulk modulus of the matrix
ℓ	length of fibre
	or
ℓ	fibre embedded length
ℓ_c	critical length = fibre length at which the fibre first breaks instead of pulling out
ℓ'_c	length related to critical length
$\ell(min)$	minimum value of embedded fibre length at which debonding initiates
ℓ_m	fibre length at the transition between catastrophic debonding and progressive debonding
ℓ_r	charcteristic fibre length beyond which fibre fracture starts to take place
ℓ_{cs}	characteristic length
L	beam span
L_c	shortest length at which fibre fracture is observed
ℓ/d	fibre aspect ratio
LEFM	linear elastic fracture mechanics
LMC	latex modified cement
LOP	limit of proportionality
m	mass
M_{cc}	moment of resistance at the first crack in bending
M_{pc}	bending moment resistance in the post cracking zone
M_u	ultimate bending moment of a beam
p	void content (%)
	or
p	perimeter of multiple filament strand
P	pull-out load
P'	load carried by fibre
$pb(t)$	total load recorded in the striking tup during an impact event
p_{crit}	pull-out load at which debonding starts
P_e	force required to start breaking the elastic fibre–matrix bond
P_{eq}	equivalent post-peak load in a beam
P_f	load generated by frictional resistance to slip
P_{max}	maximum load reached after the first crack in a beam
$P(max)$	maximum pull-out load
$P(\theta)$	peak pull-out load for a fibre at orientation θ
$Pi(t)$	inertial load in a beam during an impact test
P_{res}	residual load corresponding to deflection δ_d in a beam
P_u	breaking load of fibre
P_1	first crack load of SFRC specimen
PIC	polymer impregnated concrete
PMMA	polymethylmethacrylate
PWc_{crit}	critical % (by weight) of fibres to just make FRC unworkable

q	shear flow (shear force per unit length)
q_0	rate of water flow through idealized crack
R	distance from the fibre = effective radius of the matrix around the fibre
r	fibre radius
r_e	external radius of concrete ring
r_i	internal radius of concrete ring
RPC	reactive powder concrete
$R_{5...30}$	residual strength in bending test
SEM	scanning electron microscope
SIFCON	slurry infiltrated fibre concrete
SIMCON	slurry infiltrated mat concrete
SFRC	steel fibre reinforced concrete
SG_c	specific gravity of concrete matrix
SG_f	specific gravity of fibres
SIC test	strand in cement test
T	temperature
	or
T	compressive toughness
T_c	compressive toughness factor
u	crack opening
U	total energy in an elastic system subjected to external loads
	or
U	toughness of the composite due to straining of the fibres
U_e	strain energy stored in a system
U_f	energy contribution to toughness
ΔU_{db}	the debonding energy due to the slip between the fibres and the matrix
ΔU_f	increase in elastic strain energy of the fibres after matrix cracking
ΔU_{f-mu}	increase in fibre strain energy as the result of bridging
ΔU_{f-mc}	increase in fibre strain energy as the result of fibre bridging in the multiple cracking stage
ΔU_m	increase in elastic strain energy of the fibres after matrix cracking
	or
ΔU_m	reduction in elastic strain energy of the matrix after cracking
U_{mc}	energy of multiple cracking
U_s	surface energy absorbed in the creation of new crack surfaces
	or
U_s	work done by frictional slip after debonding
V_{cd}	shear resistance of a member without shear reinforcement
V_{fd}	contribution to shear of the steel fibres
V_f	fibre volume content
$V_{f(crit)}$	critical fibre volume content
$(V_f)_{ef-citr}$	effective critical fibre volume content

V_m	matrix volume content
V_{wd}	contribution to shear by the shear reinforcement
w	crack width
W	weight % of fibres
W_a	weight of aggregate fraction
W_L	work due to applied load
W_m	weight of mortar fraction
$\overline{W}_p$	average work done in pull-out per fibre
w/c	water:cement ratio
ΔW	additional work done by applied stress
dW	additional work on the system generated by displacement δ_{ss}
WF	specific work of fracture
WF_{db}	energy of fibre–matrix debonding
WF_t	total specific work of fracture
WF_{PO}	energy involved in fibre pull-out
x	crack spacing
y	distance from the neutral axis to the bottom of the beam
γ_{db}	work done by debonding the fibre from the matrix
γ_m	surface energy of the matrix
γ_s	surface energy of a material
δ	first crack deflection
	or
δ	'misfit' between fibre radius and radius of the hole in the 'free' matrix, due to volume changes, external stresses and Poisson effects
	or
δ	deflection of a fibre projecting over a cracked surface
δ_d	ultimate design deflection for SFRC beam in bending
δ_0	misfit value due only to matrix shrinkage and external stresses
	or
δ_0	crack opening/displacement at which debonding is complete along the full length of the embedded fibre segment
δ_{ss}	displacement
δ_{tc}	deformation in compression corresponding to a specified strain
δ_u	deflection in a bean corresponding to maximum load P_{max} in bending
Δ	relative slip of the fibre after pull-out debonding
Δ_1	deflection corresponding to first crack load P_1 in a beam in bending
Δ_{crit}	slip at which debonding of a fibre starts
Δ_0	relative slip of the fibre at the end of pull-out debonding
ε_c	tensile strain in the composite
ε_{cu}	ultimate strain of the composite
ε_f	tensile strain in the fibre
ε_{fs}	free shrinkage of FRC
ε_{fu}	ultimate (failure) fibre strain

ε_m	tensile strain in the matrix
ε_{mc}	strain in the fibre at the end of the multiple cracking process
ε_{mu}	strain in fibre at the start of multiple cracking of the matrix (or ultimate matrix strain)
ε_{no}	normal strain associated with the misfit, δ
ε_{os}	free shrinkage of plain concrete
$\varepsilon_{\rho,o}$	strain due to prestressing of the reinforcement
ε_v	Poisson strain of the fibre
$\dot{\varepsilon}$	strain rate
γ_m	matrix fracture energy, $G_m/2$
η	combined efficiency factor for orientation and length
η_ℓ	length efficiency factor of fibres
η_θ	orientation efficiency factor of fibres
η_τ	strength efficiency for bond
φ	reinforcing bar diameter
μ	coefficient of friction between fibres and matrix
ν	kinematic viscosity
ν_f	Poisson ratio of fibre
ν_m	Poisson ratio of matrix
θ	angle of fibre orientation
ρ	reinforcement ratio within the tension area of the cross-section
σ_b	flexural strength of the composite
σ_{bm}	flexural strength of the matrix
σ_{cc}	first cracking strength of the composite
σ_{cu}	tensile strength of the composite (in the post cracking zone)
σ_f	tensile stress in the fibre
$\overline{\sigma}_f$	average stress in the fibre
σ_f'	stress in the fibre at the first crack strain
$\sigma_{f(max)}$	maximum tensile stress in the fibre
σ_{fu}	tensile strength of the fibre
σ_{mu}	stress in the matrix at first crack
σ_{mu}'	tensile strength of the matrix in the absence of fibres
σ_n	fibre–matrix normal stress (contact pressure)
σ_{no}	radial compressive stress generated by matrix shrinkage
σ_{pc}	tensile post–cracking strength
σ_t	tensile stress in FRC
σ_r	radial stress in a restrained ring test
σ_θ	circumferential stress in a restrained ring test
$\overline{\sigma}_R$	average radial stress in the concrete in the vicinity of the reinforcement
$\overline{\sigma}_{pc}$	tensile stress of a rectangular block stress distribution in a reinforced beam in bending
ω	deformation of the fracture zone
τ	shear stress at the interface

$\bar{\tau}$	average interfacial shear stress
τ_{au}	adhesional bond shear strength
τ_c	critical bond strength
	or
τ_c	average shear stress in conventionally reinforced SFRC beams
τ_d	dynamic frictional resistance to sliding
τ_f	shear stress at failure in RC beam
$\tau_{fd}(\Delta)$	decaying frictional stress at slip Δ
τ_{fu}	frictional shear bond strength
τ_m	bonding due to mechanical interlocking
τ_s	frictional shear stress
	or
τ_s	static frictional resistance to sliding
τ_v	shear stress at the interface at the onset of strain softening
τ_v'	shear stress at onset of pull-out
$\tau(\max)$	maximum elastic shear stress at the interface
$\tau'(\max)$	interfacial elastic shear stress at the end of the debonded zone along a fibre
ψ	fibre perimeter
ξ	damage coefficient
	or
ξ	tortuosity factor of a crack

Chapter 1

Introduction

The use of fibres to strengthen materials which are much weaker in tension than in compression goes back to ancient times. Probably the oldest written account of such a composite material, clay bricks reinforced with straw, occurs in *Exodus* **5**: 6–7:

> And Pharaoh commanded the same day the task-masters of the people, and their officers, saying:
> 'Ye shall no more give the people straw to make bricks, as heretofore: let them go and gather straw for themselves.'

At about the same time period, approximately 3500 years ago, sun-baked bricks reinforced with straw were used to build the 57 m high hill of Aqar Quf (near present-day Baghdad) [1].

The first widely used manufactured composite in modern times was asbestos cement, which was developed in about 1900 with the invention of the Hatschek process. Now, fibres of various kinds are used to reinforce a number of different materials, such as epoxies, plastics and ceramics. Here we will concentrate on the use of fibre reinforcement in materials made with hydraulic cement binders.

For the purposes of this book, we will define *fibre reinforced cement* as a material made from a hydraulic cement and discrete, discontinuous fibres (but containing no coarse aggregate). *Fibre reinforced concrete* is made with hydraulic cement, and aggregates of various sizes, incorporating discrete, discontinuous fibres. In this book, when referring to the general class of these composites, regardless of the exact nature of the matrix (paste, mortar or concrete), the term FRC (fibre reinforced cementitious material) will be used.

Since the early use of asbestos fibres, a wide variety of other fibres have been used with hydraulic cements: conventional fibres such as steel and glass; new fibres such as carbon or kevlar; and low modulus fibres, either man-made (polypropylene, nylon) or natural (cellulose, sisal, jute). These types of fibres vary considerably both in properties, effectiveness and cost. Some common fibres, and their typical properties, are listed in Table 1.1. In addition to their mechanical properties, fibres

Table 1.1 Typical properties of fibres

Fibre	Diameter (μm)	Specific gravity	Modulus of elasticity (GPa)	Tensile strength (GPa)	Elongation at break (%)
Steel	5–500	7.84	200	0.5–2.0	0.5–3.5
Glass	9–15	2.6	70–80	2–4	2–3.5
Asbestos					
Crocidolite	0.02–0.4	3.4	196	3.5	2.0–3.0
Chrysolite	0.02–0.4	2.6	164	3.1	2.0–3.0
Polypropylene	20–400	0.9–0.95	3.5–10	0.45–0.76	15–25
Aramid (kevlar)	10–12	1.44	63–120	2.3–3.5	2–4.5
Carbon (high strength)	8–9	1.6–1.7	230–380	2.5–4.0	0.5–1.5
Nylon	23–400	1.14	4.1–5.2	0.75–1.0	16.0–20.0
Cellulose	—	1.2	10	0.3–0.5	—
Acrylic	18	1.18	14–19.5	0.4–1.0	3
Polyethylene	25–1000	0.92–0.96	5	0.08–0.60	3–100
Wood fibre	—	1.5	71.0	0.9	—
Sisal	10–50	1.5	—	0.8	3.0
Cement matrix (for comparison)	—	1.5–2.5	10–45	0.003–0.007	0.02

may also differ widely in their geometry. The steel and glass fibres that were used in the early work on FRC in the 1950s and 1960s were straight and smooth. Since then, however, more complicated geometries have been developed, mainly to modify their mechanical bonding with the cementitious matrix. Thus, modern fibres may have profiled shapes, hooked or deformed ends, they may occur as bundled filaments or fibrillated films, or they may be used in continuous form (mats, woven fabrics, textiles). Here, we will deal primarily with discontinuous fibres; materials reinforced with mats and woven fabrics will be addressed within the context of new developments in textile reinforcement of cement and fine aggregate concrete matrices.

Most of the developments with FRC involve the use of ordinary Portland cements. However, high alumina cement, gypsum and a variety of special cements have also been used to produce FRC, generally to improve the durability of the composite, or to minimize chemical interactions between the fibres and the matrix. Recent developments also include specially formulated mortar and concrete matrices with controlled particle size distributions.

Plain, unreinforced cementitious materials are characterized by low tensile strengths, and low tensile strain capacities; that is, they are *brittle* materials. They thus require reinforcement before they can be used extensively as construction materials. Historically, this reinforcement has been in the form of continuous reinforcing bars, which could be placed in the structure at the appropriate locations to withstand the imposed tensile and shear stresses. Fibres, on the other hand, are discontinuous, and are most commonly randomly distributed throughout the cementitious matrix. They are therefore not as efficient in withstanding the tensile stresses. However, because they tend to be more closely spaced than conventional reinforcing bars, they are better at controlling cracking. Thus, conventional reinforcing bars are used to increase the load-bearing capacity of concrete; fibres are more effective for crack control.

Because of these differences, there are certain applications in which fibre reinforcement is better than conventional reinforcing bars. These include:

1   Thin sheet components, in which conventional reinforcing bars cannot be used, and in which the fibres therefore constitute the *primary reinforcement.* In thin sheet materials, fibre concentrations can be relatively high, typically exceeding 5% by volume. In these applications, the fibres act to increase both the strength and the toughness (i.e. strain hardening) of the composite, and can be classified as high performance FRC (Figure 1.1). A new generation of high performance FRC is currently based on advanced formulations of fibres and matrices to achieve strain hardening behaviour even with modest fibre contents of ~ 2% by volume (Figure 1.1).

2   Components which must withstand locally high loads or deformations, such as tunnel linings, blast resistant structures, or precast piles which must be hammered into the ground.

3   Components in which fibres are added primarily to control cracking induced by humidity or temperature variations, as in slabs and pavements. In these applications, fibres are often referred to as *secondary reinforcement.* In this case the fibres provide post-cracking ductility, but the stresses are smaller than the first crack stress, that is a strain softening material (Figure 1.1). This type of composite is referred to as conventional FRC.

It is important to recognize that in general, fibre reinforcement is not a *substitute* for conventional reinforcement. Fibres and steel bars have different roles to play

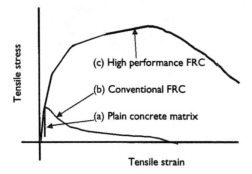

Figure 1.1 Typical stress–strain curves for conventional and high performance FRC (after ACI 544 [2]).

Figure 1.2 Industrial engineering building, Technion, Haifa, Israel, made with complex shaped panels of glass fibre reinforced cement (Architect Y. Rechter).

in modern concrete technology, and there are many applications in which both fibres and continuous reinforcing bars should be used together.

In applications (2) and (3), the fibres are not used to improve the strength (either tensile or other) of concrete, though a small improvement in strength may sometimes result from their use. Rather, the role of fibres is to control the cracking of FRC, and to alter the behaviour of concrete once the matrix has cracked, as shown by the schematic load vs. deflection curves of Figure 1.1. Thus, the fibres improve the 'ductility' of the material or more properly, its energy absorption capacity. Indeed, if an attempt is made to optimize the FRC for strength alone, this often leads to reduced toughness and more brittle behaviour. In addition, there is often an improvement in impact resistance, fatigue properties and abrasion resistance.

The applications of FRC are as varied as the types of fibres that have been used. Asbestos fibres have long been used in pipes and in corrugated or flat roofing sheets. Glass fibres are used primarily in precast panels (nonstructural), as shown in Figure 1.2. Steel fibres have been used in pavements (Figure 1.3), in shotcrete (Figure 1.4), in dams and in a variety of other structures (Figure 1.5). Increasingly, polypropylene fibres are being used as secondary reinforcement, to control plastic shrinkage cracking, and a newer generation of 'structural polymer fibres' may be applied for crack control in the hardened concrete. Vegetable fibres have been used in low-cost building materials. New fibres and new applications seem to go hand in hand.

New production technologies have evolved as new fibres have been developed, and new applications found. Clearly, in order to produce useful FRC, the production techniques must be compatible with the particular fibres and matrix. This depends not only upon the fibre type, but also on the fibre geometry. In particular,

Figure 1.3 Taxiway pavement at John F. Kennedy International Airport, New York, made with steel fibre reinforced concrete. Photograph courtesy of Bekaert Corporation.

Figure 1.4 Vancouver Public Aquarium whale pool: artificial West Coast rockscape, made with steel fibre reinforced silica fume shotcrete. Photograph courtesy of Amec, Vancouver, Canada.

Figure 1.5 Dolosse made with steel fibre reinforced concrete, for use by the US Army Corps of Engineers at Eureka, California. Each dolosse weighs about 38 tonnes. Photograph courtesy of Bekaert Corporation.

there is an inherent contradiction between the fibre geometry required to allow easy handling of the fresh FRC, and that required for maximum efficiency in the hardened composite. Longer fibres of smaller diameter will be more efficient in the hardened FRC, but will make the fresh FRC more difficult to handle.

To overcome this difficulty, there are a number of possible alternatives:

1   modification of the fibre geometry, to increase bonding without an increase in length (e.g. hooked fibres, deformed fibres or fibrillated networks);
2   chemically treating the fibre surface to improve its dispersion in the fresh matrix;
3   modifying the rheological properties of the matrix, through the use of chemical admixtures (mainly high range water reducers) and mineral admixtures (e.g. silica fume and fly ash) as well as optimization of the matrix particle size distribution;
4   using special production techniques to ensure that a sufficiently large volume of fibres can be dispersed in the mix.

The production technologies that are currently available may be classified as:

1   *Premix process* In this method, the fibres are combined with the cementitious matrix in a mixer. They are treated simply as an extra ingredient in the most common method of producing a cementitious mix. However, because the fibres reduce the workability, only up to about 2% fibres by volume can be introduced in the mix by this method.
2   *Spray-up process* This technique is used primarily with glass fibre reinforced cement. Chopped glass fibres and cement slurry are sprayed simultaneously on to the forming surface, to produce thin sheets. With this technique, substantially higher fibre volumes, up to about 6%, can be incorporated into the FRC.
3   *Shotcreting* Using a modification of normal shotcreting techniques, it has been found possible to produce steel and polypropylene fibre shotcretes, for use particularly for lining of tunnels, and for stabilization of rock slopes. With this method, too, relatively high volumes of fibres can be added to the mix.
4   *Pulp type processes* For asbestos cement replacements (cellulose or other fibres are used as a replacement for the asbestos), the fibres are dispersed in a cement slurry, which is then dewatered to produce thin sheet materials. These can be built up to the required thickness by layering. This process yields fibre contents of typically from 9% to over 20% by volume.
5   *Hand lay up* In this method, layers of fibres in the form of mats or fabrics can be placed in moulds, impregnated with a cement slurry, and then vibrated or compressed, to produce dense materials with very high fibre contents. This technique can also be used with glass fibre rovings already impregnated with a cement slurry.

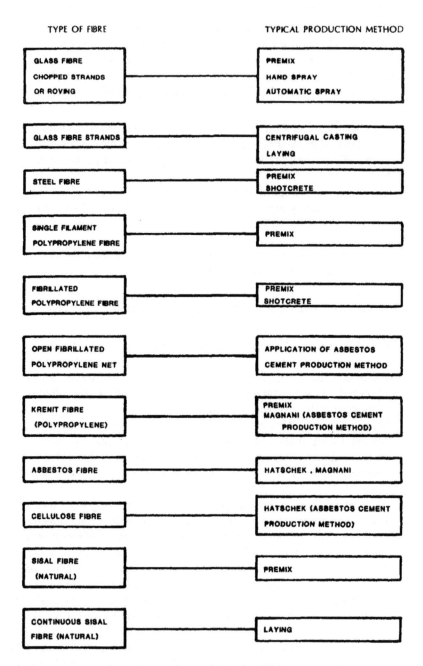

TYPE OF FIBRE

TYPICAL PRODUCTION METHOD

GLASS FIBRE CHOPPED STRANDS OR ROVING	PREMIX HAND SPRAY AUTOMATIC SPRAY
GLASS FIBRE STRANDS	CENTRIFUGAL CASTING LAYING
STEEL FIBRE	PREMIX SHOTCRETE
SINGLE FILAMENT POLYPROPYLENE FIBRE	PREMIX
FIBRILLATED POLYPROPYLENE FIBRE	PREMIX SHOTCRETE
OPEN FIBRILLATED POLYPROPYLENE NET	APPLICATION OF ASBESTOS CEMENT PRODUCTION METHOD
KRENIT FIBRE (POLYPROPYLENE)	PREMIX MAGNANI (ASBESTOS CEMENT PRODUCTION METHOD)
ASBESTOS FIBRE	HATSCHEK , MAGNANI
CELLULOSE FIBRE	HATSCHEK (ASBESTOS CEMENT PRODUCTION METHOD)
SISAL FIBRE (NATURAL)	PREMIX
CONTINUOUS SISAL FIBRE (NATURAL)	LAYING

Figure 1.6 Typical production methods for FRC (after [3]).

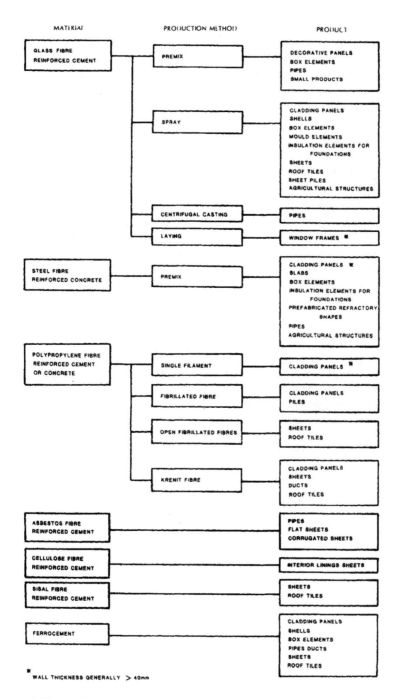

MATERIAL	PRODUCTION METHOD	PRODUCT
GLASS FIBRE REINFORCED CEMENT	PREMIX	DECORATIVE PANELS BOX ELEMENTS PIPES SMALL PRODUCTS
	SPRAY	CLADDING PANELS SHELLS BOX ELEMENTS MOULD ELEMENTS INSULATION ELEMENTS FOR FOUNDATIONS SHEETS ROOF TILES SHEET PILES AGRICULTURAL STRUCTURES
	CENTRIFUGAL CASTING	PIPES
	LAYING	WINDOW FRAMES ✳
STEEL FIBRE REINFORCED CONCRETE	PREMIX	CLADDING PANELS ✳ SLABS BOX ELEMENTS INSULATION ELEMENTS FOR FOUNDATIONS PREFABRICATED REFRACTORY SHAPES PIPES AGRICULTURAL STRUCTURES
POLYPROPYLENE FIBRE REINFORCED CEMENT OR CONCRETE	SINGLE FILAMENT	CLADDING PANELS ✳
	FIBRILLATED FIBRE	CLADDING PANELS PILES
	OPEN FIBRILLATED FIBRES	SHEETS ROOF TILES
	KRENIT FIBRE	CLADDING PANELS SHEETS DUCTS ROOF TILES
ASBESTOS FIBRE REINFORCED CEMENT		PIPES FLAT SHEETS CORRUGATED SHEETS
CELLULOSE FIBRE REINFORCED CEMENT		INTERIOR LININGS SHEETS
SISAL FIBRE REINFORCED CEMENT		SHEETS ROOF TILES
FERROCEMENT		CLADDING PANELS SHELLS BOX ELEMENTS PIPES DUCTS SHEETS ROOF TILES

✳ WALL THICKNESS GENERALLY > 40mm

Figure 1.7 Typical thin-walled products made from FRC (after [3]).

6   *Continuous production process* Continuous production of a composite mix using special machinery with the output being a continuous composite which has a thin, shaped geometry. Such processes can be based on the use of continuous reinforcement (fabrics, mats) in processes such as pultrusion or a mix with discrete short fibres which is extruded into the desired shape.

Figure 1.6 shows the typical production methods of FRC for various types of fibres. To illustrate the variety of applications, Figure 1.7 shows the range of thin-walled products that can be produced using FRC.

The object of this book is twofold: to develop, in some detail, the fundamental concepts which govern the performance of FRC, and to describe the production processes and properties of these composites and their applicability to engineering practice. To achieve these aims, the traditional material science approach is used: (i) characterization of the microstructure; (ii) relationship of the microstructure to engineering properties; (iii) relationships between the microstructural development and the processing techniques and (iv) engineering properties and design concepts.

References

1. R.N. Swamy, 'Prospects of fibre reinforcement in structural applications, in Advances in Cement-Matrix Composites', Proc. Symp. L, Materials Research Society General Meeting, Boston, MA, Nov. 1980, Materials Research Society, University Park, PA (now Pittsburgh, PA), 1980, pp. 159–169.
2. ACI Committee 544, *State of the Art Report on Synthetic Fiber-Reinforced Concrete*, American Concrete Institute, Farmington Hills, MI, 2005.
3. Anon, 'Thin-walled concrete units', *Concr. Int.* 7, 1985, 66–68.

Behaviour of fibre reinforced cementitious materials

Chapter 2

Structure of fibre reinforced cementitious materials

The properties of fibre reinforced cementitious materials are dependent on the structure of the composite. Therefore, in order to analyse these composites, and to predict their performance in various loading conditions, their internal structure must be characterized. The three components that must be considered are:

1　The structure of the bulk cementitious matrix.
2　The shape and distribution of the fibres.
3　The structure of the fibre–matrix interface.

2.1　Matrix

The bulk cementitious matrix is not significantly different from that in other cementitious materials, and it can be divided into two types depending on the particulate filler (aggregate) which it contains: paste/mortar (cement/sand–water mix) and concrete (cement–sand–coarse aggregate–water mix) [1–3].

Fibre reinforced cement pastes or mortars are usually applied in thin sheet components, such as cellulose and glass fibre reinforced cements, which are used mainly for cladding. In these applications the fibres act as the primary reinforcement and their content is usually in the range of 5–15% by volume. Special production methods need to be applied for the manufacturing of such composites.

In fibre reinforced concretes, the fibre volume is much lower (<2% by volume) and the fibres act as secondary reinforcement, mainly for the purpose of crack control. The production of such reinforced concretes is carried out by conventional means. Higher contents of fibres can be incorporated by relatively simple mixing technologies, but using advanced matrix formulations which are based on sophisticated control of the rheology and microstructure of the mix. Such formulations combine dispersants and fillers (e.g. DSP, RPC and DUCTAL® [4–6]). The dense microstructure in these composites, as well as their improved rheology can enable the incorporation and uniform dispersion of 2–6% by volume of short fibres, which can provide effective reinforcement.

2.2   Fibres

A wide range of fibres of different mechanical, physical and chemical properties have been considered and used for reinforcement of cementitious matrices, as outlined in Chapter 1. The fibre-reinforcing array can assume various geometries and in characterizing its nature two levels of geometrical description must be considered: (i) the shapes of the individual fibres and (ii) their dispersion in the cementitious matrices (Figure 2.1) [7].

The individual fibres may be subdivided into two groups: discrete monofilaments separated one from the other (e.g. steel – Figure 2.2) and fibre assemblies, usually made up of bundles of filaments, each with a diameter of $10\mu m$ or less. The bundled structure is typical of many of the man-made fibres, whether inorganic (e.g. glass – Figure 2.3(a) and (b)) [8] or organic (e.g. carbon, kevlar), and it also shows up in some natural fibres (e.g. asbestos). The bundled fibres frequently maintain their bundled nature in the composite itself (Figure 2.3(c)), and do not disperse into the individual filaments. The monofilament fibres which are used for cement reinforcement rarely assume the ideal cylindrical shape, but are deformed into various configurations (Figure 2.2), to improve the fibre–matrix interaction

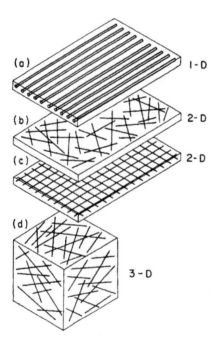

Figure 2.1 Classification of fibre arrangements in one, two and three dimensions and as continuous (a,c) or discrete, short fibres (b,d) (after Allen [7]). (a) 1D arrangement; (b,c) 2D arrangement; (d) 3D arrangement.

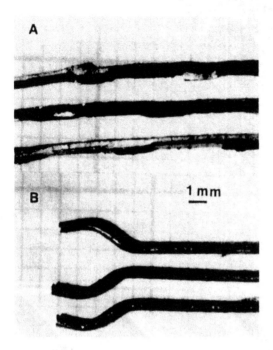

Figure 2.2 Various shapes of steel fibres (a) deformed; (b) hooked.

through mechanical anchoring. A range of complex geometries, ranging from twisted polygonal cross sections to ring type fibres have been evaluated, to provide effective anchoring, while maintaining adequate workability (e.g. [9–11]).

There are two distinctly different types of fibre-reinforcing arrays: (i) *continuous reinforcement* in the form of long fibres which are incorporated in the matrix by techniques such as filament winding or by the lay-up of layers of fibre mats; and (ii) *discrete short fibres*, usually less than 50 mm long, which are incorporated in the matrix by methods such as spraying and mixing. The reinforcing array can be further classified according to the dispersion of the fibres in the matrix, as 1D, 2D or 3D (Figure 2.1).

In the continuous form, the fibres can be aligned in a preferred orientation, which is controlled by the production process (orientation of winding, or lay-up direction of the mat) and the structure of the mat. This type of fibre reinforcement bears some resemblance to ferrocement applications; it is less common in FRC composites which are usually reinforced by discrete, short fibres, but has recently been the focus of intense development efforts (see Chapter 13 for details). In the case of dispersed fibres the dispersion in the matrix is more uniform, and the short fibres tend to assume a more random orientation. However, even in these systems the fibre distribution is rarely completely uniform, and their orientation is not

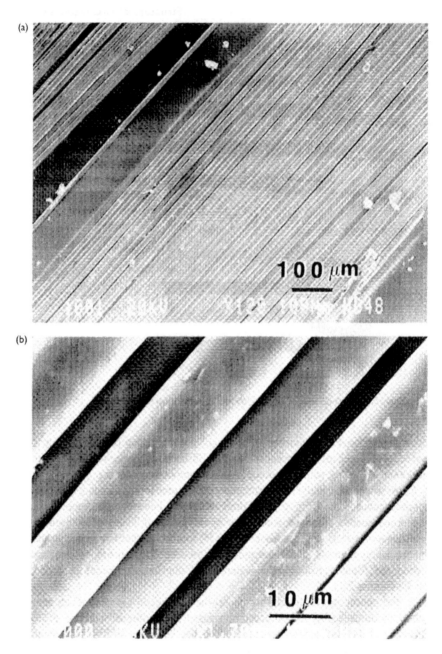

Figure 2.3 The bundled structure of glass filaments (after Bentur [8]). (a) Strands, each composed of 204 individual filaments grouped together. (b) Higher magnification of (a), showing the individual filaments in a strand. (c) The structure of the glass fibres in the cement composite, showing the bundled nature of the strand which does not disperse into the individual filaments.

(c)

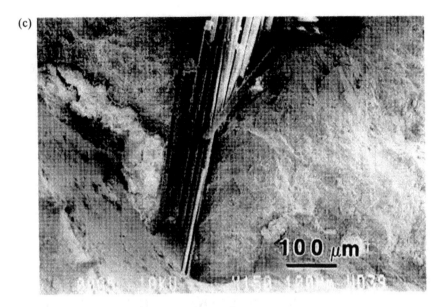

$100\,\mu m$

Figure 2.3 Continued.

ideally random. If the ratio of the fibre length to the thickness of the composite is sufficiently large, the fibres will assume a 2D distribution (Figure 2.1(b)), which is usually the case in thin components or thin cast overlays. A preferred 2D distribution can also be promoted in thick components due to vibration. This will give rise to anisotropic behaviour.

The uniformity of volume distribution of the fibres is very sensitive to the mixing and consolidation process, and in practice a uniform distribution is rarely achieved (Figure 2.4). The analytical treatment of fibre distribution can be based on various stereological models [12–15].

A geometrical parameter which is of significance in controlling the performance of the composite is the distance (spacing) between the fibres. Assuming a uniform fibre distribution, and using various statistical concepts, the average fibre–fibre spacing has been calculated, and several expressions have been derived. For cylindrical fibres, some of these equations take the form [16,17]:

$$S = \frac{K \cdot d}{V_{\mathrm{f}}^{1/2}} \tag{2.1}$$

where S is the fibre spacing; K, a constant; d, the fibre diameter; V_{f}, the fibre volume content; and K varies in the range of 0.8–1.12 depending on the orientation (1D, 2D or 3D) and the assumptions made in the calculation.

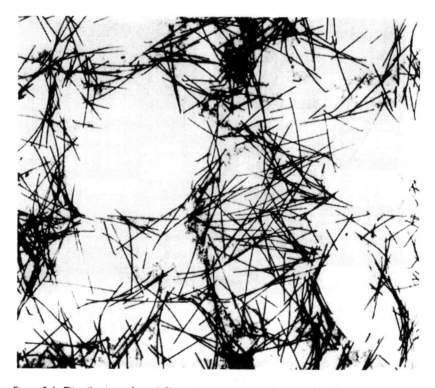

Figure 2.4 Distribution of steel fibres in concrete as observed by X-ray, showing non-uniform distribution (after Stroeven and Shah [12]).

To illustrate the relationship between fibre diameter and fibre spacing, Figure 2.5 [18] is a nomograph which yields either the fibre count (number of fibres per unit volume of FRC), or the surface area of fibres per unit volume of FRC, for unit length of fibres. If the specified volume of fibres is entered along the abscissa, then the number of fibres (or surface area) per unit volume may be found on the ordinate for a given fibre diameter.

Another way of quantifying the geometry of fibres is by using the *denier* unit common in the textile industry. A denier is the weight in grams of a 9000 m long staple. The relationship between fibre diameter and denier is shown in Figure 2.6 [19]. The fibre count and the surface area of fibres per unit volume of FRC can be expressed as functions of the weight, volume, specific gravity, denier, length and diameter as shown in the equations of Table 2.1 [19].

2.3 The structure of the fibre–matrix interface

Cementitious composites are characterized by an *interfacial transition zone* (ITZ) in the vicinity of the reinforcing inclusion, in which the microstructure of the

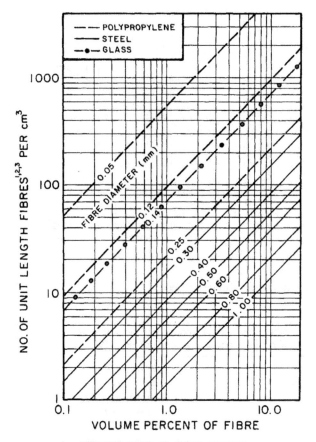

Figure 2.5 Number of fibres per unit volume, or surface area of fibres per unit volume, as a function of the volume per cent of fibres and the fibre geometry [18].

paste matrix is considerably different from that of the bulk paste, away from the interface. The nature and size of this transition zone depends on the type of fibre and the production technology; in some instances it can change considerably with time. These characteristics of the fibre–matrix interface exert several effects which should be taken into consideration, especially with respect to the fibre–matrix bond, and the debonding process across the interface (see Chapters 3 and 4).

The special microstructure of the transition zone in cementitious composites is closely related to the particulate nature of the matrix. The matrix consists of

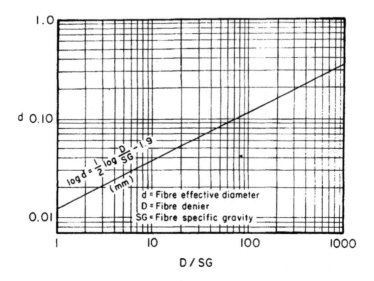

Figure 2.6 Fibre diameter vs. denier relationship [19].

Table 2.1 Fibre count (FC) and surface area (SS) of fibres per unit volume (cm³) of FRC [19]

FC = 0.077	$WT/(\ell)(d)^2(SG)$
FC = 0.077	$(V)/(\ell)(d)^2$
FC = 3.112	$WT(10)^5/(\ell)(D)$
FC = 3.112	$V(10)^5/(\ell)(D)$
SS = 0.244	$(WT)/(d)(SG)$
SS = 0.244	$(V)/(d)$
SS = 48.81	$WT/(D)^{1/2}(SG)^{1/2}$
SS = 48.81	$V/(D)^{1/2}(SG)^{1/2}$

Note
WT = weight; V = volume; SG = specific gravity; D = fibre denier; d = fibre diameter; ℓ = fibre length.

discrete cement particles ranging in diameter from ~1 to ~100 μm (average size of ~ 10 μm) in the fresh mix, which on hydration react to form mainly colloidal CSH particles and larger crystals of CH. The particulate nature of the fresh mix exerts an important influence on the transition zone, since it leads to the formation of water-filled spaces around the fibres due to two related effects:

1   bleeding and entrapment of water around the reinforcing inclusion and
2   inefficient packing of the ~ 10 μm cement grains in the 20–40 μm zone around the fibre surface.

Thus, the matrix in the vicinity of the fibre is much more porous than the bulk paste matrix, and this is reflected in the development of the microstructure as hydration advances: the initially water-filled transition zone does not develop the dense microstructure typical of the bulk matrix, and it contains a considerable volume of CH crystals, which tend to deposit in large cavities.

When considering the development of the microstructure in the transition zone, a distinction should be made between discrete monofilament fibres separated one from the other (e.g. steel), and bundled filaments (e.g. glass). With monofilament fibres, the entire surface of the fibre can be in direct contact with the matrix; with bundled filaments only the external filaments tend to have direct access to the matrix.

2.3.1 Monofilament fibres

The microstructure of the transition zone around monofilament fibres has been studied primarily in steel fibre reinforced cement pastes [20–25]. It was observed that the transition zone in the mature composite is rich in CH (usually in direct contact with the fibre surface), and is also quite porous, making it different from the microstructure of the bulk paste. These characteristics are probably the result of the nature of the fresh mix, as discussed above. The CH layer can be as thin as 1 μm (duplex film), or it can be much more massive, several μm across [23]. The porous nature of the transition zone is the result of pores formed between the CSH and the ettringite in a zone which backs up the CH layer. A schematic description of the transition zone showing the different layers (duplex film, CH layer, porous layer consisting of CSH and some ettringite) is presented in Figure 2.7(a), along with some micrographs which demonstrate the microstructure of each of the layers (Figure 2.7(b)–(d)). The formation of a CH rich zone at the fibre surface is probably the result of its precipitation from the solution in the space around the fibre, with the fibre surface being a nucleation site. The CH layer adjacent to the fibre surface is not necessarily continuous and it contains some pockets of very porous, needle-like material (Figure 2.7(c)) consisting also of CSH and some ettringite. The thin duplex film can usually be observed in the vicinity of the porous zone (Figure 2.7(d)) but not around the massive CH.

The microstructure in Figure 2.7 clearly indicates that the weak link between the fibre and the matrix is not necessarily at the actual fibre–matrix interface; it can also be in the porous layer, which extends to a distance of $\sim$ 10–40 μm from the interface, between the massive CH layer and the dense bulk paste matrix. This is consistent with characteristics of the mechanical properties of the transition zone determined by microhardness testing [26–28], showing lower values in the paste matrix in the immediate vicinity of the inclusion (aggregate, fibre) than in the bulk paste away from the inclusion surface, Figure 2.8. This is reflected in observations reported in [29] showing that during pull-out of a fibre high shear displacements occurred in an interfacial zone which appeared to be 40–70 μm wide.

(a)

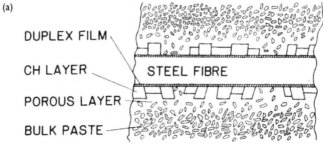

DUPLEX FILM

CH LAYER

POROUS LAYER

BULK PASTE

STEEL FIBRE

(b)

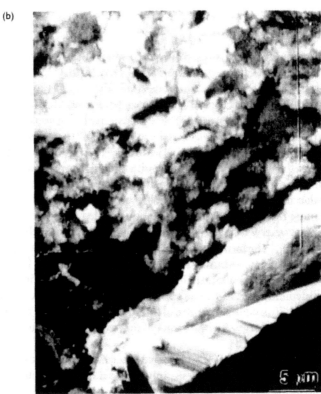

5 μm

Figure 2.7 The transition zone in steel fibre reinforced cement (after Bentur *et al.* [23]). (a) schematic description; (b) SEM micrograph showing the CH layer, the porous layer and the bulk paste matrix. (c) SEM micrograph showing discontinuities in the CH layer and (d) SEM micrograph showing the duplex film backed up by porous material.

(c)

5 μm

(d)

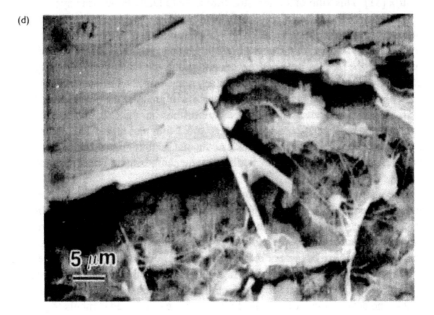

5 μm

Figure 2.7 Continued.

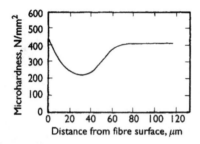

Figure 2.8 The microhardness of the cement paste matrix in contact with a steel fibre (after Wei *et al.* [26]).

It should be noted that the interfacial zone is sensitive to the processing and to the nature of the matrix. Intensive processing, which involves higher shear stresses in the fresh mix will result in a denser and smaller transition zone [30]. In the case in which the matrix is made of a well-graded mix, with fine fillers of the size of cement grains and smaller, and the fibre cross section is sufficiently small, the transition zone can be almost completely eliminated, resulting in a high bond matrix [31]. This kind of a microstructure is more likely to occur in systems such as RPC and DSP discussed previously [4–6], and in systems where the fibres are particularly small in diameter, a few tens of microns or less. In this range, the size of the fibre cross section is similar to that of the cement grains and fillers, and efficient packing of the fibre in between the cement grains can take place, resulting in an extremely dense microstructure, without any transition zone, as seen in Figure 2.9. Fibres in this size range, which is characteristic of many of the polymer and glass fibre filaments, are often referred to as microfibres, to make the distinction from macrofibres with cross-section diameters of 0.1 mm and more. The potential of getting the dense microstructure, such as the one seen in Figure 2.9, is dependent on efficient dispersion of the microfibres in the composite, to break their original bundled morphology.

Processing of softer fibres by special means, such as extrusion, can result in marked interfacial changes which are associated with abrasion of the fibre and its fibrillation, resulting in enhanced bonding [32]. Interfacial microstructural changes can occur during the pull-out of fibres induced during the loading of the composite, resulting in damage to the fibre or to the surrounding matrix, depending, to a large extent, on their relative stiffness [33]. These characteristics will be given special attention in Chapter 3. Interface tailoring is thus becoming an important tool in the development of high performance fibre reinforced cements (e.g. [34]).

2.3.2 Bundled fibres

In fibres consisting of bundled filaments, which do not disperse into the individual filaments during the production of the composite, the reinforcing unit is not a single

Figure 2.9 A dense interfacial microstructure formed around a microfibre (carbon) which was well dispersed as monofilament in the cement matrix (after Katz and Bentur [31]).

filament surrounded by a matrix, but rather a bundle of filaments [11,35–37] as shown in Figure 2.3(c) for glass fibres. The filaments in the fibre bundles of this kind are quite small, with diameters of $\sim$ 10 μm or less. The size of the spaces between the filaments does not exceed several μm, and as a consequence it is difficult for the larger cement grains to penetrate within these spaces. This is particularly the case with glass fibres, which have much less affinity for the cement slurry than does asbestos. The resulting microstructure after several weeks of hydration is characterized by vacant spaces between the filaments in the strand or limited localized formation of hydration products in some zones between the filaments (Figure 2.10). As a result, the reinforcing bundle remains as a flexible unit even after 28 days of curing, with each filament having a considerable freedom of movement relative to the others. Some stress transfer into the inner filaments may occur through frictional effects, aided by the point contacts formed by the hydration products and the sizing applied during the production of the glass fibre strands. In such a bundle, the bonding is not uniform, and the external filaments are more tightly bonded to the matrix.

The spaces between the filaments can be gradually filled with hydration products if the composite is kept in a moist environment. This process involves nucleation and growth stages, and the filament surfaces can serve as nucleation sites. Mills

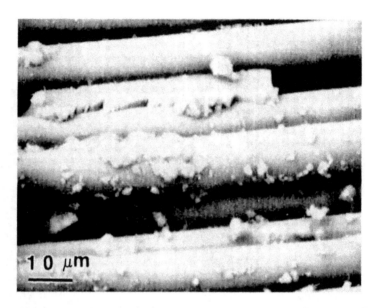

Figure 2.10 The spaces between the filaments in the reinforcing strand in a young (28 days old) glass fibre reinforced cement composite (after Bentur [37]).

[38] has demonstrated the affinity of an alkali-resistant glass fibre (AR) for nucleation and growth of CH crystals on its surface, when it was in contact with a Portland cement pore solution. This affinity is evident in the aged composite, prepared with high zirconia AR glass, where massive deposits of CH crystals were observed between the filaments [35,37], cementing the whole strand into a rigid reinforcing unit (Figure 2.11(a)). The nature of the deposited products can change, depending on the surface of the fibre. In newer generations of AR glass fibres (Cem FIL-2), in which the surface was treated by special coating [39], the hydration products deposited tend to be more porous, presumably CSH, rather than the massive crystalline CH (Figure 2.11(b)). Also, the rate of deposition is much slower [40]. This is a demonstration of the effect that the fibre surface may have on the microstructure developed in its vicinity.

The absence of CH-rich zones in the vicinity of the fibres was reported by Akers and Garrett [41] for asbestos–cement composites and by Bentur and Akers [42] for cellulose FRC composites produced by the Hatscheck process. This may be the result of the affinity of these fibres for the cement particles, and the processing treatment which involves dewatering, both of which lead to a system with very little bleeding, and probably reduce the extent of formation of water-filled spaces around the fibres in the fresh mix. This is reflected in the nature of the fibre–matrix bond failure; in asbestos composites, the cement matrix was sometimes seen to be sticking to the asbestos fibre bundle. This suggests that a strong interface was

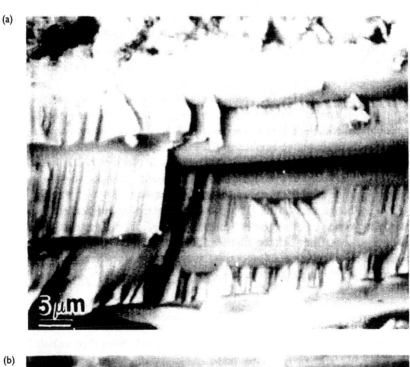

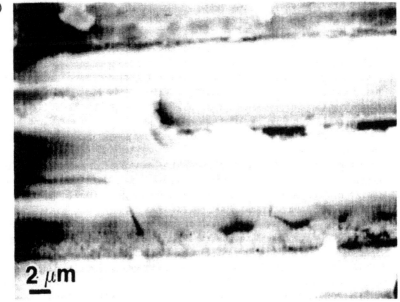

Figure 2.11 The spaces between the filaments in an aged glass fibre reinforced cement, showing them to be filled with massive CH crystals in the case of CemFIL-1 fibres (a) and more porous material in CemFIL-2 fibres; (b) (after Bentur [37]).

formed, and that failure occurred preferentially in the matrix away from the fibre bundle. The bundled nature of the asbestos fibres occasionally gave rise to another mode of failure, which involved fibre bundle failure due to separation between the filaments which make up the bundle [43]. This mode of failure was more likely to occur if, during the production of the composite, the bundle was not sufficiently opened to allow penetration of cement particles between the filaments in the bundle. Although this bears some resemblance to the observations with glass fibre strands, it should be emphasized that there is a considerable size difference between the two systems: the asbestos bundle is much smaller, consisting of fibrils of ~ 0.1 μm diameter or even less, with a fibre bundle diameter being ~ 5 μm; in the glass system each filament is ~ 10 μm in diameter.

Thus, although many of the FRC systems develop a transition zone which is porous and rich in CH, this may not generally hold true for all systems. Substantial changes in the affinity of the fibre for the matrix, combined with rheological modification of the mix or its processing, may have a major effect on the interface, and consequently on the fibre–matrix bond.

References

1. S. Mindess, J.F. Young and D. Darwin, *Concrete*, Second edition, Prentice Hall, Upper Saddle River, NJ, 2002.
2. A.M. Neville, *Properties of Concrete*, fourth Edition, John Wiley and Sons, London, 1996.
3. P.K. Mehta, *Concrete, Structure, Properties and Materials*, Prentice-Hall, Englewood Cliffs, NJ, 1986.
4. H.H. Bache, 'Principles of similitude in design of reinforced brittle matrix composites, Paper 3', in H.W. Reinhardt and A.E. Naaman (eds) *High Performance Fiber Reinforced Cement Composites*, Proc. RILEM Symp., E&FN SPON, London and New York, 1992, pp. 39–56.
5. P. Richard and M. Cheyrezy, 'Composition of reactive powder concretes', *Cem. Concr. Res.* 25, 1995, 1501–1511.
6. G. Orang, J. Dugat and P. Acker, P., 'DUCTAL: A new ultrahigh performance concrete, Damage resistance and micromechanical analysis', in P. Rossi and G. Chanvillard (eds) *Fiber Reinforced Concrete*, Proc. 5th RILEM Symp. (BEFIB 2000), RILEM Publications, Bagneux, France, 2000, pp. 781–790.
7. H.G. Allen, 'The purpose and methods of fibre reinforcement, in Prospects of Fibre Reinforced Construction Materials', in Proc. Int. Building Exhibition Conference, Sponsored by the Building Research Station, London, 1971, pp. 3–14.
8. A.E. Naaman, 'Fiber reinforcements for concrete: looking back, looking ahead', in P. Rossi and G. Chanvillard (eds) *Fiber Reinforced Concrete*, Proc. 5th RILEM Symp. (BEFIB 2000), RILEM Publications, Bagneux, France, 2000, pp. 65–86.
9. P. Rossi and G. Chanvillard, 'A new geometry of steel fibre for fibre reinforced concretes', in H.W. Reinhardt and A.E. Naaman (eds) *High Performance Fiber Reinforced Cement Composites*, Proc. RILEM Symp., E&FN SPON, London & New York, 1992, pp. 129–139.
10. O.C. Choi and C. Lee, 'Flexural performance of ring-type steel fiber-reinforced concrete', *Cem. Concr. Res.* 33, 2003, 841–849.

11. A. Bentur, 'Interfaces in fibre reinforced cements', in S. Mindess and S.P. Shah (eds) *Bonding in Cementitious Composites*, Proc. Conf. Materials Research Society, Materials Research Society, Pittsburgh, PA, 1988, pp. 133–144.

12. P. Stroeven and S.P. Shah, 'Use of radiography-image analysis for steel fibre reinforced concrete', in R.N. Swamy (ed.) *Testing and Test Methods for Fibre Cement Composites*, Proc. RILEM Conf., The Construction Press, Lancaster, England, 1978, pp. 275–288.

13. P. Stroeven and R. Babut, 'Wire distribution in steel wire reinforced concrete', *Acta Stereol.* 5, 1986, 383–388.

14. P. Stroeven, 'Morphometry of fibre reinforced cementitious materials Part II: inhomogeneity, segregation and anisometry of partially oriented fibre structures', *Mater. Struct.* 12, 1979, pp. 9–20.

15. J. Kasparkiewiez, 'Analysis of idealized distributions of short fibres in composite materials', *Bull. Pol. Acad. Sci.* 27, 1979, 601–609.

16. H. Krenchel, 'Fibre spacing and specific fibre surface', in A. Neville (ed.) *Fibre Reinforced Cement and Concrete*, Proc. RILEM Conf., The Construction Press, Lancaster, England, 1975, pp. 69–79.

17. J.P. Romualdi and J.A. Mandel, 'Tensile strength of concrete affected by uniformly distributed and closely spaced short lengths of wire reinforcement', *J. Amer. Concr. Inst.* 61, 1964, 657–670.

18. R.F. Zollo, 'An overview of the development and performance of commercially applied steel fibre reinforced concrete', Presented at USA - Republic of China Economic Councils 10th Anniversary Joint Business Conference, Taipei, Taiwan, Republic of China, December 1986.

19. R.F. Zollo, 'Synthetic fibre reinforced concrete: some background and definitions', Presented at World of Concrete, 189, Atlanta, Georgia, Feb. 21, 1989.

20. M.N. Al Khalaf and C.L. Page, 'Steel mortar interfaces: microstructural features and mode of failure', *Cem. Concr. Res.* 9, 1979, 197–208.

21. C.L. Page, 'Microstructural features of interfaces in fibre cement composites', *Composites.* 13, 1982, 140–144.

22. D.J. Pinchin and D. Tabor, 'Interfacial phenomena in steel fibre reinforced cement I. Structure and strength of the interfacial region', *Cem. Concr. Res.* 8, 1978, 15–24.

23. A. Bentur, S. Diamond and S. Mindess, 'The microstructure of the steel fibre-cement interface', *J. Mater. Sci.* 20, 1985, 3610–3620.

24. A. Bentur, S. Diamond and S. Mindess, 'Cracking processes in steel fibre reinforced cement paste', *Cem. Concr. Res.* 15, 1985, 331–342.

25. A. Bentur, S. Mindess and N. Banthia, 'The interfacial transition zone in fibre reinforced cement and concrete', in M.G. Alexander, G. Arliguie, G. Ballivy, A. Bentur and J. Marchand (eds) *Engineering and Transport properties of the Interfacial Transition Zone in Cementitious Composites*, RILEM Publications, Bagneux, France, Report 20, 1999, pp. 89–112.

26. S. Wei, J.A. Mandel and S. Said, 'Study of the interface strength in steel fibre reinforced cement-based composites', *J. Amer. Concr. Inst.* 83, 1986, 597–605.

27. P. Trtik and P.J.M. Bartos, 'Micromechanical properties of cementitious composites', *Mater. Struct.* 32, 1999, 388–393.

28. J. Nemecek, P. Kabele and Z. Bittnar, 'Nanoindentation based assessment of micromechanical properties of fiber reinforced cementitious composite', in M. Di Prisco, R. Felicetti and G.A. Plizzari (eds) *Fibre Reinforced Concrete – BEFIB 2004*, Proc. RILEM Symposium, PRO 39, RILEM, Bagneux, France, 2004, pp. 401–410.

29. Y. Shao, Z. Li and S.P. Shah, 'Matrix cracking and interface debonding in fiber-reinforced cement-matrix composites', *Advanced Cement Based Materials*. 1, 1993, 55–66.

30. S. Igarashi, A. Bentur and S. Mindess, 'The effect of processing on the bond and interfaces in steel fiber reinforced cement composites', *Cem. Concr. Compos.* 18, 1996, 313–322.

31. A. Katz and A. Bentur, 'Mechanisms and processes leading to changes in time in the properties of carbon fiber reinforced cement', *Advanced Cement Based Materials*. 3, 1996, 1–13.

32. A. Peled and S.P. Shah, 'Parameters related to extruded cement composites', in A.M. Brandt, V.C. Li and I.H. Marshall (eds) *Brittle Matrix Composites 6*, Proc. Int. Symp., Woodhead Publications, Warsaw, 2000, pp. 93–100.

33. Y. Geng and C.K.Y. Leung, 'Damage evolution of fiber/mortar interface during fiber pullout', in S. Diamond, S. Mindess, F.P. Glasser, L.W. Roberts, J.P. Skalny and L.D. Wakeley (eds) *Microstructure of Cement-Based Systems/Bonding and interfaces in Cementitious Materials*, Materials Research Society Symp. Proc. Vol. 370, Materials Research Society, Pittsburgh, PA, 1995, pp. 519–528.

34. V.C. Li, C. Wu, S. Wang, A. Ogawa and T. Saito, 'Interface tailoring for strain-hardening polyvinyl alcohol-engineered cementitious composite (PVA-ECC)', *ACI Mater. J.* 99, 2002, 463–472.

35. M.J. Stucke and A.J. Majumdar, 'Microstructure of glass fibre reinforced cement composites', *J. Mater. Sci.* 11, 1976, 1019–1030.

36. A. Bentur, 'Microstructure and performance of glass fibre-cement composites', in G. Frohnsdorff (ed.) *Research on the Manufacture and Use of Cements*, Proc. Eng. Found. Conf., Engineering Foundation, New York, 1986, pp. 197–208.

37. A. Bentur, 'Mechanisms of potential embrittlement and strength loss of glass fibre reinforced cement composites', in S. Diamond (ed.) *Proceeding – Durability of Glass Fiber Reinforced Concrete Symposium*, Prestressed Concrete Institute, Chicago, IL, 1986, pp. 109–123.

38. R.H. Mills, 'Preferential precipitation of calcium hydroxide on alkali resistant glass fibres', *Cem. Concr. Res.* 11, 1981, 689–698.

39. B.A. Proctor, D.R. Oakley and K.L. Litherland, 'Developments in the assessments and performance of GRC over 10 years', *Composites*. 13, 1982, 173–179.

40. A. Bentur, M. Ben-Bassat and D. Schneider, 'Durability of glass fibre reinforced cements with different alkali resistant glass fibres', *J. Amer. Ceram. Soc.* 68, 1985, 203–208.

41. S.A.S Akers and G.G. Garrett, 'Observations and predictions of fracture in asbestos cement composites', *J. Mater. Sci.* 18, 1983, 2209–2214.

42. A. Bentur and S.A.S. Akers, 'The microstructure and aging of cellulose fibre reinforced cement composites cured in normal environment', *International Journal of Cement Composites and Lightweight Concrete*. 11, 1989, 99–109.

43. S.A.S. Akers and G.G. Garrett, 'Fibre-matrix interface effects in asbestos-cement composites', *J. Mater. Sci.*, 18, 1983, 2200–2208.

Chapter 3

Fibre–cement interactions

Stress transfer, bond and pull-out

3.1 Introduction

The effectiveness of fibres in enhancing the mechanical performance of the brittle matrix is dependent to a large extent on the fibre–matrix interactions. Three types of interactions are particularly important:

1 physical and chemical adhesion;
2 friction;
3 mechanical anchorage induced by deformations on the fibre surface or by overall complex geometry (e.g. crimps, hooks, deformed fibres).

The adhesional and frictional bonding between a fibre and cementitious matrix are relatively weak. They have however significant contribution and practical significance in the case of composites having high surface area fibres (e.g. thin man-made synthetic filaments such as carbon, referred to sometimes as microfibres, with diameters in the range of 10 μm), and advanced cementitious matrices which are characterized by an extremely refined microstructure and very low porosity (i.e. water/binder ratios lower than about 0.3). In conventional fibre reinforced concretes, where the matrix water/binder ratio is 0.40 and above, and the fibres are of a diameter in the range of 0.1 mm or bigger, efficient reinforcement cannot be induced by adhesional and frictional bonding, and mechanical anchoring is required. For this purpose a variety of fibre shapes have been developed and are used commercially.

An additional element that should be considered is the orientation angle of the fibre, relative to the load direction. A range of competing processes needs to be considered here, in particular in the brittle cementitious matrix, where much of the fibre contribution comes at the stage where the matrix has cracked and the fibre is bridging across the crack.

These elements will be analysed in this chapter, starting with the modelling of the pull-out and bonding of straight and smooth fibres, either aligned or oriented, going through the modes of bonding of deformed fibres, where mechanical anchoring is induced. In these treatments, distinction will be made between the processes prior

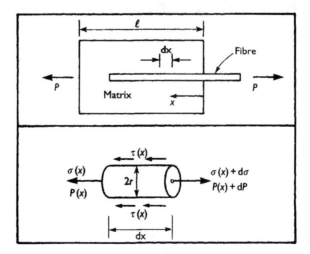

Figure 3.1 Pull out geometry to simulate fibre–matrix interaction.

to matrix cracking, and in the post-cracking stage, where the fibres bridge across cracks. The role of the fibres is particularly crucial in the latter stage.

A common element in all of these treatments is the quantification of the bonding by the simple pull-out geometry shown in Figure 3.1. This geometry is the one used in testing, to obtain a pull-out vs. slip relation, and in the modelling of the fibre–matrix interactions to provide analytical simulation for these curves. Ideally, one would like to provide a constitutive relation for the interfacial interactions by means of a characteristic curve of interfacial shear stress vs. pull-out displacement, Figure 3.2(a). Curves of this kind demonstrate a range of behaviours, from slip softening to slip hardening. Many of the models assume a constant frictional interfacial shear behaviour, Figure 3.2(b). Such relations cannot be developed by direct testing and are obtained indirectly from pull-out tests and assumptions which are at the basis of the analytical models applied to interpret these curves. More than that, interpretations in terms of interfacial shear stress are only valid from a physical point of view for well-defined geometry, such as straight and smooth fibres. As already indicated, this treatment is of direct practical significance to the more advanced cementitious composites, employing a dense matrix reinforced with thin filaments. For a conventional concrete matrix, deformed fibres are usually used, and their mechanics will be considered in Section 3.4.

3.2　Straight fibres

3.2.1　Elastic and frictional stress transfer

Qualitative and quantitative models have been developed to account for the fibre–matrix stress transfer and crack bridging, by analysing the shear stresses that

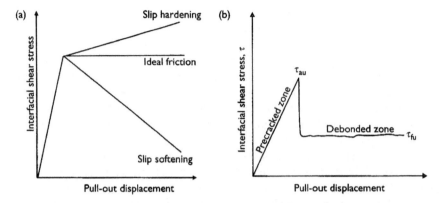

Figure 3.2 Ideal presentation of interfacial shear stress–slip curves: (a) range of behaviours demonstrating slip softening to slip hardening; (b) ideal presentation of a sharp transition from elastic stress transfer to a constant frictional stress transfer.

develop across the fibre–matrix interface in straight fibres. The models usually provide an analytical solution to a simulation of the fibre–matrix interaction, which is based on the simple pull-out geometry shown in Figure 3.1. These analytical treatments set the basis for predicting the efficiency of the fibres in the actual composite, which is usually made with short and randomly oriented fibres.

The processes involved in the fibre–matrix interaction take place mainly in a relatively small volume of the matrix surrounding the fibre. The microstructure of the matrix in this zone can be quite different from that of the bulk matrix (Chapter 2), thus invoking effects which are not always predicted by the analytical models which assume a uniform matrix down to the fibre surface. Its influence on the fibre–matrix interaction will be discussed in Section 3.5.

An understanding of the mechanisms responsible for stress transfer provides the basis for prediction of the stress–strain curve of the composite and its mode of fracture (ductile vs. brittle). Such understanding and quantitative prediction may also serve as a basis for developing composites of improved performance through modification of the fibre–matrix interaction. This might be achieved, for example, through changes in the fibre shape, or treatment of the fibre surface.

In brittle matrix composites, the stress-transfer effects should be considered for both the pre-cracking stage and the post-cracking stage, since the processes can be quite different in these two cases. Before any cracking has taken place, elastic stress transfer is the dominant mechanism, and the longitudinal displacements of the fibre and matrix at the interface are geometrically compatible. The stress developed at the interface is a shear stress which is required to distribute the external load between the fibres and matrix (since they differ in their elastic moduli), so that the strains of these two components at the interface remain the same. This elastic shear transfer is the major mechanism to be considered for predicting the limit of proportionality and the first crack stress of the composite. The elastic shear stress distribution along the fibre–matrix interface is non-uniform.

At more advanced stages of loading, debonding across the interface usually takes place, and the process controlling stress transfer becomes one of frictional slip. In this case relative displacements between the fibre and the matrix take place. The frictional stress developed is a shear stress, which is assumed in many models to be uniformly distributed along the fibre–matrix interface. This process is of greatest importance in the post-cracking zone, in which the fibres bridge across cracks. Properties such as the ultimate strength and strain of the composite are controlled by this mode of stress transfer.

The transition from elastic stress transfer to frictional stress transfer occurs when the interfacial shear stresses due to loading exceed the fibre–matrix shear strength. This will be referred to as the adhesional shear bond strength, τ_{au}. As this stress is exceeded, fibre–matrix debonding is initiated, and frictional shear stress will act across the interface in the debonded zone. The maximum frictional shear stress (i.e. the frictional shear strength) that can be supported across the interface will be called τ_{fu}. The values of τ_{fu} and τ_{au} are not necessarily the same. The value of τ_{fu} is very sensitive to normal stresses and strains; in most analytical treatments it is assumed to be constant over the entire pull-out range, implying the ideal interfacial shear stress–displacement curve shown in Figure 3.2(b). However, in practice, τ_{fu} may be reduced at advanced stages of loading (slip softening) or increased (slip hardening), depending on the nature of the interaction and the damage developed across the interface during the slip process (for more details see Section 3.5).

The transition from elastic stress transfer prior to debonding, to frictional stress transfer after debonding, is a gradual process, during which both types of mechanisms are effective. Debonding may even take place prior to the first cracking of the matrix [1], and thus, the combined effect of these two mechanisms may influence the shape of the stress–strain curve prior to matrix cracking. The occurrence of such a sequence of events depends upon the fibre–matrix adhesional shear bond strength and on the tensile strength of the matrix. If the latter is high, one may expect debonding to occur prior to matrix cracking, when the elastic shear stress exceeds the adhesional shear bond strength. Bartos [1] has argued that in such instances, the limit of proportionality may be reached before first cracking. He thus questioned the commonly held concept that matrix cracking and the deviation of the stress–strain curve from linearity (i.e. limit of proportionality) must always coincide.

In composites with a low tensile strength matrix, cracking may precede fibre debonding. In this case, fibre debonding, being the result of the interaction of an advancing crack and a fibre placed in its path, assumes a different nature. The analytical treatment of such an event can be based on fracture mechanics concepts, taking into account the stress field in front of the advancing crack as it approaches the reinforcing inclusion (fibre). In an analysis of this kind the differences in the moduli of elasticity of the matrix and the fibre should be considered, as well as the special nature of the matrix properties in the vicinity of the interface, which may be different from that of the matrix in which the crack was initiated and advanced (Chapter 2).

The shear stresses developed parallel to the fibre–matrix interface are of prime importance in controlling the fibre–matrix stress-transfer mechanism, as discussed previously. Yet, one should also consider the effect of strains and stresses that develop normal to the fibre–matrix interface. Such strains and stresses may be the result of the Poisson effect, volume changes, and biaxial or triaxial loading. They may cause weakening of the interface and premature debonding, and may also induce considerable variations in the resistance to frictional slip, which is sensitive to normal stresses.

A comprehensive approach to model the stress transfer requires simultaneous treatment of all the above-mentioned effects: elastic shear transfer, frictional slip, debonding, and normal stresses and strains. Unfortunately, such a unified approach is complex. Therefore, in this chapter, each of these effects will first be discussed separately, based on models developed for fibres of a simple shape, usually straight fibres with a circular cross section. The stress transfer in uncracked and in cracked composites will also be dealt with separately.

It should be borne in mind that although most analytical models were developed for smooth, straight fibres, as outlined in this section, the fibres used in practice have more complex shapes, such as profiled steel fibres, networks of fibrillated polypropylene film or bundles of filamentized glass fibre strands (Section 2.2). These complex shapes are the result of the production process of the fibres (e.g. fibrillated glass fibres) or the need to improve the bond characteristics by providing mechanical anchoring effects (e.g. crimped and hooked steel fibres). The latter need is a manifestation, as noted before, of the fact that the shear bonding stresses between the fibre and the matrix are relatively low, and are insufficient to generate composites with reasonable mechanical properties. Mechanical anchoring is a practical solution to compensate for the short anchorage length and low bond strength. Thus, many of the analytical models developed for smooth, straight fibres may not be applicable quantitatively (or even qualitatively) to the complex shaped fibres. These characteristics are dealt with in Section 3.3. The significance of the stress transfer in controlling the reinforcing efficiency of short fibres will be discussed in Chapter 4.

3.2.2   Stress transfer in the uncracked composites

3.2.2.1   Elastic stress transfer

During the early stages of loading, the interaction between the fibre and the matrix is elastic in nature. The first analytical model to describe the stress transfer in the elastic zone was developed by Cox [2]. Later models were based on similar concepts; they differed only in some of their numerical parameters. These models are usually referred to as *shear lag theories*. They are based on the analysis of the stress field around a discontinuous fibre embedded in an elastic matrix. A schematic representation of the deformations around such a fibre, before and after loading, is provided in Figure 3.3(a). In calculating the stress field developed due to these deformations, several simplifying assumptions are made:

1 The matrix and the fibre are both elastic materials.
2 The interface is infinitesimally thin.
3 There is no slip between the fibre and the matrix at the interface, that is 'perfect' bond exists between the two.
4 The properties of the matrix in the vicinity of the fibre are the same as those of the bulk matrix.
5 The fibres are arranged in a regular, repeating array.
6 The tensile strain in the matrix, ε_m, at a distance R from the fibre, is equal to the tensile strain of the composite, ε_c.
7 No stress is transmitted through the fibre ends.
8 There is no effect of the stress field around one fibre on neighbouring fibres.

Based on these assumptions, Cox [2] derived the following equations for the tensile stress, $\sigma(x)$, in the fibre, and the elastic shear stress at the interface, $\tau(x)$,

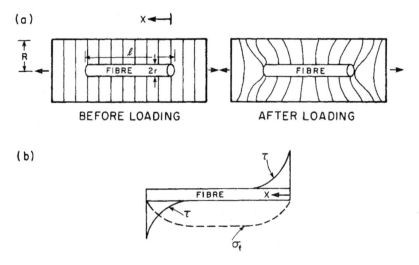

Figure 3.3 Schematic description of a fibre embedded in a matrix, and the deformation and stress fields around it: (a) geometry of the fibre and the deformation in the matrix around the fibre prior to and after loading; (b) elastic shear stress distribution at the interface (τ) and tensile stress distribution in the fibre (σ).

at a distance x from the fibre end:

$$\sigma_f(x) = E_f \varepsilon_m \left[\frac{1 - \cosh \beta_1((\ell/2) - x)}{\cosh(\beta_1 \ell/2)} \right] \tag{3.1}$$

$$\tau(x) = E_f \varepsilon_m \left[\frac{G_m}{2E_f \ell n(R/r)} \right]^{1/2} \frac{\sinh \beta_1((\ell/2) - x)}{\cosh(\beta_1 \ell/2)} \tag{3.2}$$

$$\beta_1 = \left[\frac{2G_m}{E_f r^2 \ell n(R/r)} \right]^{1/2} \tag{3.3}$$

where, with reference to Figure 3.3(a): R is the radius of the matrix around the fibre; r, the radius of the fibre; ℓ, the length of the fibre; E_f, the modulus of elasticity of the fibre and G_m, the shear modulus of the matrix, at the interface.

The value R/r depends upon the fibre packing and the fibre volume content of the composite. The following equations for long fibres of circular cross section have been derived [3]:

Square packing:

$$\ell n \, R/r = \frac{1}{2} \ell n \, (\pi/V_f) \tag{3.4}$$

Hexagonal packing:

$$\ell n \, R/r = \frac{1}{2} \ell n \, [2\pi/(3V_f)^{1/2}] \tag{3.5}$$

where V_f is the fibre volume content in the composite.

The calculated shear stress distribution at the interface, and the tensile stress distribution in the fibre are presented graphically in Figure 3.3(b). The shear stress has a maximum value at the ends of the fibre and drops to zero at the centre. It is in this end zone that the stress is transferred from the matrix to the fibre, gradually building up tensile stress within the fibre. The tensile stress increases from the fibre end moving inwards, reaching a maximum at the centre.

More detailed analyses have given similar results, and they usually differ only in the value of β_1 (e.g. [4]). In all cases, β_1, is a function of $(G_m/E_f)^{1/2}$. Experimental measurements [4], however, usually indicate much higher shear values at the fibre end than those predicted by equations such as Eq. 3.2. This is due in part to stress concentrations that may arise at the fibre ends, and in part to the difference between the moduli of elasticity of the fibre and the matrix.

The efficiency of fibre reinforcement depends to a large extent on the maximum tensile stress that can be transferred to the fibre. The maximum value would, of course, be the yield strength or tensile strength of the fibre. The shear lag theory (Eqs 3.1–3.3) provides an analytical tool to predict the shear stresses that will develop at the interface in order to achieve this maximum tensile stress. An estimate of the maximum elastic shear stress developed for different levels of tensile stress in the fibre can be obtained by calculating the ratio between the maximum elastic shear stress at the interface, $[\tau(\max) = \tau(x = 0)]$, and the maximum tensile stress within the fibre itself, $[\sigma_f(\max) = \sigma_f(x = \ell/2)]$, taking the expression in Eqs 3.2 and 3.1, respectively:

$$\frac{\tau(\max)}{\sigma_f(\max)} = \left[\frac{G_m}{2E_f \ell n(R/r)}\right]^{1/2} \coth\frac{\beta_1 \ell}{4} \tag{3.6}$$

For long fibres, Eq. 3.6 reduces to:

$$\frac{\tau(\max)}{\sigma_f(\max)} = \left[\frac{G_m}{2E_f \ell n(R/r)}\right]^{1/2} \tag{3.7}$$

For a typical steel fibre–cement composite with a 2% fibre volume, and assuming hexagonal packing, the ratio in Eq. 3.7 would be about 0.06. Thus, for a typical cementitious matrix, where the adhesional shear bond strength would not normally exceed $\sim$ 15 MPa, the maximum tensile stress that could develop in the fibre would be about 200 MPa. This value is well below the strengths of commercial fibres, which usually are in excess of 700 MPa (Table 1.1). It should also be noted that, because of the simplifying assumptions involved in the derivation of these equations, the maximum calculated shear stress values may be considered to be the lower bound. The ratio of $\tau(\max)/\sigma_f(\max)$ would probably be larger than that calculated from Eq. 3.7. Thus, in an FRC composite, the adhesional shear bond strength at the interface would normally be exceeded long before a significant tensile stress could develop in the fibre.

3.2.2.2   Combined elastic and frictional stress transfer

The earlier analysis of elastic stress transfer implies that debonding may occur at an early stage, before the fibre has been utilized effectively. If debonding occurs prior to matrix cracking, two alternatives should be considered:

1   Complete loss of bond and failure of the composite.
2   Slip in the debonded zone, resulting in activation of the frictional slip resistance mechanism.

Only the second alternative, which is presented schematically in Figure 3.2, will be discussed further, since it is most likely to occur in FRC composites. Assuming that in the debonded zone the frictional resistance to slip results in a uniform distribution of a frictional shear stress, τ_{fu}, at the interface (Figure 3.2(b)), then the shear stress distribution at the interface and the tensile stress distribution in the fibre can be described by the curves in Figure 3.4. At early stages of loading, before the maximum elastic shear bond stress, $\tau(\max)$, exceeds the adhesional shear bond strength, τ_{au} (P_0 in Figure 3.4), the distributions are similar to those presented in Figure 3.3. When the adhesional shear strength has been exceeded at

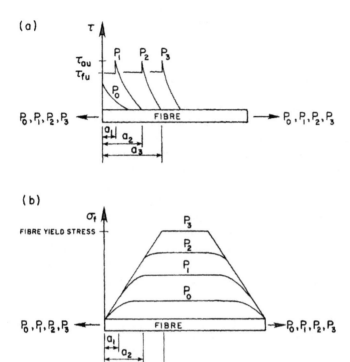

Figure 3.4 Distribution of interfacial shear stresses, τ, and fibre tensile stresses, σ_f, in zones of combined elastic and frictional shear stress transfer. $P_0, P_1, P_2,$ and P_3 indicate the distribution curves at increased stages of loading. a_1, a_2, a_3 indicate the a location of the interfacial zone where debonding occurred.

load P_1, a debonded zone of length a_1 is formed, and the shear stress distribution is uniform in this zone (with a shear stress value of τ_{fu}). Beyond zone a_1, the interfacial shear stress decreases, following the shear lag theory relationships. As a result, the tensile stress build-up in the fibre is linear throughout the debonded zone ($0 < x < a_1$) and then it increases, following the shear lag theory equation. As the external load increases (P_2) so does the length of the debonded zone (a_2), and the tensile stress distribution in the fibre changes accordingly. In the extreme case, where the debonded length is such that a tensile stress equal to the strength of the fibre can be developed, the stress distribution becomes linear (P_3).

For $\tau_{au} = \tau_{fu}$ the stress distributions presented in Figure 3.4 are similar to those developed for fibres in a plastic matrix [4], in which a constant shear stress develops at the interface as the matrix yields (i.e. the shear strain at the fibre–matrix interface exceeds its yield strain). Yet, it should be emphasized that in the case of FRC composites, in which the matrix is more brittle than the fibres, a different mechanism is responsible for inducing a constant shear stress at the interface at advanced stages of loading, that is frictional slip rather than plastic yielding.

3.2.3   Stress transfer in cracked composites

In practice, the major effect of the fibres in FRC composites occurs in the post-cracking zone, where the fibres bridge across cracks that have propagated in the brittle matrix, and thus prevent catastrophic failure. Whereas the stress-transfer mechanisms described in the previous section control the stress–strain curve of the composite prior to cracking, the mechanisms discussed in this section are the ones which influence the ultimate strength and deformation properties of the FRC composite and its mode of failure. The load transfer induced by the fibre which bridges across the crack is usually simulated by pull-out tests, in which either a single fibre or an array of fibres is pulled out of a matrix. Various pull-out test configurations are discussed in Section 6.3.8. Because of the significance of these aspects of fibres in brittle matrix composites, the pull-out of a fibre from such matrices has been treated extensively, both analytically and experimentally. These treatments can serve a double purpose:

1   They provide the basis for predicting the overall behaviour of the composite in the post-cracking zone.
2   They provide a tool for analysing the results of pull-out tests, in order to resolve the bonding mechanisms and to determine the relative contributions of elastic and frictional shear stress transfer components.

An assumption implicit in most analytical models and pull-out test configurations is that the faces of the crack across which the fibre is bridging are smooth and straight, with single fibres protruding normal to the faces. However, in an actual composite, the situation is much more complex:

1   The interaction of a crack, propagating in the matrix, with a fibre lying in its path, can be quite complex, often resulting in extensive microcracking in the vicinity of the fibre, and debonding at some distance away from the actual fibre–matrix interface. These characteristics are dealt with in Section 3.5, taking into account the stress field at the crack tip and the microstructure of the matrix at the fibre–matrix interface.
2   In practice, most FRC composites are made with fibres having a more or less random orientation, and therefore, many of the fibre bridging across the crack assume an angle different from 90° with respect to the crack surfaces. The pull-out mechanisms invoked with randomly oriented fibres can be quite different from those with fibres normal to the crack. Orientation effects will be discussed in Section 3.3.

The subject of debonding and pull-out has been studied extensively and a variety of models have been presented [5–14]. Many of them are based on common concepts which were reviewed in references [1,5,6]. A significant portion of this section will be based on these three papers.

The stress-transfer mechanisms to be considered during pull-out or bridging over an opening crack are essentially the same as those discussed in Section 3.2.1 for the uncracked composite: elastic bonding (shear lag theory) and frictional slip. The main difference is that whereas in the previous case the maximum interfacial shear stress values occur at the ends of the fibre, in the cracked composite they occur at the point at which the fibre enters the matrix. If debonding has previously occurred at this intersection (during interaction with the propagating crack or due to loading in the pre-cracked zone), the shear stress distribution will be of a combined mode, with frictional shear adjacent to the crack, and decreasing elastic shear stresses away from it (Figure 3.5(a)). In the event that no debonding preceded cracking, the interfacial shear stress distribution at the fibre–crack intersection will initially be elastic in nature, following the shear lag equations (Figure 3.5(b)), and only at advanced stages of loading will it combine frictional shear and elastic shear.

Several models (e.g. [7,11,13,14]) have analysed the pull-out of a single fibre from a matrix, based on the shear lag theory up to debonding and frictional slip

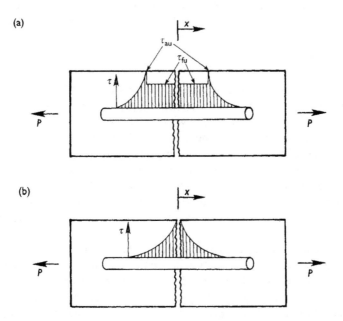

Figure 3.5 Interfacial shear stress distribution along a fibre intersecting a crack immediately after cracking: (a) debonding preceded cracking; (b) no debonding prior to cracking.

thereafter. Assuming the configuration of Figure 3.1, Greszczuk [7] derived the following equations:

$$\tau(x) = \frac{P\beta_2}{2\pi \ r}[\sin \ \beta_2 x - \coth(\beta_2\ell) \cosh(\beta_2 x)] \tag{3.8}$$

$$\beta_2 = \left[\frac{2G_\mathrm{m}}{b_\mathrm{i}rE_\mathrm{f}}\right]^{1/2} \tag{3.9}$$

where r is the radius of the fibre; b_i, the effective width of the interface; E_f, the modulus of elasticity of the fibre; G_m, the shear modulus of the matrix at the interface and ℓ, the embedded length.

In this treatment, no reference is made to the width of the interfacial zone, and it is implied that it is a very thin layer of binder.

The stress distribution predicted by Eq. (3.8) is essentially similar to that of the shear lag theory, with a maximum elastic shear stress at the point at which the fibre enters the matrix (Figure 3.5(b)). The value of this stress is:

$$\tau(\mathrm{max}) = \tau(x = 0) = \frac{P\beta_2}{2\pi \ r} \coth(\beta_2\ell) \tag{3.10}$$

In many simplified treatments of the stress-transfer problem, reference is made to an *average* interfacial shear stress value, $\bar{\tau}$, assuming a uniform interfacial shear stress distribution along the whole fibre length; for a pull-out load P,

$$\bar{\tau} = \frac{P}{2\pi \ r\ell} \tag{3.11}$$

This value has no physical significance, especially when the shear transfer is solely elastic. It can easily be shown that $\bar{\tau}$ is considerably smaller than $\tau(\mathrm{max})$, and the difference between the two increases with the length of the fibre. Thus, the use of average values can be misleading. For example, the $\bar{\tau}$ value determined from the pull-out test of a fibre of a specific length cannot be used to predict the load that can be supported in a similar system with a fibre of different length, just by multiplying by the fibre length ratio. Thus, one should avoid interpreting the results of pull-out tests (i.e. the load required to pull the fibre out of the matrix in a test configuration such as that shown in Figure 3.1) by calculating average stresses.

Once the elastic shear strength has been exceeded (at the point of entry of the fibre into the matrix) debonding will occur. Assuming that the debonding is limited to the zone in which the elastic shear stress exceeds the adhesional shear bond strength, then the load transfer process will be made up of frictional slip at the debonded end and elastic shear transfer in the rest of the fibre. This implies that when the elastic shear bond strength is exceeded, catastrophic failure will not necessarily occur. This state of combined stress-transfer mechanisms was treated analytically by Lawrence [8], and was later reviewed and extended by others

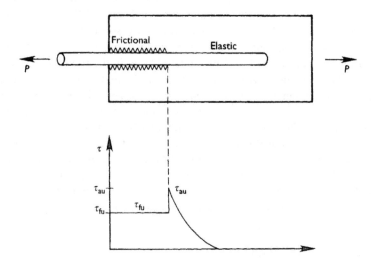

Figure 3.6 Partially debonded fibre configuration and the interfacial shear stresses to calculate combined elastic and frictional resistance to pull-out.

(e.g. [1,5,10,11,13,15]). Lawrence also considered frictional slip in the debonded zone. The debonded fibre configuration is presented schematically in Figure 3.6, assuming frictional slip in the debonded zone, with a constant shear stress of τ_{fu}. At the point at which debonding is terminated, the load, P', carried by the fibre, will be equal to the external load P, minus the load provided by the frictional resistance over the debonded zone:

$$P' = P - 2\pi\, r\, b\, \tau_{fu} \qquad\qquad (3.12)$$

where r is the fibre radius and b, the length of the debonded zone;

Beyond this point, the stress distribution in the bonded zones is governed by elastic considerations, and can be calculated by equations such as Eqs 3.8 and 3.9, assuming a fibre with a length of $(\ell - b)$ and a pull-out load of P' (Eq. 3.12).

Thus, the interfacial elastic shear stress at the end of the debonded zone is:

$$\tau'(\text{max}) = \frac{P'\beta_2}{2\pi\, r}\,\coth[\beta_2(\ell - b)] \qquad\qquad (3.13)$$

If b increases and $\tau'(\text{max})$ remains equal to or smaller than τ_{au}, no catastrophic debonding will take place; that is in order to advance the debonded zone the load P must be increased. This condition will occur when the increase in the term $\coth[\beta_2(\ell - b)]$ in Eq. 3.13 due to the increase in b, is compensated for by a decrease in P'. If this compensating effect does not take place, bond failure will occur at once without progressive debonding.

A similar analytical model, using the same concepts but somewhat different assumptions, was developed by Gopalaratnam and Shah [10], Naaman *et al.* [11,12] and Lin and Li [14] for cement matrices and by Budiansky and Hutchinson [13] for ceramic matrices, all of them representing elastic–brittle matrices. Naaman *et al.* [11,12] considered in their model a decay in the frictional resistance with increase in slip. This is demonstrated in Figure 3.7 in which the characteristic shear stress-slip diagram is presented alongside the load-slip diagram. The overall behaviour in Figure 3.7 [16] assumes interfacial shear stresses which are elastic to start with, with a gradual follow-up of debonding, which starts at load P_{crit} where the slip is Δ_{crit}. The gradual debonded zone is extended and at a slip of Δ_0 the whole fibre is debonded and the interfacial shear becomes frictional. In between Δ_{crit} and Δ_0 the stress transfer includes a mixed mode of adhesional and frictional stresses. Naaman *et al.* [11,12] developed a model for the decaying frictional slip, $\tau_{f_d}(\Delta)$ for slip Δ (the subscript 'd' indicates damage or decay):

$$\tau_{f_d}(\Delta) = \tau_{fi} \frac{e^{-(\Delta-\Delta_0)^n - \xi e^{(-1)^n}}}{1 - \xi e^{-(1-\Delta+\Delta_0)^n}}$$
$$\cdot \frac{1 - \exp\left[-2v_f \mu(1 - \Delta + \Delta_0)/E_f r_f \left((1 + v_m)/E_m\right) + ((1 - v_f)/E_f)\right]}{1 - \exp\left[-2v_f \mu l/E_f r_f \left((1 + v_m)/E_m\right) + ((1 - v_f/E_f)\right]}$$

$$(3.14)$$

where Δ is the relative slip of the fibre after pull-out debonding; Δ_0, the relative slip of the fibre at end of pull-out debonding, as a first approximation it can be taken equal to the slip at maximum load; ξ, the damage coefficient, a dimensionless constant to give the analytical descending branch of the bond shear stress vs. slip

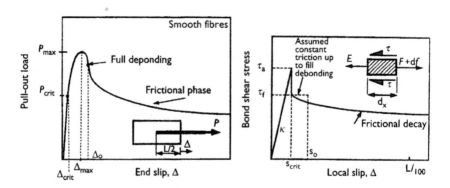

Figure 3.7 Pull-out of aligned and straight fibres with elastic response and a decaying frictional slip: (a) typical pull-out load vs. slip response for steel fibre embedded in cement-based matrix; (b) bond shear stress vs. slip relationship with frictional decay (after Bentur *et al.* [6])

curve the same decaying trend as the experimental one; μ, the friction coefficient of the fibre–matrix interface; ν, the Poisson's ratio, with subscript f for fibre and m for matrix and η, the coefficient describing the experimental shape of the descending branch of the bond shear stress vs. slip curve; for smooth steel fibres, a value of 0.2 is recommended by Naaman et al. [11,12].

The decaying frictional function (i.e. slip softening) suggested by Naaman et al. [11,12] was based on the study of pull-out of steel fibres from a cement matrix. In this system, the matrix interface is abraded by the higher modulus steel fibre, and the abrasion damage accumulates with slip, leading to the decay. However, this trend may change or even be reversed for different fibres and matrices, as demonstrated in Figure 3.8 [17]. Behaviours of this kind can be explained on the basis of microscopical observations and micro-compositional characterization of the interface [18]. When the fibre is softer than the matrix (e.g. low modulus polymer), the pull-out results in damage to the fibre, showing up as fibrillation and peeling of its surface, leading to some mechanical anchoring between the fibre and the matrix, with the consequence being an ascending, slip hardening pull-out curve [19] (for more details see Section 3.4).

Lin and Li [14] modelled the pull-out resistance of a slip hardening fibre by assuming a linear relation between the interfacial shear stress τ and the slip Δ:

$$\tau = \tau_0(1 + \beta\Delta/d) \tag{3.15}$$

where: τ is the interfacial shear stress after slip Δ; τ_0, the interfacial shear at the tip of the debonding zone where no slip occurs ($\Delta = 0$); β, the non-dimensional hardening parameter and d, the fibre diameter.

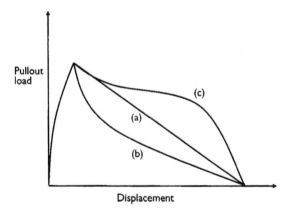

Figure 3.8 Three modes of pull-out behaviour showing ideal friction (a), slip softening; (b) slip hardening (c) (adapted from Katz and Li [17]).

The calculated pull-out load (P)-displacement (δ) relation (fibre end displacement in pull-out, or the analogous crack opening in a fibre bridging a crack) is given by the following equations:

1   For the fibre debonding stage ($0 \leq \delta \leq \delta_0$)

$$P = \frac{\pi d^2 \tau_0 (1 + \eta)}{\omega} \sqrt{\left(1 + \frac{\beta \delta}{2d}\right)^2 - 1} \tag{3.16}$$

2   For the fibre post-debonding (pull-out) stage ($\delta_0 \leq \delta \leq \ell$):

$$P = \frac{\omega d_f^2 \tau_0 (1 + \tau)}{\omega} \left[\sinh\left(\frac{\omega L}{d_f}\right) - \sinh\left(\frac{\omega(\delta - \delta_0)}{d_f}\right)\right]$$
$$+ \pi \tau_0 \beta (1 + \eta)(\delta - \delta_0)[\ell - (\delta - \delta_0)] \tag{3.17}$$

and $P = 0$ for $\delta > \ell$ where

$$\eta = (V_f E_f)/(V_m E_m)$$

$$\omega = \sqrt{4(1 + \eta)\beta \tau_0 / E_f}$$

$$\delta_0 = \frac{2d}{\beta}\left[\cosh\left(\frac{\omega \ell}{d}\right) - 1\right]$$

where V_f, V_m is the volume fraction of fibre and matrix, respectively; d, the fibre diameter; δ_0, the crack opening/displacement at which debonding is complete along the full length of the embedded fibre segment; τ_0, the the value obtained for a constant shear model, for $\beta \to 0$ in Eq. 3.16; ℓ, the embedded length.

3.2.4 Frictional slip and normal stresses

The discussion in the previous section emphasized the important role of frictional slip resistance in the stress-transfer mechanisms. The magnitude of this effect is a function of the *normal* stresses that develop across the interface, and the apparent coefficient of friction:

$$\tau_{fu} = \mu \sigma_n \tag{3.18}$$

where τ_{fu} is the frictional shear stress; μ, the coefficient of friction; and σ_n, the fibre–matrix normal stress (contact pressure); the sign of σ_n is negative for compression.

The normal stresses (sometimes called clamping stresses) are generated by a misfit, δ, between the radius of the fibre and the radius of the hole in the 'free matrix', that is the matrix in the absence of the fibre [11]. A reduction in the radius of the matrix hole for any reason would induce normal compressive stresses which increase the frictional resistance. The normal strain, ε_{no} associated with this misfit would be

$$\varepsilon_{no} = \frac{\delta}{r} \tag{3.19}$$

If the value of δ is negative (i.e. the reduction in the radius of the hole in the free matrix is greater than the reduction of the fibre radius), the resulting normal compressive stress will lead to enhanced frictional resistance. If, however, the value of δ is positive compressive normal stresses will not develop, and the frictional resistance to slip will be greatly reduced. Four effects should be considered when predicting the magnitude of the fibre–matrix misfit and the normal stresses:

1 *Volume changes*   Drying shrinkage of the matrix is an example of matrix contraction in excess of that of the fibres. Although this effect can be simply analysed in terms of shrinkage strains, it may involve some additional complexities arising from cracks generated during shrinkage.
2 *External stresses*   External pressure on the composite could generate normal compressive stresses across the interface.
3 *Poisson effects*   If the Poisson's ratio of the fibre is smaller than that of the matrix, then under tensile loading the misfit δ would be negative, resulting in normal compressive stresses and increased frictional resistance. However, it should be noted that the tensile strain of the fibre where it enters the matrix, in the vicinity of a crack, is much greater than that of the matrix, Figure 3.9 [20]. This may result locally in a Poisson contraction in the fibre greater than that of the matrix, even though the Poisson's ratio of the fibre is smaller. In this case, tensile normal stresses may be generated at the interface, and reduce or even eliminate the frictional resistance to slip.
4 *Plastic deformation*   Yielding of fibres during loading may result in a large plastic radial contraction of the fibre, which may reduce the frictional slip resistance.

Pinchin and Tabor [21,22] considered the first three effects, namely volume change, external pressure and the Poisson effect. They derived a general equation for the load build-up in a fibre, P_f, at a distance x from its edge:

$$P_f = -\frac{\pi r \delta_0 E_f}{v_f} \left\{ 1 - \exp \left[\frac{-2v_f \mu x}{E_f r \left((1 + v_m)/E_m + (1 - v_f)/E_f \right)} \right] \right\} \tag{3.20}$$

where δ_0 is the misfit value due to matrix shrinkage and external pressure (negative sign); ε_v, the Poisson contraction strain of the fibre; v_f, v_m, the Poisson's ratio of the fibre and the matrix, respectively; μ, the coefficient of friction and E_f, the modulus of elasticity of the fibre.

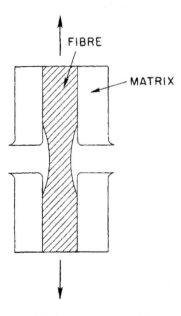

Figure 3.9   Representation of fibre and matrix in the vicinity of a crack. The Poisson contractions have been grossly exaggerated for clarity (after Kelly and Zweben [20])

For a constant embedded length, ℓ, and coefficient of friction, μ, the term in brackets in Eq. 3.20 is constant and the tensile stress in the fibre, σ_f, can be described simply as a function of the misfit, δ_0:

$$\sigma_f = -K\delta_0 \tag{3.21}$$

where

$$K = \frac{E_f}{r\nu_f}\left[1 - \exp\frac{-2\nu_f\mu\ell}{E_f r\left((1+\nu_m)/E_m + (1-\nu_f)/E_f\right)}\right] \tag{3.22}$$

Similar relations were applied by Beaumont and Aleszka [23] who expressed them somewhat differently:

$$\sigma_f = A[1 - \exp(-B\ell)] \tag{3.23}$$

where

$$A = -(E\sigma_{no})/k$$

$$B = 4k\mu/d$$

$$k = \nu_f\mu/E_f[(1 + (\nu_m/E_m)) - (1 - (\nu_f/E_f))]$$

where σ_{no} is the the radial compressive stress generated by matrix shrinkage.

A detailed analysis of the Poisson effect alone was presented by Kelly and Zweben [20]. For a simple system the normal stress across the interface can be described by Eq. 3.24, assuming that the entire composite undergoes an axial strain ε, and using elastic relations for isotropic fibres and the matrix:

$$\sigma_n = \frac{2\varepsilon(\nu_f - \nu_m)V_m}{[(V_m/K_f) + (V_f/K_m) + (1/G_m)]} \tag{3.24}$$

where K_f, K_m are the plain strain bulk moduli of the fibre and the matrix, respectively.

The sign of the normal stress will depend on $(\nu_f - \nu_m)$; if $\nu_f < \nu_m$, the normal stress will be negative (i.e. compression), resulting in frictional resistance that can be calculated by combining Eqs 3.24 and 3.18.

The coupling between shrinkage (i.e. misfit value in Eq. 3.19) and the modulus of elasticity, as predicted from Eq. 3.20, has been the subject of several experimental studies to determine the frictional bond, reported by Pinchin and Tabor [21,22], Beaumont and Aleszka [23] and Stang [24].

Pinchin and Tabor [21,22] studied the effect of normal stresses (confining pressures) on pull-out specimens. The effect of the level of normal stress was evaluated, up to a maximum of $28.5\,N/mm^2$. They calculated a fibre–matrix radius misfit value, δ_0, to be about $-0.2\,\mu m$ (Eq. 3.20), in the case of steel FRC specimens. Obviously, this value would be sensitive to matrix shrinkage, which is dependent on its composition and curing. The absolute misfit value was found to decrease during pull-out, which was suggested to be the result of local matrix compaction in the vicinity of the pulled-out fibre.

Beaumont and Aleszka [23] studied pull-out configurations with varying steel fibre lengths, using both a plain concrete matrix and a concrete matrix impregnated with PMMA. The values of A and B, and subsequently the values of μ and σ_{no} in Eq. 3.23, were obtained empirically, by curve fitting of the results of the characteristic pull-out loads obtained for various embedded lengths. The coefficients of friction were 0.6 for both the normal mortar and polymer impregnated mortar; the σ_{no} values were -19.2 and -32.6 MPa (i.e. compression) for the normal mortar and the polymer impregnated mortar, respectively. The -19.2 MPa value is of the same order of magnitude as one can obtain from Pinchin and Tabor's results.

Stang [24] developed a novel technique to measure the clamping stresses induced around fibres. He tested fibres of different moduli in a normal Portland cement matrix and in a microsilica–cement matrix which underwent autogenous curing (the microsilica matrix shrinks much more than the Portland cement matrix). The test method and the interpretation of the results were based on the concept of a matrix shrinking against a rigid inclusion, that is restrained shrinkage of a matrix against a fibre core. In the calculation of the bond stress, assumptions had to be made regarding the coefficient of friction: for steel, values in the range 0.05–0.25 were used (based on data in studies by Geng and Leung [25], Pinchin and Tabor [21]); for polypropylene, values in the range 0.45–0.75 (based on friction

Table 3.1 Predicted clamping stresses and frictional bond in fibre–cement interface with two different matrices: Portland cement (PC) and Portland cement with 10% microsilica (MS) (after Stang [24])

Fibre	Modulus (GPa)	μ	Clamping stress (MPa)		Frictional bond (MPa)	
			PC	MS	PC	MS
Steel	210	0.05–0.25	6.4	19	0.3–1.6	1–4.8
Carbon	240	0.05–0.1	5.9	17	0.3–0.6	0.9–1.7
Polypropylene	1	0.3–0.5	1.1	3.1	0.3–0.6	0.9–1.6

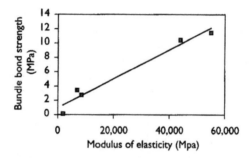

Figure 3.10 Relation between the bond strength of the surface of bundled filaments and their modulus of elasticity (after Peled and Bentur [26]).

measurements between concrete and butyl sheets), and 0.05–0.10 for carbon. The results of the study, shown in Table 3.1, clearly demonstrate the inherent influence of the bulk modulus of the fibre and the shrinkage of the matrix on the bond: a higher modulus fibre is expected to induce more restraint during shrinkage and the expected result is higher clamping stresses for a matrix which experiences greater shrinkage. In a recent study of the bond of various filaments, Peled and Bentur [26] demonstrated a significant relation between the measured pull-out bond strength and the modulus of the fibre (Figure 3.10) confirming the significance of the relations between bulk properties of the matrix and the fibre in controlling the frictional bond.

Kelly and Zweben [20] indicated that although Eq. 3.20 may predict a compressive normal stress due to the Poisson effect, this may not occur in some special cases, in which fibres are being pulled out of a matrix (i.e. a pull-out test, or fibres bridging across a crack in the composite). In these instances, the tensile strain in the fibre as it enters the matrix is high, while that of the matrix is low. At the cracked composite surface (or at the matrix surface in the pull-out test) the matrix is practically stress-free. Therefore, in these regions the normal stress across the

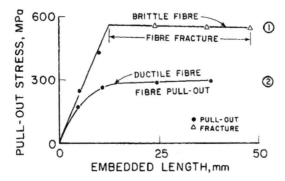

Figure 3.11 Schematic description of the pull out curves of brittle and ductile fibres embedded in a brittle matrix: (a) maximum pull-out load vs. embedded length; (b) pull
out load vs. fibre displacement. Nickel wires in cement matrix (after Morton
and Groves [27]).

interface will be tensile, resulting in debonding with no frictional resistance, that
is a localized unstable debonding failure at the interface. This may lead to irre-
producible results in pull-out tests which are carried out to determine frictional
resistance. Kelly and Zweben suggested that pull-out tests will yield consistent
results only if some matrix contraction has occurred prior to the pull-out test (e.g.
by matrix shrinkage during curing). Thus, in the pull-out configuration or in crack
bridging, the absence of residual stresses may lead to catastrophic debonding
without any frictional resistance, even though compressive normal stresses might
have been predicted on the basis of the Poisson effect (Eq. 3.20). This situation
is represented schematically in Figure 3.9, demonstrating the strains in a cracked
composite with a fibre bridging across the crack.

Kelly and Zweben extended their analysis to show that there may be significant
differences when considering an array of fibres in a composite. In such cases, lateral
restraint may be provided under certain circumstances, preventing the debonding
that is predicted to occur when a single fibre is considered.

Morton and Groves [27] have pointed out that in considering Poisson effects,
special attention should be given to the nature of the fibre, in particular whether
it can undergo plastic deformation and yielding prior to failure. With a brittle
fibre, the maximum pull-out load vs. embedded length relation assumes the shape
shown by curve (1) in Figure 3.11. When the embedded length exceeds the critical length, fibre fracture will occur. If, however, the fibre is sufficiently ductile
it will always undergo pull-out, even with a long embedded length (curve (2)
in Figure 3.11). This behaviour can be explained in terms of the plastic yielding induced in the ductile fibre, when it has a sufficient embedded length. This
fibre yielding is accompanied by a large lateral contraction. This contraction leads
to a release in the grip originally provided by matrix shrinkage. This 'plastic

release' propagates along the fibre, with only its residual length (distant from the crack) being strongly gripped. Thus, there is continuous pull-out of the fibre under a constant force, which shows up as a horizontal line in the maximum pull-out load vs. embedded length curve (curve (2) in Figure 3.11). In a subsequent work, Bowling and Groves [28] developed a model to account for these effects, taking into consideration parameters such as fibre roughness and the minimum bonded length of the fibre portion remaining in contact with the matrix.

3.2.5   Modelling and analysis of pull-out

The analysis in Section 3.2.4 provides the basis for the modelling and interpretation of pull-out tests (Figure 3.1). The loads and displacements in the course of the test are measured, and the analysis of the curves obtained can serve to evaluate the nature of the fibre–matrix interaction (elastic, frictional or both), and the characteristic bond strength values.

3.2.5.1   Modelling of pull-out curves

It is possible to model pull-out behaviour by predicting two types of pull-out curves:

1   Maximum pull-out load vs. embedded fibre length obtained after testing several specimens, each with a different fibre embedded length.
2   Pull-out load vs. fibre displacement, obtained in a single test.

Such curves can be modelled by considering three important parameters related to the fibre geometry and fibre–matrix interfacial interaction:

1   The embedded length of the fibre, ℓ
2   The adhesional shear bond strength, τ_{au}
3   The frictional shear bond strength, τ_{fu}.

The model presented here is based on the work of Lawrence [8], with further extensions by Gray [5] and Bartos [1].

The nature of the pull-out and its quantitative behaviour will depend on the ratio between τ_{fu} and τ_{au}. Thus, three different cases should be considered.

(i) $\tau_{fu}/\tau_{au} \geq 1$

After the maximum adhesional shear stress has been exceeded, debonding will initiate and gradually develop along the entire fibre length, without

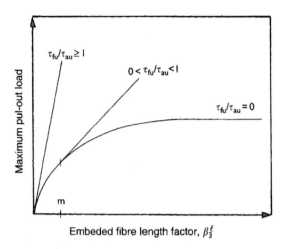

Figure 3.12 Maximum pull-out load vs. embedded length factor for different ratios of τ_{fu}/τ_{au} (after Lawrence [8]).

catastrophic failure. The maximum pull-out load will be:

$$P(\text{max}) = 2\pi \, r \, \tau_{fu} \qquad (3.25)$$

The shape of the maximum pull-out load vs. embedded length curve is linear (Figure 3.12). The shape of the pull-out load vs. fibre displacement curve during the test is presented in Figure 3.13(a). The pull-out load decreases linearly from the maximum load, as the fibre is extracted from the matrix. It is assumed that the frictional resistances before and after sliding (i.e. static and dynamic) are the same.

(ii) $0 < \tau_{fu}/\tau_{au} < 1$

In this case, after the adhesional bond strength has been exceeded, debonding will initiate and may continue without any further increase in load (i.e. catastrophic bond failure). This will take place when the length of the embedded fibre is less than a minimum value, $\ell(\text{min})$:

$$\ell(\text{min}) = \frac{1}{\beta_3} \cosh^{-1} \left(\frac{\tau_{fu}}{\tau_{au}} \right)^{1/2} \qquad (3.26)$$

where β_3 is the term $[(2\pi \, G_m/\ell n(R \cdot r)(1/A_f E_f - 1/A_m E_m)]^{1/2}$; A_f, A_m, the cross-sectional areas of the fibre and the matrix, respectively and E_f, E_m, the elastic moduli of the fibre and the matrix, respectively.

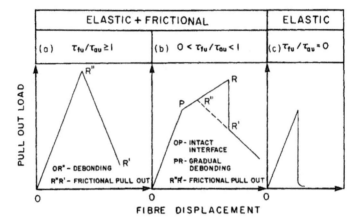

Figure 3.13 Pull-out load vs. fibre displacement for various interfacial friction conditions (after Gray [6]). (a) $\tau_{fu}/\tau_{au} \geq 1$; (b) $0 < \tau_{fu}/\tau_{au} < 1$; (c) $\tau_{fu}/\tau_{au} = 0$.

The maximum load supported in this case is:

$$\ell < \ell(\min); \qquad P(\max) = \frac{\tau_{fu}\pi\, d}{\beta_3} \tanh(\beta_3\ell) \qquad (3.27)$$

If the embedded fibre length is greater than $\ell(\min)$, catastrophic bond failure will not occur, and the fibre will gradually be extracted from the matrix after the maximum pull-out load has been reached. The maximum load supported by the fibre in this case is:

$$\ell > \ell(\min); \quad P(\max) = \frac{\tau_{fu}\pi\, d}{\beta_3} \tanh \beta_3\ell(\min) + 2\tau_{fu}\pi r\,[\ell - \ell(\min)] \qquad (3.28)$$

Once the entire fibre has been debonded, the pull-out force will drop to a level of $2\tau_{fu}\pi\, r\ell$, and will continue to decline as the fibre is extracted. These characteristics are clearly demonstrated in Figure 3.13(b).

The shape of the maximum load vs. fibre embedded length curve for this case is shown in Figure 3.12. The point marked m represents $\ell = \ell(\min)$. For shorter embedded lengths catastrophic bond failure will occur, while for longer lengths pull-out will continue, with frictional slip gradually developing over the entire embedded length. This is represented by the linear portion of the curve.

(iii) $\tau_{fu}/\tau_{au} = 0$

In this case the only mechanism of stress transfer is an elastic one, governed by shear lag equations, such as those of Greszczuk [7] and Lawrence [8].

Eq. 3.27 applies in this case. The maximum load vs. fibre embedded length for this condition is shown in Figure 3.12, and the pull-out load vs. fibre displacement curve is presented in Figure 3.13(c), showing a sharp drop of the load to zero after the maximum has been reached.

3.2.5.2   Interpretation of pull-out curves

The analysis in the previous section demonstrates the complexity of the pull-out process. The curves in Figure 3.12 indicate, for example, that increasing the embedded length will not necessarily result in a proportional increase in the pull-out load. Such proportionality, which is implicitly assumed in predictions and calculations based on average values, is only valid for the case where $\tau_{fu}/\tau_{au} \geq 1$. Therefore, evaluation of pull-out tests should not be based on the determination of a limited number of numerical parameters (maximum pull-out load, embedded fibre length and fibre cross-sectional geometry). Rather, it should include analysis of the curves obtained during such tests.

PULL-OUT LOAD VS. FIBRE DISPLACEMENT CURVES

The curves presented in Figure 3.13 provide a guide for the interpretation of curves of pull-out load vs. fibre displacement. A more detailed modelling of such curves was provided by Laws [15], who showed that the curves can assume the shapes shown in Figure 3.14. At early stages of loading, when the stress-transfer mechanism is elastic (no debonding), the curve is linear, up to point A, where debonding is initiated (i.e. the adhesional shear bond strength is exceeded). The debonding can be catastrophic (full line) or progressive (dashed line). In the latter case an increase in load is required to overcome additional frictional resistance. Whether

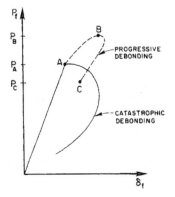

Figure 3.14 Theoretical pull out load vs. fibre extension/displacement curves derived by (Laws [15]).

failure is catastrophic or progressive will depend on the ratio τ_{fu}/τ_{au} and on the embedded length. After the maximum load has been reached in the progressive failure case (point B), the load and extension decrease until debonding is complete (point C). In actual tests, this reduction in extension is not observed, since pull-out tests are usually carried out at a constant rate of crosshead travel, which produces a curve with the shape shown in Figure 3.13(b). The load drops suddenly to the level corresponding to the frictional slip resistance generated when the fibre is debonded all along its length.

If points A and C on curves such as those shown in Figure 3.14 can be determined, one can then calculate the elastic interfacial shear bond strength (τ_{au}) and the interfacial frictional shear bond strength (τ_{fu}) using the equations developed by Laws:

$$\tau_{au} = \frac{P_A \beta_3}{2\pi r} \cdot \frac{1}{\tanh(\beta_3 \ell)} \tag{3.29}$$

$$\tau_{fu} = \frac{P_C}{2\pi r \ell} \tag{3.30}$$

A simplified calculation of the average bond strength, $\bar{\tau}$, using the definition in Eq. 3.11, and the maximum pull-out load yields:

$$\bar{\tau} = \frac{P_B}{2\pi r} \tag{3.31}$$

The average shear bond strength is an underestimate of the adhesional shear bond strength and an overestimate of the frictional shear bond strength. For short embedded lengths $\bar{\tau}$ approaches τ_{au} and for long ones $\bar{\tau}$ approaches τ_{fu}. Thus, the ratios $\bar{\tau}/\tau_{au}$ and $\bar{\tau}/\tau_{fu}$ depend on the embedded fibre length. This is still another demonstration that average bond strength has no physical significance.

A schematic description and analysis of the pull-out load vs. pull-out displacement curve and its interpretation was presented by Bartos [1] (Figure 3.15). This diagram would be obtained in a test in which the load and crosshead movement are recorded. (This is the more common test, since accurate determinations of displacements, extensions or strains of the thin fibres is a difficult task.) The following data are provided from such a diagram:

1   The maximum pull-out force P(max) at which the fibre is pulled out or broken.
2   The force P_e required to start breaking the elastic bond. At this load level there is a change in the slope of the diagram (Point E).
3   The sudden decline in the curve after the maximum has been reached indicates the termination of the debonding process. The load P_f is generated by the frictional resistance to slip which at this stage is effective over the entire fibre length:

$$P_f = \tau_{fu} \tau \, r \, \ell \tag{3.32}$$

If τ_{fu} exceeds τ_{au} the steep drop occurring at P(max) can be eliminated.

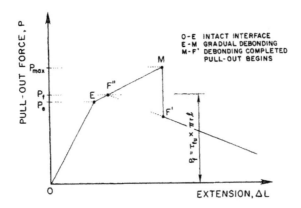

Figure 3.15 A simplified pull-out force–displacement diagram (after Bartos [1]).

Equation (3.32) presents a simplistic approach to frictional pull-out resistance, which ignores the Poisson effect. When this effect is considered (Section 3.2.4), more complex and realistic relations can be developed.

4   The decreasing curve commencing at point F′ is the result of the frictional resistance while the fibre is being extracted from the matrix, with its embedded length continuously decreasing. This zone is characterized by a 'stick-slip' behaviour, which occurs when the ductile fibre is drawn from the brittle matrix. As a result, the curve in this zone is usually not smooth.

MAXIMUM PULL-OUT LOAD VS. FIBRE EMBEDDED LENGTH CURVES

A detailed analysis of maximum pull-out load vs. fibre embedded length curves has been presented by Bartos [1,29], based essentially on the concepts discussed in Section 3.2.5.1. Bartos classified the curves into three types (Figure 3.16):

1   Elastic shear stress without any debonding (i.e. $\tau_{fu}/\tau_{au} = 0$), Figure 3.16(a). This elastic curve is shown also as dotted lines in Figures 3.16(c) and (d), and is independent of the fibre strength.
2   Frictional shear stress without any elastic stress transfer (i.e. $\tau_{fu}/\tau_{au} = \infty$), Figure 3.16(b).
3   Combined elastic and frictional stress transfer (i.e. $0 < \tau_{fu}/\tau_{au} < \infty$) Figure 3.16(c) and (d).

Bartos described the shear stress transfer in terms of shear flow, q (i.e. shear force per unit length), rather than shear stress, τ, in order to avoid considering the

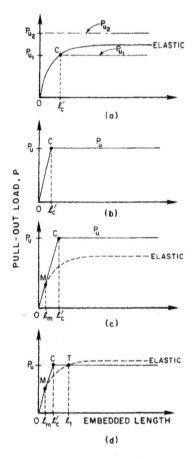

Figure 3.16 Maximum pull-out force vs. embedded length curves (after Bartos [1,29]).
(a) elastic bond only; (b) frictional bond only; (c) elastic and frictional bond
combined, strong fibre; and (d) elastic and frictional bond combined, weak
fibre. The dashed lines in (c) and (d) are identical to the elastic curve in
(a). P_u is the fibre fracture load.

perimeter of the fibre which is sometimes not well defined. The breaking load of
the fibre, P_u, plots on these curves as a horizontal line. Its intersection with the
pull-out curves (points C) defines a length ℓ'_c, which is related to the critical length,
ℓ_c. The term critical length, ℓ_c which will be discussed in detail in Chapter 4, refers
to the length of the fibres in the composite at which the fibre first break rather than
pull out. In this analysis $\ell'_c = \ell_c/2$.

For elastic shear stress transfer (Figure 3.16(a)), and $\ell < \ell'_c$, the mode of
failure will be instantaneous debonding and fibre extraction. A longer embedded

length will lead to failure by fibre breaking, without any prior debonding. For higher strength fibres (line P_u, – Figure 3.16(a)) no fibre breaking will occur, regardless of the embedded length. The mode of failure will always be catastrophic debonding.

In the frictional stress transfer case (Figure 3.16(b)) where no elastic bonding is effective, the shape of the curve is linear, up to length ℓ_c'. The slope depends on the frictional resistance; a higher resistance will result in a higher slope and shorter critical length. When $\ell < \ell_c'$, the mode of failure will be frictional slip and fibre extraction. For a longer embedded length ($\ell > \ell_c'$) fibre failure will precede debonding.

These two cases are extremes. In practice, FRC composites are characterized by combined elastic and frictional shear stress transfer (Figures 3.16(c) and (d)). The characteristic points on this curve are point C (describing ℓ_c' and P_u) and point M, which is the dividing point between catastrophic debonding failure (when $\ell < \ell_m$) and progressive debonding (when $\ell > \ell_m$). The change in the shape at this point from a curved line to a straight line was described in Section 3.2.5.1, in which the value of ℓ_m was calculated (ℓ (min) in Eq. 3.26).

In addition, the failure process during pull-out tests should be considered not only in terms of the length of the fibre, but also in terms of the strength of the fibre (breaking load). Two cases, of strong and weak fibres, are considered:

Strong fibre (Figure 3.16(d)):

1   $\ell_m > \ell$: the fibre is extracted, preceded by catastrophic debonding.
2   $\ell > \ell_m$: the fibre is extracted, preceded by progressive debonding.

Weak fibre (Figure 3.16(d)): For weaker fibres, there is another characteristic point, T, at which the elastic curve exceeds the fibre strength.

1   $\ell_m > \ell$: the fibre is extracted, preceded by catastrophic debonding.
2   $\ell_c' > \ell > \ell_m$: the fibre is extracted, preceded by progressive debonding.
3   $\ell_t > \ell > \ell_c'$: the fibre is broken, preceded by some progressive debonding.
4   $\ell > \ell_t$: the fibre is broken, without any prior debonding.

Bartos [29] also developed a graphical procedure from which the maximum load vs. embedded fibre length curve (Figure 3.16(c) and (d)) can readily be constructed on the basis of the results of a single pull-out test, in which the pull-out load vs. displacement is recorded.

3.2.6 Characteristic pull-out stress values

An interpretation of pull-out tests in terms of interfacial shear bond strength values is only feasible when the fibre geometry is constant and well defined along its length. This is usually the case when using single fibres or an array of parallel fibres, which are straight and circular in cross section. The results of such tests will be evaluated in this section. Testing of fibres with varying and complex geometries (deformed or tapered steel fibres, filamentized glass fibre strands and fibrillated polypropylene) does not lend itself readily to interpretation in terms of interfacial shear bond stresses.

In most cases, average interfacial shear bond strength values have been reported in the range 0.8–13 MPa [10,30–33]. As pointed out earlier, the physical signifi-cance of such shear bond strength values is limited. If the test results were based on short embedded lengths, the average interfacial shear bond strength would approach the adhesional shear bond strength, τ_{au}. For long embedded lengths, it would approach the frictional shear bond strength, τ_{fu}. Only a few attempts were made to determine τ_{au} and τ_{fu} from test results. These attempts were based on three different methods of analysis:

1 Interpretation of fibre pull-out load vs. fibre displacement curves [10,15,23].
2 Interpretation of maximum fibre pull-out load vs. fibre embedded length curves [29,32].
3 Interpretation of test results of the actual composite (stress–strain curves and crack spacing) to determine the interfacial shear strength values indirectly using analytical models which accounted for the composite behaviour in ten-sion and flexure. Results of such interpretations will be discussed separately in Section 4.5.

Determination of τ_{au} and τ_{fu} on the basis of the fibre pull-out load vs. fibre dis-placement curves requires identification of the characteristic points on the curves, as discussed in Section 3.2.5, using shear lag and frictional shear equations to cal-culate τ_{au} and τ_{fu}. Laws [15] has pointed out the limitation of such an approach, which requires the use of a value of G_m which is the shear modulus of the matrix in the interfacial zone. Since the matrix in the interfacial zone is different from the bulk matrix (see Chapter 2), the shear modulus of the bulk matrix should not be substituted for G_m.

Comparison of interfacial shear strength values obtained by different methods of analysis of pull-out curves of steel fibres from cementitious matrices are presented in Table 3.2. The results in this table clearly indicate the large range in interfacial shear bond strength values, from 4 to 100 MPa for τ_{au} and from 1 to 10 MPa for τ_{fu}. This order of magnitude range probably reflects differences in the materials system (fibre surface and matrix composition), but it is very likely that it also has much to do with the experimental difficulties inherent in the pull-out test: it would be very sensitive to the method of sample preparation and the pull-out test itself (gripping, alignment, rate of pull-out, environmental conditions). As a result, the variability of such tests is large, leading to coefficients of variation of 30% and even more [30]. In addition, the shapes of the curves do not always lend themselves easily to definition of the characteristic points:

1 The point of change in slope of the fibre pull-out load vs. fibre displacement curve (point E in Figure 3.15) is not always readily resolved [15].

Table 3.2 Characteristic interfacial shear strength values reported for steel fibre–cement systems tested in pull-out

Method of analysis	Reference	τ_{au} (MPa)	τ_{fu} (MPa)	Curing conditions	Remarks
Curves of fibre pull-out load vs. fibre displacement	Laws [15]	15.5 7.4	7.7 4.9	Air Water	Analysis of curves
	Beaumont and Aleszka [23]	9.0 19.5	11.5[a] 19.6[a]	Water	Mortar Polymer impregnated mortar
	Gopalaratnam and Shah [10]	4.13	1.9		Analysis of curves reported by Naaman and Shah [33]
Curves of maximum pull-out load vs. embedded length	Gray [30]	94.7	1.2	Water	Analysed after Laws [15]
		45.0	1.3	Water	Analysed after Lawrence [8]
		38.4	1.2	Water	Analysed after Bartos [29]

Note
a Determined from [23] by multiplying σ_{no} by the value reported there for coefficient of friction (0.6).

2   The point of change in slope of the maximum pull out load vs. fibre embedded length curves (point M in Figure 3.15) may occur at very small length values of 2–3 mm [31], where the experimental curve is not well defined.

It is interesting to note that when the same results are analysed by different models (Gray's results in Table 3.2), the characteristic values obtained are usually similar. This should not come as a total surprise, since the different models are based on the same concepts: shear lag combined with frictional resistance. It should also be noted that τ_{au} reflects local adhesional strength calculated on the basis of shear stress concentration profiles derived by various models, whereas τ_{fu} is an average value. Therefore the τ_{au} values can be quite high, and exceed the tensile strength of the matrix.

Some of the differences in the values in Table 3.2 may be explained on the basis of environmental conditions. The higher bond values calculated by Laws [15] for air-cured specimen may be the result of the matrix shrinkage, which leads to higher normal compressive stresses across the interface, and higher shear resistance. Yet, this may not be always the case: Pinchin and Tabor [32] obtained lower pull-out loads in sealed specimens that underwent some shrinkage, compared to water-cured specimens which had swelled. They suggested that microcracking, which

Table 3.3 Bond strength values of fibres of different moduli of elasticity [34]

Fibre	Fibre modulus (GPa)	Fibre diameter (mm)	τ_{ave} (MPa)	τ_a (MPa)	τ_f (MPa)	Reference
Steel	210	0.1–1.0	—	7.4–94.7	1.2–4.9	[34][a]
Steel	210	0.1–1.0	0.95–4.2	—	—	[35][a]
Steel	210		2	—	1.2	[37]
Steel	210	0.40,0.76	1.5	—	—	[36]
Steel	210	0.19	1.9	—	—	[38]
Steel	210		—	1.49	1.49	[39]
Steel	210	0.20	—	0.78–1.12	0.43–1.05	[40]
Poly-propylene	0.40	0.51	0.45	—	—	[25]
Poly-ethylene	0.89	0.25	0.11	—	—	[41]
Nylon	6	0.027	0.16	—	—	[42]
Kevlar 49		0.012	4.5	—	—	[42]
Poly-ethylene	120	0.038	1.02	—	—	[42]
spectra		0.042	0.40–0.63	—	—	[43]
Carbon	240	0.010	0.52–0.66	—	—	[43]

Note
a Review of data in literature.

can be induced during shrinkage, might have counteracted the increased gripping effect that one would expect to occur due to matrix shrinkage.

In a review of the interfacial properties of fibre–cement composites, Bentur *et al.* [34] compiled bond strength values reported in a variety of studies and analysed the effect of fibre composition and matrix properties. A tendency for higher bond for higher modulus fibres could be observed (Table 3.3), [35–43] as expected from the analysis presented in Section 3.2.4. The fact that the data in Table 3.3 demonstrate only a trend and not a clear-cut relationship may be associated with the different production methods and curing, leading to differences in clamping stresses, which are not the result of differences in the moduli of the fibres.

The effect of the matrix composition for fibres of different type and shape (steel macrofibres) compiled in [34] can highlight some of the parameters which control the bond level (Table 3.4) [44,45].

1 Curing   Continuous curing results in densification of the ITZ microstructure and in improvement in the bond. Although this general trend is expected, its influence was found to be greater on the bond characteristics than on the bulk properties. For example, a modest increase in the curing time from 14 to 28 days resulted in almost doubling of the bond, which is much greater than the modest increase in the strength of the matrix due to this extended curing. This reflects

Table 3.4 Effect of several variables on bond strength values of macrofibres [34]

Diameter (mm)	Matrix	τ_{ave} (MPa)	τ_a (MPa)	τ_f (MPa)	Reference
Effect of age					
0.2	OPC 14 days	1.5	—	0.84	[37]
	OPC 28 days	2.0	—	1.2	
	OPC 14 days	—	1.12	1.05	[44]
	OPC 28 days	—	2.74	1.97	
Effect of silica fume					
0.19	OPC	2.0	—	1.20	[37]
	OPC+10%SF	2.5	—	1.68	
	OPC+20%SF	2.8	—	2.57	
	OPC	1.9	—	—	[39]
	DSP	4.4	—	—	
Effect of polymeric additive					
0.40–0.76	OPC	1.5	—	—	[38]
	OPC+ PVA	2.5–2.8	—	—	
0.5	OPC	—	1.49	1.49	[12]
	OPC +Latex	—	9.80	1.82	
Effect of processing					
0.6	no processing	0.7	—	—	[45]
	20 min processing	1.0	—	—	
	40 min processing	1.3	—	—	

the fact that the ITZ microstructure is much more porous to start with, and therefore the influence of extended curing is much greater; longer time periods are needed for diffusion and deposition of hydration products in the large pores within the ITZ.

2 Modification of the cementitious binder with silica fume The influences of silica fume in densification of the ITZ is well documented. The positive effect of the silica fume stems from its physical characteristics (where its small-sized particles can pack more readily at the surface of the reinforcing inclusion to diminish the wall effect as well as to reduce bleeding) and chemical nature which can promote pozzolanic reactions within the ITZ. Interfacial bond can be enhanced by 35% and more.

3 Modification of the cementitious binder with polymers The use of polymer dispersions (latex) as an additive to cementitious binders is well established. The polymer particles are smaller by orders of magnitude than the cement grains, and they coalesce to form a continuous film. The use of water dispersed acrylics and PVA polymers (which can more readily be placed at the fibre interface and later on coalesce into a film which could be interlaced within the gel particles) resulted in a fine interfacial structure with much higher bond strength.

4 Intensive processing The ITZ microstructure is expected to be dependent on the rate of shear applied during the processing of the composite. This intensity of processing will influence the density of placement of cement particles in the vicinity of the inclusion surface as well as the rate of bleeding. Controlled tests with steel fibres demonstrated the significance of this parameter. Extended mixing of paste matrix around a steel fibre resulted in reduction in porosity and enhancement in the microhardness values of the ITZ. These changes were accompanied by enhanced bond. These trends are of practical significance when considering the optimal production processes, but they also indicate that one should be careful when considering test results of bond and ITZ microstructural characteristics based on evaluation of composite specimens in which a fibre is simply placed in a fluid matrix.

5 Surface treatment of the fibres (not shown in Table 3.4) Several surface treatments of polymer fibres have been reported to be effective in improving the bond. Such treatments are intended to enhance the chemical affinity between the fibre and the matrix and to cause some roughening of the surface to generate mechanical anchoring. Plasma treatments have been applied to enhance chemical activity of the polymer surface, replacing hydrogen on the backbone of the macromolecule with polar groups [46]. An increase by 20–100% was reported, depending on the fibre type. Treatments by various solutions were reported to be effective in enhancing the bond of polyethylene fibres, but most effective was roughening of the fibre surface [47].

3.3 Oriented fibres

For analysis of the contribution of fibres oriented at an angle θ with respect to the load direction, two different geometrical details must be considered: fibres maintaining a constant angle along their length (Figure 3.17(a)), and fibres exhibiting local flexure at the crack surface due to geometrical constraints (Figure 3.17(b)). The constant angle is valid primarily in the pre-cracking case, while the local flexure occurs mainly in the post-cracking case. These two cases will be treated separately.

3.3.1 Orientation effect assuming a constant fibre angle

The analytical treatment of this problem is different for uncracked and cracked composites. The first case is applicable to the contribution of the fibre to the modulus of elasticity; the second is more relevant to strength.

(a) Uncracked composite Krenchel [48] has treated the orientation efficiency in two ways, the first based on the assumption that the composite is constrained (i.e. subject to deformation only in the direction of the applied stress) and the second based on the assumption that deformation occurs in the other directions too. For the first case he showed that the pull-out load for fibres oriented at an angle θ is $\cos^4 \theta$ that of the aligned fibre. Thus, for fibres oriented at an angle of 45° the pull-out resistance is 0.25 that of aligned fibres. For fibres distributed at various orientations, the load-bearing capacity relative to a similar volume of

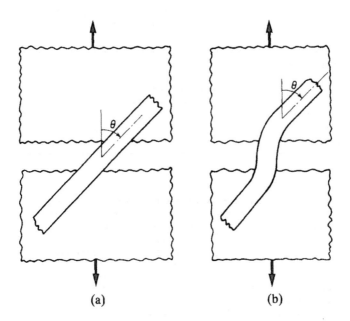

Figure 3.17 The intersection of an oriented fibre with a crack assuming (a) constant fibre orientation across the crack; (b) local fibre bending around the crack.

aligned fibres, η_θ, can be calculated as:

$$\eta_\theta = \sum a_\theta \cos^4 \theta \qquad (3.33)$$

where a_θ is the proportion of fibres oriented at an angle θ. The calculated value of η_θ is by definition the orientation efficiency, ranging from 0 to 1 (for more details see Section 4.2).

The equations for the unrestrained case are more complex, since the orientation efficiency depends on the fibre volume content, Poisson's ratio of the matrix and the ratio between the elastic moduli of the fibre and matrix. They may be found in refs. [48] and [2]. Krenchel noted that the differences in the results obtained using the two assumptions are small. This is demonstrated in Table 3.5 which compares the efficiency factors derived by Krenchel for the constrained assumption and those derived by Cox [2] for the unconstrained case.

(b) Cracked composite   A number of investigators have calculated the efficiency factors in the post-cracking case, using different assumptions. The efficiency values calculated were found to be in the range of 1/3 to 2/π for a random 2D arrangement and 1/6 to 1/2 for a random 3D arrangement (Table 3.6) [49–51].

Table 3.5 Orientation factors derived for constrained and unconstrained composites (after Krenchel [48] and Cox [2])

Fibre orientation	Orientation efficiency factor	
	Unconstrained	Constrained
Aligned, 1D	1	1
Random, 2D	1/3	3/8
Random, 3D	1/6	1/5

Table 3.6 Orientation efficiency factors in the post-cracking case

Reference	Orientation efficiency factor	
	Unconstrained	Constrained
Laws [49]	1/3	1/6
Aveston et al. [50]	$2/\pi$	1/2
Allen [51]	1/2	—

3.3.2   Influence of local bending in the post-cracking zone

It has been pointed out [50,52–58] that when analysing the effect of orientation, one should take into consideration the local bending of the fibre around the crack, which may be induced by geometrical considerations (Figure 3.17(b)). The orientation effects determined by the 'conventional' composite material approach includes an implied assumption of an orientation with a constant fibre angle across the crack (Figure 3.17(a)). However, local bending in the fibre around the crack will induce flexural stresses in the fibre, and at the same time will lead to local compressive stresses in the matrix. A complex state of stress will be developed (Figure 3.18) and the overall behaviour will depend to a large extent on the overall balance between the rigidity and ductility of the matrix and the fibre, as seen schematically in Figure 3.19. Within this context two different situations should be considered:

1   Ductile fibres bridging the cracks in a more brittle matrix.
2   Brittle fibres bridging the cracks in a more ductile matrix.

If the fibre is ductile and of low modulus it will easily bend and a dowel action may be induced leading to additional pull-out resistance. This effect may compensate for the reduced efficiency when considering only the inclination angle in a straight fibre [33,42,53,55–57,59]. If the fibre is brittle and of higher modulus

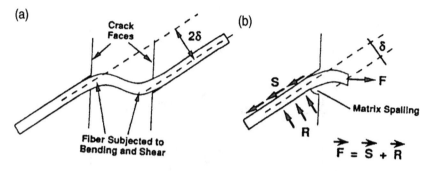

Figure 3.18 Bending of fibre across a crack (a) and components of crack bridging force; (b) (after Leung and Chi [57]).

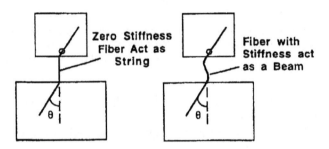

Figure 3.19 Difference in behaviour between fibre with zero stiffness and finite stiffness (after Leung and Ibanez [58]).

of elasticity, there is a build-up of local flexural stresses in the fibre, which are superimposed on the axial tensile stress, which may lead to premature failure of the fibre. This, in turn, may result in efficiency which is lower than that expected on the basis of consideration of the inclination angle only in a straight fibre, assuming only axial stresses [60–63]. The response is also dependent on the properties of the matrix in the vicinity of the fibre and its ability to withstand the additional local flexure without cracking.

These concepts can account for a range of orientation effects observed in a variety of fibre–cement systems, as demonstrated in Figure 3.20. Clearly, an increase in the orientation angle in a composite with brittle and rigid fibres (e.g. carbon) results in the reduction of pull-out resistance, whereas for ductile and low modulus fibres (e.g. polypropylene) the opposite occurs: the pull-out resistance increases with orientation, which is opposite to the trend expected in fibres which remain straight and do not bend. At this stage a qualification should be made, emphasizing that the trends in Figure 3.20 are for the values relative to pull-out

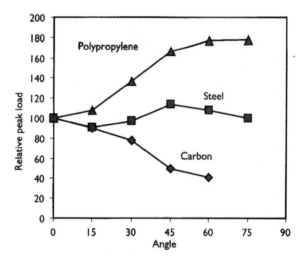

Figure 3.20 Effect of orientation on the pull-out of ductile fibres (polypropylene and steel) and a brittle carbon fibre (adopted from [6], based on data compiled from [57], [33] and [61,62] for polypropylene, steel and carbon, respectively).

resistance of aligned fibres; the latter value is usually higher for the high modulus fibres, as discussed in Section 3.2.4.

Several models have been developed to quantify these effects, by considering the following mechanisms:

1   Beam bending on elastic foundation to describe the interaction of the fibre and the matrix near the crack [56,58,60,63].
2   Matrix crushing and crumbling which occur readily in a brittle matrix which is relatively porous [60–63].
3   Energy consumed in bending and plastic deformation of a ductile fibre [55,56].
4   Frictional pulley effect at the exit point of the fibre (referred to as 'snubbing effect' by Li *et al.*) [42].

Based on these processes, models accounting for the overall behaviour were developed for cement composites with brittle fibres such as carbon and glass [60–62] and ductile ones, such as steel, polyethylene and polypropylene [55–58]. Some of these will be briefly reviewed here.

Aveston *et al.* [50] developed a model which considered the sum of the bending stress and the axial tensile stress at the point at which the fibre intersects the crack surface. They calculated the ratio between this total local stress and the axial stress that can be supported by the fibre, as presented graphically in

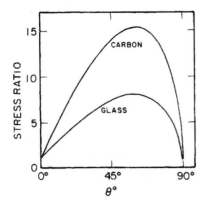

Figure 3.21 Total stress (tensile + bending) in a fibre crossing a crack at an angle θ, relative to the stress in the aligned fibre (stress ratio) for various cement composites (after Aveston *et al.* [50]).

Figure 3.21 for cementitious composites with different fibres. The maximum is obtained at intermediate values of θ; it ranges from about 7–15 for glass and carbon fibres, respectively. This implies that the stress that could be supported by inclined fibres at such an angle would be smaller by about an order of magnitude than that calculated assuming a constant fibre angle across the crack, that is, the fibre orientation efficiency factor would be considerably smaller than that calculated in Section 3.3.1.

Aveston *et al.* [50] suggested that this is not necessarily always the case, and that the overall orientation efficiency would also depend on the response of the matrix to the local flexural stresses. If the matrix is sufficiently weak, it will crumble, and the flexural stresses will be effectively relaxed. They thought this to be the case in the carbon fibre reinforced cement tested in their work. Stucke and Majumdar [52] applied this mechanism to account for the embrittlement of glass fibre reinforced cement and suggested that the densification of the ageing matrix around the fibres leads to a build-up of flexural stresses in the fibres in the cracked zone, which in turn results in premature failure. In the younger composite, the matrix interface is more porous and weaker, and crumbles before any significant flexural stress can develop in the fibres.

Morton and Groves [53,54] and Brandt [55,56] extended the analysis of flexural stresses in inclined fibres and considered a ductile fibre, such as steel, which will yield rather than break due to the local bending. In their modelling they considered either that the fibre remains in the elastic zone or that it develops plastic hinges when it is stressed beyond the elastic zone. Morton and Groves [53] showed that for short fibres, the pull-out force and pull-out work can be modelled by summing the contributions of two mechanisms: (i) pull-out resistance due to interfacial shear; and (ii) plastic deformation of the fibres which lie at an oblique angle to the crack and undergo yielding due to the local bending. The contribution of the first mechanism declines as θ increases, while the contribution of the second process has its maximum at an angle of about 45°. The contribution of the bending mechanism to the pull-out force and pull-out work was determined analytically by a simple theory based on the calculation of the force needed to produce a plastic hinge in the portion of the fibre bridging the crack. Both experimental and analytical results demonstrating this effect are shown in Figure 3.22.

When considering the practical effect of the bending mechanism of the inclined, ductile fibre, one should compare its contribution with that generated by the 'conventional' effect of simple interfacial shear resistance. Morton and Groves [53] showed that the magnitude of the pull-out force and energy induced by bending

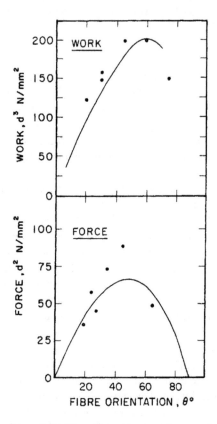

Figure 3.22 The effect of orientation on the pull-out force and work generated by the bending across the crack of a ductile nickel fibre in a brittle polyester matrix (after Morton and Groves [53]).

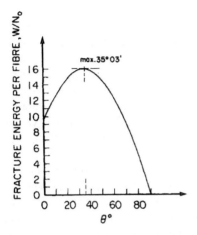

Figure 3.23 Effect of fibre orientation on the total calculated energy involved in pull-out of a ductile fibre, exhibiting plastic yielding (after Brandt [56]).

of the oriented fibre is a function of the aspect ratio of the fibre. This contribution is comparable with that induced by simple interfacial shear only when the fibre is much shorter than its critical length. Morton [54] applied these concepts to steel FRC, and showed that for randomly distributed fibres, which were either short or had poor bond, a substantial part of the composite work of fracture was due to plastic deformation of that portion of the fibres which were not perpendicular to the cracked matrix surface.

Brandt [55,56] extended the analysis of flexural stresses in ductile fibres by summing the energy consumed in five processes, as a function of the fibre orientation, θ: W_1 is the debonding of the fibres from the matrix; W_2, the pulling out of fibres, against the interfacial friction; W_3, the plastic deformation of the fibres; W_4, the crumbling ('yielding') of the matrix in compression near the exit points of the fibres and W_5, the complementary friction between the fibres and the matrix due to local compression at the bending points.

The results of his analysis are plotted in Figure 3.23, for values typical of steel fibres in a cement matrix, showing the energy contribution of each of these processes as a function of orientation angle θ.

It is interesting to note the maximum observed at an intermediate angle of about 35–50° [53–56] (Figures 3.22 and 3.23), which is in apparent contradiction to the results of Aveston *et al.* [50]. In the latter case, the efficiency is reduced at intermediate angles (Figure 3.21). The difference is that the analysis of Morton and Groves [53,54] and Brandt [55,56] was developed for a yielding type of fibre (i.e. steel) which will undergo plastic deformation when the local flexural stress is sufficiently high, rather than break as implied in Aveston *et al.*'s analysis. Naaman and Shah [33] also demonstrated that an increase in the angle of the fibre with respect to the direction of the pull-out load may not necessarily decrease the pull-out load and pull-out work. They observed a maximum in pull-out work at an intermediate angle, as predicted by the analyses of Morton and Groves [53] and Brandt [55]. Naaman and Shah suggested that the pull-out process of an inclined fibre may involve at least two mechanisms, one decreasing and the other increasing with fibre angle. The mechanism that decreases with fibre angle may be due to the effect of orientation on the axial stress contribution; the mechanism that increases with fibre angle may be associated with local yielding of the fibre at the cracked surface, when the fibre is ductile and can undergo plastic deformation.

Extensive modelling of these effects was developed by Leung, Li and their co-workers in a series of studies [19,42,57,58,63]. Two types of models were developed, for low modulus–high rupture strain fibres and for brittle fibres–brittle matrix systems: (i) the low modulus–high ultimate strain fibre systems were modelled by assuming a 'string-like' fibre which can be considered as a flexible string passing over a small frictional pulley at the exit point from the crack surface (Figure 3.24) and (ii) the brittle fibre–brittle matrix modelling was based on the concept of simulation of the fibre as a beam bending on an elastic foundation; this is similar to the concept of Bowling and Grove [28], but the beam was considered to be elastic–brittle, taking into account crumbling and spalling, which is a better simulation of a cementitious matrix than the yielding matrix assumed in Bowling and Groves' model [28].

The pulley effect for flexible high ultimate strain fibres can be considered in terms of a frictional force , F (Figure 3.24) which is the result of a normal stress, N exerted at the exit point which induces the frictional resistance [42,58]. The peak pull-out load for a fibre at an orientation θ, $P(\theta)$, is related to the load in an aligned fibre, $P(0)$:

$$P(\theta) = P(0)\exp(f\theta) \tag{3.34}$$

where f is the snubbing friction factor. This model is referred to as the snubbing friction model.

Li *et al.* [42] verified this relation experimentally and found the snubbing factor to be 0.994 and 0.702 for nylon and polypropylene fibres, respectively, indicating a continuous increase in the pull-out resistance with orientation angle. It should be noted that in the experimental derivation of f the maximum load $P(\theta)$ was normalized for the actual embedded length, which was obtained by subtracting from the original length the spalled length at the exit point, where the wedged matrix fails [42]. Leung and Ybanez [58] extended the snubbing friction model by analytically considering the matrix spalling as well as the whole process of bending: the fibre is straight at the beginning of the pull-out and it is gradually being curved and kinked at the wedge, implying that the snubbing friction is changing with increase in slip displacement (Figure 3.25). They found that no spalling

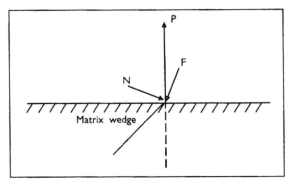

Illustration of fibre pull-out at an angle

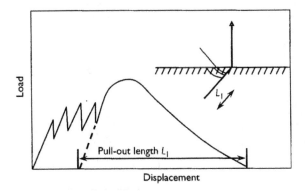

Displacement

Figure 3.24 Concepts of modelling of the behaviour of a string-like fibre pulled at an angle θ by considering a small frictional pulley at the point of exit of the fibre from the matrix (after Li *et al.* [42]).

is predicted up to about 45° angle. This is consistent with their experimental results showing spalling taking place only at a 60° angle and above, and this occurs only after a small slip displacement has taken place. Before this spalling, at a small displacement, fibre bending rather than fibre snubbing is the dominant mechanism. From a practical point of view, the high pull-out resistance at these small crack openings should be neglected, and the characteristic pull-out curve for predicting the macroscopic behaviour should be based on the post-spalling part of the pull-out curve. This extended model can account for the pull-out–crack-opening curve of a string-like fibre, showing that at a small angle the resistance is continuously reduced with increase in opening, while at higher angle it is maintained over a crack opening range of a few millimetres before being reduced (Figure 3.26).

Katz and Li [61,62] modelled the flexural behaviour by simulating the fibre partially as a beam resting on an elastic foundation and partially as a cantilever

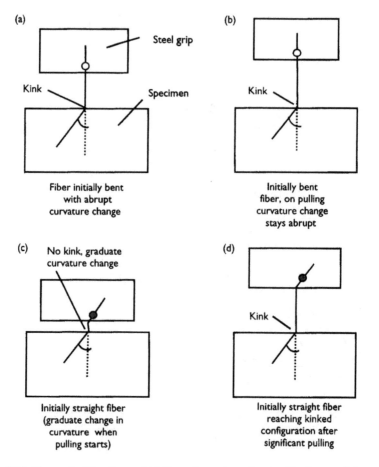

Figure 3.25 The gradual bending and kinking of an oriented string-like fibre with increase in crack opening (after Leung and Ybanez [58]).

sticking out into the crack and bridging it (Figure 3.27). The relations between the geometrical characteristics are as follows:

$$\delta = 0.5u \sin \theta \qquad (3.35)$$

$$l = 0.5d \tan \theta + 0.5u \cos \theta \qquad (3.36)$$

where δ is the deflection; l, the cantilever length; u, the crack opening; θ, the orientation angle and d, the fibre diameter.

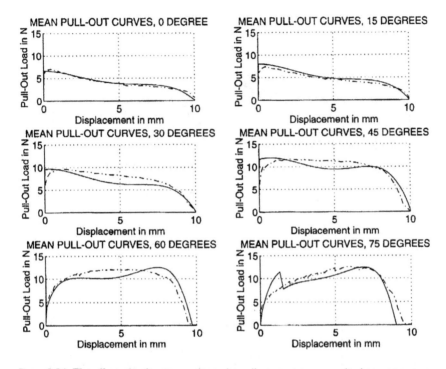

Figure 3.26 The effect of inclination angle on the pull-out resistance vs. displacement curves in pull-out tests of a flexible fibre (0.508 mm diameter polypropylene in a 0.5 w/c mortar) (after Leung and Ybanez [58]).

The loads on the cantilever part are axial load, N, and bending load, P. The supported beam part is under axial load, N, shear load, V, bending moment, M. The shear load V and the bending moment M lead to the following deflection, y_s, of the supported part, which can be calculated as:

$$y_s = \frac{2V\lambda}{k} e^{-\lambda x_s} \cos(\lambda x_s) \tag{3.37}$$

$$y_s = \frac{2M\lambda^2}{k} e^{-\lambda x_s} [\cos(\lambda x_s) - \sin(\lambda x_s)] \tag{3.38}$$

where

$$\lambda = \left(\frac{k}{4E_f I_f}\right)^{1/4}$$

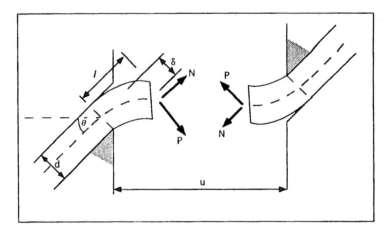

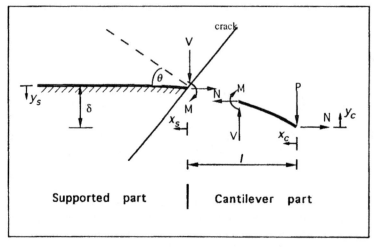

Figure 3.27 Schematic description of the fibre composed of one part which is a beam resting on an elastic foundation and the other being a cantilever sticking into the crack to bridge it: (a) overall description, (b) scheme of loads (after Katz and Li [61,62]).

where k is the matrix foundation stiffness and E_f, I_f are the modulus and moment of inertia of the fibre, respectively.

The relation between the bending load, P, and the axial load, N, acting at $x_c = 0$, and the moment on the cantilever part are:

$$M = -E_f I_f \frac{d^2 y_c}{dx_c^2} = -N y_c + P x_c \qquad (3.39)$$

The differential equation in (3.39) can be solved for boundary conditions of $y_c = 0$ at $x_c = 0$ to determine the deflection equation of the cantilever:

$$y_c = 2C \sinh(mx_c) + \frac{P}{N}x_c \tag{3.40}$$

where

$$m = \left(\frac{N}{E_f I_f}\right)^{1/2}$$

Reducing the axial load increases the deflection to infinity as N approaches 0 and in such a case the theory of large deflection must be applied.

In order to maintain the geometrical considerations for continuity, the angle of the slope of the fibre, ϕ, at the surface of the crack ($x_s = 0$) must be equal to the angle of the cantilever at $x_c = l$. These conditions are presented in the following equations:

$$\phi(x_s = 0) = \phi_c(x_c = l) \tag{3.41}$$

$$y_s(x_s = 0) + y_c(x_c = l) = \delta \tag{3.42}$$

Based on these equations the moment, M, and the shear force, V, can be obtained, and the total deflection can be calculated:

$$\delta = K_2 P \tag{3.43}$$

where

$$K_2 = -\frac{4\lambda}{k} K_1 N [m \cosh(ml) + \lambda \sinh(ml)] + 2K_1 \sinh(ml) + \frac{l}{N} \tag{3.44}$$

$$\frac{1}{K_1} = -\frac{4\lambda^2 N^2}{k} [m \cosh(ml) + \lambda \sinh(ml)] - 2Nm \cosh(ml) \tag{3.45}$$

The load N can be calculated assuming fibre–matrix frictional bond τ, and taking a pull-out length of l_N, which is only the right term in Eq. 3.36, $0.5\mu \cos \theta$ (the other term in Eq. 3.36, is related to the part of the fibre which separates from the matrix on bending, Figure 3.27):

$$N(l_N) = \begin{cases} \pi \left(\frac{1}{2} E_f d^3 \tau l_N\right)^{1/2} & \text{for } l_N < l_0 \\ \pi \tau L d \left(1 - \frac{l_N - l_0}{L}\right) & \text{for } l_0 < l_N < L \\ 0 & \text{for } l_N > L \end{cases} \tag{3.46}$$

where

$$l_0 = \frac{2L^2\tau}{E_f d};$$

$$l_N = 0.5u\cos\theta.$$

The matrix stiffness k in these equations is a function of the matrix modulus, E_m and the foundation depth below the fibre, which increases as one moves from the surface of the crack into the supported beam. Leung and Li [63] determined the ratio k/E_m by finite element analysis for E_f/E_m of 1, and found it to be about 0.20 close to the crack surface; it increases to about 0.45 as one moves from the crack into the matrix. The ratio k/E_m increases slightly for fibres stiffer than the matrix, up to 0.23 and 0.55 at the crack surface and away from it. However, it does not change much for $E_f/E_m > 6$. An average value for k/E_m of 0.25 was used in the analysis.

From these equations it is possible to calculate the reaction on the fibre. Dividing the reaction by the fibre diameter is the pressure which the fibre exerts on the matrix. It was assumed that matrix spalling occurs when this stress exceeds the spalling strength of the matrix which is taken as $\sigma_m = \varepsilon_m E_m$. The spalling length should be added to the free length of the fibre.

An additional load was considered in this analysis, which is the friction load, N_f, applied from the matrix into the fibre at the bending point. This load is a function of the reaction of the matrix into the fibre, R, and a frictional coefficient, μf:

$$N_f = \mu f R \tag{3.47}$$

The reaction R is calculated by integrating the reaction of the matrix along the fibre. The function describing this reaction decays rapidly from the crack surface towards the supported part of the fibre. The decay distance is equal to approximately the fibre diameter.

The final bridging load, F_{brdg}, can be calculated by summing all the effects considered above:

$$F_{brdg} = P\sin\theta + (N + N_f)\cos\theta \tag{3.48}$$

The tensile stress in the fibre can be calculated from this equation, and its maximum is approximately at the exit point of the fibre. To determine the total stress of the fibre at this point, the stress induced by the bending moment, M, should be added:

$$M = -2K_1 PN\sinh(ml) \tag{3.49}$$

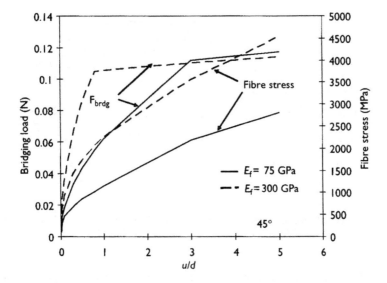

Figure 3.28 Effect of fibre modulus on bridging load and fibre stresses in oriented fibre (45°) in a 30 GPa modulus matrix (after Katz and Li [61,62]).

The maximum tensile stress in the fibre at the exit point is:

$$\sigma = \sigma_b + \sigma_p = \frac{Md}{I_f 2} + \frac{4(N + N_f)}{\pi d^2} \qquad (3.50)$$

Some interesting implications resulting from calculation based on this model are shown in Figure 3.28, demonstrating the effect of increase in fibre modulus (75 and 300 GPa) on the bridging load and fibre stresses in a 30 GPa modulus matrix with 45° oriented fibre: higher modulus results in a significantly higher stress in the fibre, but the bridging loads at crack openings which are about twice the fibre diameter or bigger, are similar. A similar relation demonstrates the effect of matrix modulus (30 and 20 GPa) for 75 GPa fibre (Figure 3.29): the reduction in matrix modulus did not have an influence on the bridging load, but resulted in reduction in the fibre stress. Such trends clearly indicate that the bulk properties of the fibres and matrix may exert considerable influence on fibre–matrix interactions in the case of oriented fibres. Also, lowering the modulus of the matrix or the fibre may in some cases reduce the stresses in the fibres without drastically affecting their bridging capacity. The combined effect of orientation and crack width using the Katz and Li model [61,62] is shown in Figure 3.30. It can be seen that the stress build-up in the fibres at larger crack opening may lead to their premature fracture.

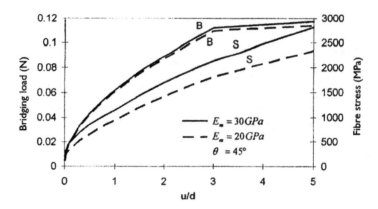

Figure 3.29 Effect of matrix modulus on bridging load, *B*, and fibre stress, *S*, in oriented fibre (45°) of 70 GPa modulus (after Katz and Li [61,62]).

It should be noted that this analytical treatment and its outcome (Figures 3.28–3.30) assume brittle fibres. The situation might be different when ductile fibres are considered, as was indeed demonstrated by Brandt for steel fibres [55,56].

3.4   Deformed fibres

In the case of conventional fibres, with diameter of the order of 0.1 mm or bigger (called sometimes macrofibres), the adhesional and frictional bond with the cementitious matrix is not sufficient for developing adequate reinforcing efficiency. This is true for steel as well as polymeric fibres. To overcome this limitation, it is common to induce deformations in the fibre to provide it with a complex shape which will provide anchoring effects. The bonding achieved by the anchoring mechanism has been shown to be much greater than the one achieved by interfacial effects.

Several experimental studies were carried out to assess the contribution of the anchoring mechanisms [e.g. 16,64–73]. Typical results are shown in Figure 3.31 for polymer fibres. Similar trends are seen for crimped and hooked steel fibres.

It can be assumed that the anchoring mechanisms involve processes by which (i) energy is dissipated as the fibre undergoes plastic deformation while being pulled out or (ii) the stressing and cracking of the matrix which is within the sphere of influence of the fibre, which extends deeper into the matrix than that of a straight fibre. Both of these mechanisms would suggest that the fibre and matrix bulk properties play a role in controlling the bonding. Indeed, in the case of steel fibre reinforced concrete, higher strength fibres provided improved bonding behaviour, as seen in Figure 3.32 for end deformed fibres where the steel properties were controlled by changes in the carbon content. Higher strength steel provided improved pull-out behaviour.

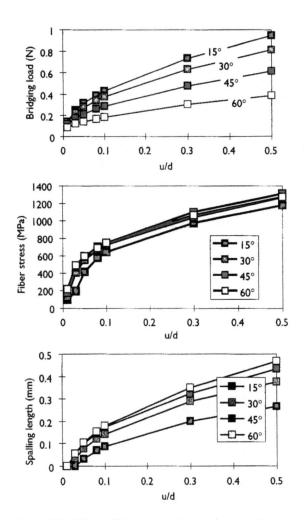

Figure 3.30 Effect of fibre orientation on the stress in oriented fibre as a function of crack opening (after Katz and Li [61,62]).

Attempts were made to resolve some of the influences of the shape of fibres by experimental evaluations, based on comparison of the pull-out behaviour of deformed fibres with similar fibres of straight geometry [16,64–70]. Naaman [16] applied an approach in which the adhesional and frictional contribution in the deformed fibre were neutralized by application of an agent (e.g. grease, wax) on the fibre surface. The effect of such treatments on the debonding load and post-peak load in pull-out tests of end hooked steel fibre are shown in Figure 3.33. The trends observed indicate a high level of contribution of the anchoring effect in the

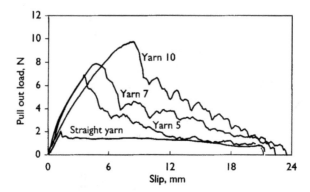

Figure 3.31   Effect of crimping (crimps per cm) of polymer fibre on the pull-out curves (after Peled and Bentur [66]).

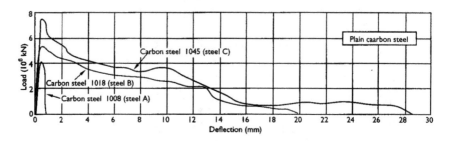

Figure 3.32   Influence of steel fibre properties on the pull-out curves of end hooked fibres (after Krishnadev *et al.* [67]).

post-peak zone, which is mainly the result of the shape of the fibre with only a small contribution of interfacial shear. Weiler *et al.* [64] reported a similar evaluation, but in their case the neutralization of the interfacial bonding was achieved in a model system of a waxed hooked fibre in an epoxy matrix. The pull-out curves of this system, along with straight and hooked fibres in a concrete mix were compared. This set of experiments also demonstrated the overwhelming effect of the hooks.

A similar approach is currently taken for polymeric fibres, which are being developed for semi-structural reinforcement of concrete, to achieve performance equivalent to that of steel fibres. The effect of the crimped shape of polymeric fibres was empirically analysed in terms of the amplitude of each crimp and the density of crimps (i.e. the number of crimp waves along the embedded length). It was shown that the pull-out resistance of each wave was very sensitive to the amplitude of the crimp (Figure 3.34(a)) and the pull-out resistance of each wave ('crimp') was

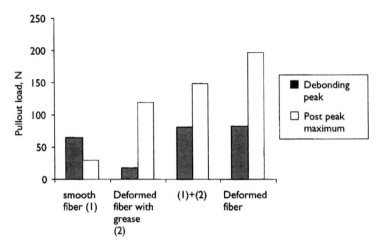

Figure 3.33 Effect of treatment of the fibre surface with grease on the characteristic pull-out loads in pull-out tests of straight and end hooked steel fibre of 0.45 mm diameter and 25.4 mm length (adapted from Naaman [16]).

linearly related to the crimp amplitude (Figure 3.34(b)). Thus, the overall pull-out resistance of the fibre, P_{max}, could be expressed in terms of the number of crimps along its embedded length (first term in Eq. 3.51) and the amplitude of each crimp (second term in Eq. 3.51):

$$P_{max} = (8Amp + 0.18)\left(\frac{l}{\lambda}\right) \qquad (3.51)$$

where P_{max} is the maximum pull-out load, N; Amp, the amplitude of each crimp, that is its length perpendicular to the fiber, mm; l, the embedded length of crimped fibre, mm and λ, the wavelength of individual crimp, that is its projected length along the fibre axis, mm.

This implies that it is not the length of the fibre which controls its bonding, but rather the number of crimps. Similarly, for hooked steel fibres, it was resolved that the contribution of the hook is independent of fibre length, and it is much greater than the contribution of the interfacial bond of a straight fibre of a similar length [68].

An alternative approach to enhance bond without changing the overall shape of the fibre is to indent or roughen the surface. This approach has been taken with steel as well as with polymer fibres [47,69] and it can result in significant influences. Singh et al. [70] demonstrated a fourfold increase in bond strength of structural polymer fibres (polypropylene with ~ 10 GPa modulus), from about 0.5 MPa to about 2 MPa in the indented fibre. At this bond level the critical length

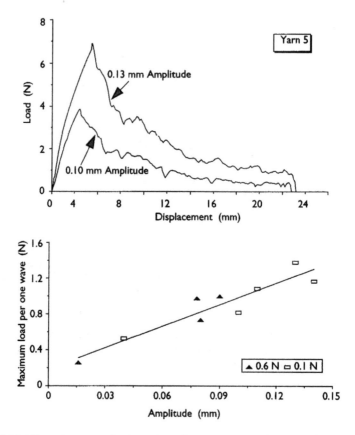

Figure 3.34 Effect of wave amplitude on the pull-out resistance of crimped fibre: (a) Pull-out load–slip curves for fibres with different crimp (wave) amplitude, and (b) maximum pull-out load per one crimp vs. the crimp (wave) amplitude (after Peled and Bentur [66]).

of the fibre is about 50 mm, which implies efficient use of a relatively short fibre that can be incorporated in the concrete by mixing.

Several models have been developed to account for the behaviour of deformed fibres. Most of them address the contribution of the hook in an end hooked fibre, assuming that the fibre is 'extracted' from the groove by undergoing plastic deformation as it is being pulled out. Such a mechanism was demonstrated in experimental studies by Van Gysel [65] and Banthia *et al.* [70]. Microscopic observations [70] indicated that there were no signs of matrix crushing or cracking in the groove left after the fibre pull-out. It was also demonstrated that the microhardness of the pulled-out fibre was higher than the virgin fibre, consistent with the work hardening induced during the plastic deformation of the extracted

Table 3.7 Influence of steel properties on pull-out behaviour
(after Banthia et al. [69])

Heat treatment	Fibre Vickers hardness (virgin fibre)	Relative peak load
Cold rolled	285	1.0
600°C, 5 min	208	0.75
600°C, 10 min	176	0.65
600°C, 15 min	162	0.57

fibre. This is in further agreement with the observations that the pull-out resistance was independent of concrete properties [68,70] over a wide range of compressive strength (33–83 MPa) and w/c ratio (0.5 and 0.9), but highly dependent on the bulk fibre properties. These characteristics are demonstrated in Figure 3.32 and also in Table 3.7 for steel fibres of different strength, which was achieved through heat treatments. This mode of fibre pull-out is consistent also with the report of Weiler *et al.* [64] who observed fibre extraction when carrying out pull-out in a model system consisting of a transparent epoxy matrix.

An analytical model for describing such extraction of deformed steel fibre was presented by Alwan *et al.* [68] and Chanvillard and Aitcin [71]. In both cases the contribution of the deformation is simulated by the formation of a plastic hinge which is generated as the fibre slips through the groove during the pull-out process. The model of Alwan *et al.* [68] will be described in some detail for end hooked steel fibre. They considered several mechanisms which are effective during the process: elastic stretching of the fibre and build-up of interfacial stress, debonding at the interface (partial and full) and mechanical clamping which is induced by cold work leading the hook to deform and be extracted through the groove (Figure 3.35). The latter action was described by means of plastic hinges (Figure 3.36(b)), and the pull-out resistance induced by such hinges was calculated by means of a frictional pulley model (Figure 3.36(a)) and the characteristics of the plastic hinge (Figure 3.36(b)) . The additional load resistance in pull-out, when the two pulleys are effective (T_1) and when only one is effective (T_2) at stage c (Figure 3.35), were calculated in terms of the yield strength f_y of the fibre:

$$\Delta P' = T_1 = \frac{(f_y \pi r f^2 / 3 \cos \theta) [1 + \mu \cos \beta / (1 - \mu \cos \beta)]}{(1 - \mu \cos \beta)} \tag{3.52}$$

$$\Delta P'' = T_2 = \frac{(f_y \pi r_f^2 / 6 \cos \theta)}{(1 - \mu \cos \beta)} \tag{3.53}$$

For the steel fibre studied in this work ($d_f = 0.5$ mm, $\theta = 45°$, $\beta = 67.5°$, $f_y = 896$ MPa, $\mu = 0.5$), the values of the contribution of the hook for two hinges and

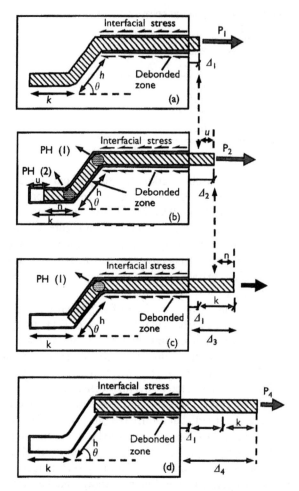

Figure 3.35 Modelling of the pull-out of hooked end steel fibre when being extracted from the matrix: (a) hooked steel fibre at the onset of complete debonding, (b) hooked steel fibre during mechanical interlock with two plastic hinges, (c) mechanical interlock with one plastic hinge, (d) hooked steel fibre at onset of frictional bond (after Alwan et al. [68]); PH is plastic hinge.

one hinge (T_1 and T_2, respectively), were 125 and 51 N, respectively. These values are consistent with the experimental pull-out, where the T_2 value is much higher than the resistance to pull-out of a straight fibre of similar diameter and length (25 and 44 N for 12.7 and 25.4 mm long fibres, respectively).

An alternative approach to develop mechanical anchoring in a fibre was recently suggested by Naaman [72]. It is based on twisting of polygonal fibres (e.g. triangular, square cross section), which allows ribs to be developed. The greater the

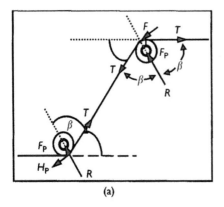

(a)

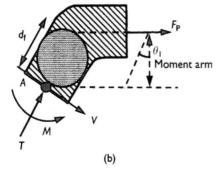

(b)

Figure 3.36 Principles of calculation of pull-out of a hooked end steel fibre: (a) line sketch of the frictional pulley model and (b) free body diagram of the fibre plastic hinge (after Alwan *et al.* [68]).

number of twists per unit length, the greater the number of ribs, resulting in a more significant post-peak slip hardening. The experimental results in this study suggest that the post-peak pull-out resistance is proportional to the density of the 'ribs' per unit length, similar to the observations for the crimped polymer fibres. Such pull-out curves could also be interpreted in terms of an apparent slip hardening bond, which is defined in a simplified way as the average bond stress at a given slip value (i.e. the pull-out load at the slip value divided by the fibre perimeter and effective embedded length at this point).

The influence of the shape of the deformed fibre is usually addressed from the point of view of pull-out resistance of the aligned fibre. A shape which may lead to improved pull-out resistance of the aligned fibre may alter its sensitivity to the effect of orientation. The effect of shape on orientation was studied by Banthia and Trottier [73,74], and the compilation of their results in terms of relative peak load is presented in Figure 3.37. It clearly shows a considerable difference with

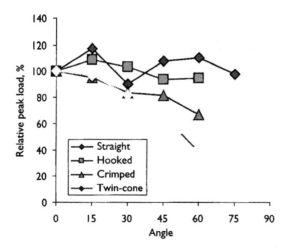

Figure 3.37 Influence of orientation angle on the pull-out resistance of steel fibres of different geometries, adopted from the data of Banthia and Trottier [73,74].

respect to orientation. At present there are no in-depth models to account for such influences. Yet, the practical implications with regard to the efficiency of the fibres are important, and need to be considered in the evaluation and development of new fibre shapes.

3.5   Interfacial microstructure and processes

The special interfacial microstructures of the matrix in the vicinity of the fibre and the complex geometry of some fibres and their surface properties can cause the mode of debonding and pull-out to be quite different from that predicted by the models based on the simple pull-out geometry presented in Figure 3.1, which is characterized by symmetrical pull-out and debonding at the actual fibre–matrix interface.

The complexity of the processes for deformed fibres has been discussed in Section 3.4. However, even for fibres which have the 'ideal' straight shape and circular cross section, the pull-out processes in the actual composite may be more complex than predicted by the analytical models. This section deals with special processes at the interface which are induced by (i) the special microstructure of the cement matrix in the vicinity of the fibres; (ii) the special characteristics of the fibre surface itself and (iii) the geometrical and microstructural characteristics of reinforcement by strands consisting each of a large number of thin filaments.

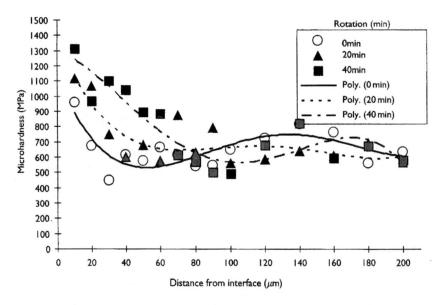

Figure 3.38 Effect of processing time on the microhardness profile in the ITZ around a steel macrofibre (after Igarashi *et al.* [45]).

3.5.1 Effects of matrix interfacial microstructure

In cementitious composites the microstructure around an inclusion (e.g. fibre, aggregate) is often different from the bulk microstructure. This zone is referred to as the interfacial transition zone (ITZ) and is described in Section 2.3.1 (Figure 2.7). The presence of this zone results in a gradient of microstructure, and as a consequence a gradient in mechanical properties which shows up in microhardness characteristics (Figure 3.38). The structure of the ITZ is not a 'pure' material property and it is affected by the processing of the composite. Intensive mixing can result in the densification of the ITZ, by reducing bleeding and forcing a better packing of the cement particles at the fibre surface. Such an effect is clearly exhibited by the change in the microhardness profile around the fibre (Figure 3.38).

The special microstructure has two consequences: (i) a porous and weak interface which will cause an overall reduction in bonding, (ii) a weak interface which is not at the fibre surface, but rather in the porous layer of the ITZ, somewhat away from the fibre surface. The first characteristic directly affects the bond and the pullout resistance, whereas the second one will have an indirect effect by influencing the mode of debonding when cracks develop in the matrix and propagate towards the fibre.

3.5.1.1 Effect on bond and pull-out resistance

Means taken to reduce the size and inhomogeneity of the ITZ will lead to its strengthening and the expected result is an overall increase in bond, as demonstrated by Igarashi *et al.* [45]: the bond of steel fibre to a paste matrix increased with mixing time: 0.7, 1.0 and 1.3 MPa of average bond for 0, 20 and 40 min of agitation of the mortar matrix against the fibre (Table 3.4).

In view of the presence of the special ITZ microstructure several models were developed, assuming a three phase material: bulk, ITZ and fibre. In these models specific characteristics (i.e. modulus of elasticity) were assigned to the ITZ. Li *et al.* [75] calculated a parameter related to the stiffness of the ITZ to resolve its value for different systems. Mobasher and Li [76] extended a fracture mechanics model to describe the pull-out curves in terms of adhesional and frictional bond, stiffness of the interface and interface toughness. They demonstrated that all of these parameters influence the pull-out behaviour, and their change with age could account for the variation in the pull-out behaviour, as discussed further in Chapter 5.

3.5.1.2 Effect on mode of debonding and cracking

As long as the interfacial microstructure inhomogeneity remains, which is frequently the case in conventional concrete reinforced with fibres, a complex mode of cracking and debonding may be induced, as outlined below.

The loading of FRC composites involves initiation and propagation of cracks in the brittle matrix, and their subsequent advance across the fibres. The fibres have a dual role: to suppress the initiation and propagation of cracks, that is to increase the matrix first cracking stress, and to bridge across the cracks once they have advanced, to prevent them from leading to a catastrophic failure. The interaction of an advancing crack with the fibres determines whether the mode of failure will be brittle or ductile.

In many of the monofilament and bundled systems this interaction results in a deviation of the crack path as it crosses the fibre, and leads to its separation into a number of microcracks [77–82], as seen in Figure 3.39(a). Observations at higher magnification show that in many instances the crack seemed to be arrested in the matrix, just ahead of the fibre, which it never quite reached: rather, it typically turned 90° on a new course parallel to the fibre, about 10–40 μm in front of the actual paste–fibre interface (Figure 3.39(b)). In such instances the debonding was not at the actual interface.

These observations can be explained on the basis of the weak interface in the transition zone (porous layer in Figure 2.5) and the Cook–Gordon crack arrest mechanism [82] shown in Figure 3.40. The stress field ahead of the elliptical crack tip consists of the familiar stress concentration profile in the direction of the external tensile stress (y direction), and a tensile stress field in the

(a)

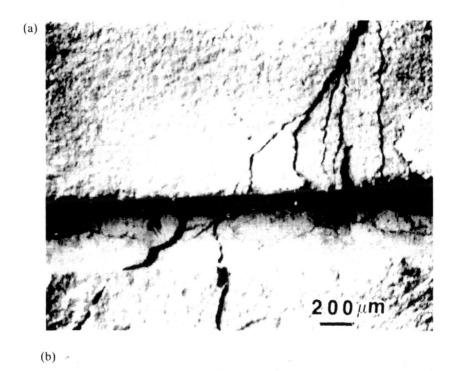

(b)

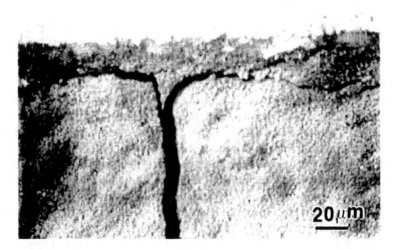

Figure 3.39 Interaction of a crack and a steel fibre (after Bentur *et al.* [81]). (a) deviation and branching of the crack (the crack was moving upwards and the fibre was removed to observe the crack path in the groove around it); (b) debonding and crack blunting in the matrix, away from the actual fibre-matrix interface (the crack was moving upwards).

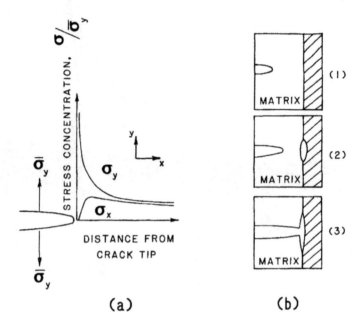

Figure 3.40 The Cook-Gordon crack arrest mechanism in a composite with a weak inter-face [82]. (a) Stress field in front of the crack tip; (b) schematic description of the crack arrest; (1) crack approaches a weak interface; (2) interface fails ahead of the main crack; (3) T-shape crack stopper. In practice the crack is usually diverted.

perpendicular direction (x direction). The latter field has its maximum not at the crack tip, but slightly ahead of it. As the propagating crack approaches the fibre, the perpendicular tensile stress may cause debonding at the fibre–matrix interface, before the crack has reached the interface (Figure 3.40(b–2)). This will occur if the interfacial bond strength is low, less than about 1/5 of the tensile strength of the matrix. When the crack finally reaches the interface, its tip will be blunted by the already present debonding crack (Figure 3.40(b–3)). The stress concentration will be reduced, and the forward propagation of the crack will be arrested. Thus, the crack path will be diverted, and instead of continuing straight across the fibre it will choose the path of least resistance which involves debonding along the weak interface away from the actual fibre–matrix interface. As a result, crack arrest, debonding and crack branching will take place (Figure 3.41).

Thus, the special microstructure of the transition zone leads to a complex crack pattern. In particular, two characteristics of this cracking pattern should be considered further, since they may produce fibre–matrix interactions which are not predicted by most analytical models.

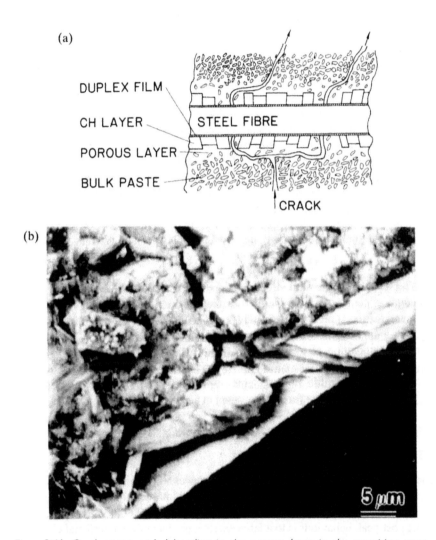

(a)

DUPLEX FILM

CH LAYER

STEEL FIBRE

POROUS LAYER

BULK PASTE

CRACK

(b)

5 μm

Figure 3.41 Crack arrest and debonding in the porous layer in the transition zone. (a) schematic description; and (b) SEM micrograph showing a crack arrested and diverted in the porous layer which is backing a massive CH rim (after Bentur *et al.* [81]).

1 Debonding away from the actual interface   In the case of monofilament rein-forcement, such as steel fibres, the debonding in the composite will tend to take place in the porous layer of the transition zone. This is quite different from the process taking place in a pull-out test. In the latter case, the stress field is such that it will promotedebonding at the actual interface. Thus, the nature of the bond measured in a pull-out test may be different from that controlling the performance in an actual composite, and the significance of this test should be re-evaluated.

2 Asymmetry and local flexure at the crack–fibre interaction   The asymmetry of the crack with respect to the fibre, caused by its deviation and branching as it crosses the fibre, leads to local flexure. Therefore, pull-out of the fibre in the actual composite cannot be considered only in terms of the axial load in the fibre and uniform interfacial shear resistance. The locally high flexural stress may lead either to premature failure of the fibre if it is brittle, or to local yielding and stretching of a ductile fibre, thus providing an additional energy absorbing mechanism. The asymmetric crack may also create an improved bonding effect, which bears some resemblance to anchoring. Such effects were discussed in greater detail, for fibres orientated at some angle with respect to the load (Section 3.3). This discussion suggests that such local flexure can also occur in the actual composite, even when the fibres are oriented parallel to the load.

The level of the flexural stress developed in the fibre will also depend on the strength of the matrix in its vicinity. If it is weak and porous, as may be the case in the transition zone, the matrix may crumble under the local flexural stress, thus relieving the flexural stress that may develop in the fibre.

The combination of local flexure, flexibility of the reinforcing unit and the density and tightness of the matrix grip around the fibre is significant in controlling the durability of brittle fibre systems, especially glass fibre reinforced cement [52,82], and this will be further discussed in Chapter 5.

3.5.2   Effect of fibre properties

The pull-out curve in the post-peak zone can exhibit a decaying nature (slip softening), ascending nature (slip hardening) and an intermediate linear decline which is characteristic of an 'ideal' frictional behaviour (Figure 3.8). The nature of the post-peak behaviour is to a large extent a product of the interfacial changes occurring during the pull-out process, which are affected by the characteristics of the ITZ but also by the rigidity of the ITZ relative to the fibre. Such influences were resolved by Geng and Leung [18]. For a fibre which is more rigid and harder than the matrix (e.g. steel) the pull-out results in damage to the ITZ, leading to abrasion and grinding and decline in the frictional resistance with increased slip (Figure 3.42). When the fibre is softer than the ITZ (e.g. nylon), the pull-out results in damage to the fibre, showing up as fibrillation and peeling of its surface, leading to some mechanical anchoring between the fibre and the matrix with the consequence of an ascending post-peak pull-out curve. Such an effect was reported to occur more readily in a hydrophilic fibre (e.g. nylon) which tends to swell as moisture penetrates into the abraded fibre (i.e. nylon would show an ascending

A. Before Debonding   B. Debonding

Steel
Fiber

CH

Crack   Porous C-S-H

C. Pullout (Short Distance)   D. Pullout (Long Distance)

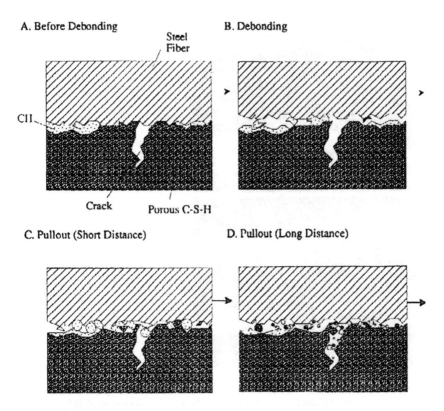

Figure 3.42 Schematic description of the evolution of damage during pull-out at the interface between a steel fibre and a cementitious matrix (after Geng and Leung [18]).

post-peak curve whereas polypropylene will exhibit a slight decay). Peeling during pull-out of a low modulus polymer fibre was observed also in crimped fibres (Figure 3.43) [84] and it was suggested that it is one of the mechanisms promoting an ascending post-peak behaviour in pull-out.

When dealing with fibres of soft surface, attention should be given to the processing of the composite. It was shown that PVA composite produced by extrusion exhibited higher strength and strain hardening behaviour than a similar composite prepared by casting [85]. The difference in behaviour was attributed to fibre roughening and peeling in the case of the extruded composite.

Special effects may occur in low modulus fibres due to the Poisson effect. The pre-tensioning of the fibre to position it in the mould, for specimen preparation, can affect the load–slip curves, as shown in Figure 3.44 for 890 MPa modulus polyethylene fibre [86]. At a relatively low level of pre-tension (0.8% of fibre

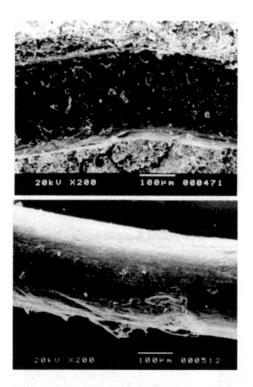

Figure 3.43 Roughening and fibrillation of the surface of a crimped low modulus polyethylene fibre, as seen after pull-out (after Bentur *et al.* [84]).

strength; 0.1 N pre-tension load in a 0.25 mm diameter fibre), the load–slip curve is insensitive to the time of pre-tension release (Figure 3.44(a)). However, at higher pre-tension (4.8% of strength; 0.6 N pre-tension load), considerable differences were observed for pre-tension release at 1 and 6 days. The lower bonding characteristics at 1 day release could be accounted for by damage to the interfacial zone due to the lateral expansion of the fibre, on the release of pre-tension. The higher bond performance at 6 days release could be explained by the presence of a much stronger ITZ, which could better resist the normal pressure generated during the release of pre-tension. It should be noted that the pre-tension of 0.6 N is relatively small (4.8% of the strength of the fibre), and the release would create a compressive stress of about 0.3 MPa, assuming a full restraint. Although this stress is low, it is perhaps sufficient to cause damage to a porous ITZ, at the very early stage of 1 day of curing. SEM observations supported this explanation, showing damage in the ITZ for the 0.6 N pre-tensioning which was released at 1 day.

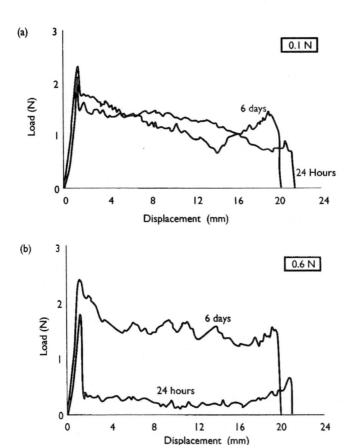

Figure 3.44 Load–slip curves of low modulus polyethylene fibre, pre-tension and released at different sequences: (a) pre-tension at 0.8 MPa and release after 1 and 6 days, and (b) pre-tension at 4.8 MPa and release after 1 and 6 days (after Peled *et al.* [86]).

3.5.3 Bundled fibres

Most of the modern man-made fibres are produced in the form of bundles of filaments (called strands), with each bundle consisting of several hundred (e.g. glass) to several thousand (e.g. carbon) filaments. The filaments in the strand are usually held together by means of a sizing, which is a thin polymeric treatment applied on the filaments during their production process. In the production process of the composite, the strand can be separated into the individual filaments to obtain a uniform distribution in the matrix, or the filaments can remain bundled, resulting

Table 3.8 Effect of bundle structure on average bond values determined by pull-out tests

Fibre	Filament diameter (μm)	Number of filaments in bundle	τ_{single} (MPa)	τ_{bundle} (MPa)	$\tau_{bundle}/\tau_{single}$	Reference
Nylon	27	220	0.16	0.051	0.321	[42]
Kevlar	12	1000	4.50	0.198	0.044	[42]
Polyethylene (spectra)	38	20	1.02	0.328	0.322	[42]
		40	1.02	0.502	0.492	
		57	1.02	0.505	0.495	
		118	1.02	0.352	0.352	
Glass[a]	12.5	204	1.1	0.38	0.350	[87]

Note
a Based on microscopic observations.

in a reinforcing unit which is a strand surrounded by the matrix (Figures 2.3 and 2.10). Which microstructure is obtained is to a large extent is dependent on the production process of the composite (shear rate, to break the bundle structure) and the tendency of the sizing to dissolve in water during the mixing process.

The interfacial microstructure is quite different when the filaments remain bundled. The cement grains which are of the same size as the filaments (and usually bigger than the spaces between the filaments) can hardly penetrate in between them. Thus, the bonding of each of the filaments is different, with the inner ones being less tightly gripped to the matrix. The overall bond of the bundle will depend on the extent to which cement grains can penetrate into the bundle.

Li *et al.* [42] formulated this case mathematically by calculating the exposed surface area of filaments in the bundle relative to the total surface area, assuming hexagonal close packing. The calculation indicated that the fraction of exposed area is 0.08 for a bundle with 200 filaments, and it reduces to 0.04 for a bundle with 2000 filaments. Experimental data of pull-out resistance of a single filament and a filament in a bundle are given in Table 3.8. The data indicate that as expected, the average bundle bond strength is considerably smaller than that of the single filament. However, in most of the systems it was in the range of 0.3–0.4, considerably greater than the calculated value of 0.04–0.08. This suggests that in practice, the bundle is somewhat opened up, allowing some penetration of cement grains or hydration products. This is supported by microscopical observations of cement reinforced with glass strands [87].

The stress transfer in a bundled reinforcement cannot be described in simple terms of interfacial stresses. Bartos [88] proposed a bonding mechanism involving 'telescopic' behaviour, in which the external filaments in the bundle are tightly bonded to the matrix and may fracture, while the internal ones may be sliding. He suggested that this mode of pull-out may provide unique advantages, if controlled, as it can induce at the same time high bonding and enhanced strength (due to the external filaments), as well as energy absorption and ductility (due to the slippage of the inner filaments). Studies of the bonding of individual filaments in a strand using micro-indentation technique confirmed that the bond of the inner filaments is considerably smaller than the outer ones [89]. It should be noted that the differences between external and internal bonding of a reinforcing unit was also reported by Sadatoshi and Hannant [90]. They showed that slip could occur at the fibre–matrix interface, but also within the fibre, in the case of fibrillated polypropylene. They developed a model of a fibre consisting of a core and external sleeve, and considered situations where slip can occur in both.

The bundled structure may also provide a reinforcing unit which is flexible and can be engaged in some local bending as it bridges a crack. Such local bending in a brittle matrix composite may produce premature fracture when the fibre is brittle; but with a bundled reinforcement, even with brittle filaments, local bending capacity can be provided by the inner filament that are only loosely bonded to the matrix and can slide one relative to the other [91,92]. These special characteristics play an important role in the case of glass fibre reinforced cements and will be further discussed in Chapters 5 and 8.

References

1. P. Bartos, 'Review paper: bond in fibre reinforced cements and concretes', *Int. J. Cem. Comp. & Ltwt. Concr.* 3, 1981, 159–177.
2. H.L. Cox, 'The elasticity and strength of paper and other fibrous materials', *Br. J. Appl. Phys.* 3, 1952, 72–79.
3. H.R. Piggott, *Load Bearing Fibre Composites*, Pergamon Press, Oxford, 1980.
4. A. Kelly, *Strong Solids*, Oxford University Press, Oxford, 1973.
5. R.J. Gray, 'Analysis of the effect of embedded fibre length on fibre debonding and pull-out from an elastic matrix, Part 1: review of theories', *J. Mater. Sci.* 19, 1984, 861–870.
6. A. Bentur, S.T. Wu, N. Banthia, R. Baggott, W. Hansen, A. Katz, C.K.Y Leung, V.C. Li, B. Mobasher, A.E. Naaman, R. Robertson, P. Soroushian, H. Stang and L.R. Taerwe, 'Fiber-matrix interfaces', in A.E. Naaman and H.W. Reinhardt (eds) *High Performance Fiber Reinforced Cement Composites - 2*, Proc. Int. RILEM Conf., Ann Arbor, MI, June 11–14, 1995, E&FN SPON, London and New York, 1996, 149–192.
7. L.B. Greszczuk, 'Theoretical studies of the mechanics of the fibre-matrix interface in composites', in *Interfaces in Composites*, American Society for Testing and Materials, ASTM STP 452, Philadelphia, PA, 1969, 42–58.
8. P. Lawrence, 'Some theoretical considerations of fibre pull-out from an elastic matrix', *J. Mater. Sci. 7*, 1972, 1–6.
9. V. Laws, P. Lawrence and R.W. Nurse, 'Reinforcement of brittle matrices by glass fibres', *J. Phys. D: Appl. Phys.* 6, 1973, 523–537.
10. V.S. Gopalaratnam and S.P. Shah, 'Tensile failure of steel fibre reinforced mortar', *J. Eng. Mech. ASCE.* 113, 1987, 635–652.

11. A.E. Naaman, G.G. Namur, J.M. Alwan and S. Najim, 'Fiber pull-out and bond slip, Part I: analytical study', *ASCE Journal of Structural Engineering.* 117, 1991, 2769–2790.

12. A.E. Naaman, G.G. Namur, J.M. Alwan and S. Najim, 'Fiber pull-out and bond slip, part II: experimental validation', *ASCE Journal of Structural Engineering* 117, 1991, 2791–2800.

13. B. Budiansky, J.W. Hutchinson and A.G. Evans, 'Matrix fracture in fiber reinforced ceramics', *Journal of the mechanics and physics of solids.* 34, 1986, 167–189.

14. Z. Lin and V.C. Li, 'Crack bridging in fiber reinforced cementitious composites with slip-hardening interfaces', *J.Mech. Phys. Solids* 45, 1997, 763–787.

15. V. Laws, 'Micromechanical aspects of the fibre-cement bond', *Composites.* 13, 1982, 145–151.

16. A.E. Naaman, Personal contribution to ref. [6] (ref. 28 in [6]).

17. A. Katz and V. Li, 'Bond properties of micro-fibers in cementitious matrix', in S. Diamond, S. Mindess, F.P. Glasser, L.W. Roberts, J.P. Skalny and L.D. Wakeley (eds) *Microstructure of Cement-Based Systems/Bonding and interfaces in Cementitious Materials*, Proc. Materials Research Society Symp., Vol. 370, Materials Research Society, Pittsburgh, PA, 1995, 529–537.

18. Y. Geng and C.K.Y Leung, 'Damage evolution of fiber/mortar interface during fiber pullout', in S. Diamond, S. Mindess, F.P. Glasser, L.W. Roberts, J.P. Skalny and L.D. Wakeley (eds) *Microstructure of Cement-Based Systems/Bonding and interfaces in Cementitious Materials*, Proc. Materials Research Society Symp., Vol. 370, Materials Research Society, Pittsburgh, PA, 1995, pp. 519–528.

19. Y. Wang, V.C. Li and S. Backer, 'Modeling of fibre pull-out from cement matrix', *Int. J. Cem. Comp. & Ltwt. Conc.* 10, 1988, 143–150.

20. A. Kelly and C. Zweben, 'Poisson contraction in aligned fibre composites showing pull-out', *J. Mater. Sci.* 11, 1976, 582–587.

21. D.J. Pinchin and D. Tabor, 'Inelastic behavior in steel wire pull-out from portland cement mortar', *J. Mater. Sci.* 13, 1978, 1261–1266.

22. D.J. Pinchin and D. Tabor, 'Interfacial contact pressure and frictional stress transfer in steel fibre cement', in R.N. Swamy (ed.) Proc. RILEM Conference Testing and Test Methods of Fibre-Cement Composites, The Construction Press, England, 1978, pp. 337–344.

23. P.W.R Beaumont and J.C. Aleszka, 'Polymer concrete dispersed with short steel wires', *J. Mater. Sci.* 13, 1978, 1749–1760.

24. H. Stang, 'Significance of shrinkage-induced clamping pressure in fiber-matrix bonding in cementitious composite materials', *Advanced Cement Based Materials* 4, 1996, 106–115.

25. Y. Geng and C.K.Y. Leung, 'A microstructural study of fiber/mortar interfaces during fibre debonding and pull-out', *J. Mater. Sci.* 31, 1996, 1285–1294.

26. A. Peled and A. Bentur, 'Cementitious composites reinforced with textile fabrics', Paper 35 in H.W. Reinhardt and A.E. Naaman (eds) Proceedings of the Third International Workshop on High Performance Fiber Reinforced Cement Composites (HPFRCC3), Proc. RILEM, PRO 6, RILEM Publications, Bagneux, France, 1999, 31–40.

27. J. Morton and G.W. Groves, 'Large work of fracture values in wire reinforced brittle matrix composites', *J. Mater. Sci.* 10, 1975, 170–172.

28. J. Bowling and G.W. Groves, 'The debonding and pull-out of ductile wires from a brittle matrix', *J. Mater. Sci.* 14, 1979, 431–442.

29. P. Bartos, 'Analysis of pull-out tests on fibres embedded in brittle matrices', *J. Mater. Sci.* 15, 1980, 3122–3128.
30. R.J. Gray, 'Analysis of the effect of embedded fibre length on the debonding and pull-out from an elastic matrix, Part 2: application to steel fibre-cementitious matrix composite system', *J. Mater. Sci.* 19, 1984, 1680–1691.
31. R.C. De Vekey and A.J. Majumdar, 'Determining bond strength in fibre reinforced composites', *Mag. Concr. Res.* 20, 1968, 229–234.
32. D.J. Pinchin and D. Tabor, 'Interfacial phenomena in steel fibre reinforced cement II. Pull-out behaviour of steel wires', *Cem. Concr. Res. 8*, 1978, 139–150.
33. A.E. Naaman and S.P. Shah, 'Pull-out mechanism in steel fibre reinforced concrete', *J. Struct. Div. ASCE* 102, 1976, 1537–1548.
34. A. Bentur, S. Mindess and N. Banthia, 'The interfacial transition zone in fibre reinforced cement and concrete', in M.G. Alexander, G. Arliguie, G. Ballivy, A. Bentur and J. Marchand (eds) *Engineering and Transport properties of the Interfacial Transition Zone in Cementitious composites*, RILEM Publications, Bagneux, France, Report 20, 1999, 89–112.
35. A. Bentur and S. Mindess, *Fibre Reinforced Cementitious Composites,* First edition, Elsevier Applied Science, London and New York, 1990.
36. P.N. Balaguru and S.P. Shah, *Fiber Reinforced Cement Composites*, McGraw Hill, New York, 1992.
37. M. Kawamura and S. Igarashi, 'Fluorescence microscopy study of fracture process of the interfacial zone between a steel fiber and the cementitious matrix under pull-out loading', in O. Buyukozturk and M. Wecheratana (eds) Proc. ACI Special Publication on Interface, Fracture and Bond, American Concrete Institute, Farmington Hills, MI, 1995.
38. H. Najm, A.E. Naaman, T.-J. Chu and R.E. Robertson, 'Effect of poly(vinyl alcohol) on fiber cement interfaces. Part I: bond stress-slip response', *Advn. Cem. Bas. Mat.* 1, 1994, 115–121.
39. W. Hansen, Personal contribution to ref. [6] (ref. 31 in [6]).
40. S. Wei, J.A Mandel and S. Said, 'Study of the interface strength in steel fiber reinforced cement-based composites', *J. Amer. Conc. Inst.* 83, 1986, 597–605.
41. A. Peled, D. Yankelevsky and A. Bentur, 'Bonding and interfacial microstructure in cementitious matrices reinforced by woven fabric', in S. Diamond, S. Mindess, F.P. Glasser, L.W. Roberts, J.P. Skalny and L.D. Wakely (eds) *Microstructure of Cement Based Systems, Bonding and Interfaces in Cememtitious Materials*, Proc. Materials Research Society Symp., Vol. 370, Materials Research Society, Pittsburgh, PA, 1995, 549–558.
42. V.C. Li, Y. Wang and S. Backer, 'Effect of inclining angle, bundling and surface treatment on synthetic fibre pull-out from cement matrix', *Composites.* 21, 1990, 132–140.
43. A. Katz and V.C. Li, 'Bond properties of micro-fibers in cementitious matrix', in S. Diamond, S. Mindess, F.P. Glasser, L.W. Roberts, J.P. Skalny and L.D. Wakeley (eds) *Microstructure of Cement-Based Systems/Bonding and interfaces in Cementitious Materials*, Proc. Materials Research Society Symp., Vol. 370, Materials Research Society, Pittsburgh, PA, 1995, 529–537.
44. Z. Li, B. Mobasher and S.P. Shah, 'Characterization of interfacial properties in fiber-reinforced cementitious composites', *J. Amer. Ceram. Soc.* 74, 1991, 2156–2164.
45. S. Igarashi, A. Bentur and S. Mindess, 'The effect of processing on the bond and interfaces in steel fiber reinforced cement composites', *Cem. Concr. Compos.* 18, 1996, 313–322.

46. V.C. Li, H.-C. Wu and Y.-W. Chan, 'Effect of plasma treatment of polyethylene fibers on interface and cementitious composite properties', *J. Amer. Ceram. Soc.* 79, 1996, 700–704.

47. A. Peled, H. Guttman and A. Bentur, 'Treatments of polypropylene fibres as a means to optimize for the reinforcing efficiency in cementitious composites', *Cem. Concr. Compos.* 14, 1992, 277–285.

48. H. Krenchel, *Fibre Reinforcement*, Akademick Forlag, Copenhagen, 1964.

49. V. Laws, 'The efficiency of fibrous reinforcement of brittle matrices', *J. Phys. D: Appl. Phys.* 4, 1971, 1737–1746.

50. J. Aveston, R.A. Mercer and J.M. Sillwood, 'Fibre reinforced cements – scientific foundations for specifications, in *Composites – Standards, Testing and Design*, Proc. National Physical Laboratory Conference, England, 1974, 93–103.

51. H.G. Allen, 'The strength of thin composites of finite width, with brittle matrices and random discontinuous reinforcing fibres', *J. Phys. D: Appl. Phys.* 5, 1972, 331–343.

52. M.J. Stucke and A.J. Majumdar, 'Microstructure of glass fibre reinforced cement composites', *J. Mater. Sci.* 11, 1976, 1019–1030.

53. J. Morton and G.W. Groves, 'The cracking of composites consisting of discontinuous ductile fibres in a brittle matrix-effect of orientation', *J. Mater. Sci.* 9, 1974, 1436–1445.

54. J. Morton, 'The work of fracture of random fibre reinforced cement', *Mater. Struct.* 12, 1979, 393–396.

55. A.M. Brandt, 'On the optimal direction of short metal fibres in brittle matrix components', *J. Mater. Sci.* 20, 1985, 3831–3841.

56. A.M. Brandt, 'On the optimization of the fibre orientation in cement based composite materials', in G.C. Hoff (ed.) *Fibre Reinforced Concrete*, ACI SP-81, American Concrete Institute, Detroit, MI, 1984, 267–286.

57. C.K.Y. Leung and J. Chi, 'Crack bridging force in random ductile fiber reinforced brittle matrix composites', *J. Eng. Mech. ASCE.* 121, 1995, 1315–1324.

58. K.Y. Leung and N. Ybanez, 'Pull-out of inclined flexible fiber in cementitious composite', *J.Eng. Mech. Div. ASCE.* 123, 1997, 239–246.

59. A.K. Maji and J.L. Wang, 'Noninvasive diagnosis of toughening mechanisms in fiber reinforced concrete', in S. Mindess and J.P. Skalny (eds) *Fiber Reinforced Cementitious Materials*, Proc. Materials Research Society Symp., Vol. 211, *Materials Research Society*, Pittsburgh, PA, 1991, pp. 169–174.

60. J. Aveston, R.A. Mercer and J.M. Sillwood, 'Fibre reinforced cement – scientific foundations for specifications', in *Composites – Standards, Testing and Design*, Proc. National Physical Laboratory Conference, UK, 1974, 93–103.

61. A. Katz and V.C. Li, Inclination Angle Effect of Carbon Fibers in Cementitious Composites, UMCEE Report No. 94–27, Advanced Civil Engineering Materials Research Laboratory, Department of Civil and Environmental Engineering, University of Michigan, 1994.

62. A. Katz and V.C. Li, 'Inclination angle effect of carbon fibers in cementitious composites', *J. Eng. Mech. ASCE.* 121, 1995, 1340–1348.

63. C.K.Y. Leung and V.C. Li, 'Effect of fiber inclination on crack bridging stress in brittle fiber reinforced brittle matrix', *J. Mech. Phys. Solids.* 40, 1992, 1333–1362.

64. B. Weiler, C. Grosse and H.W. Reinhardt, 'Debonding behavior of steel fibres with hooked ends', Paper 35 in H.W. Reinhardt and A.E. Naaman (eds) Proceedings of

the Third International Workshop on High Performance Fiber Reinforced Cement Composites (HPFRCC3), Proc. RILEM, PRO 6, RILEM Publications, Bagneux, France, 1999, 423–433.

65. A. Van Gysel, 'A pullout model for hooked end steel fibres', Paper 29 in H.W. Reinhardt and A.E. Naaman (eds) Proceedings of the Third International Workshop on High Performance Fiber Reinforced Cement Composites (HPFRCC3), Proc. RILEM, PRO 6, RILEM Publications, Bagneux, France, 1999, 351–360.

66. A. Peled and A. Bentur, 'Fabric efficiency and its reinforcing efficiency in textile reinforced cement composites', *Composites Part A.* 34, 2003, 107–118.

67. M.R. Krishnadev, S. Berrada, N. Banthia and J.F. Fortier, 'Deformed steel fiber pull-out mechanics: influence of steel properties', in R.N. Swamy (ed.) *Fibre Reinforced Cement and Concrete,* Proc. RILEM Symp., E&FN SPON, 1992, London and New York, 390–399.

68. J.M. Alwan, A.E. Naaman and P. Guerrero, 'Effect of mechanical clamping on pull-out response of hooked steel fibers embedded in cementitious matrices', *Concr. Sci. Eng.* 1, 1999, 15–25.

69. S. Singh, A. Shukla and R. Brown, 'Pullout behavior of polypropylene fibers from cementitious matrix', *Cem. Concr. Res.* 34, 2004, 1919–1925.

70. N. Banthia, J.F. Trottier, M. Pigeon and M.R. Krishnadev, 'Deformed steel fiber pull-out: material characteristics and metallurgical processes', in H.W. Reinhardt and A.E. Naaman (eds) *High Performance Fiber Reinforced Cement Composites,* Proc. RILEM Symp., E&FN SPON, London and New York, 1992, 456–466.

71. G. Chanvillard and P.C. Aitcin, 'Micro-mechanical modeling of the pull-out behavior of corrugated wire drawn steel fibres from cementitious matrices', in Proc. Materials Research Society Symp., Vol. 211, *Materials Research Society*, Pittsburgh, PA, 1991, 107–202.

72. A.E. Naaman, 'Engineered steel fibers with optimal properties for reinforcement of cement composites', *J. Advanced Concrete Technology.* 1, 2003, 241–252.

73. N. Banthia and J.F. Trottier, 'Concrete reinforced with deformed steel fibers, Part I: bond-slip mechanisms', *ACI Mater. J.* 91, 1994, 435–446.

74. N. Banthia and J.F. Trottier, 'Concrete reinforced with deformed steel fibers, Part II: toughness characterization', *ACI Mater. J.* 92, 1995, 146–154.

75. Z. Li, B. Mobasher and S.P. Shah, 'Characterization of interfacial properties in fiber reinforced cement-based composites', *J. Amer. Ceram. Soc.* 74, 1991, 2156–2164.

76. B. Mobasher and C.Y. Li, 'Modeling of stiffness degradation of the interfacial transition zone during fibre debonding', *J. of Composite Engineering.* 5, 1995, 1349–1365.

77. R.N. Swamy, 'Influence of slow crack growth on the fracture resistance of fibre-cement composites', *Int. J. Cem. Comp.* 2, 1980, 43–53.

78. W.A. Patterson and H.C. Chan, 'Fracture toughness of glass reinforced cement', *Composites.* 6, 102–104.

79. A. Bentur and S. Diamond, 'Crack patterns in steel fibre reinforced cement paste', *Mater. Struct.* 18, 1985, 49–56.

80. A. Bentur and S. Diamond, 'Effect of aging of glass fibre reinforced cement on the response of an advancing crack on intersecting a glass fibre strand', *Int. J. Cem. Comp & Ltwt. Concr.* 8, 1986, 213–222.

81. A. Bentur, S. Diamond and S. Mindess, 'Cracking processes in steel fibre reinforced cement paste', *Cem. Concr. Res.* 15, 1985, 331–342.

82. J. Cook and J.E. Gordon, 'A mechanism for the control of crack propagation in all brittle systems', *Proc. Roy. Soc.* 282A, 1964, 508–520.
83. A. Bentur, 'Mechanisms of potential embrittlement and strength loss of glass fibre reinforced cement composites', S. Diamond (ed.) Proc. Durability of Glass Fiber Reinforced Concrete Symp., Prestressed Concrete Institute, Chicago, IL, 1985, pp. 109–123.
84. A. Bentur, A. Peled and D. Yankelevsky, 'Enhanced bonding of low modulus polymer fibre-cement matrix by means of crimped geometry', *Cem. Concr. Res.* 27, 1997, 1099–1111.
85. A. Peled and S.P. Shah, 'Parameters related to extruded cement composites', in A.M. Brandt, V.C. Li and I.H. Marshall (eds) *Brittle Matrix Composites* 6, Woodhead Publications, Warsaw, 2000, pp. 93–100.
86. A. Peled, A. Bentur and D. Yankelevsky, 'The nature of bonding between monofilament polyethylene yarn and cement matrices', *Cem. Concr. Compos.* 20, 1998, 319–327.
87. D.R. Oakley and B.A. Proctor, 'Tensile stress-strain behavior of glass fiber reinforced cement composites', in A. Neville (ed.) *Fibre Reinforced Cement and Concrete*, The Construction Press, Lancaster, England, 1975, pp. 347–359.
88. P. Bartos, 'Brittle matrix composites reinforced with bundles of fibres', in J.C. Maso (ed.) *From Material Science to Construction Materials*, Proc. RILEM Symp., Chapman and Hall, Dordrecht, The Netherlands, 1987, 539–546.
89. W. Zhu and P. Bartos, 'Assessment of interfacial microstructure and bond properties in aged GRC using novel microindentation method', *Cem. Concr. Res.* 27, 1997, 1701–1711.
90. O. Sadatoshi and D.J. Hannant, 'Modeling of stress-strain response of continuous fiber reinforced cement composite', *ACI Mater. J.* 91, 1994, 306–312.
91. P. Trtik and P.J.M Bartos, 'Assessment of glass fibre reinforced cement by in-situ SEM bending test', *Mater. Struct.* 32, 1999, 140–143.
92. A. Bentur, 'Long term performance of fiber reinforced cementitious composites', in J.P. Skalny and S. Mindess (eds) *Materials Science in Concrete – V*, The American Ceramic Society, Westerville, Ohio, 1998, 513–536.

Chapter 4

Mechanics of fibre reinforced cementitious composites

4.1 Introduction

In FRC composites, the major role played by the fibres occurs in the post-cracking zone, in which the fibres bridge across the cracked matrix. In a well-designed composite the fibres can serve two functions in this zone:

1   They may increase the strength of the composite over that of the matrix, by providing a means of transferring stresses and loads across cracks. This implies an ascending stress–strain curve after first cracking, and this behaviour is referred to as strain hardening.
2   More importantly, they increase the toughness of the composite by providing energy absorption mechanisms, related to the debonding and pull-out processes of the fibres bridging the cracks. This occurs even when the stress–strain curve is descending after first crack (referred to as strain softening).

The sequence of events following first cracking in the composite determines whether these strengthening and toughening effects will take place. As cracking occurs in the brittle matrix, the load is transferred to the fibres; if failure is to be prevented at this stage, the load-bearing capacity of the fibres, $\sigma_{fu} V_f$, in the case of aligned and continuous fibres, should be greater than the load on the composite at first crack. This relation can be quantified on the basis of the elastic stresses at the cracking strain of the matrix in the composite, ε_{mu} [1]:

$$\sigma_{fu} V_f > E_m \varepsilon_{mu} V_m + E_f \varepsilon_{mu} V_f \tag{4.1}$$

The right-hand term in Eq. 4.1 represents the first crack load of a composite. When Eq. 4.1 is satisfied (i.e. when the fibre content, V_f, is sufficiently high), the first crack to occur in the composite will not lead to catastrophic failure, but will result in redistribution of the load between the matrix and the fibres. That is, the load carried by the matrix in the cracked zone will be imposed on the bridging fibres, and the matrix at the edges of the crack will become stress-free. Additional loading will result in additional cracks, until the matrix is divided into a number of segments, separated by cracks. As shown in Figure 4.1 [2], loading

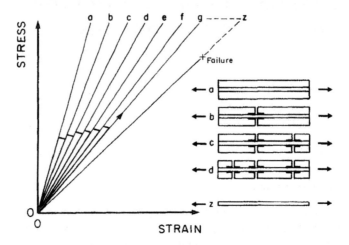

Figure 4.1 Simplified description of the multiple cracking process and the resulting curve in a brittle matrix–fibre composite (after Allen [2]).

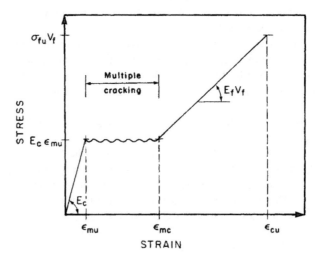

Figure 4.2 Schematic description of the stress–strain curve, based on the ACK model [1].

would initially be along path 'a'; after initial cracking, the loading path would deviate to 'b' and then 'c' and so on. This process is known as *multiple cracking*. It occurs at an approximately constant stress, which is equal to the first crack stress $\varepsilon_{mu} \cdot E_c$, where E_c is the modulus of elasticity of the composite. This region of the stress–strain curve is approximately horizontal or slightly ascending (Figure 4.2).

When there is no further multiple cracking, and the matrix is divided by parallel cracks, any additional tensile load will cause stretching or pull-out of the fibres. At this stage, the rigidity of the reinforcing array is the reason for the ascending nature of the curve in the post-multiple cracking zone. Assuming that the composite consists of aligned and continuous fibres which stretch under load, the slope in this range is $E_f V_f$ and failure will occur when the fibres reach their load-bearing capacity, at a composite stress of a $\sigma_{fu} V_f$ for aligned and continuous fibres (Figure 4.2). Thus, the overall mechanical behaviour of the FRC composite can usually be described in terms of the three stages of the tensile stress vs. strain curve of Figure 4.2 (the ACK model [1]):

1 'Elastic range', up to the point of first crack; the matrix and the fibres are both in their linear, elastic range.
2 'Multiple cracking' range, in which the composite strain has exceeded the ultimate strain of the matrix.
3 'Post-multiple cracking' range, during which the fibres are being stretched or pulled out of the cracked matrix.

Various models and analytical treatments have been proposed to account for the overall shape of the tensile stress–strain curve and to predict the salient points of the curve: modulus of elasticity in the elastic zone, E_c, first crack stress and strain (σ_{mu}; ε_{mu}), the strain at the end of the multiple cracking zone (ε_{mc}), and the ultimate stress and strain of the composite, σ_{cu} and ε_{cu} respectively. Specific attention has been given to the analytical estimation of the energy involved in the fracture of the composite. The concepts used include the composite materials approach, multiple cracking and fracture mechanics. While each of these concepts may be useful for describing some of the characteristics of the overall mechanical behaviour, none is capable of describing the entire response of the material to load. For example, the composite materials approach may best be used to model the elastic range and the strength of the composite, while fracture mechanics concepts may be used to predict the first crack strain, and in some instances the fracture energy. At present, only a combination of these concepts permits quantification of at least most of the characteristics of the mechanical behaviour.

4.2 Efficiency of fibre reinforcement

The actual composite usually consists of short fibres dispersed in the matrix, of which at least some are at an angle with respect to the load orientation. The contribution of such short, inclined fibres to the mechanical properties of the composite is smaller than that of long fibres oriented parallel to the load assumed in Eq. 4.1, that is the efficiency of the short and inclined fibres is less. The efficiency of fibre reinforcement can be judged on the basis of two criteria: the enhancement in strength and the enhancement in toughness of the composite, compared with the brittle matrix. These effects depend upon the fibre length, the orientation of the fibres and the fibre–matrix shear bond strength. These three factors are not independent, since the effects of both fibre length and orientation are highly sensitive to the bond. A further complication in the treatment of fibre efficiency is that parameters which enhance tensile strength do not necessarily lead to higher toughness.

In many engineering applications, the fibre efficiency is expressed in terms of an efficiency factor, which is a value between 0 and 1, expressing the ratio between the reinforcing effect of the short, inclined fibres and the reinforcement expected from continuous fibres aligned parallel to the load. These factors, η_ℓ and η_θ for length and orientation efficiency, respectively, can be determined either empirically, or on the basis of analytical calculations. They are frequently used in combination with properties that can be accounted for by the rule of mixtures (Section 4.3). In this section, various analytical treatments to account for efficiency factors will be reviewed, with special emphasis on the effects of bond. The effects of length and orientation will be described separately.

4.2.1   Length effects

The effect of length can be analysed in terms of the stress transfer mechanisms discussed in Chapter 3. A critical length parameter, ℓ_c, can be defined as the minimum fibre length required for the build-up of a stress (or load) in the fibre which is equal to its strength (or failure load), as shown in Figure 4.3. The calculated value of ℓ_c depends on the assumptions made regarding the stress transfer mechanism, as demonstrated in Figure 4.3(b). Curve (1) represents a frictional stress transfer mechanism; Curve (2) represents an elastic stress transfer mechanism (see Section 3.2). For $\ell < \ell_c$ (Figure 4.3(a)) there is insufficient embedded length to generate a stress equal to the fibre strength and the fibre is not utilized efficiently. Only if the length of the fibre considerably exceeds ℓ_c does the stress along most of the fibre reach its yield or tensile strength (Figure 4.4(c)), thus mobilizing most of the potential of the fibre reinforcement.

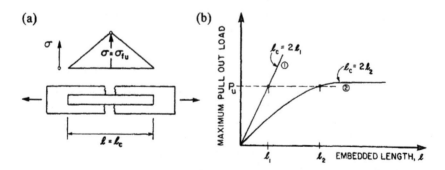

Figure 4.3 Definition of critical length. (a) Stress distribution, assuming frictional stress transfer; (b) intersection of fibre breaking load P_u, with pull-out load vs. length curves.

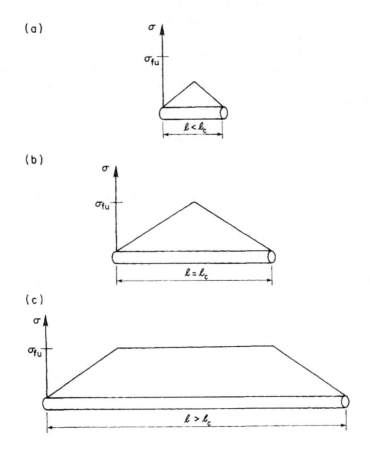

Figure 4.4 Stress profile along a fibre in a matrix as a function of fibre length (after Kelly [3]).

For a frictional shear stress transfer (or linear stress transfer), the calculated value of ℓ_c (Figure 4.3(a)) is:

$$\ell_c = \frac{\sigma_{fu} \cdot r}{\tau_{fu}} \tag{4.2}$$

The efficiency factor is defined as the average stress along the fibre, $\bar{\sigma}_f$, relative to its strength, σ_{fu}, that is $\eta_\ell = \bar{\sigma}/\sigma_{fu}$. For the triangular or trapezoidal profiles in Figure 4.4, the following efficiency factors are obtained [3]:

for $\quad \ell > \ell_c \quad \eta_\ell = 1 - \frac{\ell_c}{2\ell}$ $\qquad$ (4.3)

for $\quad \ell < \ell_c \quad \eta_\ell = \frac{\tau_{fu}}{2r\sigma_{fu}} = \frac{1}{2}\frac{\ell}{\ell_c}$ $\qquad$ (4.4)

The stress distribution profiles in Figure 4.4 and the efficiency factors of Eqs 4.3 and 4.4 are the ones generally used in composite materials theory. They are valid for the zone in which the matrix is not cracked, and the interfacial shear stress is constant. This may be the case for polymer or metal matrices, but does not always hold true for the whole range of loading in FRC composites. Therefore, in these systems, different treatments are required for the pre-cracking and post-cracking zones. These two will be considered below.

4.2.1.1   Efficiency in the pre-cracking zone

Laws [4] has indicated that in the elastic zone, the contribution of the fibre, and thus its efficiency, is a function of the strain $\varepsilon_c(x)$ in the composite. For elastic stress transfer (Figure 3.3(b), Section 3.2), the average stress in the fibre, $\overline{\sigma}_f$, as a function of $\varepsilon_c(x)$ is:

$$\overline{\sigma}_f = E_f \varepsilon_c(x) \left(1 - \frac{\tanh \beta_1 \ell/2}{\beta_1 \ell/2} \right) \tag{4.5}$$

This relationship is derived from Eqs 3.1–3.3. For frictional stress transfer, with a constant interfacial shear stress, τ_{fu}, the average stress in the fibre is:

$$\overline{\sigma}_f = \left(1 - \frac{\ell_x}{2\ell} \right) E_f \varepsilon_c(x) \tag{4.6}$$

where $\ell_x/2$ is the length required for a stress of $E_f \varepsilon_c(x)$ to be built up in the fibre, and therefore:

$$\ell_x = \frac{E_f \varepsilon_f(x) r}{\tau_{fu}} \tag{4.7}$$

The length efficiency factor in the pre-cracking zone for the frictional stress transfer mechanism can be calculated from Eq. 4.6, and its value is a function of the strain in the composite:

$$\eta_\ell = 1 - \frac{\ell_x}{2\ell} \tag{4.8}$$

where ℓ_x is given by Eq. 4.7. At the ultimate strain of the matrix, ε_{mu}, which denotes the end of the pre-cracking (elastic) zone, the length efficiency factor can be calculated as:

$$\eta_\ell = 1 - \frac{\ell_c}{2\ell} \frac{\varepsilon_{mu}}{\varepsilon_{fu}} \tag{4.9}$$

where ℓ_c is defined by Eq. 4.2. Equation 4.9 indicates that the efficiency factor of $1 - \ell_c/2\ell$, as derived in Eq. 4.3, is valid only if the ultimate strains of the fibre and the matrix are the same, which is obviously not the case in FRC composites.

4.2.1.2 Efficiency in the post-cracking zone

Krenchel [5] and Laws [4] have analysed this zone, assuming that the fibres are intersected by the crack, so that the shorter embedded lengths of each fibre in the matrix are uniformly distributed between 0 and $\ell/2$. At a strain of $\varepsilon_c(x)$ fibres whose shorter embedded length is less than $\ell_x/2$ will slip, and will not contribute to strength, where ℓ_x is given by Eq. 4.7. The probability that a fibre will slip is ℓ_x/ℓ and the average stress supported by the fibres at a composite strain of $\varepsilon_c(x)$ is:

$$\bar{\sigma}_f = \left(1 - \frac{\ell_x}{\ell}\right) E_f \varepsilon_c(x) = \left(1 - \frac{\ell_c}{\ell}\frac{\varepsilon_c(x)}{\varepsilon_{fu}}\right) E_f \varepsilon_c(x) \tag{4.10}$$

where ℓ_c is defined by Eq. 4.2.

On this basis Laws [4] derived the following equations for the efficiency factors related to composite strength:

$$\text{for} \quad \ell \gg 2\ell_c \quad \eta_\ell = 1 - \frac{\ell_c}{\ell} \tag{4.11}$$

$$\text{for} \quad \ell \gg 2\ell_c \quad \eta_\ell = \frac{1}{4} - \frac{\ell}{\ell_c} \tag{4.12}$$

For $\ell \gg 2\ell_c$ sufficient embedded length is available to develop a stress equal to the fibre strength, and the failure will be predominantly by fibre fracture. For $\ell \ll 2\ell_c$ the fibres are so short that they will pull out before sufficient stress is developed to cause fibre failure. These strength efficiency factors for a cracked matrix (Eqs 4.11 and 4.12) are smaller than those derived by Kelly (Eqs 4.3 and 4.4) for a composite with a ductile, uncracked matrix.

In the derivation of Eqs 4.11 and 4.12, it was assumed that the portion of the fibres which slip and pull out do not contribute to strength. These relations would change if one were to assume that some frictional resistance is effective after pullout and sliding are initiated, and the fibres are being extracted from the matrix. Laws modified Eqs 4.11 and 4.12 to include such effects which are considered in terms of static frictional resistance prior to sliding (τ_s) and dynamic frictional resistance during sliding (τ_d):

$$\text{for} \quad \ell \ll 2\ell_c' \quad \eta_\ell = \frac{\ell}{2\ell_c(2 - \tau_d/\tau_s)} \tag{4.13}$$

$$\text{for} \quad \ell \ll 2\ell_c' \quad \eta_\ell = 1 - \frac{\ell_c}{2\ell}(2 - \tau_d/\tau_s) \tag{4.14}$$

where $\ell_c' = 1/2\ell_c(2 - \tau_d/\tau_s)$.

Assuming the ratio of τ_d/τ_s be in the range of 0–1, the range of length efficiency factors will be as follows:

$$\text{for} \quad \ell \ll 2\ell_c \quad \frac{1}{4}\frac{\ell_c}{2\ell} < \eta_\ell < \frac{1}{2}\frac{\ell_c}{2\ell} \tag{4.15}$$

$$\text{for} \qquad \ell \gg 2\ell_c \qquad 1 - \frac{\ell_c}{2\ell} < \eta_\ell < 1 - \frac{\ell_c}{2\ell} \tag{4.16}$$

Usually, ℓ_c is calculated from Eq. 4.2. The higher limits of the efficiency factors in Eqs 4.15 and 4.16 are similar to those derived by Kelly (Eqs 4.3 and 4.4); when using these upper bounds for FRC composites it is implied that $\tau_{fu} = \tau_s = \tau_d$. However, Laws [4] estimated τ_d to be 2/3 of τ_s on the basis of the results of de Vekey and Majumdar [6].

Equations 4.15 and 4.16 show the sensitivity of the efficiency to the fibre length. In order to achieve 90% strength efficiency, the fibre must be 5–10 times longer than its critical length. This can be achieved by controlling the geometry of the fibre (higher aspect ratio ℓ/d) or by enhancing the fibre–matrix interaction (i.e. higher τ_{fu} values for the case of frictional bond). The effect of length and bond on the strength efficiency are presented graphically in Figures 4.5(a) and (b), (curves marked P_f) based on Eqs 4.2, 4.11 and 4.12.

The increase in fibre length, from less than ℓ_c to greater than ℓ_c is associated not only with an increase in efficiency but also with a change in the mode of fracture. A low fibre length is associated with pull-out failure; above ℓ_c fibre fracture will precede fibre pull-out. Kelly [3] has treated this problem quantitatively, assuming that the energy consumed in fibre fracture is negligible compared with the energy consumed during fibre pull-out. Thus, the change in the mode of failure, from fibre pull-out to fibre fracture, results in a marked reduction in the energy involved in the failure of the composite, leading to a more *brittle* material. Kelly calculated the average work done in pull-out per fibre, $\overline{W}_p$, for short fibres ($\ell \ll \ell_c$), assuming

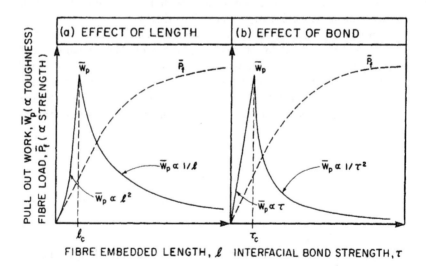

Figure 4.5 Effect of length (a) and bond (b) on the contribution (efficiency) of the fibre to strength, $\bar{P}_f$, and toughness (pull out work, $\overline{W}_p$), $\bar{P}_f$, is proportional to the fibre efficiency factor, η.

that: (i) all the fibres pull out; (ii) the pull-out lengths are uniformly distributed between 0 and $\ell/2$ and (iii) the same frictional shear stress is effective throughout the entire pull-out process (i.e. $\tau_s = \tau_d = \tau_{fu}$):

$$\text{for} \qquad \ell << \ell_c \qquad \overline{W}_p = \frac{1}{24}\pi\, d\tau_{fu}\ell^2 \qquad (4.17)$$

For longer fibres ($\ell >> \ell_c$), some of the fibres will break prior to pull-out, and their contribution to the fracture energy will be negligible. Only the fraction of fibres that pull out (ℓ_c/ℓ) will contribute to the fracture energy through their frictional pull-out work, $\overline{W}_p$:

$$\text{for} \qquad \ell >> \ell_c \qquad \overline{W}_p = \frac{\ell_c}{\ell}\frac{1}{24}\pi\, d\tau_{fu}\ell_c^2 \qquad (4.18)$$

A graphical presentation of Eqs 4.17 and 4.18 is given in Figure 4.5(a) (curve marked W_p) showing the effect of length on *toughness* (i.e. pull-out work). The maximum toughness is obtained when $\ell = \ell_c$. As the length increases, more fibres break rather than pull out, and the energy consumed in failure is reduced, leading to a more brittle composite. However, the increase in length is accompanied by an improvement in the load-bearing capacity of the fibres, which enhances their strength efficiency (Eq. 4.11). Thus, in the range of $\ell > \ell_c$ there is a contradiction between the requirements for strength and toughness, implying that design of the composite for higher strength by increasing the fibre length can lead to a loss of toughness and embrittlement.

Equations 4.15–4.18 were developed to describe the strength efficiency of the fibres, and their contribution to toughness as a function of their length. These equations can be recalculated to determine the effect of bond strength, τ_{fu}, on the strength and toughness of a composite with fibres of a constant length, ℓ, assuming a frictional stress transfer mechanism:

Strength efficiency for bond, η_τ:

$$\tau < \tau_c \quad \eta_\tau = 1/2\tau/\tau_c \qquad (4.19)$$

$$\tau > \tau_c \quad \eta_\tau = 1 1/2\tau_c/\tau \qquad (4.20)$$

Pull-out work, $\overline{W}_p$:

$$\tau < \tau_c \quad \overline{W}_p = \frac{1}{12}\ell\sigma_f V_f(\tau/\tau_c) \qquad (4.21)$$

$$\tau > \tau_c \quad \overline{W}_p = \frac{1}{12}\ell\sigma_f V_f(\tau_c/\tau)^2 \qquad (4.22)$$

The critical bond strength, τ_c, is:

$$\tau_c = \frac{\sigma_f r}{\ell} \qquad (4.23)$$

The results of this calculation are presented in Figure 4.5(b), showing that an increase in frictional bond strength (τ_{fu}) is beneficial for strength, but can be detrimental to toughness when increased beyond a critical value, τ_c. This dependency on bond strength is of interest in many FRC composites, which consist of short fibres whose bond to the matrix can be changed by various surface treatments, or during the ageing of the composite. For instance, densification of the cement matrix during ageing can lead to improved bond, which if too high, can result in embrittlement.

Equations 4.15–4.23 and the curves in Figure 4.5 were derived on the basis of somewhat simplified concepts, assuming frictional stress transfer only. Yet, they do provide semi-quantitative guidelines for the design and development of fibre–cement composites; strength and toughness can be optimized by modifying the fibre length and the fibre–matrix bond.

4.2.2   Orientation effects

In most of the FRC, the fibres are randomly oriented, either in two or three dimensions. The effect of orientation was considered within the context of fibre pull-out in Section 3.3, and orientation efficiencies were evaluated and calculated for a variety of assumptions. Characteristic values calculated on the basis of various approaches and models are provided in Tables 3.5 and 3.6.

4.2.3   Combined length and orientation efficiency

Laws [4] concluded that the combined efficiency factors, due to both length and orientation, cannot be simply calculated as the product of the length efficiency factor and the orientation efficiency factor. That is, the orientation efficiency factor is also a function of the length in the case of short fibres. For a random, 2D fibre array, Laws [4] derived the following relations:

$$\ell << 5/3\ell_c \quad \eta = 3/8 \text{x} 3/10(\ell/\ell_c) \tag{4.24}$$

$$\ell >> 5/3\ell_c \quad \eta = 3/8(1 - 5/6\ell_c/\ell) \tag{4.25}$$

The combined effect of orientation (aligned and random arrays) and length on the overall efficiency, as calculated by Laws [4] is shown in Figure 4.6. She demonstrated that the efficiency is dependent on the frictional bond before slip (static) and the frictional bond after fibre slip (dynamic). The higher efficiency values for the systems where dynamic slip is present (dashed line in Figure 4.5) are the result of the contribution of this type of frictional resistance to the strength of the short fibre composite.

The combined efficiency of length and orientation is inherently included in the outcome of the analysis of inclined fibres which are induced to bend locally (Section 3.3.2).

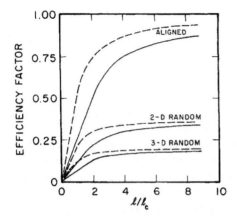

Figure 4.6 The total efficiency factor, η, as a function of the ℓ/ℓ_c ratio for frictional shear stress transfer (dashed lines $\tau_d = \tau_{fu}$; full lines $\tau_d = 0$) after Laws [4].

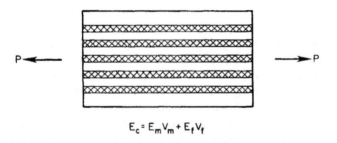

$$E_c = E_m V_m + E_f V_f$$

Figure 4.7 The rule of mixtures (the parallel model).

4.3   Composite materials approach

The composite materials approach is usually based on *the rule of mixtures*, which models the composite as shown in Figure 4.7. It states that the properties of the composite are the weighted average of the properties of its individual components. For mechanical properties such as strength and modulus of elasticity, the concept of the rule of mixtures is valid only if the two components are linear, elastic and the bond between them is perfect. Therefore, the rule of mixtures can only be applied in the elastic, pre-cracked zone of the fibre reinforced cement, and even in this zone it should be considered as an upper limit, since in practice the bond is not perfect. Taking into account the effects of fibre orientation and length (using the efficiency factors described in Section 4.2), the rule of mixtures can be used to predict the modulus of elasticity, E_c and the first crack stress of the composite, σ_{mu}, in tension:

$$E_c = E_m V_m + \eta_\ell \eta_\theta E_f V_f \qquad (4.26)$$

$$\sigma_{mu} = \underset{\text{(matrix)}}{\sigma'_{mu}} + \underset{\text{(fibre)}}{\eta_\ell \eta_\theta \sigma'_f V_f} \qquad (4.27)$$

where σ'_f is the stress in the fibre at the first crack strain and σ'_{mu} is the tensile strength of the matrix in the absence of fibres.

Substituting values typical for fibre reinforced cements into Eqs 4.26 and 4.27 (V_f of 5% or less, and E_f of 210 GPa or less) readily demonstrates that the improvement in the modulus of elasticity and first crack stress over the matrix values ($E_m = 20 - 30$ GPa and $\sigma'_{mu} = 3-5$ MPa) cannot exceed 10–20%, which, from a practical point of view, is not very significant. This has been confirmed experimentally [7,8]. Thus, as discussed previously, the main influence of the fibres is in the post-cracking zone, where the contribution of the matrix is small or even negligible, because of its multiple cracking. Therefore, in this zone, Eq. 4.27 should be modified, by neglecting the contribution of the matrix. It can then be rewritten to calculate the tensile strength of the composite, σ_{cu} in the post-cracking zone:

$$\sigma_{cu} = \eta_\ell \eta_\theta \sigma_{fu} V_f \qquad (4.28)$$

It should be noted that the efficiency factors in the post-cracking zone can be different from those of the pre-cracked zone (Section 4.2).

Equation 4.28 is valid for fibre volumes which exceed a critical value, $V_{f(\text{crit})}$. Below this value, the load-bearing capacity of the fibres, as expressed by the strength σ_{cu} in Eq. 4.28, is smaller than the first crack stress expressed by Eq. 4.27. Kelly [3] has applied these equations to describe the strength of the composite as a function of fibre volume, assuming efficiency factors of 1, that is continuous and aligned fibres (Figure 4.8). Equation 4.27 applies for lower fibre contents, and Eq. 4.28 for the higher fibre contents. The intersection of the two equations is at the critical fibre volume content, which can be readily calculated [9] by substituting the elastic range relations of $\sigma'_f = \varepsilon_{mu} E_f$ and $\sigma_{mu} = \varepsilon_{mu} E_c$:

for continuous aligned fibres: $\qquad V_{f(\text{crit})} = \dfrac{E_c}{E_m} \dfrac{\sigma_{mu}}{\sigma_{fu}} \qquad (4.29)$

where E_c is given by Eq. 4.26 with $\eta_\ell = \eta_\theta = 1$. Since, in a typical fibre reinforced cementitious composite, E_c is not much greater than E_m and $\sigma_{mu} \cong \sigma'_{mu}$ Eq. 4.29 may be approximately expressed as:

$$V_{f(\text{crit})} = \dfrac{\sigma'_{mu}}{\sigma_{fu}} \qquad (4.30)$$

The fibres will contribute markedly to strength only when $V_f > V_{f(\text{crit})}$.

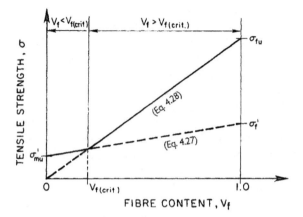

Figure 4.8 Relations between composite strength, σ_{cu}, and fibre volume content as predicted by Eqs (4.27) and (4.28), for continuous and aligned fibres (after Kelly [3]).

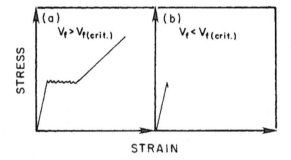

Figure 4.9 Schematic presentation of the stress–strain curves for composites with (a) $V_f > V_{f(crit)}$ and (b) $V_f < V_{f(crit)}$.

In this range, the mode of fracture is characterized by multiple cracking of the matrix. After first cracking, the load carried by the matrix is transferred to the fibres, which due to their sufficiently large volume can support this load without failure. Additional loading leads to more matrix cracking, which is still not accompanied by failure of the composite. In contrast, at $V_f < V_{f(crit)}$ the mode of failure will be by the propagation of a single crack, since there is an insufficient volume of fibres to support the load that was carried by the matrix as it cracked. A schematic description of the stress–strain curves of composites with $V_f > V_{f(crit)}$ and $V_f < V_{f(crit)}$ is given in Figure 4.9.

Critical fibre volumes calculated from Eqs 4.29 and 4.30 are in the range of 0.3% to 0.8% for steel, glass and polypropylene reinforced cements [9]. However, Eqs 4.29 and 4.30 were derived for continuous and aligned reinforcement. If the orientation and length efficiency are taken into account, the critical fibre volumes will be considerably greater; increases by factors of about 3 or 6 are expected due to random orientation in two or three dimensions, respectively. Thus, in practice, the critical fibre volume is generally greater than 1–3%. This is in the range of the maximum fibre content that can be incorporated by conventional mixing procedures. Therefore, in many fibre reinforced concretes produced on site, the presence of the fibres results in some improvement in the post-cracking ductility, but not in the post-cracking load-bearing capacity. The stress–strain curves for such matrices would be similar to the one presented in Figure 4.9(b) and they are known as strain softening composites. In recent years, with the advent of new means to control the rheology of mortars and concretes, and the use of new fibres with controlled interfacial properties, composites with strain hardening behaviour (e.g. enhanced load-bearing capacity in the post-cracking zone – Figure 4.9) were developed.

For short fibres ($\ell << \ell_c$), failure of the composite in the post-cracking zone occurs primarily by fibre pull-out. Equation 4.28 can be adapted to account for this by substituting the resistance of the fibres to pull-out for the average fibre stress at fracture (σ_{fu}). Assuming a constant frictional bond strength, τ_{fu}, and a uniform distribution of the embedded length of the shorter ends of the fibres bridging the crack between 0 and 1/2, the average resistance of a fibre to pull out is $\tau_{fu}\pi d\ell/4$. The number of aligned fibres per unit area is $4V_f/\pi d^2$, and the composite tensile strength σ_{cu} will therefore be:

for aligned short fibres: $$\sigma_{cu} = V_f \tau_{fu} \frac{\ell}{d} \qquad (4.31)$$

Linear relationships between tensile strength and the parameter $V_f \ell/d$ have indeed been reported by several investigators [10,11], Figure 4.10. It is interesting to note that the curves in Figure 4.10(b) extrapolate back close to the origin, as expected from Eq. 4.31, while in Figure 4.10(a) the extrapolated intersection with the strength axis is at a value equal to the matrix strength. Although curves of both types have been reported in the literature, there is no contradiction between the two. Equation 4.31 applies to systems where $V_f > V_{f(crit)}$ and the data in Figure 4.10(b) were obtained for such systems. On the other hand, the data in Figure 4.10(a) were obtained for low fibre volumes, apparently in the range of $V_f < V_{f(crit)}$ (the highest V_f value in the data in Figure 4.10(b) is 1.8%).

Thus, the data in Figure 4.10(a) are representative of the first portion of the curve in Figure 4.8, where the strength of the composite is equal to its first crack stress. For the case of short fibres, the second term in Eq. 4.27 should be modified to:

for $V < V_{f(crit)}$ $$\sigma_{cu} = \sigma_{mu} = \sigma'_{mu}V_m + \eta_\theta V_f \ell/d \qquad (4.32)$$

Thus, Eq. 4.32 can be used to describe the data in Figure 4.10(a). Data such as those presented in Figure 4.10, however, emphasize the need to evaluate the fibre content of the composite, $V_f > V_{f(crit)}$ or $V_f < V_{f(crit)}$, before attempting to develop general relations to describe its behaviour.

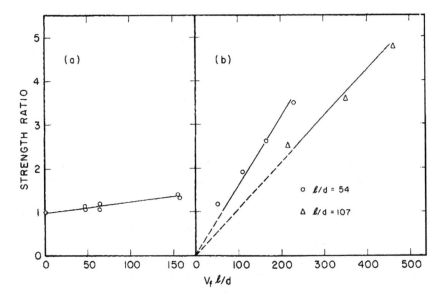

Figure 4.10 Experimental relations between composite ultimate tensile strength, σ_{cu}, and the parameter $V_f l/d$ for steel FRC: (a) adapted from the data of Johnston and Gray [10] for low fibre volume content; and (b) adapted from the data of Ng et al. [11] for $V_f > V_{f(crit)}$.

It should also be noted that even for $V_f > V_{f(crit)}$, not all data can be described by a single line going through the origin. Rather, there are a number of lines, each with a different slope. This may reflect the influence of the orientation efficiency, that is, Eq. 4.31 should be rewritten in a more general way,

$$\sigma_{cu} = \eta_\theta V_f \tau_{fu} l/d \qquad (4.33)$$

For the data in Figure 4.10(b), the different slopes may be associated with different η_θ values, which might be a function of the fibre aspect ratio.

It has been shown [12–14] that for ductile fibres the pull-out resistance is not necessarily reduced with an increased orientation angle (Section 3.3). Aveston et al. [15] have suggested that the only orientation efficiency effect which needs to be taken into account is the change in the number of fibres per unit area, which for 2D and 3D random distributions has been calculated as $(2/\pi 4 V_f/\pi d^2)$ and $(1/2 \cdot 4V_f/\pi d^2)$, respectively. Thus, the tensile strength of the composite for short, pulled-out fibres of random orientation is [9]:

for 2D random distribution: $\qquad \sigma_{cu} = \dfrac{2}{\pi} V_f \tau_{fu} \dfrac{l}{d} \qquad (4.34)$

for 3D random distribution: $\qquad \sigma_{cu} = \dfrac{1}{2} V_f \tau_{fu} \dfrac{l}{d} \qquad (4.35)$

Equations 4.30, 4.34 and 4.35 can then be used to calculate the approximate critical fibre volumes by equating σ_{cu} with σ'_{mu}:

for 1D aligned fibres:
$$V_{f(crit)} \approx \frac{\sigma'_{mu}}{\tau_{fu}} \frac{1}{\ell/d} \tag{4.36}$$

for 2D random distribution:
$$V_{f(crit)} \approx \frac{\pi}{2} \frac{\sigma'_{mu}}{\tau_{fu}} \frac{1}{\ell/d} \tag{4.37}$$

for 3D random distribution:
$$V_{f(crit)} \approx 2\frac{\sigma'_{mu}}{\tau_{fu}} \frac{1}{\ell/d} \tag{4.38}$$

Equations 4.36–4.38 indicate that for short fibres, the critical fibre volume is a function of both the aspect ratio of the fibre and the fibre–matrix bond. Measured values of τ_{fu} are in the range of 1–10 MPa. Figure 4.11 demonstrates the influence of bond (Figure 4.11(a)) and orientation (Figure 4.11(b)). For typical aspect ratios of 50–100, $V_{f(crit)}$ is in the range of 1–3%. This is the same order of magnitude as the value derived from Eq. 4.30.

The strength and modulus of elasticity relationships presented in this section are based on several simplifying assumptions, especially those involved in the parameters preceding the terms $\sigma_{fu}V_f$, $\tau_{fu}\ell/d$ and $E_f V_f$. However, these equations have been found to be the most useful for the development of empirical relationships from which these parameters could be determined experimentally, using the following general equations:

$$\sigma_{mu} = K_1\sigma'_{mu}V_m + K_2\sigma_{fu}V_f \tag{4.39}$$

$$\sigma_{cu} = K_3\sigma_{fu}V_f \tag{4.40}$$

$$\sigma_{cu} = K_4V_f\ell/d \tag{4.41}$$

where $K_{1,2,3,4}$ are constants. The constants in these equations can be obtained by a linear regression analysis of experimentally obtained curves of strength vs. $\sigma_{fu}V_f$ or $V_f\ell/d$. These constants are related to efficiency factors and bond strength, and the empirical values have sometimes been used to calculate these basic parameters.

Some investigators [7,16–18] have shown that the rule of mixtures can also be applied to account for the flexural strength, σ_b, taking into account the matrix as well as the fibre contribution.

$$\sigma_b = A\sigma_{bm}(1 - V_f) + BV_f\ell/d \tag{4.42}$$

where σ_{bm} is the flexural strength of the matrix and A and B are constants.

A regression analysis of flexural strength results [7,18,19] indicates that this equation provides a good description of the experimental data (Figure 4.12). Laws [19] has questioned the validity of Eq. 4.42, since the cracked matrix cannot provide any contribution to strength. She demonstrated that while one can derive Eq. 4.42 to describe the flexural strength, this has nothing to do with the rule of mixtures (see Section 4.7).

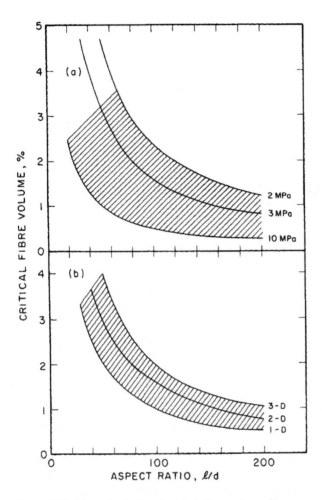

Figure 4.11 Plots of calculated critical fibre volume, $V_{f(crit)}$ vs. aspect ratio l/d for short fibres: (a) composites with different fibre–matrix bond strength τ_{fu}, with $\sigma_{mu} = 3$ MPa and random 2D fibre array; and (b) composites of different fibre orientations, with $\sigma_{mu} = 5$ MPa and $\tau_{fu} = 3$ MPa.

4.4 Fracture mechanics and interactions across cracks

Fracture mechanics refers to the analysis of the fracture of materials by the rapid growth of pre-existing flaws or cracks. Such rapid (or even catastrophic) crack growth may occur when a system acquires sufficient stored energy that, during crack extension, the system releases more energy than it absorbs. Fracture of this type (often referred to as *fast fracture*) can be predicted in terms of an energy

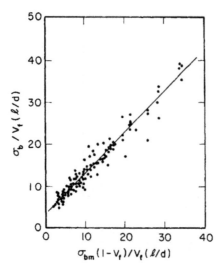

Figure 4.12 Correlation between composite flexural strength and matrix strength based on Eq. (4.42) (after Swamy and Mangat [17]).

criterion. If we consider an elastic system containing a crack and subjected to external loads, the total energy in the system, U, is

$$U = (-W_L + U_E) + U_s \tag{4.43}$$

where $-W_L$ is the work due to the applied loads L; U_E, the strain energy stored in the system and U_S, the surface energy absorbed in the creation of new crack surfaces.

A crack will propagate when $dU/d_c < 0$, where d_c is the increase in the crack length. Using this theory, one can derive the *Griffith equation*, which gives the theoretical fracture strength for brittle, linearly elastic materials:

$$\sigma_f = \left(\frac{2E\gamma_s}{\pi c}\right)^{1/2} \tag{4.44}$$

where c is the one-half the crack length; γ_s, the surface energy of the material. This is the basic equation of *linear elastic fracture* mechanics (LEFM).

If we define a parameter $G_c = 2\gamma_s$ = *critical strain energy release rate*, then we may write the criterion for catastrophic crack growth as

$$\sigma_f\sqrt{\pi c} = \sqrt{EG_c} \tag{4.45}$$

That is, fracture will occur when, in a stressed material, the crack reaches a critical size (or when in a material containing a crack of some given size, the stress reaches a critical value).

Alternatively, we may define a parameter $K_c = \sigma\sqrt{\pi c}$ = *critical stress intensity factor.*

$$K_c^2 = EG_c \tag{4.46}$$

K_c has the units of $N \cdot m^{-3/2}$, and is often referred to as the *fracture toughness* (not to be confused with the term 'toughness', which is used in this book to refer to the area under the stress–strain curve).

The LEFM parameters, G_c and K_c, are one-parameter descriptions of the stress and displacement fields in the vicinity of a crack tip. In much of the early work on the applications of fracture mechanics to cement and concrete, it was assumed that they provided an adequate failure criterion. However, later research showed that even for these relatively brittle materials, LEFM could only be applied to extremely large sections (e.g. mass concrete structures such as large dams). For more ordinary cross-sectional dimensions, non-linear fracture parameters provide a much better description of the fracture process.

In Section 3.5, the simple concepts of fibre–crack interaction were discussed, in relation to the microstructure of the interfacial transition zone. Fibres enhance the strength and, more particularly, the toughness of brittle matrices by providing a crack arrest mechanism. Therefore, fracture mechanics concepts have also been applied to model some of the properties of fibre reinforced concrete. Mindess [20] has reviewed the difficulties in modelling the behaviour of fibre reinforced concrete using the fracture mechanics approach. LEFM might be adequate to predict the effects of the fibres on first cracking. However, to account for the post-cracking behaviour (which is responsible for the enhanced toughness), it is essential to resort to elastic–plastic or non-linear fracture mechanics. A crude measure of the toughness (i.e. the energy absorbed during fracture) of the composite, can be obtained from the area under the stress–strain curve in tension. The fracture mechanics approaches which should provide a more precise description of the behaviour of FRC include the crack mouth opening displacement (CMOD), R-curve analysis, the fictitious crack model (FCM), and various other treatments, all of which model (either implicitly or explicitly) a zone of discontinuous cracking, or *process zone,* ahead of the advancing crack. These approaches provide fracture parameters which are, at least, dependent on the fibre content, whereas the LEFM parameters (G_c or K_c) are most often insensitive to fibre content. It might be added here that, while the J-integral has often been used to describe these systems, theoretically it cannot be applied to composite systems such as fibre reinforced concrete, where there is substantial stress relaxation in the micro-cracked region in the vicinity of the crack tip.

When considering the fibre–crack interactions using fracture mechanics concepts, three distinct processes must be treated:

1   *Crack suppression* this is the increase in stress, because of the presence of the fibres, required for crack initiation (i.e. the increase in the first cracking stress).

2   *Crack stabilization* this refers to the arrest of the cracks generated upon first cracking of the matrix, and which have begun to propagate across the fibres.
3   *Crack bridging* the action of fibres across the crack, which cannot be considered a discontinuity, but rather as an element that can transfer loads, but it is characterized by a unique relation of stress vs. crack opening (width), $\sigma-\omega$ curve. This approach, which is frequently referred to as the smeared crack approach, can be considered within the framework of the fictitious crack model.
4   *Fibre–matrix debonding* this process can be modelled as crack propagation along the fibre–matrix interface.

4.4.1   Crack suppression

Romualdi and Batson [21,22] were the first to use LEFM concepts to analyse the mechanics of crack suppression in a cement matrix induced by the presence of fibres. In the absence of any cracks, the extensions of both the matrix and the fibre under tensile loading are the same. However, when cracks are present, the matrix tends to extend more than the fibres, because of the stress concentrations just ahead of the crack tip. The fibres oppose this tendency. Through interfacial shear bond stresses, they apply 'pinching forces' which effectively reduce the stress intensity factor of the crack. As a result, higher applied stresses are now required to produce a stress field ahead of the crack tip such that the maximum stress exceeds the critical stress intensity factor, K_c, of the cement matrix. The shear bond stress distribution which causes this pinching effect is shown in Figure 4.13; this permits a determination of the contribution of neighbouring fibres to the stress intensity factor. Their work led to the introduction of the concept of the *spacing factor*, S; the stress required to cause matrix cracking was found to be inversely proportional to S.

Romualdi and Mandel [23] calculated the effective spacing factor in a 3D random short fibre reinforced concrete to be:

$$S = 2.76r(1/V_f)^{1/2} \tag{4.47}$$

where r is the fibre radius. They used this relationship to demonstrate the validity of the spacing factor concept for predicting the improvement in first crack stress due to fibres, as shown in Figure 4.14. A number of other numerical expressions for the spacing factor have since been developed, either by considering the distances between the centroids of individual fibres [24] or by considering the distances between fibres crossing a given plane in the composite [25]. Though the various spacing equations are similar in form, they do not properly account for the chief geometrical factors which define a fibre, that is length and diameter, since the different equations lead to quite different values.

There are a number of limitations to the application of spacing factor equations for predicting first crack strains. For instance, some experimental results considered by Edgington *et al.* [8] showed considerable deviation from the theoretical

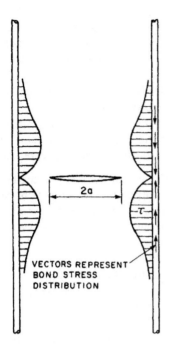

Figure 4.13 The 'pinching effect' and interfacial shear stress distribution predicted by Romualdi and Batson [22] in the arrest of crack propagation in a matrix, between the fibres.

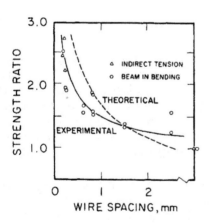

Figure 4.14 Effect of fibre spacing on first crack stress ratio (after Romualdi and Mandel [23]).

predictions of Romualdi *et al.* [21–23]. It appears that in calculations of *S*, one must account not only for length and diameter, but also for the effects of fibre orientation and the nature of the fibre–matrix bond. Romualdi and Batson assumed 'perfect' bond, which is simply not realistic for short fibre–cement composites. Swamy *et al.* [16] have thus suggested the concept of 'effective' spacing, which takes into account modifications due to both geometrical and bond considerations.

An alternative approach to the calculation of the first crack stress was proposed by Aveston *et al.* [1], based on energy balance considerations. For a crack to form under conditions of a fixed tensile stress, the energy changes to be considered are:

1   Work, ΔW, done by the applied stress, since the body increases in length.
2   Work, γ_{db}, done in debonding the fibre from the matrix. This term can be cal-
     culated by a fracture mechanics analysis of fibre–matrix debonding, assuming
     that the debonding energy at the fibre–matrix interface, G_{db} [26] is less than
     or equal to the surface energy, γ_m, of the matrix.
3   Work, U_s done by frictional slip after debonding.
4   The reduction in elastic strain energy of the matrix after cracking, ΔU_m.
5   The increase in elastic strain energy of the fibres after matrix cracking, ΔU_f,
     due to load transfer from the cracked matrix to the fibre.

Hence, if the work expended in creating a new crack (i.e. the surface energy of the crack area) is γ_m, then a crack will only form if:

$$2\gamma_m V_m + \gamma_{db} + U_s + \Delta U_f \leq \Delta W + \Delta U_m \tag{4.48}$$

On this basis, the first cracking strain of the matrix is:

$$\varepsilon_{mu} = \left[\frac{12\tau_{fu}\gamma_m E_f V_f^2}{E_c E_m^2 r V_m} \right]^{1/3} \tag{4.49}$$

Equation 4.49 predicts an effective increase in the matrix cracking strain for a high fibre volume content, a high interfacial frictional shear strength, and a small fibre diameter. In essence, this is another way of predicting the degree of crack suppression (i.e. increase in first cracking stress), in accordance with the spacing factor concepts. A significant increase in the matrix cracking strain should occur with well-bonded, small diameter filaments. This has been observed with asbestos and glass fibres, which generally consist of filaments (often in bundles) less than ~ 15 μm in diameter.

Equation 4.49 was developed with the assumption of longitudinal and aligned fibres. Tjiptobroto and Hansen [27] extended the treatment of ACK, and based on the same principles of energy considerations, they developed a relation for the more realistic composite which consists of short and oriented fibres. The differ-ence between the revised model and that of ACK was in (i) the separation between the terms associated with friction which are taken into account at the time of their occurrence; (ii) the effect of discontinuity in the fibres (leading to a different and more complex stress distribution along their length) and (iii) the orientation (randomness), which results in a smaller number of fibres bridging the crack. In this revised model, calculations of increase in fibre strain energy, decrease in matrix strain energy and change in external work were carried out, and based on energy balance considerations, the first cracking strain of the matrix, ε_{mu}, was determined for several cases (assumptions):

(i) *Approximation assuming that the stress transfer length is equal to the fibre length, ℓ_f, and a linear strain distribution along the fibre*:

$$\varepsilon_{mu} = \left(\frac{2\gamma_m V_m}{[(3/4)E_c - (7/24)E_f V_{ef}(1+\alpha)]\alpha\ell_f} \right)^{1/2} \quad (4.50)$$

(ii) *Approximation taking into account that the actual stress transfer length ℓ_{tr}, is smaller than the fibre length, ℓ_f, and assuming a linear strain distribution along the fibre*:

The stress transfer length can be calculated from the approximate value of ε_{mu}, using for that purpose ε_{mu} derived in Eq. 4.49, and assuming a linear stress transfer along the fibre:

$$\ell_{tr} = \frac{d_f \varepsilon_{mu}(1+\alpha)E_f}{4\tau} \quad (4.51)$$

Assuming a linear stress transfer:

$$\varepsilon_{mu} = \left(\frac{2\gamma m V m}{[(3/4)E_c - (7/24)E_f V_{ef}(1+\alpha)]\alpha(\beta\ell_f)} \right)^{1/2} \quad (4.52)$$

where

$$\beta = \frac{\ell_{tr}}{\ell_f/2}$$

(iii) *Effect of the assumption of linear stress transfer*   If the stress distribution is assumed to follow the shear lag model (i.e. non-linear) rather than the frictional model (i.e. linear), the terms describing the strain energy in the fibre will have to be modified. The ratio between the two terms (frictional–linear)/(shear lag–non-linear) was calculated and its influence was analysed as a function of fibre content for a low w/c ratio matrix [27]. Since the strain energy is included in Eqs 4.50 and 4.52 in the second part of the denominator with a negative sign and within a square-rooted term, the over or underestimation was shown to be relatively small. For example, for V_f of 12% in a low w/c matrix, the overestimation of the energy term is about 60%, but the effect on ε_{mu} is only about 14%.

4.4.2   Crack stabilization

Once first cracking has taken place in the brittle matrix, the fibres serve to inhibit unstable crack propagation. At this stage, the cracking patterns are complex, with discontinuous micro-cracks present ahead of the principal crack. This can be deduced from the various analytical models [28–30] and has also been observed microscopically (Figure 4.15) [31]. Thus, in the cracked composite, it is difficult to define the 'true' crack tip. The simplistic definition of a traction-free crack (as assumed in LEFM) is not applicable to FRC. Stress is transferred across the crack by a variety of mechanisms, as can be seen from the idealization of a crack proposed by Wecharatana and Shah [32], in Figure 4.16. Three distinct zones can be identified:

1   Traction-free zone.
2   Fibre bridging zone, in which stress is transferred by frictional slip of the fibres.
3   Matrix process zone, containing micro-cracks, but with enough continuity and aggregate interlock to transfer some stress in the matrix itself.

Many of the analytical treatments of these effects involve either the assumption of a specific stress field around the apparent crack tip, or the consideration of a

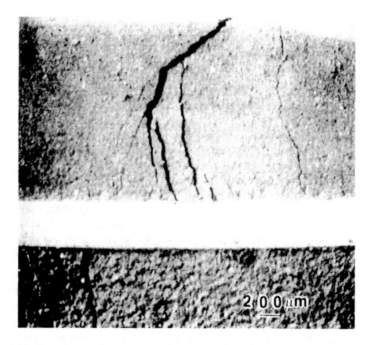

Figure 4.15 Complex crack patterns at the intersection of an advancing crack and a fibre lying in its path (after Bentur and Diamond [31]).

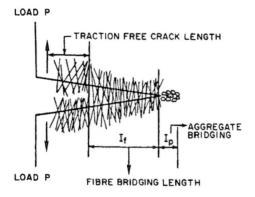

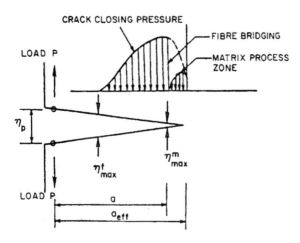

Figure 4.16 Idealized representation of an advancing crack and the stress field around it, in a fibre reinforced cement (after Wecharatana and Shah [32]).

traction-free crack surface which is subjected to a closing pressure (Figure 4.17) [33].

There are, in addition, a number of other more complicated models. Bazant and his co-workers [34–38] have developed a *smeared crack model*, in which fracture is modelled as a blunt smeared crack band. The fracture properties are characterized by three parameters: fracture energy, uniaxial strength, and width of the crack band. This approach lends itself particularly to computer-based finite element modelling of the cracks. For very large structures, this theory becomes equivalent to the LEFM approach. For smaller structures, however, the theory predicts a lower critical strain energy release rate, because the fracture process zone cannot develop fully. This theory has been shown to provide a good fit to experimental data from the literature.

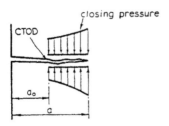

Figure 4.17 Schematic description of a traction-free crack with a closing pressure, to model the fracture behaviour of fibre reinforced cement (after Jenq and Shah [33]).

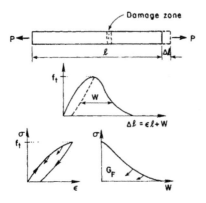

Figure 4.18 Analysis of tensile loading by the fictitious crack model (after Hillerborg [28]).

Hillerborg [39–42] has developed the *fictitious crack model* to describe the fracture of both, plain and fibre reinforced concretes. In this model, the deformation of a specimen is given in terms of two diagrams (Figure 4.18):

1   The stress–strain (σ–ε) diagram (including the unloading branch).
2   The stress-crack opening (σ–ω) diagram for the fracture zone itself.

The σ–ω curve gives, then, the additional deformation in a test specimen due to the presence of a damage zone. Moreover, the area under the σ–ω curve (called the fracture energy, G_F), equals the energy absorbed per unit (projected) area during the fracture process due to the additional deformation of the damage zone:

$$G_F = \int_0^{\omega_c} \sigma\,(\omega)d\omega \qquad (4.53)$$

4.4.3 Crack bridging by fibres

In recent years the mechanisms of fibre bridging across a crack have been investigated extensively, to provide a tool for the design and characterization of strain softening and strain hardening composites [43–46]. A distinction was made between the Griffith type crack and a steady-state flat crack (Figure 4.19). The action of the fibres across the crack surfaces is simulated by non-linear springs, whose response is described by the σ–ω curve (Figure 4.20). The characteristic points on the curve are the maximum bridging stresses which can be supported by the fibres, σ_{cu}, and the crack opening after which a decline in the stress takes place, δ_p. In the case of the Griffith type crack behaviour, the propagation of the crack is associated with the 'springs' at the middle portion of the crack being broken (Figure 4.19(a)): either δ_p or σ_{cu} are exceeded (strong interface will lead to low δ_p while a weak interface will result in a low σ_{cu}). This behaviour is characteristic of a strain-softening composite, and the 'springs' in the middle are unloaded as the crack propagates. However, if by adjusting the interface the δ_p and σ_{cu} are made sufficiently high, the 'springs' can maintain their bridging effect as the crack propagates (Figure 4.19(b)), and a quasi-plastic (multiple cracking) or even strain hardening behaviour may be generated.

The concept of the fictitious crack model has been extended by Hillerborg to fibre reinforced cement [28], as shown schematically in Figure 4.16(b). Here, the effect of fibres in bridging over the crack is taken into account, in addition to the stress transfer across the process zone. The crack width across which stress transfer occurs becomes much larger, due to the effect of the fibres which can slip over relatively large distances. In this case, the significant parameter for characterization and design is no longer G_F, but rather the crack bridging function which describes the change in stress, σ_ω over the whole range of ω, that is crack opening as large as 2 mm. This function includes in it the contribution of the fibre

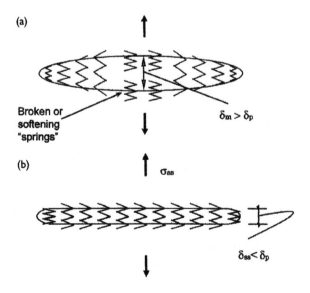

(a)

Broken or softening "springs"

$\delta_m > \delta_p$

(b)

σ_{ss}

$\delta_{ss} < \delta_p$

Figure 4.19 (a) Griffith crack; (b) steady-state crack (after Li *et al.* [44]).

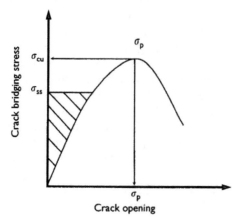

Figure 4.20 Bridging stress-crack opening curve, σ_b–δ, for fibre bridging over a crack (after Leung [47]).

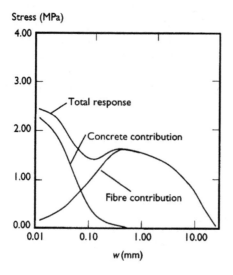

Figure 4.21 σ–ω curve and the contribution of the matrix (aggregates) and fibre, for 2% steel reinforced concrete (after Li *et al.* [45]).

and the matrix (aggregate interlock and process zone), as shown in the example in Figure 4.21.

These concepts were quantified in a series of studies [44–48] aimed at identifying the conditions required to obtain strain hardening or quasi-plastic behaviour, which is manifested also as a steady-state cracking behaviour. The characteristic σ–ω curve was modelled for randomly distributed discontinuous fibres by averaging the contribution of those individual fibres that cross the matrix crack plan, by making several assumptions [49–52]. One of the simpler relations is presented here [50]:

$$\sigma(\delta) = \begin{cases} \sigma_{cu}[2\delta/\delta_p)^{1/2} - (\delta/\delta_p)] & \text{for } \delta \le \delta_p \\ \sigma_{cu}(1 - 2\delta/L_f)^2 & \text{for } \delta_p \le \delta \le L_f/2 \\ 0 & \text{for } L_f/2 \le \delta \end{cases} \tag{4.54}$$

The extension of a crack, Δa, initiated at the first cracking stress, σ_{cr}, was considered, outlining the conditions for energy balance (Figure 4.22). The additional work on the system generated by displacement δ_{ss}, dW, should be equal to the change in the surface energy due to the creation of new crack surfaces, dE_s, and the additional strain energy consumed in stretching the fibres across the crack, dU:

$$dW = dU + dE_s \tag{4.55}$$

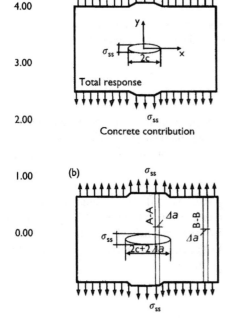

Figure 4.22 Crack extension under steady-state condition (after Leung [47]).

The individual terms in Eq. 4.55 can be calculated:

$$dW = (2\Delta a)\sigma_{ss}\delta_{ss} \tag{4.56}$$

$$dU = (2\Delta a)\left[\int_0^{\delta_{ss}} \sigma(\delta)d\delta\right] \tag{4.57}$$

$$dE_s = (2\Delta a)G_{tip} \tag{4.58}$$

where G_{tip} is the crack tip critical energy release rate of the composite.

Substituting the individual terms in Eqs 4.56–4.58 in the energy balance Eq. 4.55, yields the following relation:

$$\sigma_{ss}\delta_{ss} - \int_0^{\delta_{ss}} \sigma(\delta)d\delta = G_{tip} \tag{4.59}$$

The left-hand side of Eq. 4.59 represents the shaded area in Figure 4.20 which is the complementary energy of the σ–ω curve. In order to obtain steady-state cracking, the maximum complementary energy (i.e. its value when it reaches the peak value of σ_{cu}) should be greater than the energy required for crack propagation [53,54]:

$$\sigma_{cu}\delta_p - \int_0^{\delta_p} \sigma(\delta)d\delta \geq G_{tip} \tag{4.60}$$

To meet the conditions in Eq. 4.60, for a crack bridging function described in Eq. 4.59, the fibre content V_f should exceed a critical volume, V_f^{crit}:

$$V_f \geq V_f^{crit} \equiv \frac{12 J_{tip}}{g\tau(L_f/d_f)\delta_p} \tag{4.61}$$

The implications of Eq. 4.61 in setting the requirements for the matrix and interface properties in order to achieve pseudo strain-hardening behaviour was discussed in [54]. Graphical presentation of the combinations leading to such behaviour is presented in Figure 4.23 [55].

For the purpose of design, to facilitate using finite element calculations, several σ–ω relations have been suggested by RILEM [56]. They include a multilinear relation (Figure 4.24, Eq. 4.62) and a bilinear relation (Figure 4.25, Eq. 4.63). The multilinear relation can allow for strain-hardening behaviour, while the bilinear relation is always strain softening. The bilinear relation requires a total of four material parameters to describe the σ–ω curve.

$$\sigma_\omega = \sigma_i - \alpha_i\omega \quad \text{for} \quad \omega_i - 1 < \omega \leq \frac{\sigma_{i+1} - \sigma_i}{\alpha_{i+1} - \alpha_i} \tag{4.62}$$

$$\omega_0 = 0 \quad \sigma_1 = f_1 \quad \alpha_1 > 0$$

$$\sigma_\omega = \begin{cases} \sigma_1 - \alpha_1\omega & \text{for} \quad 0 \le \omega \le \omega_1 = \dfrac{\sigma_2 - \sigma_1}{\alpha_2 - \alpha_1} \quad \alpha_1 > 0 \quad \sigma_1 = f_1 \\[3mm] \sigma_2 - \alpha_2\omega & \text{for} \quad \omega_1 < \omega \le \omega_c = \dfrac{\sigma_2}{\alpha_2} \quad \alpha_2 > 0 \end{cases}$$

$$(4.63)$$

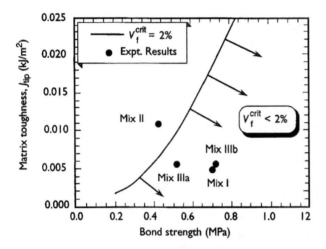

Figure 4.23 Relations between matrix properties (J_{tipp}) and interfacial bond to achieve strain hardening behaviour; adapted from Li et al. [55]. ($E_f = 117$ GPa, $L_f = 12.7$ mm, $d_f = 38$ μm, $E_m = 25$ GPa).

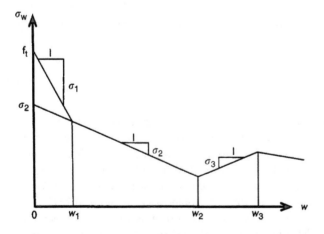

Figure 4.24 Multilinear stress-crack opening curve, adapted from [56].

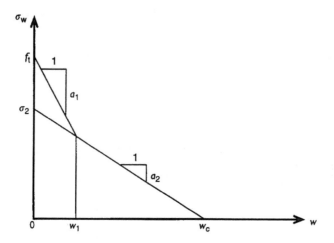

Figure 4.25 Bilinear stress-crack opening curve, adapted from [56].

4.4.4 Fibre–matrix debonding

A fracture mechanics approach has also been applied to the problem of fibre-debonding and pull-out [57,58], as an alternative to the treatment based on the analysis of elastic and frictional shear stresses (Chapter 3). The object is to develop material parameters to account for debonding which are more reliable and easier to evaluate experimentally than the interfacial shear bond strength values. In this treatment, it is assumed that the debonded region is stress-free ($\tau_f = 0$ in Figure 4.26), and this zone is treated as an interfacial crack of length b. Using the classical Griffith theory (or LEFM), the conditions leading to the propagation of this crack, and to spontaneous debonding, can be calculated. Outwater and Murphy [26] calculated the tensile stress in the fibre required for catastrophic debonding as

$$\sigma = \left(\frac{8E_f G_{db}}{d} \right)^{1/2} \tag{4.64}$$

where G_{db} is the energy required to debond a unit surface area of fibre.

The earlier analysis considers only the energy balance in the fibre itself. Subsequently, Stang and Shah [58] developed a solution which takes into account the compliance of the entire pull-out system. Morrison *et al.* [57] further extended the analysis by taking into account also the frictional resistance in the debonded zone, which is the more realistic case for FRC composites. The critical strain energy release rates for debonding calculated by Morrison *et al.* [57], 2.5 N/m, were similar to those calculated by Mandel *et al.* [59], of 4–7.4 N/m. These values are, however, lower than typical values of the critical strain energy release rate of plain

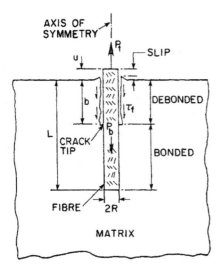

Figure 4.26 Schematic description of the model used to consider the pull-out problem in terms of fracture mechanics concepts, with a propagating debonding crack of length *b* (after Morrison *et al.* [57]).

mortars, 5–12 N/m [60]. Thus, for a crack to follow the path of least resistance, it should propagate along the interface rather than through the matrix.

This fracture mechanics approach tends to confirm the conclusions reached previously. That is, the interface in fibre cement composites is relatively weak. This leads to preferential crack propagation along the fibre–matrix interface.

4.5 Multiple cracking

In FRC composites, as stated earlier, the fibres are effective primarily in the post-cracking zone. Immediately following first cracking, a process of multiple cracking is initiated, in which the matrix cracks successively, and divides into segments of similar length, as shown schematically in Figure 4.1. This process occurs at an approximately constant stress, which is equal to the first crack stress, $\varepsilon_{mu}E_c$. This region of the stress–strain curve is horizontal, or slightly ascending, resembling plastic yielding (Figure 4.2). Therefore, this behaviour is sometimes referred to as 'pseudo-plastic'. It should be emphasized that this pseudo-plasticity is associated with successive cracking: continuity is not maintained across the composite. The multiple cracking process is extremely important, since it controls the *toughening* of the cementitious composite. The control of crack spacing and width at this stage also has a considerable influence on the serviceability of the composite.

4.5.1   Multiple cracking in a composite with continuous and aligned fibres

The first detailed analytical treatment of multiple cracking was presented by Aveston, Cooper and Kelly [1], for continuous and aligned fibre reinforcement, assuming frictional shear stress transfer. They showed that the matrix will eventually fracture into segments of lengths between x and $2x$, where x is determined by the rate of stress transfer between the fibre and the matrix (Figure 4.27(a)). The length $2x$ is such that the stress transferred to the matrix at a distance x from the crack does not exceed its cracking stress, that is $\sigma_m = \varepsilon_{mu}E_m$ (Figure 4.27(b)). For a constant frictional shear transfer with stress τ_{fu}, the stress build-up in the matrix is linear, and by a simple force balance, the value of x can be derived (assuming fibres of circular cross section):

$$\sigma'_{mu}V_m = 2\pi r \frac{V_f \tau_{fu}}{\pi r^2} x,$$

therefore

$$x = \left(\frac{V_m}{V_f}\right) \frac{\sigma'_{mu}r}{2\tau_{fu}} \tag{4.65}$$

The strain distribution in the matrix at the final stage of multiple cracking is linear, with a value of zero at the crack surface, going to values of ε_{mu} and $\varepsilon_{mu}/2$ at the centres of blocks with length of $2x$ and x, respectively.

This redistribution of the stresses and strains in the matrix must be accompanied by appropriate adjustments of the stresses and strains in the fibre, leading also to linear stress and strain redistributions. The load originally carried by the matrix at the crack surface ($\sigma'_{mu}V_{mu}$) will be fully transferred to the fibre bridging the crack, at the point of its entry into the matrix. This additional load will gradually be reduced along the fibre; it will be fully transferred to the matrix at the centre of the $2x$ long block, but only half of this additional load will be transferred to the

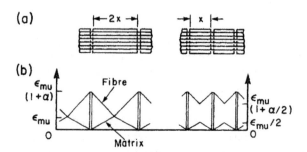

Figure 4.27 Multiple cracking according to the ACK model [1]: (a) crack spacing (x and $2x$ spacings) and bridging fibres; and (b) strain distributions in fibres and matrix.

matrix at the centre of the block with length x. These additional stresses will result in an increase in fibre strains at the cracked surface. This increase (compared with the original strain, ε_{mu}, at the start of multiple cracking) will be:

$$\Delta \varepsilon_f = \alpha \varepsilon_{mu} \tag{4.66}$$

where $\alpha = E_m V_m / E_f V_f$.

The total strain in the fibre at the crack will thus be:

$$\varepsilon_{mu} + \Delta \varepsilon_f = (1 + \alpha)\varepsilon_{mu} \tag{4.67}$$

The strain distributions in the fibre and matrix are presented in Figure 4.27(b).

According to this model, the multiple cracking process will continue at a constant stress ($E_c\varepsilon_{mu}$) until the composite is completely separated into segments with lengths between x and $2x$. The increase in the composite strain during this stage of multiple cracking can be calculated from the increase in the average strain in the fibre. This increase is $\alpha\varepsilon_{mu}/2$ for the case of cracking into $2x$ long segments and $3\alpha\varepsilon_{mu}/4$ for segments of length x. Thus, the upper and lower limits for the strain at the end of the multiple cracking process, ε_{mc} are:

$$\varepsilon_{mu}(1 + \alpha/2) < \varepsilon_{mc} < \varepsilon_{mu}(1 + 3\alpha/4) \tag{4.68}$$

Beyond the multiple cracking stage, no more cracking can take place, and additional loading will result in stretching of the fibres until they break. Assuming stretching of continuous and aligned fibres, the modulus of elasticity in the post multiple cracking zone is $E_f V_f$ (i.e. no contribution from the cracked matrix) until the fibre load-bearing capacity, $\sigma_{fu} V_f$, is reached. The ultimate composite strain at this point, ε_{cu}, depends on the ultimate strain of the fibre and the crack spacing, giving lower and upper limits of:

$$\left(\varepsilon_{fu} - \frac{\alpha\varepsilon_{mu}}{2}\right) < \varepsilon_{cu} < \left(\varepsilon_{fu} - \frac{\alpha\varepsilon_{mu}}{4}\right) \tag{4.69}$$

The shape of the resulting stress–strain curve is presented in Figure 4.2. The composite strength value of $\sigma_{fu} V_f$, and Eq. 4.69 are based on the assumption that failure by fibre fracture will precede failure by fibre pull-out. This is less likely to occur in short fibre-reinforced composites.

Aveston and Kelly [61] have extended the ACK model, and developed an analytical treatment for elastic bonding, assuming that the elastic strains in the matrix and the fibre at the interface are the same. In this case the matrix remains bonded to the fibre after it has cracked and, except for the deformation at the crack surface, the whole system remains linearly elastic. The earlier model presented in Eqs 4.66–4.69 assumed frictional slip after cracking of the matrix. In this later treatment, the additional stress on the fibre at the crack is transferred back to the matrix at a rate controlled by elastic considerations. Their calculations, based on concepts similar to those of the shear lag theory, indicates that the shape of the stress–strain curve is not sensitive to the assumption regarding the nature of bond (Figure 4.28).

The gradual cracking of the matrix results in an increase in the interfacial elastic shear stress, until the adhesional shear strength value is exceeded. At this stage debonding begins, and combined elastic and frictional shear stress transfer must be considered. Assuming the adhesional elastic shear strength to be equal to the matrix tensile strength, Aveston and Kelly [61] showed that elastic bonding can be maintained during the entire multiple cracking process only if the fibre volume contents exceed 36%, 38% and 50% in cementitious composites reinforced with carbon, steel and glass fibres, respectively. Obviously, these values are well outside the practical range for FRC. Thus, in practice, pure elastic bonding is highly unlikely, and the multiple cracking model should be based on combined elastic and frictional shear. Additional analysis [61] has indicated that the crack spacing in typical FRC composites, calculated using this assumption, will not differ by more than 15% from the value determined by the simplified model, which assumes that frictional shear is the only stress transfer mechanism (Eq. 4.65). Thus, it appears that the simple theory adequately describes the behaviour of FRC composites, as long as the elastic shear bond strength is not much greater than the matrix tensile strength.

A similar treatment was later presented by Laws [62], based on Lawrence's [63] model for combined elastic and frictional fibre–matrix stress transfer. Laws

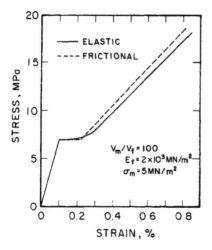

Figure 4.28 Idealized stress–strain curves for Portland cement reinforced with 1% by volume of steel fibres, assuming elastic and frictional stress transfer (after Aveston and Kelly [61]).

calculated the minimum crack spacing, x, as a function of fibre volume content for a typical glass FRC using three different assumptions:

1   frictional stress transfer only;
2   elastic stress transfer only;
3   combined elastic and frictional stress transfer.

The results of this calculation are shown in Figure 4.29. Curve A, (combined elastic and frictional stress transfer), is considered to be the most realistic. Curve C (frictional stress transfer), is ~ 20% higher than curve A, in the range typical for FRC composites ($V_f = 1$–10%; $V_m/V_f = 90$–100). Thus, considering only frictional shear can provide a reasonable estimate of the behaviour of the composite, as suggested by Aveston and Kelly [61]. Curve B (elastic stress transfer) yields much lower values for crack spacing.

    Equation 4.65 can be used to predict the crack spacing under tensile loading, if the frictional shear bond strength, τ_{fu}, is known. Alternatively, observations of the crack spacing during loading can be used to calculate τ_{fu} from this equation. Aveston et al. [15] confirmed the validity of this equation for continuous and aligned steel fibres in cementitious composites, by plotting the experimentally observed crack spacing x, as a function of $\varepsilon_{mu} V_m/V_f$, and showing that the relationship was reasonably linear (Figure 4.30). The slope of this curve was used to · calculate τ_{fu}, which was found to be 6.8 MPa. This is in reasonable agreement with average shear bond strength values reported for a steel fibre in a cementitious matrix. Similar analyses have been carried out to find the relationship between

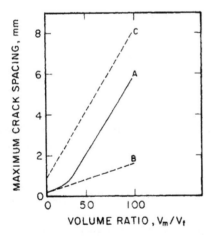

Figure 4.29 Calculated minimum crack spacing as a function of V_m/V_f for glass fibre reinforced cement composite (after Laws [62]): Frictional stress transfer only, curve C; elastic stress transfer only, curve B; combined elastic and frictional stress transfer, curve A.

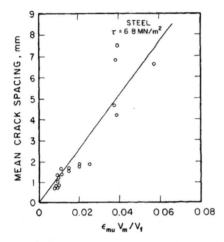

Figure 4.30 Mean crack spacing in continuous and aligned steel fibre reinforced cement as a function of $\varepsilon_{mu}/V_m/V_f$ (after Aveston *et al.* [15]).

bond and crack spacing for carbon [15], glass [64] and polypropylene [65] fibres in a cementitious matrix.

The analysis described in this section indicates the significance of bond in controlling multiple cracking. However, it has been found in practice that bonding in an actual composite may be different, and sometimes more effective, from that predicted by assuming interfacial shear only.

4.5.2   Effect of fibre orientation

Aveston and Kelly [61] later extended the ACK model to include the effects of fibre orientations: random 2D and 3D fibre distributions were considered. They treated this problem in two stages:

1   calculation of the number of fibres that actually bridge across a crack;
2   determination of the distance, in the direction of the applied stress, over which the fibres bridging the crack transfer sufficient load to cause further cracking of the matrix.

They showed that the numbers of fibres crossing the crack are $V_f/\pi r^2$, $2/\pi(V_f/\pi r^2)$ and $1/2(V_f/\pi r^2)$ for aligned, and random 2D and 3D fibre distributions, respectively.

The geometry of an inclined fibre bridging a crack is shown in Figure 4.31. For simplicity, fibre bending and matrix crumbling at the crack are neglected. The force exerted by the fibres on the matrix at a distance x' from the crack, and in a direction normal to the crack surface, can be calculated by adding two components: the frictional axial force in the fibre, multiplied by $\sin\theta$, and the

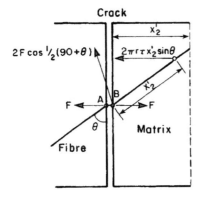

Figure 4.31   Geometry of a fibre crossing a crack (after Aveston and Kelly [61]).

force in the horizontal portion of the fibre at the crack (AB in Figure 4.31). The distance x', at which the force exerted on the matrix is sufficient to cause cracking is calculated as:

$$x'_{(2D)} = \frac{\pi}{2}x \qquad \text{(random 2D distribution)} \tag{4.70}$$

$$x'_{(3D)} = 2x \qquad \text{(random 3D distribution)} \tag{4.71}$$

In Eqs 4.70 and 4.71, x is the minimum crack spacing assuming aligned fibres (Eq. 4.65). In the derivation of $x'_{(2D)}$ and $x'_{(3D)}$, it was assumed that the only stress transfer mechanism is a frictional one. Equations 4.70 and 4.71 predict an increase in minimum crack spacing by about 50% for a planar fibre distribution, and 100% for a 3D fibre distribution.

4.5.3   Effect of fibre length

Laws [66] considered the effect of fibre length on the multiple cracking process, and on the stress–strain curve predicted by the ACK model. The analysis took into account the probability that a short fibre bridging the crack would slip, rather than transfer a sufficiently high load back to the matrix, so as to cause it to crack at a distance of x to $2x$ from the first crack. It was further assumed that the fibres intersected by a crack would have their shorter embedded lengths uniformly distributed between 0 and $\ell/2$. The results of this treatment indicated that shorter fibres would increase the strain in the multiple cracking zone, and would also change the shape of the curve beyond this stage: the curve would no longer be linear, and the ultimate strain would be reduced. These characteristics are shown schematically in Figure 4.32. The quantitative effect of length was described in terms of a length efficiency factor.

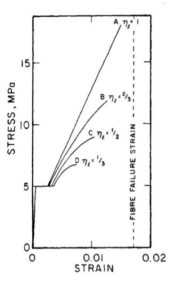

Figure 4.32 The schematic description of the effect of fibre length on the stress–strain curve (after Laws [66]).

The effect of fibre length on multiple cracking was also treated by Proctor [67], based on a modification of the ACK relationships for multiple cracking strains (Eq. 4.68) which involved multiplying the modulus of elasticity of the fibres, E_f, by an efficiency factor which accounts for the reduced rigidity of the reinforcing system. Proctor's relationships yield lower multiple cracking strains and lower ultimate strains than those of Laws [66]. However, both treatments indicate that for shorter fibre lengths, the multiple cracking strains increase. This is, essentially, due to the reduction in the effective rigidity of the fibre reinforcement (E_f in Eq. 4.66), leading to greater extension of the fibres due to the additional load imposed on them during matrix cracking.

4.5.4 *Effect of matrix strength and fibre content*

Tjiptobroto and Hansen [68] extended the ACK model and the multiple cracking concept to a strain hardening composite with a high strength matrix, in which the stress–strain curve of the composite is characterized by three stages, which they defined as elastic, strain hardening and crack opening (strain softening), Figure 4.33. They used a similar energy approach, but in their model they considered the different timing of the various energy terms, as outlined in Figure 4.33. One should especially note that the ACK model does not differentiate between the first cracking point and the end of multiple cracking with respect to energy changes, which are all combined in the ACK model at the first cracking point. At

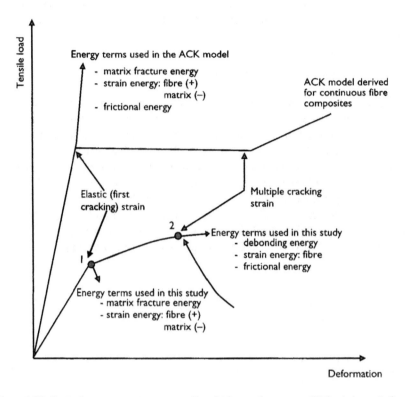

Figure 4.33 Typical stress–strain curve for high performance FRC (adapted from Tjiptobroto and Hansen [68]).

point 1 (on the curve in Figure 4.33) the first crack occurs, and it opens up at the end of the zone between points 1 and 2; at the same time additional microcracks are formed. Experimental and theoretical considerations indicated that the strain softening initiated at the end of this zone is the result of the drastic opening of the first crack, which eventually leads to failure.

In the multiple cracking zone (between points 1 and 2) the energy consumed in opening up of the first crack is:

$$E_{1-2} = \Delta U_{f-mc} + \Delta U_{fr} + U_{db} \tag{4.72}$$

where ΔU_{f-mc} is the increase in the fibre strain energy due to crack bridging in the multiple cracking stage; ΔU_{fr}, the frictional energy due to the slip between the fibres and the matrix; ΔU_{db}, the debonding energy required to destroy the elastic bond between the fibre and the matrix.

The energy required to form additional microcracks in this zone, denoted as $E_2, E_3, \ldots E_n$, is the same for each, and can be described by the following relation:

$$E_2 \approx E_3 \approx E_n \approx G_m V_m + \Delta U_{f-mu} - \Delta U_m \tag{4.73}$$

where $G_m V_m$ is the fracture energy of the matrix in the composite to create new surface; G_m, the matrix fracture energy; V_m, the matrix volume fraction; ΔU_{f-mu}, the increase in fibre strain energy as the result of bridging; ΔU_f, the decrease in matrix strain energy due to the stress relaxed when cracking occurs.

Multiple cracking will occur if the energy to open the first crack, E_{1-2} is greater than that required for formation of the subsequent microcracks, $E_2, E_3, \ldots, E_n$. The process of multiple cracking is repeated until the sum of the energies $E_2, E_3, \ldots, E_n$ becomes larger than E_{1-2}. This is consistent with the assumption that the first crack always becomes the failure crack. The number of cracks is thus E_{1-2}/E_2.

The individual energy terms in Eqs 4.72 and 4.73 were evaluated [68]: For Eq. 4.72:

$$\Delta U_{f-mc} = \frac{V_{ef}}{E_f} \left[\frac{7}{48} \frac{\tau_{fu}^2 \ell_f^3}{r^2} - (E_f \varepsilon_{mu})^2 \frac{\ell}{4} \right] \tag{4.74}$$

$$\Delta U_{fr} = \frac{V_{ef} \tau_{fu} \ell^2}{4r} \left[\frac{\tau_{fu} \ell}{3 E_f r} - \frac{1}{2} \varepsilon_{mu} \right] \tag{4.75}$$

$$U_{db} = \frac{V_{ef} \ell G_{II}}{r} \tag{4.76}$$

$$V_{ef} = 0.5 V_f \tag{4.77}$$

where V_{ef} is the effective volume fraction of fibres; V_f, the volume fraction of fibres; E_f, the modulus of elasticity of fibres; ℓ, the fibre length; r, the fibre radius; ε_{mu}, the first crack strain of the composite; τ_{fu}, the frictional interfacial stress; G_{II}, the second mode fracture energy, that is in a shearing fracture process of the fibre–matrix interface.

The terms in Eq. 4.73:

$$\Delta U_{f-mu} = \frac{1}{24} E_f V_{ef} \ell \alpha (18 + 7\alpha) \varepsilon_{mu}^2 \tag{4.78}$$

$$\Delta U_m = \frac{11}{24} E_m V_m \ell_f \varepsilon_{mu}^2 \tag{4.79}$$

where

$$\alpha = \frac{E_m V_m}{E_f V_f}$$

E_m is the modulus of elasticity of the matrix and V_m, volume fraction of fibres.

For multiple cracking to occur, the critical fibre content should be such that the following relation is satisfied:

$$E_{1-2} > E_2 \tag{4.80}$$

By introducing the appropriate terms of the equations for E_{1-2} and E_2, the critical fibre volume, $V_{f(crit)}$ is:

$$V_{f(crit)} = 2V_{ef(crit)} = \frac{\gamma_m + \sqrt{\gamma_m^2 + A}}{A} \tag{4.81}$$

where

$$A = 2\gamma_m + \frac{\ell}{r}\left(\frac{11}{48}\tau_{fu}^2\frac{\ell_f^2}{E_f r}\right);$$

γ_m is the matrix surface energy ($G_m/2$).

For typical values of conventional and high performance systems (Table 4.1), Tjiptobroto and Hansen [68] calculated the values for E_{1-2} and E_2, and determined the critical fibre volume, to be 3.3% and 15% for high performance and conventional composites, respectively.

This model and calculations explain the feasibility of using dense matrices (leading to high interfacial bond) and thin fibres for producing high performance

Table 4.1 Typical parameters for conventional and high performance fibre cement composites (after Tjiptobroto and Hansen [68])

Material	Conventional FRC	High performance FRC
Matrix		
Compressive strength, MPa	20	175
Modulus of elasticity, GPa	21	49
Interfacial frictional stress, τ_f, MPa	1	5
Fracture energy, G_m, N/m	20	120
Interfacial fracture energy, G_{II}, N/m	2.5	120
w/c ratio	0.4–0.6	0.18
Steel fibre		
Modulus of elasticity, GPa	200	200
Length, mm	25	6
Diameter, mm	0.5	0.15

FRC. These composites are characterized by high strength and ductility which is the result of a multiple cracking process, which can be achieved even with a modest content of fibres.

4.6   Modelling of tensile behaviour

4.6.1   Stress–strain curve

The tensile stress–strain curve is usually modelled using the ACK model [1,61]. The salient points on the curve presented in Figure 4.2 (initial modulus of elasticity, matrix cracking strain, strain at the end of the multiple cracking zone and the ultimate strain) can be determined, on the basis of the concepts described in the previous sections.

Laws [66,69] modified this model to take into account the effect of fibre length and orientation on the shape of the curve. In the simplest case, the stress–strain relations for the reinforcing fibre array were analysed, ignoring the influence of the matrix. This is a first approximation of the behaviour in the post-multiple cracking zone, where the influence of the matrix is small. The stress contribution of the fibre array to the composite, at any stage of loading x, is given by the product of the average stress in the fibre, $\bar{\sigma}_f(x)$ and its volume concentration, V_f. Based on the efficiency relations discussed previously (Section 4.2) the value of $\bar{\sigma}_f V_f$ can be shown to be a function of the strain $\varepsilon_f(x)$ in the reinforcing array:

$$\bar{\sigma}_f(x) \cdot V_f = \left(1 - \frac{\ell'_c}{\ell} \frac{\varepsilon_c(x)}{\varepsilon_{fu}}\right) E_f V_f \varepsilon_f(x) \tag{4.82}$$

where

$$\ell'_c = 1/2\ell_c(2 - \tau_d/\tau_s).$$

The stress–strain response of the fibre array can be calculated from Eq. 4.82, as shown in Figure 4.34. These curves demonstrate that if the frictional resistance after the initiation of sliding is small (low ratios of τ_d/τ_s), the maximum stress which the fibre array can support (point F in Figure 4.34) occurs at a strain below the failure strain of the fibre (ε_{fu}), unless the length of the fibre approaches twice the critical length ℓ_c (curve D in Figure 4.34). For the case where $\tau_{fu} = \tau_d = \tau_s$, which is more typical of FRC in general, failure will occur at the failure strain of the fibre (ε_{fu}) if the fibre length is equal to or larger than the critical length, ℓ_c. Relationships of this kind, and their sensitivity to the τ_d/τ_s ratio, emphasize the significance of the pull-out performance of the fibres, after sliding and extraction from the matrix have been initiated, in controlling the shape of the stress–strain curve. Thus, when characterizing the fibre matrix bond by a pull-out test, the significance of the shape of the pull-out load–displacement curve in the post peak zone, cannot be underestimated.

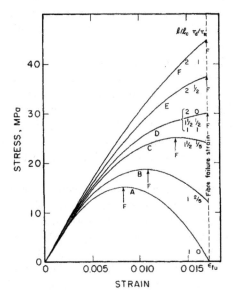

Figure 4.34 Stress–strain curves of a glass fibre array determined from Eq. 4.82 for various ratios of l/l_c and τ_d/τ_{fu}, assuming aligned short fibres held across a crack. $E_f = 70$ GPa, $\sigma_{fu} = 1200$ MPa, $V_f = 0.05$. Failure occurs at F (after Laws [66]).

Equation 4.82 was further modified by Laws to take into account the effect of orientation for a random 2D fibre array:

$$\bar{\sigma}(x) \cdot V_f = \frac{3}{8}\left(1 - \frac{5}{6}\frac{\ell_c'}{\ell}\frac{\varepsilon_c(x)}{\varepsilon_{fu}}\right) E_f V_f \varepsilon_f(x) \tag{4.83}$$

This equation is based on the efficiency factor relationships developed for the case in which the orientation and length effects are combined (Section 4.2.3). Equations 4.82 and 4.83 can then be applied to calculate the strength of the composite from the maximum value of $\bar{\sigma}_f(x) \cdot V_f$ of the fibre array:

$$\sigma_{cu} = \eta_\ell \cdot \eta_\theta \cdot \sigma_f(x)_{max} \cdot V_f \tag{4.84}$$

where η_ℓ and η_θ are the efficiency factors for length and orientation, η_θ, is equal to 1 and 3/8 for aligned and random 2D fibre arrays, respectively, whereas

$$\text{for} \quad \ell >> 2C\ell_c' \quad \eta_\ell = (1 - C\ell_c'/\ell) \tag{4.85}$$

$$\text{for} \quad \ell << 2C\ell_c' \quad \eta_l = \frac{1}{4C}\ell/\ell_c' \tag{4.86}$$

where C is equal to 1 or to 5/6 for aligned and random 2D fibre arrays, respectively.

The stress–strain relations developed for the fibre array can serve as a first estimate for the stress–strain curve of the composite in the post-cracking zone, neglecting the contribution of the matrix. However, for more accurate modelling, the constraint of the matrix has to be taken into account; some of the stress imposed on the fibre at the crack is transferred back to the coherent portions of the matrix which exist between the x to $2x$ spaced cracks (Figure 4.27). Thus, the average strain in the composite will be less than the strain in the fibre at the crack, by a value of $\Delta\varepsilon$. This value can be calculated as $\Delta\varepsilon = C \cdot \alpha \cdot \varepsilon_{mu}$, where α was defined in Eq. 4.66 and C is a numerical constant which depends on the crack spacing. These concepts can be used to derive a model describing the stress–strain curve of the composite [66,67]. The results of such an analysis are demonstrated in Figure 4.32.

The ACK model, even in its simple form, can be used to predict the effect of the properties of various fibres on the stress–strain curve of FRC. This has been demonstrated by Hannant [9], who calculated the stress–strain curve for typical FRC composites reinforced with asbestos, glass and polypropylene fibres with $V_f > V_{(crit)}$, and for a steel FRC where $V_f < V_{(crit)}$. The results are shown in Figure 4.35. For a high modulus fibre (asbestos), the multiple cracking zone is considerably reduced, giving rise to a more brittle behaviour, compared with the

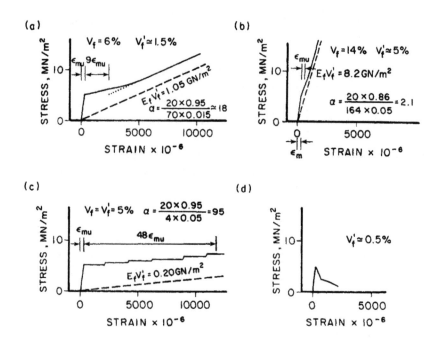

Figure 4.35 Calculated stress–strain curves of four types of fibre cement composites (after Hannant [9]).

lower modulus fibres. However, with a low modulus fibre, such as polypropylene, the strengthening potential of the fibre will be mobilized only after excessive deformation has taken place. Other investigations have also found good agreement between the tensile stress–strain curve determined experimentally, and that calculated by the ACK model (as demonstrated in Figure 4.36).

A different concept for modelling tensile behaviour was proposed by Nathan *et al.* [70], involving a trilinear stress–strain curve characterized by three linear ranges; uncracked, cracked and post-cracking (Figure 4.37). The slope in each of the ranges and the points separating them were calculated using the rule of mixtures concepts, combined with the appropriate efficiency factors. The main difference between this model and the ACK model occurs in the medium range, which in the ACK model is the multiple cracking zone and is essentially horizontal, whereas in this model the stress–strain curve continues to ascend. In this range, Nathan *et al.* [70] and Ng *et al.* [11] applied the rule of mixtures, but reduced the contribution of the matrix by a factor k to account for its reduced rigidity:

$$E_{cq} = kE_m V_m + \eta_\ell \eta_\theta E_f V_f \tag{4.87}$$

The value of k is less than 1; its value is determined empirically.

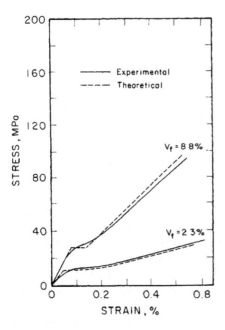

Figure 4.36 Experimental and analytical (ACK model) stress–strain curves of cement reinforced with continuous and aligned steel fibres (after Aveston *et al.* [15]). Dashed line, experimental; full line, calculated.

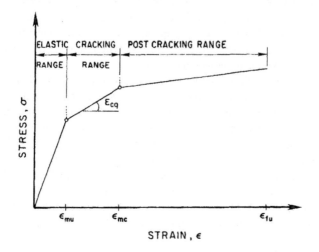

Figure 4.37 Trilinear stress–strain curve to model the behaviour of fibre reinforced cement (after Nathan *et al.* [70]).

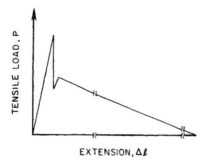

Figure 4.38 Modelling of stress–strain curve of composites with $V < V_{f(crit)}$ (after Lim *et al.* [71]).

While a good fit was obtained between the experimental results and the analytical model, the physical concepts in the Nathan *et al.* [11,70] model and the ACK model are significantly different in the multiple cracking zone (or cracking range) which accounts for the different stress–strain curve predictions.

Lim *et al.* [71] extended the Nathan *et al.* [11,70] model to account for fibre volumes smaller than $V_{f(crit)}$. In this modification the post-cracking zone was analysed by assuming that the behaviour is dominated by the widening of a single major crack. The loads and deformations in this zone were calculated by considering a bridging fibre pulling out of the matrix. The deformations were the sum of the elastic strain in the uncracked matrix and the widening of the crack. The stress–strain curve predicted by these concepts is shown in Figure 4.38.

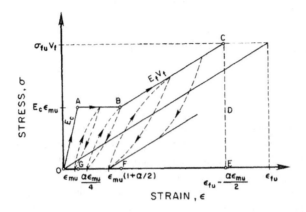

Figure 4.39 Residual strains and hysteresis during unloading–loading cycles in the post-cracking zone, as predicted from the ACK model [1].

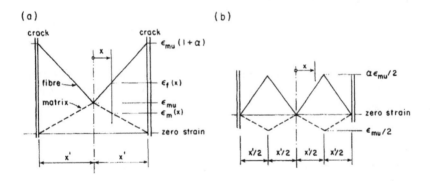

Figure 4.40 Strain distribution in the matrix and fibres in multiple cracked composite with 2x crack spacing during loading (a) and after unloading (b) (after Keer [73]).

4.6.2 Cyclic behaviour

The ACK model has also been used to predict stress–strain curves during cycles of loading and unloading [1,72,73]. In the post-cracking zone, unloading results in a residual strain, and the loading–unloading moduli are different, leading to hysteresis in each unload–load loop (Figure 4.39). The initial slopes on both unloading and reloading have the limiting value of the modulus of the composite, E_c, but the modulus decreases continuously as increasing lengths of the fibres slip through the matrix. The final slopes vary between E_c and $E_f V_f$, depending on the crack spacing and maximum specimen strain. The residual strain distribution in a composite that has been unloaded from the post-multiple cracking zone is shown in Figure 4.40(b). It is a product of the different strains generated in the fibres and matrix during multiple cracking (Figure 4.40(a)), which prevent complete relaxation of the composite strain during unloading. Analytical relations were developed to account for the stress–strain behaviour in the unloading and reloading branches, and the energy consumed in each cycle. It has been demonstrated [73] that different relations should be used in the multiple cracking zone, and in the post-multiple cracking zone.

4.6.3 Toughness

The toughness of a composite in tension has been calculated for both short ($\ell << \ell_c$) and long ($\ell >> \ell_c$) fibres. For short fibres, most of the energy involved in failure is associated with fibre pull-out, and it has been calculated by Hibbert and Hannant [74], using relations similar to those discussed in Section 4.2.1.2 (Eqs 4.17 and 4.18), which take into account the frictional work involved in the extraction of the fibres from the matrix. The ACK model, assuming long and aligned fibres ($\ell >> \ell_c$), was used to calculate the toughness by integrating the area under the idealized stress–strain curve [1,74]. Using an average crack spacing of $1.364x$, the toughness, U, of the composite [74] is:

$$U = E_c \varepsilon_{mu}^2 (0.5 + 0.659\alpha a) + 0.5[\varepsilon_{fu} - \varepsilon_{mu}(1 + \alpha a)][E_c \varepsilon_{mu} \cdot \sigma_{fu} V_f]$$

(4.88)

Under these conditions, the fibres reach their ultimate strain before they fracture. The energy contribution to the toughness due to this mechanism, U_f, can be calculated as:

$$U_f = 0.5\sigma_{fu} V_f$$

(4.89)

Neglecting the energy contribution of the matrix, the difference between U and U_f can be considered as the energy of multiple cracking, U_{mc} [75]:

$$U_{mc} = U - U_f = 0.159\alpha E_c \varepsilon_{mu}^2$$

(4.90)

Thus, the toughness of a composite consisting of long and aligned fibres can be calculated [74] as the sum of the energies involved in multiple cracking and in fibre fracture:

$$U = U_{mc} + U_f = 0.159\alpha E_c \varepsilon_{mu}^2 + 0.5\sigma_f \varepsilon_{fu} V_f$$

(4.91)

Though Eq. 4.91 has its limitations, due to the simplifying assumptions on which it is based, it is useful for comparing different types of fibres with a wide range of properties, from low modulus polypropylene to high modulus Kevlar. The results of Hibbert and Hannant's calculations, based on Eqs 4.89–4.91 are shown in Table 4.2.

The multiple cracking contributes only a small portion of the total energy. The multiple cracking energy can be increased by increasing the failure strain of the

Table 4.2 Theoretical energy absorbed by aligned continuous fibre composites, as calculated by Eqs 4.89 to 4.91 (after Hibbert and Hannant [74])

	E_f (Gnm^{-2})	σ_{fu} (MNm^{-2})	$\epsilon_{fu} = \sigma_{fu}/E_f$	$\sigma = \frac{E_m V_m}{E_f V_f}$	U_{mc} (kJm^{-3})	U (kJm^{-3})
Fibrillated Polypropylene	2	200	0.1	135	36.5	1036
Steel wire	200	1000	0.005	1.35	0.63	250
Glass strand	70	1250	0.018	3.85	1.3	1125
Kevlar 49	130	2900	0.022	2.08	0.83	3190

matrix (ϵ_{mu} in Eq. 4.90). This would be accompanied by a reduction in the ultimate strain of the composite (Eq. 4.69), even though the area under the stress–strain curve would increase (Eq. 4.89). However, this increase could be modest, whereas the reduction in ultimate strain would be considerable. Thus, an increase in ϵ_{mu}, which can be obtained by improved fibre–matrix bond (Eq. 4.49), or by use of high strength matrices, may lead to a marked reduction in the ultimate strain of the composite, and result in apparent embrittlement. This is one of the processes that may occur during the ageing of composites (e.g. glass fibre reinforced cement), where the matrix surrounding small diameter fibre densifies (Chapter 5). The reduced ultimate strain may not be associated with a loss in toughness in tension, but the analytical treatment of flexural behaviour in Section 4.7 will show that it can lead to a loss in flexural toughness.

It is interesting to note, from Table 4.3, that this simple model predicts that by going from polypropylene to Kevlar, that is, increasing the modulus of elasticity by two orders of magnitude and the strength by one order of magnitude, the toughness would increase only by a factor of ~3. Thus, although the high quality of Kevlar would obviously be reflected in a marked improvement in composite strength, its advantage with respect to toughness is not as dramatic.

4.6.4 The crack bridging approach for calculation of tensile parameters

The crack bridging approach, presented in Section 4.4.3 can serve as a basis for calculation of tensile parameters, by integrating the relevant stress-crack opening function over the crack opening to obtain the toughness, or determine the maximum stress that can be supported, to serve as an estimate of strength. In doing so, the effects of fibre orientation, length and rupture of individual fibres are inherently considered, as they are already included as considerations in the modelling of the stress-crack opening function. This concept was used by Maalej et al. [51] to calculate the strength and the toughness of a composite, as a function of fibre length, for randomly distributed short fibre composite. In their analytical treatment, they

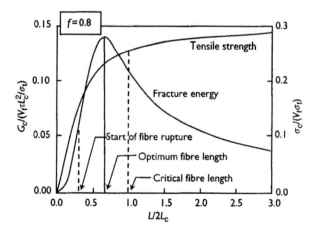

Figure 4.41 Effect of fibre length on composite tensile strength and toughness (after Maalej et al. [51]).

identified a characteristic fibre length ℓ_r (Eq. 4.92) beyond which fibre fracture starts to take place, and they modified the stress-crack width function (Eqs 4.54) to take this into account:

$$\ell_r = 2\ell_c e^{-f\pi/2} \tag{4.92}$$

where ℓ_c is the critical fibre length defined in Eq. 4.2; f, snubbing factor defined in 3.34.

The strength of the composite (i.e. the peak on the stress-crack displacement curve) was calculated:

$$\overline{\sigma}_{cu} = m_1(m_2 - \overline{\ell}^{-1}) \tag{4.93}$$

where

$$\hat{\sigma}_{cu} = \sigma_{cu}/V_f\sigma_{fu};$$
$$m_1 = (1/125) + (19/80)e^{-(9/4)f};$$
$$m_2 = (g/4m_1)e^{-f(\pi/2)} + e^{f(\pi/2)}$$
$$\overline{\ell} = \overline{\ell}/2\overline{\ell}_c.$$

The toughness of the composite was obtained by integration of the stress-crack opening curve:

$$\hat{G}_c = \begin{cases} \dfrac{1}{3}\left[g(\theta_b)\overline{\ell}_f^2 + h(\theta_b)\overline{\ell}_f^{-1}\right] & \text{for} \quad \ell_r \leq \ell < 2\ell_c \\ \dfrac{1}{3}g_2\overline{\ell}_f^{-1} & \text{for} \quad \ell \geq 2\ell_c \end{cases} \tag{4.94}$$

where

$$h(\theta_b) = \frac{1}{2(1+f^2)}[(f\sin 2\theta_b + \cos 2\theta_b)e^{-2f\theta_b} + e^{-f b}];$$

$$g(\theta_b) = \frac{1}{4+f^2}[(f\sin 2\theta_b - 2\cos 2\theta_b)e^{f\theta_b} + 2];$$

$$\Phi_b = -\frac{1}{f}\ln\left(\frac{\ell}{2\ell_c}\right);$$

$$g_2 = \frac{1}{2(1+f^2)}[1 + e^{-f\pi}].$$

The parameter θ_b is a critical fibre inclination angle ($0 \le \theta_b \le \pi/2$) that makes the distinction between two groups of fibres, those that break during the process ($\pi/2 > \theta > \theta_b$) and those which pull out ($0 < \theta < \theta_b$).

Curves showing the effect of fibre length on strength and toughness are presented in Figure 4.41. They are similar in nature to the curve in Figure 4.5 which were based on simplified assumption, demonstrating that the toughness curve goes through a maximum, but in this case at a fibre length which is shorter than the critical fibre length defined in Eq. 4.2.

4.7   Modelling of flexural behaviour

Simple bending theory is not really applicable to FRC, because of the post-cracking characteristics of these composites. This is demonstrated in Figure 4.42, for ideally elastic and ideally elastic–plastic materials, both having the same tensile strength. Bending theory is applicable up to the first crack stress, but it cannot account for the flexural behaviour beyond this point. The stress distributions presented in Figure 4.42(b) show that when the elastic limit in bending of an ideally elastic material is reached, it will fail. However, the elastic–plastic material can continue to support additional loads. This is accompanied by modification of the stress distribution, with the neutral axis moving upwards and the tensile stress distribution becoming rectangular in shape. As a result, the load vs. deflection curve in the elastic–plastic material will continue to ascend beyond the elastic limit (Figure 4.42(c)). Thus, the ductility associated with pseudo-plastic behaviour leads to an increase in the load-bearing capacity of the ideally plastic material, even though its tensile strength is not greater than that of the ideally elastic material.

The bending capacity is usually expressed in terms of the flexural strength, σ_b calculated from the ultimate bending moment of the tested beam, M_u, assuming a linear elastic stress distribution. For a beam with a rectangular cross section of width b and height h,

$$\sigma_b = 6M_u/bh^2 \tag{4.95}$$

For an ideally elastic material, σ_b is equal to the ultimate tensile strength. However, as can be deduced from the stress distribution in Figure 4.42(b) for

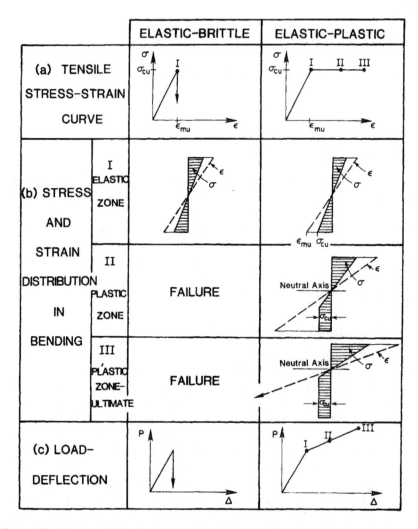

Figure 4.42 Behaviour of ideally elastic and ideally elastic–plastic materials in flexure, showing the stress and strain distribution in flexure at three different stages (I, II, III) and the resulting load–deflection curves.

an elastic–plastic material, σ_b will be greater than the ultimate tensile strength. Thus, in the flexural loading of a ductile, pseudo-plastic material, σ_b does not represent the tensile strength, and has no particular physical significance. In FRC composites, σ_b depends on both the ultimate tensile strength and the post-cracking ductility, with the latter sometimes being of greater importance.

Since FRC composites, are, in practice, frequently subjected to flexural loading and are usually tested in bending, it is important to understand the processes

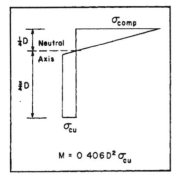

Figure 4.43 The stress distribution in flexure of an ideally elastic–plastic material at the instance of failure (after Hannant [77]).

involved in this mode of loading, and the ratios between σ_b and the tensile strength, σ_{cu}. Edgington [75] and Allen [76] have shown that at the ultimate stage of flexural loading the neutral axis in FRC composites may be as much as 0.8 of the beam height from its tensile face. Using a more conservative value of 0.75 of the height, and the stress distribution in Figure 4.43, Hannant [77] calculated σ_b as a function of the tensile strength:

$$\sigma_b = 2.44\,\sigma_{cu} \tag{4.96}$$

Thus, fibres which do not lead to an increase in tensile strength, but which enhance the post-cracking ductility, can lead to an increase in the load-bearing capacity of the composite in bending. Hannant [77] has estimated that in order to achieve the stress distribution in Figure 4.43, which gave the factor of 2.44 in Eq. 4.96, the ultimate tensile strain of the composite must be about 0.9%. In composites with a lower strain capacity, σ_b will be reduced. That is, the factor 2.44 in Eq. 4.96 is an upper limit.

The enhancement of σ_b beyond the ultimate tensile strength suggests that in bending, the critical fibre volume can be reduced by a factor whose maximum value is 2.44, that is the critical fibre volume for bending may be about 40% of the critical volume for tension. Thus, even a more limited amount of fibre reinforcement, which provides a load-bearing capacity in the post-cracking zone which is smaller than the matrix strength, can lead to improved σ_b, provided that the lower post-cracking stress can be maintained over sufficiently large strains, and its value is greater than 40% of the peak load.

The modelling of flexural behaviour and strength can be carried out by two different modes: (i) modelling the tensile curve and determining the flexural behaviour by considering a continuous beam in flexure, with different charac-teristic functions for the tensile and compressive behaviour and (ii) modelling

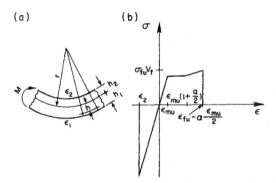

(a)   (b)

Figure 4.44 Stress distribution at failure in bending of a composite with tensile behaviour modelled by the ACK theory (after Aveston *et al.* [15]). (a) Specimen geometry (b) stress distribution.

flexural behaviour in terms of a crack which opens in mid-span at the tensile zone, using the fibre crack bridging functions. These two approaches will be dealt with in the next two sections.

4.7.1 Flexural behaviour modelled by stress–strain relations in tension

The stress–strain curve in tension, modelled for FRC, has been used to develop an analytical description of the flexural behaviour (σ_b and load vs. deflection curves) of FRC. In such treatments the improvement in flexural behaviour is the result of two effects: enhancement in post-cracking strength and in ductility.

Aveston *et al.* [15] calculated the ratio between σ_b and σ_{cu} in a composite with continuous and aligned fibres. Failure in flexure occurred when the strain at the tensile face of the beam reached the value of ε_{cu} derived from Eq. 4.69 of the ACK model. The stress distribution at this stage, based on the ACK model is shown in Figure 4.44. The ratios between σ_b and σ_{cu} ($= E_f V_f \varepsilon_{fu}$), calculated for assumed crack spacings of $2x$ and x differed only by 2–3%. The results of this calculation in which the ratio of σ_b/σ_{cu} is described in terms of the ratios between $\varepsilon_{fu}/\varepsilon_{mu}$ and the value of α in the ACK model, are shown graphically in Figure 4.45.

The upper limit in Figure 4.45 is about 3, somewhat higher than the value of 2.44 predicted by Hannant [77]. A σ_b/σ_{cu} ratio of ~ 2.6 was reported by Aveston *et al.* [15] for steel FRC and values of 1.9–2.6 were obtained by Allen [76] and Laws and Ali [69] for different types of glass FRC.

A similar approach was adopted by RILEM in a recent recommended method for calculation of flexural capacity, the σ–ε method [78]. The RILEM recommendations provide guidelines for the stress distribution as a function of the level of loading, characterized by the deflection or crack opening displacement. The

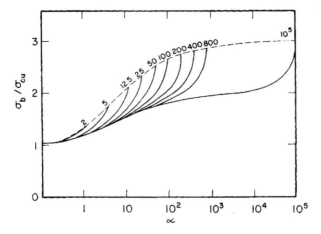

Figure 4.45 Ratio of modulus of rupture to ultimate tensile strength, σ_b/σ_{cu}, vs. $\alpha(= E_m V_m/E_f V_f)$ for composites with different $\varepsilon_{fu}/\varepsilon_{mu}$ ratios (after Aveston *et al.* [15]).

stress distribution is similar in nature to that described in Figure 4.43, with the neutral axis being at 66% and 90% of the cross-section height, for mid-span deflections of 0.46 and 3 mm of beams tested according to the RILEM recommendations (0.5 and 3.5 mm crack opening displacements, respectively). The tensile strength for the calculated tensile block diagram is the residual tensile strength, defined as:

$$\sigma_{R,i} = \frac{3 P_{R,i} L}{2 b h_{sp}^2} \tag{4.97}$$

where $\sigma_{R,i}$ is the residual strength at deflection i; $P_{R,i}$, the load at deflection i; h_{sp}, the distance between tip of the notch in the beam and the top cross section; b, the width of the beam.

Load–deflection curves in bending have been calculated from the tensile stress–strain curves, assuming that the behaviour in the post-cracking zone in tension is described by linear relations [69] or by curves derived from the ACK model [15]. In a modified treatment, the influence of the cracking on the flexural behaviour was also considered [79]. The results of such a calculation based on the ACK model in tension are shown in Figure 4.46. The flexural behaviour is expressed in terms of apparent flexural stresses and strains. These curves demonstrate the greater efficiency of the fibres in enhancing the performance in flexure. For example, at the critical fibre volume in tension (curve A in Figure 4.46), only enhanced ductility is observed in tension, whereas in bending the load-bearing capacity is also increased. Another significant effect is the 'smoother' transition across the

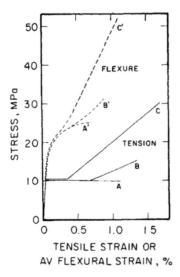

Figure 4.46 Tensile stress–strain curves for continuous and aligned fibre reinforced cement predicted by the ACK theory (full lines) and the flexural response calculated from them. E_f = 76 GPa; E_m = 25 GPa; ε_{mu} = 0.04%; ε_{fu} = 2%; V_f = 0.7% (curves A, A'); 1% (curves B, B'); 2% (curves C, C') (after Laws [19]).

first crack zone in bending, which is not accompanied by the drastic change in slope which occurs in tensile loading.

These differences suggest that the flexural test which is commonly used to characterize the properties of FRC may be difficult to interpret in an attempt to resolve parameters of basic physical significance, such as those associated with first crack stress and multiple cracking. However, since in practice most FRC composites are subjected to flexural loading, the flexural test is probably the most suitable one to use.

Laws [19] discussed some interesting relationships that may be resolved when plotting the calculated σ_b as a function of fibre volume content, V_f (Figure 4.47). The dashed line A in this figure, and its intersection with the flexural strength curves, represents systems with a fibre content equal to the critical fibre content in tension (i.e. those characterized by lines A and A' in Figure 4.46). At these intersections the σ_b of the composite is greater than that of the matrix. At higher fibre volumes (full lines in Figure 4.47), the relationships between σ_b and V_f are non-linear. However, over the practical range of fibre contents this non-linearity is small. Thus, if the best fit straight line is used over a limited fibre volume range, it will intercept the σ_b axis at a positive value, which is approximately equal to the matrix σ_b. Thus, over a limited fibre content range, the σ_b–V_f relationships can be approximately linear, and are essentially similar to those of the rule of mixtures, in spite of the fact that this rule is physically invalid in this range, in which the matrix

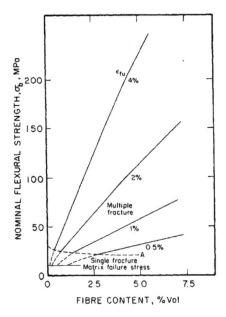

Figure 4.47 Predicted σ_b vs. fibre volume curves for continuous and aligned fibre reinforced cements for the parameters in Figure 4.46, except for ε_{fu} values which are shown. Full lines relate to volume content of fibre above the critical volume content for tensile reinforcement; dashed lines represent the relations for lower volume content (after Laws [19]).

is cracked. Laws suggested that this might explain the apparently contradictory statements and observations that the law of mixtures can be used to account for the flexural strength of the composite, but not for its tensile strength.

4.7.2 Flexural behaviour modelled by crack bridging functions

The flexural behaviour of FRC can be described in terms of the opening up of a crack in the tensile zone, with the fibres bridging over the crack, Figure 4.48. This concept can be quantified in terms of the fictitious crack concept, based on a characteristic $\sigma-\omega$ relation, in which the fibre bridging provides an important contribution, as discussed in Section 4.4.3. Various models have been developed based on this concept, and an overview of this topic and its practical implications was presented by RILEM committee TC – 162 TDF [56]. This section follows the RILEM document [56].

The overall behaviour of the beam in flexure, with a crack generated in the tensile zone, can be described in terms of a non-linear hinge (Figure 4.49), located at the centre of the beam. The hinge can be described in terms of an element of

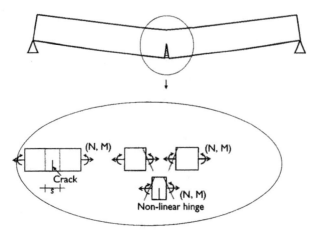

Figure 4.48 Schematic behaviour of an FRC beam loaded in flexure with fibres bridging over the crack which develops at the centre of the beam and is modelled as a non-linear hinge.

length s which connects the two sections of the beam. The width of the hinge, s, has been estimated to be twice the length of the crack [80] and half the cross-section height [81]. In order for the hinge section to connect to the rest of the beam, its two faces are assumed to remain plane and be loaded with the generalized stresses shown in Figure 4.49, with axial force N and moment M. These concepts provide the background for structural design with FRC, as outlined in Chapter 14.

A common feature underlying all the models is a treatment which is based on dividing the cross section into three parts (Figure 4.49): compressive zone up to the neutral axis (from $x = 0$ down to $x = x_0$ in Figure 4.49), tensile zone from the neutral axis ($x = x_0$) to the crack tip and the cracked zone, from the crack tip ($x = h - a$) to the full depth of the beam ($x = h$). The width of the crack at the bottom face of the beam is the crack mouth opening displacement, ω_{cmod}. The stress distributions in the compressive and tensile zones are elastic in nature, whereas in the cracked zone various models for the crack bridging are applied. Another characteristic in the modelling is the assumption regarding the geometry of the fictitious crack surface: remaining plane in some models and the assumption of deviation from a plane surface in others. In all the models the curvature of the non-linear hinge, k_m, is:

$$k_m = \varphi/s \tag{4.98}$$

In the models assuming a plane fictitious crack surface, ω_{cmod}, is a direct function of the crack opening angle, φ^*, and the length of the fictitious crack, a:

$$\omega_{cmod} = \varphi^* a \tag{4.99}$$

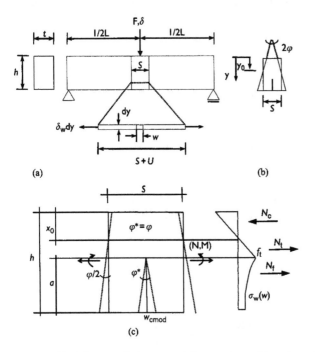

Figure 4.49 Schematic behaviour of the non-linear hinge in an FRC beam loaded in flexure (a) overall behaviour, (b) hinge section and (c) stress distribution in the hinge for strain softening composite.

In the case of the non-planar assumption, the ω_{cmod} is determined from the $\sigma-\omega$ relation, the overall curvature k_m and the length of the fictitious crack.

The simplified approach of Pederson [82] presented in the RILEM report [56] assuming that the crack has a linear profile (Eq. 4.99) results in the following expressions for the generalized stresses on the cross section (Figure 4.49):

$$N_f = \frac{1}{\varphi^*} \int_0^{\omega_{cmod}} \sigma_\omega(u)du \qquad (4.100)$$

$$M_f = \frac{1}{(\varphi^*)} \int_0^{\omega_{cmod}} \sigma_\omega(u)du \qquad (4.101)$$

$$N_c = \frac{\varphi E x_0^2}{2s} \qquad (4.102)$$

$$N_t = \frac{(f_t)^2 s}{2\varphi E} \qquad (4.103)$$

In these equations, the neutral axis is calculated based on the assumptions in Figure 4.49, as:

$$h - x_0 = \frac{1}{\varphi} \left(\frac{f_t}{E}s + \omega_{cmod} \right) \tag{4.104}$$

The moment relative to the centre line of the cross section was calculated from these equations:

$$M = \left(\frac{h}{2} - \frac{x_0}{3} \right) N_c + \left(\frac{h}{6} + \frac{x_0}{3} - \frac{2a}{3} \right) N_t + \left(\frac{h}{2} - a \right) N_f + M_f \tag{4.105}$$

Casanova and Rossi [80] applied different assumptions with regard to the internal kinematics of the hinge, and considered two curvatures: the elastic curvature of the uncracked part of the hinge, k_1, and that of the cracked zone, k_2:

$$k_1 = \frac{12M}{Eh^3} \tag{4.106}$$

$$k_2 = \frac{\varepsilon_c}{x_0} \tag{4.107}$$

where ε_c is the strain in the extreme fibre in compression.
They calculated the average curvature, k_m (Eq. 4.98), as:

$$k_m = \frac{2k_1 + k_2}{3} \tag{4.108}$$

The generalized stresses and moment derived by Casanova and Rossi [80] are similar to Eqs 4.100–4.103 and 4.105, in which φ/s is substituted by k_m.

A modified approach for the modelling of the non-linear hinge was presented by Stang and Oleson [83] in which the hinge was modelled as incremental layers of springs which act without transferring shear between each other. Assuming a bilinear σ–ω relation they described the stress distribution in the hinge at the various stages of loading (Figure 4.50) and developed expressions to describe them.

Maalej and Li [84] developed relations between flexural and tensile strength of strain softening FRC, based on these concepts and determined this strength ratio as a function of the brittleness ratio, B (Eq. 4.109, Figure 4.51) and the length of the fibre relative to its critical length (Figure 4.52).

$$B = \frac{T_b d}{E_c \omega_c} \tag{4.109}$$

The ratio in Figure 4.52 is calculated with the assumption of no fibre fracture. For the limiting case of $B \to 0$ the ratio is 3, as predicted by the mechanics of materials

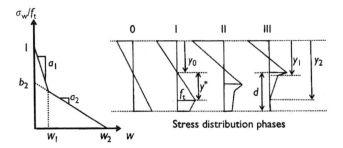

Figure 4.50 Stress distribution at the various stages of flexural loading, assuming a bilinear strain softening $\sigma - \omega$ relation (after Stang and Oleson [83]).

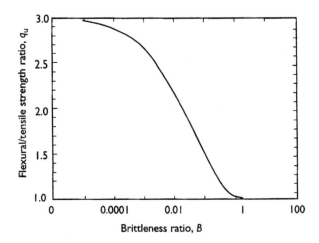

Figure 4.51 Effect of brittleness ratio, B, on the ratio between flexural and tensile strength (after Maalej and Li [84]).

approach assuming ideal elastic–plastic material. For the case where B approaches infinity the ratio is 1, characteristic of an ideal elastic–brittle material. The decline in the curve in Figure 4.52, after the maximum, is the result of fibre rupture which occurs more readily when the fiber length is approaching and exceeding the critical length.

The same concept was further extended to strain hardening composites, in which the tensile behaviour was characterized in terms of the first crack stress, which is approximately equal to the matrix tensile strength, and the ultimate strength at the end of the strain hardening portion of the stress–strain curve [85]. It was shown that the relative improvement in flexural strength was roughly twice as high as the

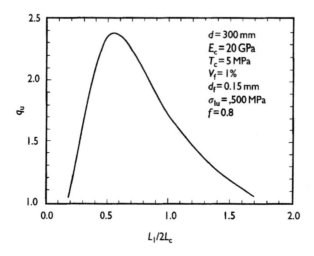

Figure 4.52 Effect of fibre length on the ratio between flexural and tensile strength (after Maalej and Li [84]).

improvement in tensile strength [85] and this can be accounted for by the marked increase in toughness.

References

1. A. Aveston, G.A. Cooper and A. Kelly, 'Single and multiple fracture', in *The Properties of Fibre Composites*, Proc. Conf. National Physical Laboratories, IPC, Science and Technology Press, England, 1971, pp. 15–24.
2. H.G. Allen, 'The purpose and methods of fibre reinforcement', in *Prospects of Fibre Reinforced Construction Materials*, Proc. Int. Building Exhibition Conference, Building Research Station, England, 1971, pp. 3–14.
3. A. Kelly, *Strong Solids*, Oxford University Press, Oxford, 1973.
4. V. Laws, 'The efficiency of fibrous reinforcement of brittle matrices', *J. Phys. D: Appl. Phys.* 4, 1971, 1737–1746.
5. H. Krenchel, *Fibre Reinforcement*, Akademick Forlag, Copenhagen, 1964.
6. R.C. De Vekey amd A.J. Majumdar, 'Determining bond strength in fibre reinforced composites', *Mag. Concr. Res.* 20, 1968, 229–234.
7. S.P. Shah and V.B. Rangan, 'Fibre reinforced concrete properties', *J. Amer. Concr. Inst.* 68, 1971, 126–135.
8. J. Edgington, D.J. Hannant and R.I.T. Williams, 'Steel fibre reinforced concrete', Current Paper CP69/74, Building Research Establishment, England, 1974.
9. D.J. Hannant, *Fibre Cements and Fibre Concretes*, John Wiley & Sons, Chichester, 1978.
10. C.D. Johnston and R.J. Gray, 'Uniaxial tensile testing of steel fibre reinforced cementitious composites', in R.N. Swamy (ed.) *Testing and Test Methods of Fibre*

Cement Composites, Proc. RILEM Symp., The Construction Press, Lancaster, England, 1978, pp. 451–461.

11. H.K. Ng, G.K. Nathan, P. Paramasivan and S.L. Lee, 'Tensile properties of steel fibre reinforced mortar', J. Ferrocement 8, 1978, 219–229.

12. A.E. Naaman and S.P. Shah, 'Pull out mechanism in steel fibre reinforced concrete', J. Struct. Div. ASCE. 102, 1976, 1537–1548.

13. J. Morton and G.W. Groves, 'The cracking of composites consisting of discontinuous ductile fibres in brittle matrix-effect of orientation', J. Mater. Sci. 9, 1974, 1436–1445.

14. A.M. Brandt, 'On the optimal direction of short metal fibres in brittle matrix composites', J. Mater. Sci. 20, 1985, 3831–3841.

15. J. Aveston, R.A. Mercer and J.M. Sillwood, 'Fibre reinforced cements-scientific foundation for specifications', in Composites, Standards Testing and Design, Proc. National Physical Laboratory Conference, England, 1974, pp. 93–103.

16. R.N. Swamy, P.S. Mangat and C.V.S.K Rao, 'The mechanics of fibre reinforcement of cement matrices', Fiber Reinfored Concrete, ACI SP-44, American Concrete Institute, Detroit, MI, 1974, pp. 1–28.

17. R.N. Swamy and P.S. Mangat, 'A theory for the flexural strength of steel fibre reinforced concrete', Cem. Concr. Res. 4, 1974, 313–325.

18. Y.W. Mai, 'Strength and fracture properties of asbestos-cement mortar composites', J. Mater. Sci. 14, 1979, 2091–2102.

19. V. Laws, 'On the mixture rule for strength of fibre reinforced cements', J. Mater. Sci. Letters. 2, 1983, 527–531.

20. S.Mindess, 'The fracture of fibre reinforced and polymer impregnated concretes: a review', in F.H. Wittmann (ed.) Fracture Mechanics of Concrete, Elsevier Science Publishers, Amsterdam, 1983, pp. 481–501.

21. J.P. Romualdi and G.B. Batson, 'Mechanics of crack arrest in concrete', J. Eng. Mech. ASCE. 89, 1963, 147–168.

22. J.P. Romualdi and G.B. Batson, 'Behaviour of reinforced concrete beams with closely spaced reinforcement', J. Amer. Concr. Inst. 60, 1963, 775–789.

23. J.P. Romualdi and J.A. Mandel, 'Tensile strength of concrete affected by uniformly dispersed and closely spaced short lengths of wire reinforcement', J. Amer. Concr. Inst. 61, 1964, 657–672.

24. D.C. McKee, The Properties of Expansive Cement Mortar Reinforced with Random Fibres, PhD thesis, University of Illinois, Urbana, IL, 1969.

25. H. Krenchel, 'Fibre spacing and specific fibre surface', in A. Neville (ed.) Fibre Reinforced Cement and Concrete, Proc. RILEM Symp., The Construction Press, Lancaster, England, 1976, pp. 69–79.

26. J.O. Outwater and M.H. Murphy, 'On the fracture energy of unidirectional laminates', in Proc. 26th Annual Conf. an Reinforced Plastics, Composites Division of Society of Plastics Industry, Washington, DC, 1969, paper 11-C-1, pp. 1–8.

27. P. Tjiptobroto and W. Hansen, 'Model for predicting the elastic strain hardening of fiber reinforced composites containing high volume fractions of discontinuous fibers', ACI Mater. J. 90, 1993, 134–142

28. A. Hillerborg, 'Analysis of fracture by means of the fictitious crack model, particularly for fibre reinforced concrete', Int. J. Cem. Comp. 2, 1980, 177–184.

29. Z.P. Bazant and L. Cedolin, 'Fracture mechanics of reinforced concrete', J. Eng. Mech. Div., ASCE. 106, 1980, 1287–1305.

30. M. Wecharatana and S.P. Shah, 'Nonlinear fracture mechanics parameters', in F.H. Wittmann (ed.) *Fracture Mechanics of Concrete*, Elsevier Science Publishers B.V., Amsterdam, 1983, pp. 463–480.

31. A. Bentur and S. Diamond, 'Crack patterns in steel fibre reinforced cement paste', *Mater. Struct.* 18, 1985, 49–56.

32. M. Wecharatana and S.P. Shah, 'A model for predicting fracture resistance of fibre reinforced concrete', *Cem. Concr. Res.* 13, 1983, 819–829.

33. Y.S. Jenq and S.P. Shah, 'Crack propagation resistance of fibre reinforced concrete', *J. Struct. Eng. ASCE.* 112, 1986, 19–34.

34. Z.P. Bazant and L. Cedolin, 'Blunt crack band propagation in finite element analysis', *J. Engr. Mech. Div., ASCE.*105, 1979, 297–315.

35. Z.P. Bazant and L. Cedolin, 'Fracture mechanics of reinforced concrete', *J. Engr. Mech. Div., ASCE.* 106, 1980, 1287–1305.

36. Z.P. Bazant and L. Cedolin, 'Fracture mechanics of reinforced concrete', in *Fracture in Concrete*, Proceedings of a session sponsored by the Committee on Properties of Materials at the ASCE National Convention in Hollywood, Florida, Oct. 1980, American Society of Civil Engineers, pp. 28–35.

37. Z.P. Bazant and B.H. Oh, 'Crack band theory for fracture of concrete', *Mater. Struct.* 16, 1983, 155–178.

38. Z.P. Bazant. 'Mechanics of fracture and progressive cracking in concrete structures', in G.C. Sih and A. DiTommaso (eds) *Fracture Mechanics of Concrete*, Martinus Nijhoff Publishers, The Netherlands, 1985, pp. 1–94.

39. A. Hillerborg, 'Determination and significance of the fracture toughness of steel fibre concrete', in S.P. Shah and A. Skarendahl (eds) *Steel Fibre Concrete*, US–Sweden Joint Seminar, Stockholm, 1985, pp. 257–271.

40. A. Hillerborg, 'Numerical methods to simulate softening of fracture of concrete', in G.C. Sih and A. DiTommaso (eds) *Fracture Mechanics of Concrete*, Martinus Nijhoff Publishers, The Netherlands, 1985, pp. 141–170.

41. A. Hillerborg, 'Dimensionless, presentation and sensitivity analysis in fracture mechanics', in F.H. Wittmann (ed.) *Fracture Toughness and Fracture Energy of Concrete*, Elsevier Science Publishers B.V., Amsterdam, 1986, pp. 413–421.

42. A. Hillerborg, 'Analysis of one single crack', in F.H. Wittmann (ed.) *Fracture Mechanics of Concrete*, Elsevier Science Publishers B.V., Amsterdam, 1983, pp. 223–249.

43. V.C. Li, 'On engineered cementitious composites (ECC): a review of the material and its application', *J. Advanced Concrete Technology.* 1, 2003, 215–230.

44. V.C. Li, H. Mihashi, H.C. Wu, J. Alwan, R. Brincker, H. Horii, C. Leung, M. Maalej and H. Stang, 'Micromechanical models of mechanical response of HPFRCC', in A.E. Naaman and H.W. Reinhardt (eds) *High Performance Fiber Reinforced Cement Composites 2*, Proc. RILEM Symp., E&FN SPON, London and New York, 1996, pp. 43–100.

45. V.C. Li, H. Stang and H. Krenchel, 'Micromechanics of crack bridging in fibre-reinforced concrete', *Mater. Struct.* 26, 1993, 486–494.

46. V.C. Li and C.K.Y. Leung, 'Steady state and multiple cracking of short random fiber composites', *ASCE J. of Engineering Mechanics.* 118, 1992, 2246–2264.

47. C.K.Y. Leung, 'Design criteria for pseudo-ductile fiber reinforced composites', *ASCE J. Engineering Mechanics.* 122, 1996, 10–18.

48. Z. Lin and V.C. Li, 'Crack bridging in fiber reinforced cementitious composites with slip-hardening interfaces', *J. Mech. Phys. Solids.* 45, 1997, 763–787.

49. V.C. Li, Y. Wang and S. Backer, 'A micromechanical model of tension softening and bridging toughening of short random fiber reinforced brittle matrix composites', *J. Mech. Phys. Solids.* 39, 1991, 607–625.

50. V.C. Li, 'Postcrack scaling relations for fiber reinforced cementitious composites', *J. Mater. Civ. Eng.* 4, 1992, 41–57.

51. M. Maalej, V.C. Li and T. Hashida, 'Effect of fiber rupture on tensile properties of short fiber composites', *ASCE J. Engineering Mechanics.* 121, 1995, 903–913.

52. V.C. Li and K.H. Obla, 'Effect of fiber length variation on tensile properties of carbon fiber cement composites', *Int. J. of Composite Engineering.* 4, 1994, 947–964.

53. D.B. Marshall and B.N. Cox, 'A J-integral method of calculation of steady-state matrix cracking stresses in composites', *Mech. Mater.* 7, 1988, 127–133.

54. V.C. Li, 'From micromechanics to structural engineering – the design of cementitious composites for civil engineering applications', *Structural engineering/Earthquake engineering.* 10, 1993, 37–48.

55. V.C. Li, D.K. Mishra and H.C. Wu, 'Matrix design for pseudo strain-hardening fiber reinforced cementitious composites', *Mater. Struct.* 28, 1995, 586–595.

56. RILEM TC 162-TDS: Test and design methods for steel fibre reinforced concrete, *Mater. Struct.* 35, 2002, 262–278.

57. J.K. Morrison, S.P. Shah and Y.S. Jenq, 'Analysis of fibre debonding and pull out in composites', *J. Eng. Mech. Div., ASCE.* 114, 1988, 277–294.

58. H. Stang and S.P. Shah, 'Failure of fibre reinforced composites by pull-out fracture', *J. Mater. Sci.* 21, 1986, 953–957.

59. J.A. Mandel, S. Wei and S. Said, 'Studies of the properties of the fibre-matrix interface in steel fibre reinforced mortar', *Amer. Concr. Inst. Mat. J.* 84, 1987, 101–109.

60. R.N. Swamy, 'Linear elastic fracture mechanics parameter of concrete', in F.H. Wittmann (ed.) *Fracture Mechanics of Concrete*, Elsevier Applied Science B.V., Amsterdam, 1983, pp. 411–461.

61. J. Aveston and A. Kelly, 'Theory of multiple fracture of fibrous composites', *J. Mater. Sci.* 8, 1973, 352–362.

62. V. Laws, 'Micromechanical aspects of the fibre-cement bond', *Composites.* 13, 1982, 145–152.

63. P. Lawrence, 'Some theoretical considerations of fibre pull out from elastic matrix', *J. Mater. Sci.* 7, 1972, 1–6.

64. D.R. Oakley and B.A. Proctor, 'Tensile stress-strain behaviour of glass fibre reinforced cement composites', in A. Neville (ed.) *Fibre Reinforced Cement and Concrete*, Proc. RILEM Symp., The Construction Press, Lancaster, England, 1975, pp. 347–359.

65. R. Baggott and D. Gandhi, 'Multiple cracking in aligned polypropylene fibre reinforced cement composites', *J. Mater. Sci.* 16, 1981, 65–74.

66. V. Laws, 'Stress-strain curve of fibrous composites', *J. Mater. Sci. Letters.* 6, 1987, 675–678.

67. B.A. Proctor, 'The stress-strain behaviour of glass fibre reinforced cement composites', *J. Mater. Sci.* 21, 1986, 2441–2448.

68. Tjiptobroto, P. and W. Hansen, Tensile strain hardening and multiple cracking in high-performance cement-based composites containing discontinuous fibers, *ACI Mater. J.* 90, 1993, 16025.

69. V. Laws and M.A. Ali, 'The tensile stress/strain curve of brittle matrices reinforced with glass fibre', in *Fibre Reinforced Cement*, Proc. Conf., Institution of Civil Engineers, London, 1977, pp. 115–123.

70. G.K. Nathan, P. Paramasivam and S.L. Lee, 'Tensile behaviour of fibre reinforced cement paste', *J. Ferrocement* 7, 1977, 59–79.

71. T.Y. Lim, P. Paramasivam and S.L. Lee, 'Analytical model for tensile behaviour of steel-fibre concrete', *Amer. Concr. Inst. Mat. J.* 84, 1987, 286–298.

72. J.G. Keer, 'Behavior of cracked fibre composites under limited cyclic loading', *Int. J. Cem. Comp. & Ltwt. Concr.* 3, 1981, 179–186.

73. J.G. Keer, 'Some observations on hysteresis effects in fibre cement composites', *J. Mater. Sci. Letters* 4, 1985, 363–366.

74. A.P. Hibbert and D.J. Hannant, 'Toughness of fibres cement composite', *Composites.* 13, 1982, 105–111.

75. J. Edgington, *Steel fibre reinforced concrete*, PhD Thesis, University of Surrey, 1973.

76. H.G. Allen, 'Stiffness and strength of two glass-fibre reinforced cement laminates', *J. Compos. Mater.* 5, 1971, 194–207.

77. D.J. Hannant, 'The effect of post cracking ductility on the flexural strength of fibre cement and fibre concrete', in A. Neville (ed.) *Fibre Reinforced Cement and Concrete*, Proc. RILEM Symp., vol. 2, Construction Press Ltd., Lancaster, England, 1975, pp. 499–508.

78. RILEM TC 162-TDF, 'Test and design methods for steel fibre reinforced concrete, σ-ε design method', *Mater. Struct.* 36, 2003, 560–567.

79. J.G. Keer and D.J. Hannant, 'The prediction of the load-deflection behaviour of a fibre reinforced cement composite', in R.N. Swamy, R.L. Wagstaffe and D.R. Oakley (eds) *Developments in Fibre Reinforced Cement and Concrete*, Proc. RILEM Symp., Sheffield, RILEM Technical Committee 49-FTR, 1986, paper 1.6.

80. P. Casanova and P. Rossi, 'Analysis and design of steel fibre-reinforced concrete beams', *ACI Structural Journal.* 94, 1997, 595–602.

81. J. Ulfkjaer, S. Krenk and R. Brincker, 'Analytical model for fictitious crack propagation in concrete beam', *ASCE J. Engineering Mechanics.* 121, 1995, 7–15.

82. C. Pederson, *New Production Processes Materials and Calculation Techniques for Fiber Reinforced Pipes*, PhD thesis, Department of Structural Engineering and Materials, Technical University of Denmark, Series R, No. 14, 1996, quoted from Ref. 56.

83. H. Stang and J.F. Olesen, 'On the interpretation of bending tests on FRC materials', in H. Mihashi and K. Rokugo (eds) *Fracture Mechanics of Concrete Structures*, Proc. FRAMCOS 3, AEDIFICATIO Publishers, Germany, 1998, pp. 511–520.

84. M. Maalej and V.C. Li, 'Flexural strength of fiber cementititous composites', *J. Mater. Civ. Eng.* 6, 1994, 390–406.

85. M. Maalej and V.C. Li, 'Flexural/tensile-strength ratio in engineered cementitious composites', *J. Mater. Civ. Eng.* 6, 1994, 513–527.

Chapter 5

Long-term performance

Chemical and physical processes

The special structure of fibre reinforced cements and the nature of the fibres incorporated may lead to long-term effects which are different from those experienced in conventional concretes and mortars. These effects can result in marked changes in the properties of the composite, either strengthening or weakening. Although each system can undergo unique changes invoked by processes characteristic to it, there are several underlying general principles which are relevant to all FRC [1] and they will be highlighted in this section. The application of these principles for each individual system will be dealt with in the relevant chapter dealing with the particular fibre system (i.e. Chapters 7–11).

A range of behaviours has been observed in the various systems, and they can be classified as outlined below and shown schematically in Figure 5.1. These changes in properties can be accounted for by several processes with each acting either on its own or in combination with others:

- fibre degradation due to chemical attack;
- fibre–matrix interfacial physical interactions;

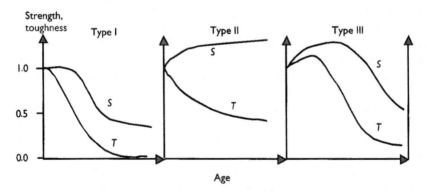

Figure 5.1 Classification of long-term behaviour of FRC composites, characterized by the influence of time on the strength (S) and toughness (T) of the composites.

- fibre-matrix interfacial chemical interactions;
- volume instability and cracking.

5.1   Fibre degradation

Degradation of fibres by chemical attack can result from two types of processes: (i) direct attack by the cementitious matrix due to reactions with the highly alkaline pore water (pH > 13) or (ii) attack by external agents which penetrate through the cementitious matrix into the fibre.

Alkaline degradation has been known to occur in glass fibres [2–4] and natural fibres [5–7] while degradation due to penetration of external agents is more characteristic of steel [8–11]. The mechanisms are different, but the outcome is the same, namely reduction in strength and toughness over time (Type I behaviour, Figure 5.1). The elimination of the fibre as a reinforcing agent due to the chemical attack will eventually lead to a decline in the strength and toughness of the composite, to the level of those of the matrix. Since the matrix tensile/flexural strength is usually in the range of 20–50% of the composite strength in its initial stages, whereas the toughness of the matrix is only a few per cent of that of the composite in its unaged state, the reductions in properties over time are different: levelling off of the strength reduction at 20–50% and loss of practically all the toughness (Figure 5.1 – type I).

A variety of degradation mechanisms have been observed due to alkaline attack. In the case of glass fibres it is the breaking of the $Si-O-Si$ bonds in the glass network [2,12,13] whereas in natural fibres the processes include alkaline hydrolysis which causes molecular chains to divide and the degree of polymerization to decrease significantly [6,7,14]. This attack is particularly harsh in the hemicellulose and lignin components of the natural fibres. The cellulose itself is less sensitive. The means developed to overcome these effects include modification of fibre composition (development of alkali resistant glass composition [12,15] and removal of lignin and hemicellulose [16,17]), modification of the fibre surface to provide it with a protective film [15,18] and the use of modified cementitious matrices which are of low alkalinity [19,20] or reduced permeability to moisture which is a necessary ingredient for corrosion [21]. In-depth analysis of chemical degradation of glass and cellulose fibres is presented in Chapters 8 and 11.

The attack by external agents is more characteristic of steel fibres, where penetration of chlorides can depassivate the steel and lead to its corrosion [8–11]. Although this type of influence might be expected to be particularly harsh in steel fibre reinforced concrete because of the small cover over the fibres, experience has indicated otherwise [8], with such systems performing even better than those of conventional reinforcement. Several mechanisms have been proposed to account for this difference [10,11] and they are addressed in Chapter 7.

5.2   Physical and microstructural effects

Changes of properties over time can occur due to microstructural changes at the fibre–matrix interface. The special microstructural characteristics of the interfacial transition zone (ITZ) have already been discussed in Chapters 2 and 3, indicating its special nature. This microstructure is prone to changes over time, due to continued hydration and densening of the ITZ, which to start with is more porous than the bulk cement paste. Thus, although the overall changes in the bulk microstructure due to continued hydration may be small, the relative changes at the ITZ may be quite considerable. The influences of these microstructural changes could be particularly large in the case of thin filaments (microfibres) whose surface area is rather large. In such systems the bond to the microfibres is likely to be close to the critical bond (Section 4.2), where transition from pull-out failure to fibre fracture may occur (Figure 4.5(b)). A simplified calculation of the critical bond level for a microfibre with 10 μm diameter, 10 mm length and a strength of 1000 MPa (Eq. 4.23) will yield a critical bond value of 0.5 MPa. This is a relatively low value, which may be exceeded as the composite ages and the ITZ becomes denser. A result of this effect, as may be predicted from the diagram in Figure 5.2, is an increase in strength and reduction in toughness over time. This may account for the type II ageing behaviour (Figure 5.1). This ageing mechanism will be addressed here as the 'bond effect'. Such a behaviour has been observed for PVA fibres [22] and may also account for the increase in the first cracking strength of cementitious composites.

Micromechanical modelling has been employed to account for the influences of bond. Kim *et al.* [23] modelled the behaviour of aged cellulose fibre–cement

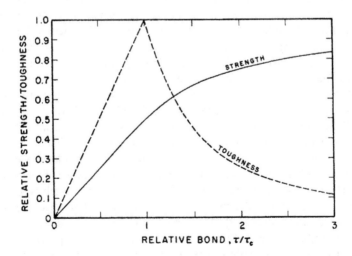

Figure 5.2 Effect of relative bond strength on the strength and toughness of a composite consisting of aligned and short fibres (based on Eqs 4.19–4.23).

composites, and demonstrated that the change in the flexural load–deflection curve over time (increase in strength and reduction in toughness) could be accounted for by an increase in the interfacial frictional bond from 0.8 MPA in the unaged state, to 3.0 MPa after ageing. Mobasher and Li [24] modelled the pull-out by taking into account the rigidity of the ITZ, and demonstrated that the ageing of glass fibre reinforced cement, leading to a pull-out curve with limited post-peak resistance, could be accounted for by an increase in the modulus of this zone by a factor of about 4. This is consistent with densification of the ITZ, which leads in the case of microfibre to drastic changes in the mechanical behaviour.

When considering the changes at the ITZ and their influence on the mechanical properties, one should also refer to its influence on the efficiency of inclined fibres, which is not identical to that of aligned ones. It should be kept in mind that in most cementitious composites the fibres are randomly dispersed in two or three dimensions, and therefore most of them are oriented with respect to the applied load. In such instances, in addition to the effect of densening of the interface on the 'bond effect', which assumes a uniaxial pull-out failure mechanism, one should consider also its influence on the bending of inclined fibres (Section 3.3) which is induced when they bridge across a crack (Figure 3.17). For a rigid and brittle fibre, the flexural stresses induced may lead to premature failure and loss of efficiency. This tendency will be greater when the surrounding ITZ becomes denser, and cannot effectively relax the flexural stresses by crumbling. The latter effect will be termed here the 'bending effect'.

Katz and Bentur [25,26] modelled the combined influence of the bond and bending effects in a composite in which an increase in interfacial density over time takes place (quantified in terms of its hardness). The influences of this change on the strength and toughness for the case of uniformly dispersed and randomly oriented reinforcing filaments were analysed. The data in Figure 5.3 demonstrate the competing nature between these processes, which can account for type III ageing behaviour (Figure 5.1). This type of behaviour was found to account for the behaviour of composites reinforced with high modulus thin carbon fibres (PAN type fibres) of 7 μm diameter, which were uniformly dispersed in a dense matrix. The detrimental effect of the 'fibre bending' which takes place as the ITZ becomes denser, is reduced when the fibres have a smaller diameter or lower modulus of elasticity (they can bend more readily without breaking), or when they are made of a ductile material. These influences have been modelled by Katz [26] and are presented in Figure 5.4. The influence of reducing the modulus (Figure 5.4(a)) can account for the observation that pitch type carbon fibres, having a lower modulus than the PAN fibres, did not undergo a reduction in properties during ageing [19]. Also, the positive influence on ageing due to a reduction in fibre diameter (Figure 5.4(b)) may provide an explanation for the reports that asbestos fibre reinforced cements usually maintain their high flexural strength over time, in spite of the fact that the fibres are brittle and the matrix is dense.

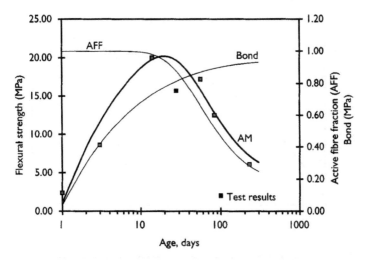

Figure 5.3 The contribution of the 'bond effect' and 'bending effect' to the change in flexural strength over time of a cement composite reinforced with brittle high modulus microfibres (after Katz and Bentur [25]).

Interfacial densening assumes very special characteristics in cement composites where the reinforcing fibre is in the form of bundled filaments surrounded by matrix, rather than individual filaments uniformly dispersed. Initially the spaces within the filaments remain largely vacant, since the cement grains are too big to penetrate into them [27,28]. Thus, the effective bond is small, and the reinforcing unit which is a strand rather than an individual filament is in practice flexible and effective in bridging cracks (Figure 5.5). However, if hydration products deposit gradually in between the filaments in the strand during ageing (i.e. deposition of calcium hydroxide precipitated from solution over time) the bond effectively becomes high, and premature fracture, leading to type I failure (Figure 5.1), may occur. Recently this mechanism was confirmed in an experimental study by Zhu and Bartos [29] using a micro-indentation technique. With this technique they could push off individual filaments in the strand and assess their bond. They showed that upon ageing the bond of the inner filaments in the strand increases considerably and approaches that of the external filaments that were in direct contact with the matrix before ageing, and thus had a high bond to start with. This type of ageing is characteristic of glass fibre reinforced cement produced by the spray method and it will be described in detail in Chapter 8.

A special case of interfacial effects controlling the durability of microfibre reinforced composite is cellulose reinforcement. The cellulose reinforcing unit obtained in the pulping process is of a hollow cylindrical shape. It is relatively flexible, and depending on the pulping process, it can be rendered immune to alkaline attack. However, upon ageing, two types of changes may occur: densification of

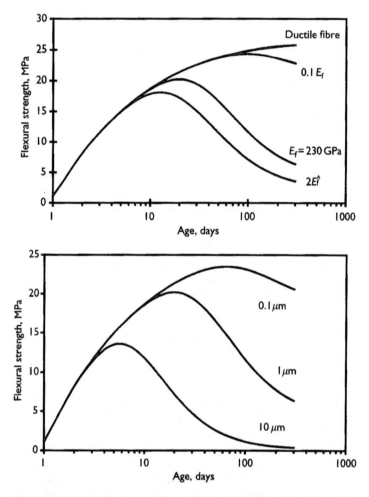

Figure 5.4 Modelling the influences of the modulus of the fibre (a) and its diameter (b) on the changes in strength of FRC composites over time, due to continued hydration leading to the densening and strengthening of the ITZ (after Katz [26]).

the matrix microstructure around the fibre and petrification of the fibre by filling of the lumen with reaction products and impregnating the cell wall with calcium silicates [30–32]. The petrification usually occurs in carbonating conditions, and its extent depends on the composition of the matrix. The interfacial densening results in loss of strength and ductility (type I behaviour), whereas the petrification results in enhanced strength and reduced ductility (type II behaviour). Both effects can be simulated in accelerated tests, based on wetting/drying cycles in carbonating

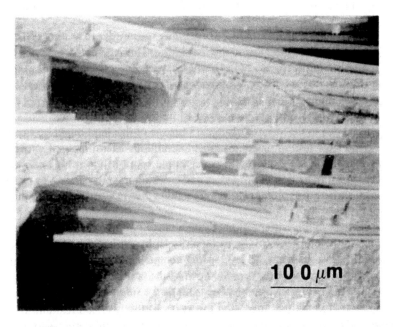

Figure 5.5 Glass fibre strand bridging over a crack, showing the bending deformation involved. Only the filaments which are loosely held together can
be flexible enough to accommodate this deformation (after Bentur and
Diamond [27]).

or non-carbonating environments [30–32]. This mode of ageing is discussed in
details in Chapter 11.

5.3 Volume stability and cracking

Many of the long-term performance problems of fibre reinforced cement composites are not the result of changes in the 'composites' properties, but are rather
induced by volume changes in the material, due to temperature and humidity
changes. It is necessary to recognize that the shrinkage potential of fibre reinforced cement thin sheets may exceed considerably the shrinkage of concrete.
Typical shrinkage strains for the latter are usually less than 0.05%, whereas for the
thin sheet composites swelling values in the range of 0.10% to almost 2% have
been measured [33]. It should be noted that this is the situation in thin sheet FRC
but not in fibre reinforced concrete where shrinkage is equal to or smaller than
that of the unreinforced concrete. The higher swelling in the thin sheet composites
is a result of a variety of influences, including higher cement content than in a
typical concrete mix (i.e. less aggregates) as well as the presence of fibres that
are moisture-sensitive, such as cellulose and wood particles. In addition, the thin

geometry of many of the thin sheet fibre reinforced cements leads to higher drying rates.

Volume changes which are induced in natural exposure due to wetting and drying may cause internal damage due to microcracking. This has been demonstrated for wood particle reinforced cement [34], where it was shown that ageing in water did not lead to strength loss, whereas drying/wetting cycles resulted in loss in strength, with the magnitude of loss becoming greater for the more extreme drying cycles (Figure 5.6). These changes were seen to be the result of internal damage in the material due to the volume changes. Damage of this kind might be expected to be greater for testing under restrained conditions where internal stresses of greater magnitude are expected to occur as the result of the restrained volume changes. Thus, reduction in properties and cracking might be observed even for composites where the volume changes are smaller than in wood particle-reinforced cement [35].

The changes in dimensions might cause a range of problems which go beyond reduction in the properties of the material itself. In applications of thin sheets of fibre reinforced cements, such as cladding components connected to a structure, the changes in dimension may lead to bowing and to micro and macrocracking of panels. The extent of such a problem depends on the volume changes characteristic to the material (which is a material property) and the type and restraint of the structure and the joints. In some applications, this type of long-term performance problem is more critical than the mere changes in the properties of the material with time [36–38]. Unfortunately, this problem has not received adequate attention in the open literature, which usually highlights durability issues related to the properties of the material itself. The approach to be taken to deal with this issue

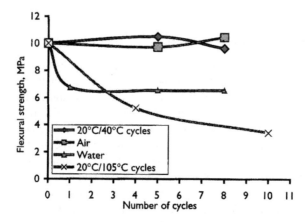

Figure 5.6 Effect of ageing in water, air and drying/wetting cycles (drying at 40°C and 105°C) on the flexural strength of wood particle reinforced cement (after Bentur and Mindess [34]).

is either to study the restrained shrinkage performance of the material itself, as reported by Becker and Laks [35], or to test full-scale panels, including their connections to the structure, as recommended in ISO standard 8336. Restrained shrinkage of fibre reinforced cements has been studied more intensively in recent years with a somewhat different motivation, which was to resolve to what extent the presence of a small content of fibre reinforcement could reduce the cracking tendency of concrete [39–41]. These studies, although not directly intended for the evaluation of long-term performance of thin sheet fibre reinforced cement, can provide a valuable contribution to the methodology and scientific background needed to deal with this problem. Standardization of tests of this kind is currently underway, using a configuration similar to that reported by See et al. [42].

5.4 Accelerated ageing tests

The long-term performance of fibre reinforced cements is of great significance in the development and evaluation of new composites. An important practical tool for this purpose is accelerated testing in the laboratory, where the properties of the composites are determined before and after exposure to the accelerated ageing conditions. In order to develop an efficient test of this kind, it is necessary first to evaluate the physical and chemical processes that may lead to changes in properties in natural exposure, and then devise the means to accelerate them in laboratory-controlled tests. In view of the variety of processes which may lead to ageing, it is difficult to devise a universal test. An illustration of this problem is presented in Table 5.1 and in [33,43], which show that composites which perform well in one type of accelerated test may do poorly in another, and the critical test is different for the different composites.

If the durability problem is expected to be the result of alkali degradation of the fibres or densening of the interfacial matrix, then immersion in hot water would provide an effective means for accelerating the process. The temperature of the water should be sufficiently high to accelerate the chemical processes involved, but not too high to cause a change in their nature. Temperatures in the range of 50°–60°C are usually accepted for this purpose. It should be noted that these processes are dependent on the presence of wet conditions. Thus in composites where the improved durability performance is based on modification of the matrix to reduce ingress of water in natural exposure, the continuous water immersion may be too harsh and the test results under such conditions may be misleading. This has been demonstrated to be the case in polymer-modified glass fibre reinforced cements, where natural weathering has shown better performance than predicted on the basis of the hot water immersion test [21,44]. In these composites, the enhanced durability is the result of the formation of a polymer film. This film may not be stable under continuously wet conditions and may not demonstrate its efficiency in reducing water ingress as it does in a natural environment. Therefore, even in the case of this kind of ageing, where the processes are of chemical degradation or

Table 5.1 Retention of flexural strength after accelerated ageing of commercial fibre cements composites made of different types of fibres (after Baum and Bentur [33])

Type of fibre	Flexural strength retention (%)	
	60°C Water	Wetting/drying
Asbestos	100	90
AR-glass	70	100
Cellulose	100	60

interfacial densening, the hot water immersion test should be applied with some caution.

In composites which are more sensitive to volume changes, such as natural fibres, cellulose and wood particle reinforced cements, a more adequate test is one based on the accelerated ageing of drying/wetting cycles. If carbonation is expected to be a contributing factor, then a carbonating environment should be superimposed on the drying cycle of the test. There are indications that the cycles in accelerated tests may have to be adjusted to the nature of the material, and for those which tend to dry at a slower rate the drying cycle should be longer [33].

It is evident on the basis of this discussion that, in order to use accelerating ageing conditions, an intelligent choice of test should be made, based on assessment of the processes which can lead to durability problems. For this purpose there is a need for in-depth understanding of the composite. This chapter provides the general background necessary to deal with this issue. The standards and specification for fibre reinforced thin sheet cement composites reflect this situation, and they specify the two types of accelerated tests (continuous immersion in hot water and drying/wetting cycles), giving the freedom to choose the more appropriate one [45,46]. These documents also deal with the situation where macrocracking and bowing may occur in a panel made of moisture-sensitive material due to restrained shrinkage, which may be induced in actual use. For that purpose, heat-rain tests are specified, in which full-scale panels, including their connections to the supporting structure, are exposed in their external face to cycles of water spray and drying. This type of test is intended to serve as a means for evaluation of a prototype of a system, since its outcome is a function of the properties of the material and the design and execution of the joints.

References

1. A. Bentur, 'The Role of Interfaces in Controlling the Durability of Fiber Reinforced Cements', *ASCE J. of Materials in Civil Eng.* 12, 2000, 2–7.
2. L.J. Larner, K. Speakman and A.J. Majumdar, 'Chemical Interaction between Glass Fibers and Cement', *J. Cryst. Solids.* 20, 1976, 43–74.

3. A. Bentur, 'Mechanisms of Potential Embrittlement and Strength Loss of Glass Fibre Reinforced Cement Composites', in S. Diamond (ed.) *Durability of Glass Fibre Reinforced Concrete,* Proc. Int. Symp., Prestressed Concrete Institute, Chicago, IL, 1985, pp.109–123.
4. L. Franke and E. Overback, 'Loss in Strength and Damage to Glass Fibers in Alkaline Solutions and Cement Extracts', *Durability Build. Mater.* 5, 1987, 73–79.
5. H.E. Gram, *Durability of Natural Fibers in Concrete,* Research Report, Swedish Cement and Concrete Research Institute, Sweden, 1983.
6. N.B. Milestone and I. Suckling, 'Interactions of Cellulose Fibre in an Autoclaved Cement Matrix', in K. Kovler, J. Marchand, S. Mindess and J. Weiss (eds) *Concrete Science and Engineering,* Proc. RILEM Symp., RILEM Publications, Evanston, IL, 2004, 153–164.
7. S.A.S. Akers and J.B. Studinka, 'Ageing Behavior of Cellulose Fibre Cement Composites in Natural Weathering and Accelerated Tests', *Int. J. Cement composites and Lightweight Concrete.* 11, 1989, 93–97.
8. G.C. Hoff, 'Durability of Fiber Reinforced Concrete in a Severe Marine Environment', in *Concrete Durability,* ACI SP-100, American Concrete Institute, Detroit, MI, 1987, pp. 997–1026.
9. K. Kosa and A.E. Naaman, 'Corrosion of Steel Fiber Reinforced Concrete', *ACI Mater. J.* 87, 1990, 27–37.
10. M. Raupach and C. Dauberschmidt, 'Corrosion behavior of steel fibres in artificial pore solution', Article 097, 15th International Corrosion Congress, Frontiers in Corrosion Science and Technology, Granada, Sept. 22–27, 2002.
11. P.S. Mangat and K. Gurusamy, 'Corrosion Resistance of Steel Fibers in Concrete Under Marine Exposure', *Cem. Con. Res.* 18, 1988, 44–54.
12. M. Chakraborty, D. Das, S. Basu and A. Paul, 'Corrosion Behavior of a ZrO_2-Containing Glass in Aqueous Acid and Alkaline Media and in a Hydrating Cement Paste', *Int. J. Cem. Compos.* 1, 1979, 103–109.
13. J. Orlowsky, M. Raupach, H. Cuypers and J. Wastiels, 'Durability Modelling of Glass Fibre Reinforcement in Cementitious Environment', *Mater. Struct.* 38, 2005, 155–162.
14. H.E. Gram, 'Durability Studies of Natural Organic Fibers in Concrete, Mortar or Cement' Paper 7.1 in R.N. Swamy, R.L. Wagstaffe, and D.R. Oakley (eds) *Development in Fibre Reinforced cement and Concrete,* RILEM Symp., Sheffield, RILEM Publications, Cachan Cedex, France, 1986.
15. A.J. Aindow, D.R. Oaley and B.A. Proctor, 'Comparison of the Weathering Behavior of GRC with Prediction made from Accelerated Aging Tests', *Cem. Concr. Res.* 14, 1984, 271–274.
16. M.D. Cambell and R.S.P. Coutts, 'Wood Fibre Reinforced Cement Composites', *J. Mater. Sci.* 15, 1980, 1962–1970.
17. S. Harper, 'Developing Asbestos Free Calcium Silicate Building Boards', *Composites.* 13, 1982, 123–138.
18. B.A. Proctor, D.R. Oakely and K.L. Litherland, 'Development in the Assessment and Performance of GRC over 10 Years', *Composites.* 13, 1982, 173–179.
19. M. Hayashi, T. Suenaga, I. Uchida and T. Takahashi, 'High Durability GFRC Using Low Alkali, Low Shrinkage CGS Cement', in R.N. Swamy (ed.) *Fibre Reinforced Cement and Concrete,* Proceedings of the 4th Int. Symp., E&FN Spon, London, 1992, pp. 888–901.
20. H.E. Gram, 'Methods for Reducing the Tendency towards Embrittlement in Sisal Fibre Concrete', Publ. No. 5, *Nordic Concrete Research,* Oslo, 1983, pp. 62–71.

21. J. Bijen, 'Improved Mechanical Properties of Glass Fiber Reinforced Cement by Polymer Modification', *Cem. Concr. Compos.* 12, 1990, 95–101.

22. J. Hikasa, T. Genba, A. Mizobe and M. Ozaka, 'Replacement for Asbestos in Reinforced Cement Products – "Kuralon" PVA Fibres, Properties and Structure', presented at the *Man Made Fibres Congress* (Dornbirn), The Austrian Chemical society, Austria, 1986.

23. P.J. Kim, H.C. Wu, Z. Lin, V.C. Li, B. deLhoneux and S.A.S. Akers, 'Micromechanics-Based Durability Study of Cellulose Cement in Flexure', *Cem. Concr. Res.* 29, 1999, 201–208.

24. B. Mobasher and C.Y. Li, 'Modeling of Stiffness Degradation of the Interfacial Zone During Fiber Debonding', *J. Composite Engineering.* 5, 1995, 1349–1365.

25. A. Katz, A. Bentur, 'Mechanisms and Processes Leading to Changes in Time in Properties of GFRC', *Advances in Cement Based Materials.* 3, 1996, 1–13.

26. A. Katz, 'Effect of Fiber Modulus of Elasticity on the Long Term Properties of Micro-Fibre Reinforced Cementitious Composites', *Cem. Concr. Compos.* 18, 1996, 389–400.

27. A. Bentur and S. Diamond, 'Effect of Aging of Glass Fibre Reinforced Cement on the Response of an Advancing Crack on Intersecting a Glass Fibre Strand', *Int. J. Cem. Comp & Ltwt. Concr.* 8, 1986, 213–222.

28. M. Stucke and A.J. Majumdar, 'Microstructure of Glass Fibre Reinforced Cement Composites', *J. Mater. Sci.* 11, 1976, 1019–1030.

29. W. Zhu and P.J.M. Bartos, 'Assessment of Interfacial Microstructure and Bond Properties in Aged GRC using a Novel Microindentation Method', *Cem. Concr. Res.* 27, 1997, 1701–1711.

30. A. Bentur and S.A.S. Akers, 'The Microstructure and Aging of Cellulose Fiber Reinforced Cement Composites Cured in Normal Environment', *Int. J. Cem. Compos. Lightweight Concr.* 11, 1989, 99–110.

31. A. Bentur and S.A.S. Akers, 'The Microstructure and Aging of Cellulose Fiber Reinforced Autoclaved Cement Composites', *Int. J. Cem. Compos. Lightweight Concr.* 11, 1989, 111–116.

32. B.J. Mohr, H. Nanko and K.E. Kurtis, 'Durability of Pulp Fiber-Cement Composites by Wet/Dry Cycling', *Cem. Concr. Compos.* 27, 2005, 435–438.

33. H. Baum and A. Bentur, *Fiber Reinforced Cementitious Materials for Lightweight Construction: Development of Criteria and Evaluation of Their Long Term Performance*, research report, National Building Research Institute, Technion, Israel Institute of Technology, Haifa, Israel, 1994.

34. A. Bentur and S. Mindess, 'Effect of Drying and Wetting Cycles on Length and Strength changes of Wood Fiber Reinforced Cement', *Durability Build. Mater.* 2, 1983, 37–43.

35. R. Becker and J. Laks, 'Cracking Resistance of Asbestos Cement Panels Subjected to Drying', *Durability Build. Mater.* 3, 1985, 35–49.

36. M.W. Fordyce and R.G. Wodehouse, *GRC and Building,* Butterworths, Seven Oaks, Kent, U.K., 1983.

37. N.W.Hansen, J.J. Roller, J.I. Daniels and T.L. Weinman, 'Manufacture and Installation of GFRC Facades', in J.I. Daniels and S.P. Shah (eds) *Thin Section Fiber Reinforced Concrete and Ferrocement,* ACI SP-124, American Concrete Institute, Detroit, MI, 1990, pp. 182–213.

38. G.R. Williamson, 'Evaluation of Glass Fiber reinforced Concrete Panels for Use in Military Construction', in S. Diamond (ed.) *Durability of Glass Fibre Reinforced Concrete,* Proc. Int. Symp., Prestressed Concrete Institute, Chicago, IL, 1985, pp. 54–63.

39. N. Banthia, M. Azzabi, and M. Piegon, 'Restrained Shrinkage Cracking in Fiber Reinforced Cementitious Composites', *Mater. Struct.* 26, 1993, 405–413.
40. M. Sarigapbuti, S.P. Shah and K.D. Vinson, 'Shrinkage Cracking and Durability Characteristics of Cellulose Fiber Reinforced Concrete', *ACI Mater. J.* 90, 1993, 309–318.
41. K. Kovler, J. Sikuler and A. Bentur, 'Restrained Shrinkage Tests of Fibre Reinforced Concrete Ring Specimens: Effect of Core Thermal Expansion', *Mater. Struct.* 26, 1993, 231–237.
42. T.H. See, E.K. Attiogbe and M.A. Miltenberger, 'Shrinkage Cracking of Concrete Using Ring Specimens', *ACI Mater. J.* 100, 2003, 239–245.
43. S.G. Bergstrom and H.E. Gram, 'Durability of Alkali-Sensitive Fibers in Concrete', *Int. J. Cem. Compos. Lightweight Concr.* 6, 1984, 75–80.
44. A. Sugiura and M. Wakasuti, 'Durability and Fire Resistance of Polymer Modified Glass Fiber Reinforced Cement', in S. Nagataki, T. Nireki, and T. Tomosawa (eds) *Durability of Building Materials and Components 6,* Proc. Int. Conf. (Japan), E&FN Spon, London, 1993, pp. 139–146.
45. UEAtc Directive, 'Directive for the Assessment of the Durability of Thin Fibre Reinforced Cement Products', European Union of Agreement, Centre Scientifique et Technique du Batiment, CSTB, Paris, France, 1989.
46. ISO Standard 8336, 'Fiber-Cement Flat Sheets,' 1993.

Chapter 6

Test methods

6.1 Introduction

The evaluation of the properties of FRC composites is of prime importance for these composites to be used effectively and economically in practice. Some of these properties are largely matrix dependent, and can be measured by the methods commonly used for conventional concrete, for example, compressive strength and freeze–thaw durability. Other properties, however, depend much more on the presence of fibres and on the fibre–matrix interactions. These properties are quite different from those of the matrix itself, and must be evaluated by test methods which are quite different from those used for plain cements and concretes. It is these properties that are of the greatest interest here, since they represent the areas in which the addition of fibres leads to significant improvements in properties such as toughness, crack control and impact resistance. This chapter deals only with this second type of tests, that is, tests which are particular to FRC. Some of these have been developed to the level of standard test methods adopted by ASTM, EN, ISO as well as other international and national agencies, while others are still used mainly for research or in special cases in practice. The significance of the various test methods will be outlined, indicating their importance and discussing their limitations.

It is beyond the scope of this chapter to provide detailed guidelines for the testing procedures, since these can be found in the references cited. Since standards are revised frequently, the most recent versions of these standards should be consulted. It should be noted that at the time of writing (early 2005) the FRC standards are still in a state of considerable flux. Indeed, it is the lack of a widely agreed upon body of standards that has inhibited the introduction of FRC into structural design codes such as ACI 318, *Building Code Requirements for Structural Concrete*.

6.2 Properties of the fresh mix

Test methods for the evaluation of the fresh FRC are primarily for the following characteristics:

1   Workability of a mix in which the fibres are dispersed by the mixing method itself.

2 Workability of the matrix in spray applications, in which the fibres and matrix are sprayed from individual nozzles.
3 Content of fibres in the fresh mix.
4 Plastic shrinkage cracking.

6.2.1 Workability

Fibres generally tend to stiffen a concrete mix, and make it seem harsh when static, though it may still respond well to vibration. Under vibration, the stiffening effect of the fibres tends to disappear, and so a properly designed FRC mix can be handled in much the same way as plain concrete in terms of mobility and ability to flow [1]. Therefore, workability tests based on static conditions, such as the slump test, are not very useful, and can be quite misleading, since the concrete is in fact workable when vibrated. Thus, in order to assess the workability of fresh FRC mixes, it is recommended that tests in which dynamic effects are involved be used [1,2].

Over the years, a great many workability tests have been devised. Koehler and Fowler [3] have described 61 different test methods, though only some of these have ever been adopted as standards, and even fewer are used for FRC. The most common test methods in current use include:

• Slump test (ASTM C143, *Standard Test Method for Slump of Hydraulic Cement Concrete)*. This test is the oldest (it first appeared as an ASTM standard in 1922) and most widely used test of concrete workability. Because it is a static test, it is not a good indicator of the workability of FRC. However, according to ACI Committee 544 (2), 'once it has been established that a particular FRC mixture has satisfactory handling and placing characteristics at a given slump, the slump test may be used as a quality control test to monitor the FRC consistency from batch to batch'.

• VeBe test: This test is described in BS 1881: Part 104: 1983 *Method for Determination of VeBe Time*. In this test (Figure 6.1), a standard slump cone is cast, the mould is removed, and a transparent disc is placed on top of the cone. This is then vibrated at a controlled frequency and amplitude until the lower surface of the transparent disc is completely covered with grout. The time in seconds for this to occur is the VeBe time. Although the VeBe test is suitable for the characterization of the workability properties of fresh FRC, it is not readily applicable for quality control on site.

• Walz flow table test (in the German DIN 1045 and 1048, and the European EN 206). This is basically a compaction test, best suited for mixes of 'medium' workability [4,5]. In this simple test, a tall metal container (200 mm × 200 mm × 400 mm) is filled with concrete without compaction. The concrete is then compacted either by rodding or by vibration, and the degree of compaction is then calculated as the height of the container divided

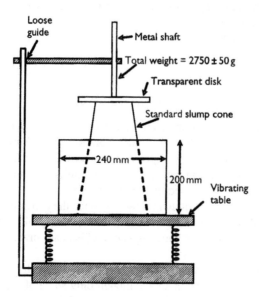

Figure 6.1 VeBe apparatus.

by the average height of the compacted concrete. The test is widely used in Europe, but not in North America.

• Inverted slump cone (ASTM C995, *Standard Test Method for Time of Flow of Fibre-Reinforced Concrete Through Inverted Slump Cone*). This test was developed specifically for FRC. A standard slump cone is inverted over a yield bucket and is then filled with concrete in three approximately equal layers without compaction. A 25 mm diameter internal vibrator is inserted vertically and centrally into the cone and is permitted to descend at a rate such that it touches the bottom of the bucket in 3 s. The time for the cone to become empty of concrete is recorded as the inverted slump-cone time. This test is sensitive to the mobility and fluidity of FRC, and is used for mixes which seem too stiff when evaluated by the slump test (slump < 50 mm). It should not be used if the time of flow is less than 8 s; such fluid mixes are better evaluated by the slump test.

The correlations obtained by the different test methods have been evaluated in several studies [1,2,6]. Relationships of this kind are presented in Figure 6.2, showing that even at a low slump the fresh FRC responds well to vibration, as estimated by the VeBe test (Figure 6.2(a)). The linear relationship between the results of the inverted slump cone and the VeBe tests (Figure 6.2(b)) suggests that both are sensitive to similar parameters. This indicates that the inverted slump cone test may be a suitable alternative to the VeBe test, in particular for site application.

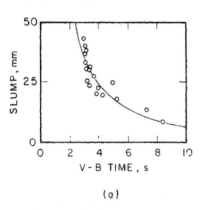 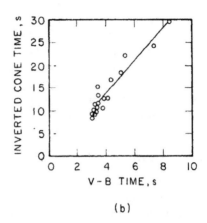

(a)   (b)

Figure 6.2 Relations between the results obtained by various workability tests of fibre reinforced cements. (a) Slump vs. VeBe; (b) VeBe vs. inverted slump cone (after Johnston [1]).

It should be noted that all workability tests mentioned above (and indeed almost all others) merely provide a comparative assessment of different FRC mixes; they do not measure parameters of physical significance.

Workability tests for the mortar matrix in sprayed fibre reinforced cement are specified mostly for glass fibre reinforced cements. They are based on the spread of a slurry which is cast into a cylindrical mould, with the mould subsequently lifted to allow the slurry to spread.

6.2.2   Fibre content in the fresh mix

The fibre content of the fresh mix is an important parameter for purposes of uniformity and quality control in the production of FRC. It is usually based on wash-out tests [7], in which the fibres are separated from the matrix. This procedure is more readily applicable when the matrix is mortar, since it can easily be washed away when placed in a mesh basket, while the long fibres are retained on the mesh. There are at least two such standard test methods for glass fibre reinforced concrete: ASTM C1229, *Standard Test Method for Determination of Glass Fibre Content of Glass Fibre-Reinforced Concrete (GFRC) (Wash-Out Test);* and BS 6432:1984, *Determining Properties of Glass Fibre-Reinforced Cement Material.* For other fibre reinforced concretes, the only published standard appears to be that published by the Japan Concrete Institute: JCI-SF 7, *Method of Tests for Fibre Content of Fibre-Reinforced Concrete,* which covers both fresh and hardened concretes. A wash-out method is recommended for fresh concrete, and a magnetic probe method is recommended for both fresh and hardened steel fibre reinforced concretes. Neither method is in common use.

Uomoto and Kobayashi [8] described an electromagnetic method which can be used to determine the steel fibre contents of both fresh and hardened concretes. The system measures the current changes induced by the ferromagnetic properties of the concrete, which are a function of the steel fibre content. For evaluation of the fibre content in fresh concrete, plastic moulds must be used (as they do not induce an electrical current by themselves). It should be noted that methods of this kind are not applicable to non-ferromagnetic fibres such as polypropylene, for which wash-out methods must be developed.

6.2.3 Plastic shrinkage and cracking

Many of the applications of fibres, in particular low modulus fibres at low fibre contents (less than ~0.3% by volume) are intended to reduce the sensitivity to cracking due to plastic shrinkage. To determine the improvement offered by the fibres, it is necessary to test the cracking sensitivity during early drying. Measurement of the plastic shrinkage by itself is not sufficient. Early age shrinkage measurements can be carried out by a variety of methods, such as ASTM C827, *Standard Test Method for Change in Height at Early Ages of Cylindrical Specimens of Cementitious Mixtures*. However, while tests of this type may show a reduction in plastic shrinkage due to the presence of fibres [9], this does not necessarily indicate an overall reduction in the cracking tendency, which is a function not only of the plastic shrinkage but also of the reinforcing effect of the fibres in the fresh concrete. The cracking tendency can only be judged on the basis of tests in which the shrinkage is *restrained* to promote tensile stresses in the concrete, and by observing the nature of the cracks developed and their time of formation. The physical significance of most of the tests that have been proposed is limited, and therefore they can be used only for a qualitative, comparative assessment of the effects of fibres, by comparing the observed nature of the cracks and their time of formation in different mixes.

One type of restrained test is the *ring test*, in which an annulus of concrete shrinks around a stiff inner ring. The concrete is cast between two steel rings, with the inner ring providing the restraint. This method can be applied to study the cracking sensitivity of fresh concrete, by immediately exposing the ring to drying conditions, or to test the cracking sensitivity of the hardened concrete by first curing in moist conditions, and then demoulding the external ring and exposing the concrete to drying [10,11]. A modified ring test was reported by Dahl [12], in which steel ribs are welded to the outer steel ring, to provide additional restraint (Figure 6.3). Ring specimens can be exposed to various drying conditions, such as low relative humidity at room temperature, mild drying in a heated oven, or a wind tunnel. The extent of cracking will obviously depend upon the restraint condition and the drying environment. The effects of these parameters in relation to field performance have not yet been evaluated satisfactorily, and thus the information obtained is assessed by comparison of the fibre-reinforced mix to a reference mix without fibres. The results can be quantified by providing the number and width of cracks developed, or by the total crack width.

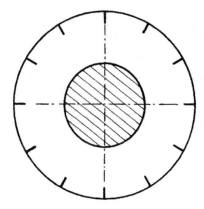

Figure 6.3 The configuration of restraining steel ring specimen to study plastic shrinkage cracking (after Dahl [12]).

Another measure of plastic shrinkage [13] is the performance of a thin slab, approximately one metre square, after warm, dry air is passed over it. The slab is restrained at the perimeter by a wire mesh, while the bottom surface is free. The large surface area, the accelerated drying and the restraint enhance plastic shrinkage cracking. This test, too, is qualitative in nature, and therefore must be applied at the same time to a reference mix and the fibre-reinforced mix; the extent of cracking in both is compared. Kraii [13] suggested a quantification of the cracking by calculating a weighted average which takes into account the crack length and width.

There are currently no standard tests in ASTM or other commonly used standards for crack control or resistance to plastic shrinkage cracking. There are currently (early 2005) two tests under consideration by ASTM specifically for assessment of plastic shrinkage cracking. The first involves using a bonded FRC overlay on a specially prepared substrate, as first suggested by Banthia *et al.* [14]; the assembly is then subjected to a specific drying environment. The hardened concrete substrate contains protuberances (Figure 6.4) obtained by casting it against a mould with depressions. The protuberances enhance the bond with the overlay and impose a uniform restraint on it. The width of each crack and the total crack area are then determined.

The second method is similar in principle. It uses a special steel form insert both to provide restraint and to provide stress risers at specific points in the specimen. Several systems of this kind have been developed [15,16], and one of these is shown in Figure 6.5. The average crack width is determined in such tests*.

---

* A test of this type has now (2006) been adopted, under the designation ASTM C1579 – *Standard Test Method for Evaluating Plastic Shrinkage Cracking of Restrained Fiber Reinforced Concrete (using a steel form insert)*.

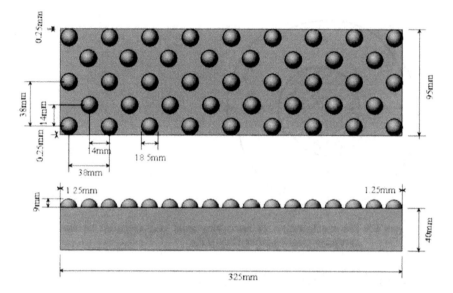

Figure 6.4 Dimensions of the substrate for measuring plastic shrinkage cracking using a bonded overlay, from N. Banthia, R. Gupta and S. Mindess, 'Development of Fiber Reinforced Concrete Repair Materials', *Canadian J. of Civil Engineering*, Vol. 33, No. 2, 2006, pp. 126–133 with permssion from NRC Research Press.

The concepts behind these various tests are simple. However, additional work is still required regarding their reproducibility and correlation with field performance.

6.3   Properties of the hardened composite

6.3.1   *Static testing in compression*

Fibres have relatively little effect on the compressive strength of concrete, and there are in general no special test methods for this property. The same tests that are used for the compressive strength of plain concrete are equally applicable to FRC. However, the Japan Concrete Institute has developed a test method for determining the compressive *toughness* of FRC, JSCE SF5: *Method of Test for Compressive Strength and Compressive Toughness of Steel Fibre-Reinforced Concrete*. The test arrangement is shown in Figure 6.6 [17]. According to the JSCE SF5, the test may be carried out using an open loop testing machine, but it has been found that this is applicable only for compressive strengths below about 60 MPa; for higher strengths, a catastrophic brittle failure occurs [17]. However, a closed-loop machine allows the test method to be used for much higher compressive strength concretes.

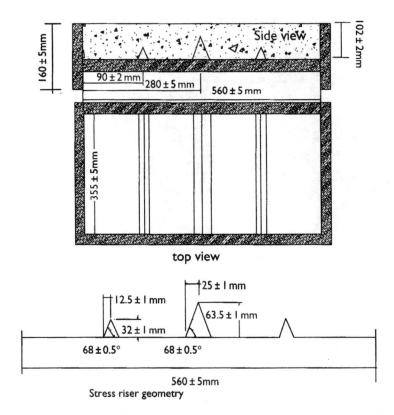

Figure 6.5 Specimen and stress riser geometry for plastic shrinkage test using a steel form [16].

The resulting load vs. deflection curves may then be analysed according to JSCE-SF5. A compressive toughness factor, T, is defined as:

$$T = 4T_c/(\pi d^2 \delta_{tc}) \qquad (N/mm^2) \tag{6.1}$$

where T_c is the compressive toughness, (J), the area under the load vs. deformation curve out to a strain of 0.75%; d, the specimen diameter (mm); δ_{tc}, the deformation corresponding to 0.75% converted to strain (mm); 0.75 mm for a specimen length of 200 mm.

A somewhat similar, though more elaborate, test method for determining the complete compressive stress–strain curve of plain concrete (which would also work for FRC) has been developed by RILEM-TC 148-SCC [18]. This method requires a servo-controlled testing machine.

Figure 6.6 Test set-up for compressive toughness according to JSCE SF5 [17].

6.3.2   Static testing in tension

Tensile tests of FRC are seldom specified in standards or specifications, and are currently used primarily in research. One of the major problems is to provide a gripping arrangement which will not lead to cracking at the grips. Various gripping methods, similar to those applied to unreinforced cement and concrete have been used successfully [19,20]. In order to get the complete load vs. deflection curves, the testing system must be equipped with strain or deflection measurement gauges. In practice, to obtain the complete curve, a servo-mechanical or servo-hydraulic (i.e. a closed loop), rigid testing machine must be used.

More recently, RILEM TC 162-TDF [21] has recommended a uniaxial tension test to be used to determine the stress vs. crack opening relationship for steel fibre reinforced concrete (SFRC), which may be used for other fibre concretes as well. The values obtained from this test are intended to be used in a design method for SFRC (see Chapter 14). The test involves a 150 mm diameter notched cylindrical specimen, 150 mm long, which is glued to the platens of the testing machine, as shown schematically in Figure 6.7. The test should be run using a closed loop mechanical testing machine. Because of the notch, the test is not suitable for the determination of tensile strength, but may be used to determine the energy dissipation between any two values of crack opening.

6.3.3   Static testing in flexure

The static mechanical tests of most interest for FRC composites are flexural (bending) tests. To characterize the flexural behaviour of FRC, it is necessary

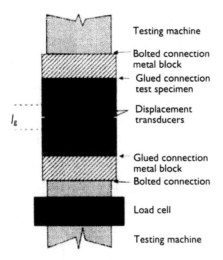

Testing machine

Bolted connection
metal block

Glued connection
test specimen

Displacement
transducers

Glued connection
metal block

Bolted connection

Load cell

Testing machine

l_g

Figure 6.7 Schematic of test set-up for uniaxial tensile testing, after RILEM TC 162-TDF [21].

to measure the complete stress vs. strain (or load vs. deflection) curves, which reflect the effect of the fibres on the toughness of the composite and its crack control potential. As with direct tension tests, in order to obtain complete load vs. deflection curves, the test system must be equipped with strain or deflection measurement gauges. Because of the importance of obtaining a reliable curve in the post-cracking zone, a rigid testing machine, and preferably a closed loop testing arrangement are necessary.

In the flexural test, the calculation of the modulus of elasticity, the first crack stress (i.e. the point at which the load vs. deflection curve first departs significantly from linearity, also sometimes known as the limit of proportionality) and the ultimate flexural stress are based on elastic bending theory, which is obviously invalid in the post-cracking zone. As was shown in Section 4.7, the calculated flexural strength is considerably higher than the direct tensile strength. Bar Shlomo [22] pointed out another limitation, due to the assumption in elastic bending theory of small deflections. In the testing of thin sheets, for instance, the deflection at maximum load can be quite high, leading to a bending moment which is larger than that calculated from the bending equation, with the difference sometimes being as high as 15%.

Of particular importance in the interpretation of the load vs. deflection curves in the post-cracking zone, which represents the strain capacity and toughness of the composite, and thus provides an indication of the quality of the material from the point of view of crack control. This characteristic is of much greater

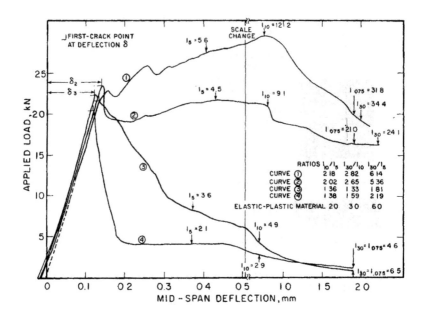

Figure 6.8 A range of load–deflection curves obtained in the testing of steel fibre reinforced concrete (after Johnston [23]).

significance than any enhancement of flexural strength that may occur. For example, the data presented in Figure 6.8 [23] show that although these composites do not differ much in their flexural strengths, there is an enormous difference in their post-cracking behaviour. Various attempts have been made over the years to quantify such curves in terms of a parameter which could be used for comparisons between different fibre types or contents, as well as for specifications and quality control. The discussion later deals with some of the approaches that have been attempted.

According to Mindess *et al.* [24], any toughness or residual strength parameter for FRC should, ideally, satisfy the following criteria:

1   It should have a physical meaning that is both readily understandable and of fundamental significance if it is to be used for the specification or quality control of FRC.
2   The 'end-point' used in the calculation of the toughness parameters should reflect the most severe serviceability conditions anticipated in the particular application.
3   The variability inherent in any measurement of concrete properties should be low enough to give acceptable levels of both within-batch and between-laboratory precision.

4   It should be able to quantify at least one important aspect of FRC behaviour (e.g. strength, toughness or crack resistance) and should reflect some characteristics of the load vs. deflection curve itself.

5   It should be as independent as possible of the specimen size and geometry.

Unfortunately, none of the standardized test methods described below can fulfil all (or even most) of these criteria, and it is important to understand the difficulties encountered when using these methods. There are at least five different tests that have been standardized:

- ASTM C1018: test method for flexural toughness and first-crack strength of fibre reinforced concrete (using beam with third-point loading).
- ASTM C1399: test method for obtaining average residual strength of fibre reinforced concrete.
- ASTM 1550: standard test method for flexural toughness of fibre reinforced concrete (using centrally loaded round panel).
- JSCE SF-4: method of test for flexural strength and toughness of fibre reinforced concrete.
- RILEM TC162-TDF [21]: test and design methods for steel fibre reinforced concrete – bending test.

Unfortunately, there is still no agreement on which (if any) of these tests best represents the toughness of FRC, and the tests often give conflicting results when compared with each other, or when carried out by different laboratories [25–28].

ASTM C1018   A small beam specimen (100 mm $\times$ 100 mm $\times$ 350 mm) is tested in flexure under 4-point loading, and 'toughness indices' are defined in terms of the ratio of the area under the load vs. deflection curve out to some specified deflection to the area under the curve out to the point of 'first crack', as shown in Figure 6.9. One or more of the toughness indices I_5, I_{10} and I_{20} are often specified as acceptance criteria.

In addition, 'residual strengths' $R_{5,10}$ and $R_{10,20}$ are usually calculated, which represent the average post-cracking load that the specimen may carry over a specific deflection interval. They are presented as a percentage of the load at first crack and are calculated from the toughness indices:

$$R_{5,10} = 20(I_5 - I_{10}) \text{ and } R_{10,20} = 10(I_{20} - I_{10}) \tag{6.2}$$

This test is probably the one most commonly used in North America, but it suffers from a number of shortcomings:

- Since the deflections out to first crack (and indeed to the peak load) are very small, it is important to determine the load vs. deflection curve precisely. In

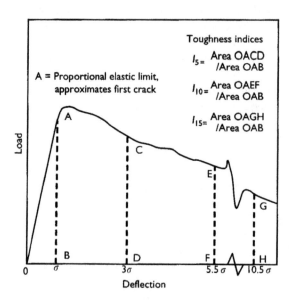

Figure 6.9 Schematic of load vs. deflection curve and definition of toughness parameters according to ASTM C1018.

particular, it is essential to correct for the 'extraneous' deflections that occur due to seating of the specimen on the supports and machine deformations; different laboratories make these corrections differently, and hence may get different results [25]. If this is not done properly (as is all too often the case in practice), then spurious values of the toughness indices will be obtained. Extraneous deformations can be accounted for in several ways, either by using a special loading frame as specified in JSCE SF4 (see later), or by mounting dial gauges or LVDTs not only at the midpoint of the specimen, but also over the specimen supports, so that the support deflections can be subtracted from the midpoint deflection. As a quick check that this is being done properly, the measured first-crack deflection for third-point loading should be compared with that calculated assuming elastic behaviour up to first crack. This deflection is given by:

$$\delta = \frac{23PL^3}{1296EI} \frac{(1 + 216D^2(1 + v))}{115L^2} \tag{6.3}$$

where P is the first crack load; L, the span; E, the elastic modulus; I, the moment of inertia; D, the specimen depth and v, Poisson's ratio.

If the measured and calculated values differ by more than a few per cent, then there is something wrong with the deflection measurements.

- The calculated toughness parameters depend on how the point of 'first crack' is defined, but since some microcracking begins to take place almost immediately after loading begins, it is difficult to define this point unambiguously.
- An *instability* often occurs in the measured load vs. deflection curve immediately after the first significant crack, particularly for low toughness FRCs, unless either a closed-loop servo-controlled testing machine or a very stiff machine is used; different loading systems can lead to quite different calculated toughness values and residual strength factors. Figure 6.10(a) [25] shows the effects of different testing machines on the measured load vs. deflection curves, with quite different shapes of the curves immediately beyond the peak load. In Figure 6.10(b) [25] the region of instability is shown more clearly; the recorded lines B–C do not represent the true load–deflection response of the beams. They are an artefact of the particular loading system.
- The toughness parameters are not independent of the specimen dimensions.
- The toughness parameters are sometimes not particularly sensitive in distinguishing between different fibre types or geometries [29] due to the variability inherent in trying to determine the point of first crack and to the difficulties induced by the instability mentioned earlier.

As a result of these problems, there is active consideration at this time (early 2005) to discontinue this test method entirely, and replacing it with a somewhat different flexural toughness test*. The proposed test (Standard test method for flexural performance of fibre reinforced concrete (using beam with third-point loading)) maintains the same procedures as ASTM C1018 for obtaining the load vs. deflection curve, but the analysis of the curve is completely different. In addition to determining the stresses at first crack and at peak load, the residual strengths at various deflection levels are determined directly from the load vs. deflection curve, as well as (optionally) the toughness out to a specified deflection calculated as the area under the load vs. deflection curve. (Note: this completely eliminates any consideration of the toughness indices of ASTM C-1018). This proposed test method appears to be sensitive to different fibre types and volumes.

ASTM 1399   In this test, a small (100 mm × 100 mm × 350 mm) FRC beam is cracked in a standard manner by loading it in bending in combination with a steel plate, as shown in Figure 6.11. The purpose of the steel plate is to prevent complete failure when the beam begins to crack. The steel plate is then removed, and the (cracked) specimen is reloaded to obtain a reload vs. deflection curve. The average residual strength of the beam over the deflection range of 0.5–1.25 mm is then determined. Banthia and Dubey [30,31] found that the load vs. deflection

---

* ASTM C1018 has now (2006) been withdrawn. It has effectively been replaced by ASTM 1609 – *Standard Test Method for Flexural Performance of Fiber-Reinforced Concrete (Using Beam with Third-Point Loading)*.

(a)

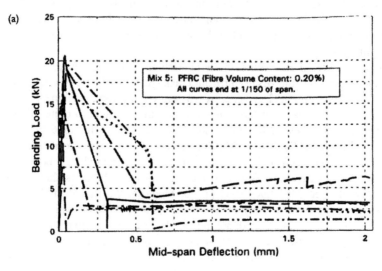

(b)

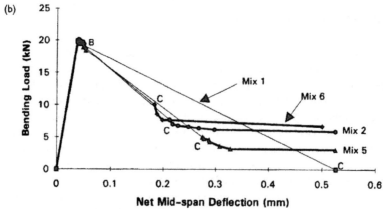

Figure 6.10 Effect of different loading machines on the load–deflection curves of (a) 0.2% polypropylene fibre beams; (b) region of instability for low fibre volume beams; mix 1 represents plain concrete [25].

curves obtained from this test agreed closely with those obtained using a closed-loop machine with proper displacement control. They also found that the variability in the test was relatively low, and that it was capable of identifying the influence of fibre characteristics such as type, geometry, volume fraction and elastic modulus. The principal problems with this test are:

• The effect of the fibres on the performance immediately after the first crack is ignored.

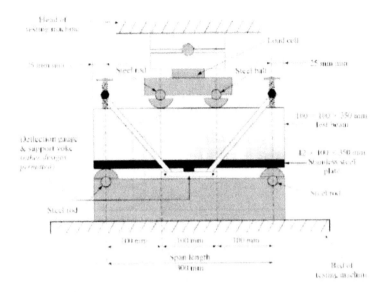

Figure 6.11 Apparatus for determining the residual strength of FRC according to ASTM
C1399. Copyright ASTM International. Reprinted with permission.

- The length of the pre-crack obtained is unknown. For different types of FRC, the pre-cracks obtained with this loading system may be different in length, making comparisons of the residual load-bearing capacities between different FRC beams difficult.
- Simple beam theory (as required in this test method) cannot be used to calculate the strength of a cracked system, so it is not clear what the calculated 'residual strengths' mean.

ASTM 1550   This relatively new test, based on the work of Bernard [32,33] involves the centre-point loading of a large circular plate, 800 mm in diameter and 75 mm thick, supported on three points. The performance of the specimen is judged on the basis of the energy absorbed in loading the plate to some selected values of central deflection. This test appears to discriminate reasonably well amongst different fibre types and volumes, though its results are not always consistent with those obtained from other toughness tests. This test has become popular with producers of fibre reinforced shotcrete, and is often used in the mining industry. A study in which it was compared with the RILEM three-point bending test [34] found that the two tests were compatible in that they led to similar conclusions, and that it had a lower scatter than the bending test. Its chief disadvantage is that the specimen itself is too large and heavy to be handled easily, and does not fit into many commonly used testing machines.

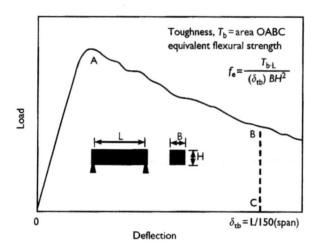

Figure 6.12 Schematic of load vs. deflection curve and definition of toughness parameters according to JSCE SF-4.

JSCE SF-4   This test was standardized by the Japan Society of Civil Engineers in 1984. This test too is a bending test of a small (100 mm × 100 mm × 350 mm) FRC specimen. In this test, the total area under the load vs. deflection curve out to a specified deflection ($\delta_{tb} = L/150$) is measured and is referred to as the *toughness*. The toughness factor (or equivalent flexural strength), which is a measure of the average residual strength, is calculated as:

$$\text{toughness factor} = \text{toughness} \times L/(BH^2 \times \delta_{tb}) \tag{6.4}$$

where the symbols are as defined in Figure 6.12.

One major advantage of this test is that the JSCE SF-4 toughness parameters are little influenced either by different loading systems or by extraneous deflections, and the instability in the load vs. deflection curves mentioned previously has only a small effect on the calculated values. However, this test too has a number of limitations:

- The toughness parameters are dependent on the specimen dimensions.
- The toughness parameters calculated cannot distinguish between pre-peak and post-peak behaviour; FRC mixes with quite different load vs. deflection curves can yield similar toughness parameters.
- Because the test has a fixed deflection end point, it cannot easily be adapted to different deflection or serviceability conditions, nor does it reflect the characteristics of the load vs. deflection curve.
- The toughness parameters are age-dependent up to an age of about 60 days.

RILEM TC162-TDF[35]   In this procedure, a *notched* FRC beam, of dimensions 150 mm × 150 mm × 550 mm is tested in centre-point loading, as shown in Figure 6.13. The notch is introduced so that the crack mouth opening displacement (CMOD) may be measured. This test is performed using a CMOD controlled machine (i.e. it is a closed loop test). The parameters determined include the limit of proportionality (referred to above as the point of first crack), the flexural strength, and residual flexural strengths at different deflections. The energy absorption capacities are also determined out to particular deflections as a function of the area under the load vs. deflection curve. This method is intended to provide values which can be used for the structural design of FRC beams, though as yet it is used primarily in research (e.g. Barros *et al.* [36]). Both the test and the design considerations have been described in detail by Vandewalle [37].

Amongst the many other test methods that have been proposed for the determination of the toughness of FRC, three are worth a brief mention here.

EFNARC   The European Federation of Producers and Contractors of Special Products for Structures (EFNARC) has proposed a plate test [38] to quantify the toughness of FRC, and in particular fibre shotcrete. The test involves a 600 mm square plate, 100 mm thick, simply supported on all four sides on a 500 × 500 mm span, and loaded at the centre. The toughness (absorbed energy) is determined from the load vs. deflection curve out to a deflection of 25 mm. Bernard [39] found a good correlation between this plate test and the round panel test. The EFNARC test is sometimes used in Europe, but rarely in North America.

'Template' approach   The template approach suggested by Morgan *et al.* [40] also utilizes a 100 mm × 100 mm × 350 mm beam tested in four-point bending. However, it does not involve a calculation of toughness *per se.* Instead, *toughness performance levels*, such as those shown in Figure 6.14, are used. Toughness performance levels I, II, III and IV represent increasing levels of performance. It would then only be necessary to compare the actual load vs. deflection curve to the template to see whether the FRC in question satisfied the required toughness performance level. The advantage of this method is that it is not sensitive to the location of the first crack, to the occurrence of extraneous deflections, or to the instability of the measured load vs. deflection curves, since the exact shape of the curve out to a deflection of 0.5 mm is not taken into consideration. Similar template approaches have been adopted in the Norwegian and Austrian codes and by EFNARC [38].

South African Water Bed   This test method is gaining popularity in the shotcrete industry because the relatively large plate specimens (1600 mm × 1600 mm × 75 mm) are thought to better represent field conditions. A schematic of the test set-up is shown in Figure 6.15 [41]. The panels are fastened in place over a water bladder, which is then filled with water to apply pressure to the specimen. The energy absorbed (i.e. the toughness) to a series of given deflections (ranging from 25 to 150 mm) is determined from the load vs. deflection curves.

(a)

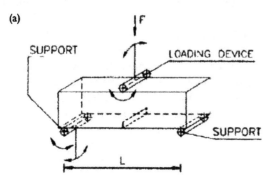

(b)

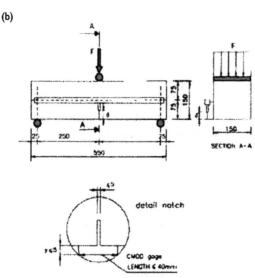

Figure 6.13 (a) Schematic of test set-up according to RILEM TC162-TDF bending test [35]; (b) arrangement of displacement monitoring gauges and details of sawn notch. Copyright RILEM. Reprinted with permission.

6.3.4   Impact testing

The impact resistance of plain concrete is relatively low. However, fibre reinforcement is particularly effective in improving the performance of concrete under dynamic loading, sometimes by more than an order of magnitude. This occurs mainly because of the higher strain capacity and residual load-bearing capacity of the FRC in the post-cracking zone. Plain concrete is known to be highly strain-rate sensitive in compression [42], tension [43,44] and flexure [45,46], though the degree of rate sensitivity varies. The situation is even more complex in FRC,

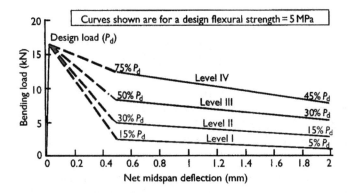

Figure 6.14 Template approach to specifying toughness in terms of residual strengths [40].

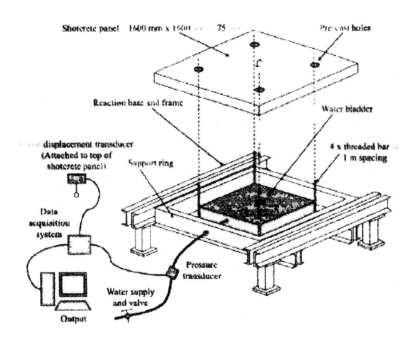

Figure 6.15 Schematic of South African Water Bed test [41].

since the concrete matrix, the fibres and the fibre–matrix bond are strain-rate sensitive to different degrees. Thus we are not able to predict the behaviour of FRC under impact on the basis of static test results, as the failure mechanisms are often quite different, depending on the particular FRC and strain rate. The problem is made more difficult by the fact that concrete systems may be subjected to very different strain rates ($\acute{\varepsilon}$) in practice, depending on the particular source of the dynamic event [47]:

Traffic	$\acute{\varepsilon} = 10^{-6}-10^{-4} \text{ s}^{-1}$
Gas explosion	$\acute{\varepsilon} = 5 \times 10^{-5}-5 \times 10^{-4} \text{ s}^{-1}$
Earthquake	$\acute{\varepsilon} = 5 \times 10^{-3}-5 \times 10^{-1} \text{ s}^{-1}$
Pile driving	$\acute{\varepsilon} = 10^{-2}-10^{0} \text{ s}^{-1}$
Aircraft landing	$\acute{\varepsilon} = 5 \times 10^{-2}-2 \times 10^{0} \text{ s}^{-1}$
Hard impact	$\acute{\varepsilon} = 10^{0}-5 \times 10^{1} \text{ s}^{-1}$
Hypervelocity impact	$\acute{\varepsilon} = 10^{2}-10^{6} \text{ s}^{-1}$

Clearly, no single test method could be expected to cover this enormous range of strain rates. Partly because of this, and because of the complexity of the FRC itself, its impact and blast properties remain poorly understood. There are a number of reasons for this state of affairs [48–50].

1   There are no generally agreed upon standard test methods with which to evaluate the impact properties of FRC (or even of plain concrete). A great many test procedures have been used in different laboratories: instrumented drop weight impact machines of various sizes; instrumented Charpy machines; split hopkinson pressure bar (SHPB); projectile impact, with projectiles of different masses, geometries and impact velocities; and blast loading with explosive charges of various intensities. As well, non-instrumented, more subjective tests are also used, such as repeated impact drop weight tests to some arbitrary degree of damage. Clearly, the test arrangements and specimen geometries have been sufficiently different that the results from the various studies are generally not comparable, and often appear to be contradictory.
2   There are no 'standard' cementitious materials that can be used as a benchmark to calibrate test techniques and equipment. That is, there are no cement-based materials whose strain-rate behaviour and fracture energy are 'known'.
3   All impact tests are very sensitive to the precise details of the particular test procedure.
4   There is still no general agreement on which parameters should be used to best characterize the response of concrete to impact loading.

Some of the available test techniques, the significance of the principal test parameters, and how best to characterize the results, are discussed below.

Instrumented drop weight test   In this test [e.g. 45,51] a mass is raised to a predetermined height and is then dropped onto the target specimen, which may be a beam, a plate or a compression specimen. Normally, the system is instrumented with some combination of load cells, accelerometers, strain gauges and

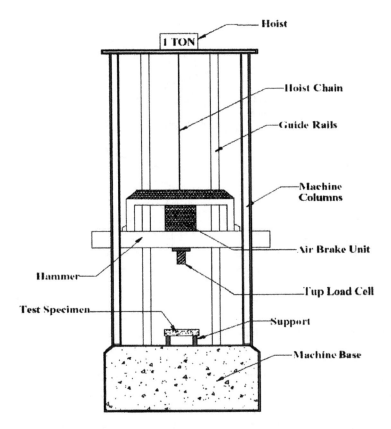

Figure 6.16 Schematic sketch of an instrumented drop weight impact machine at the University of British Columbia.

displacement transducers. Of course, since the impact event may have a duration of only a few milliseconds, a high speed data acquisition is essential. As well, high speed digital (or other) cameras can provide valuable additional information regarding the nature of the crack propagation. A schematic of such a machine at the University of British Columbia, capable of dropping a 575 kg mass from a height of up to 2.5 m is shown in Figure 6.16.

Swinging pendulum machines   These machines are generally of the form of instrumented Charpy machines, and are similar in principle to the drop weight machines. A swinging pendulum strikes the specimen [46,52], transferring momentum and inducing high stress rates.

Split Hopkinson pressure bar   In this device [42,43,53] the FRC specimen is sandwiched between two elastic bars and very high stress rates can be generated by propagating a pulse through one of the bars, using a striker bar. The device is

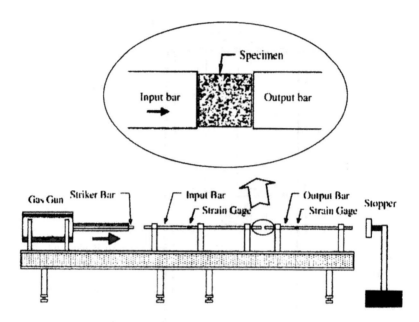

Figure 6.17 Schematic of a split Hopkinson pressure bar apparatus.

shown schematically in Figure 6.17. When the input bar is struck, a well-defined rectangular compression wave is generated. When the wave reaches the specimen, some of it reflects back through the input bar, and some transmits through the specimen to the output bar. This generates very high stresses and stress rates in the specimen. One-dimensional wave analysis is then used to obtain the stress vs. strain curves.

Projectile impact   In these tests, a projectile of some sort (frequently a round ball, sometimes a shaped projectile) is fired at an FRC specimen, using some sort of gun (e.g. [54,55]). This initiates a compressive wave which runs through the material to a free edge. The wave is then reflected and is transformed into a tensile wave that propagates in the opposite direction. The principal failure modes are scabbing on the rear side of the specimen, and scabbing on the front. In some cases, complete penetration may occur. The specimen may be instrumented, but often the results are assessed in terms of the visual damage to the specimen.

Explosive charges   To generate extremely high stress rates, an explosive charge may be attached directly to the specimen, and then detonated.

Repeated impact drop weight test   While the tests described earlier all require fairly sophisticated equipment and instrumentation, repeated impact drop weight tests are the simplest of the impact tests. Basically, a weight of some type is dropped repeatedly onto the specimen, and the number of blows required to cause a specified

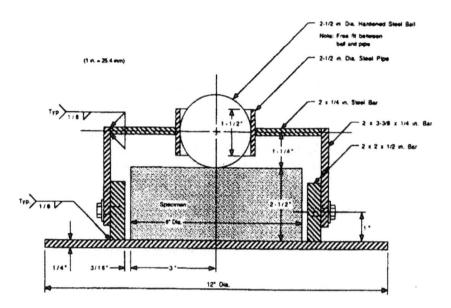

Figure 6.18 Schematic of repeated impact drop-weight apparatus (after ACI Committee 544 [2]).

amount of damage is recorded. One such test (Figure 6.18) has been described in detail by ACI Committee 544 [2]. A cylindrical specimen 152 mm in diameter and 63.5 mm thick is placed in a standard compaction hammer apparatus (ASTM D1557), and the number of blows of the 4.54 kg hammer falling through 457 mm required to cause the first visible crack and then to cause ultimate failure are recorded. This test suffers from a high variability, and is not now used very much.

Inertial loading All the instrumented impact tests, except for the split Hopkinson pressure bar, result in high specimen accelerations that manifest as inertial forces in the system. These must somehow be accounted for to derive any meaningful information, as they may account for a considerable portion of the recorded load. In the initial period of the impact event, a beam or a plate specimen is accelerated, and the inertial force induced in addition to the force required to bend the specimen. As a result during this time period, the recorded load is considerably greater than that resisted by the beam. This shows up as an oscillation in the total load vs. time curve. The duration of this inertial effect is short, and it was recommended by Server [56] that for reliable impact measurements the time to fracture should exceed the time of inertial oscillation by a factor of 3. Beyond that time, the measured load is equal to the bending load in the specimen. Unfortunately, concrete is too brittle for this condition to be met, since the time to fracture is too short.

In order to overcome this difficulty, various techniques have been developed. Suaris and Shah [46,57] reduced the inertial effect by placing a rubber pad between the striking tup and the specimen. They demonstrated that the amplitude of the inertial oscillations could be substantially reduced by this means, and the inertial effect eliminated. The application of a damping pad at the tup–beam interface, however, leads to a reduction in the effective stress rate.

A different approach, which eliminates the need for a damping pad, has been developed [51,58]. Here, the inertial effect is accounted for directly by recording the acceleration of the specimen itself during the impact event, using accelerometers attached to the specimen. From the accelerometer readings, the generalized inertial load can be derived using the principle of virtual work. Then, the load actually involved in deflecting the specimen during the impact event, $P_b(t)$, can be calculated as the difference between the total load recorded by the tup, $P_t(t)$ and the inertial load, $P_i(t)$:

$$P_b(t) = P_t(t) = P_i(t) \tag{6.5}$$

For instance, for a beam specimen, the generalized inertial load is given by:

$$P_i(t) = \rho A (\mathrm{d}^2 u / \mathrm{d}t^2)(1/3 + 8(ov)^3 / 3l^2) \tag{6.6}$$

where $u(t)$ is the mid-span deflection of the beam at time 't'; ρ, the mass density of the beam; A, the cross-sectional area of the beam; l, the clear span of the beam and ov, the length of overhang on each side of the supports.

For a square plate specimen simply supported on all four sides, the generalized inertial load is given by [59]:

$$P_i(t) = \frac{\rho h l^2}{4} \ddot{u}(x, y, t) \operatorname{cosec} \frac{\pi x}{l} \tag{6.7}$$

where l is the length of the side of the plate; ρ, the the mass density; h, the the thickness of the plate and $\ddot{u}(x, y, t)$, the acceleration at any point (x, y) on the plate at time t.

The deflection at mid-span can be calculated by integrating twice, with respect to time, the acceleration at mid-span, or by direct reading with an LVDT or laser transducer, and thus the bending load vs. time curve can be established. The area under this curve is the energy expended in straining and fracturing the beam. This approach permits a differentiation between the three types of loads, $P_b(t)$, $P_t(t)$ and $P_i(t)$. As may be seen from Figure 6.19 [51], the inertial load may be a large fraction of the total load during the initial stages of the impact event, particularly for FRC.

These results also permit an energy balance calculation, showing that the energy expended in straining and fracturing the beam may be only a small portion of the total energy (Table 6.1), thus highlighting the error involved in using the total energy to characterize the beam behaviour during an impact test. The 'unaccounted

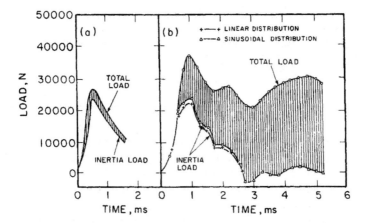

Figure 6.19 Total load, $P_t(t)$ inertial load, $P_i(t)$ vs. time curves of beams subjected to impact (a) plain concrete; (b) reinforced concrete (after Bentur et al. [61]).

Table 6.1 Energy balance during the impact test (after Bentur et al. [51])

	Plain concrete	Reinforced concrete	
	At peak load (Nm)	At peak load (Nm)	At ultimate[a] (Nm)
Total energy	23.6	66.8	1662.2
Kinetic energy	4.2	9.2	0
Beam energy	2.0	5.9	1150.7
Unaccounted energy	17.4	51.7	471.5

Note
a Calculated out to the point at which the load falls back to 1/3 of the peak load.

energy' in Table 6.1 is due, in part, to the energy 'lost' within the machine as elastic and vibrational energy. The implication of this observation is that one cannot equate the energy lost by the falling hammer to that absorbed by the specimen. The ratios between the various energies will depend upon the type of impact system, and in particular on the machine stiffness and the relative masses of the impact hammer and test specimen.

6.3.5 Effects of test parameters on impact data

As stated earlier, the precise details of the impact test system may have a considerable effect upon the resulting data. The effects of some of the principle test parameters are discussed here.

Table 6.2 Impact test data for high strength concrete beams

Test	Hammer weight (kg)	Drop height (mm)	Maximum load (kN)	Mid-span deflection at maximum load (mm)	Fracture energy (Nm)
1	578	156	75.3	12.43	318
2	578	1500	253.6	2.57	373.1
3	60	1500	190.5	2.05	258.4

Rigidity of the impact machine There have been no formal studies of the effect of the overall rigidity of the impact machine. While the rigidity probably has little effect upon the peak load, it will affect the post-peak behaviour, in particular because of the amount of energy 'lost' in the machine due to the elastic and vibrational effects mentioned earlier.

Rigidity of the specimen supports In principle, the specimen supports should be as rigid as possible. However, if the system is too rigid, such that the time required for the applied load to reach its maximum value is less than about one-half of the natural frequency of the specimen, then the stress waves set up during the impact event should be taken into account [60], which complicates the analysis enormously. To avoid having to make these corrections, Gopalaratnam *et al.* [61] used rubber pads at the specimen supports in their instrumented Charpy tests, but with the consequence of considerably reducing the loading rate.

In addition, there is inevitably some local cracking and crushing at the support points, particularly for beam tests, and so the size and shape of the supports will also have some effect on the measured values.

Impact velocity and energy of impact Both the velocity and energy have a significant effect on the material behaviour. For instance, Table 6.2 shows some test results obtained using two different instrumented drop weight machines on steel fibre-reinforced beams. The tests were carried out in the Civil Engineering laboratories at the University of British Columbia, using similar drop hammers (except for their weights) and the same data acquisition system. Tests 1 and 2 have both different velocities of impact and hence different impact energies; tests 1 and 3 have the same impact energies but different velocities; and tests 2 and 3 have the same impact velocities but different impact energies. In all cases, the beams failed completely during the impact events. It may be seen that there are no clear relationships amongst the impact velocity, the impact energy, the weight of the drop hammer, the mid-span deflection and the fracture energy. It is therefore not clear which test parameters best characterize the material.

Similarly, Table 6.3 shows the results of tests on steel fibre-reinforced beams, using the same 60 kg drop hammer as earlier, in which the drop height (and hence the impact velocities and impact energies) was varied. It may be seen that with increasing drop height, the strength increased, while the fracture energy decreased. All the specimens here also failed completely. Again, it is unclear how best to describe the material behaviour.

Table 6.3 Impact data for steel fibre-reinforced concrete beams

Drop height (mm)	Static	1000	1100	1200	1300
Maximum load (kN)	59.1	281.5	291.4	299.2	307.0
Nominal flexural strength (MPa)	17.2	81.8	84.7	87.0	89.2
Fracture energy (Nm)	263	489	450	386	353

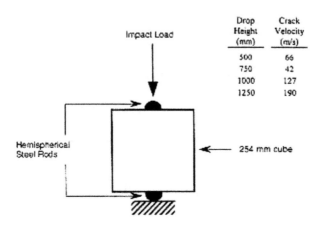

Figure 6.20 Impact of fibre-reinforced cubes in splitting tension [62].

Figure 6.20 shows the results of splitting tension (diametral compression) tests carried out on steel fibre reinforced concrete cubes [62]. All cubes failed completely in vertical splitting. In general, the crack velocity increased with increasing impact velocity (or impact energy). Again, it is quite unclear as to which drop height would best characterize the material behaviour.

Bindiganavile [63] has shown that even maintaining a condition of identical impact energy is insufficient to standardize impact tests. He tested both steel fibre and polypropylene fibre concretes, using the same instrumented drop-weight machines referred to earlier. With a large drop mass, the steel FRC appeared to be tougher; the reverse was true for tests carried out with a smaller mass (and hence a higher impact velocity). Which, then, is tougher under impact loading – steel FRC or polypropylene FRC? Similarly, Bindiganavile and Banthia [64] demonstrated that heavier hammers simulate longer pulses, while greater drop heights simulate shorter pulses. They showed that quite different flexural toughness values can be obtained for the same FRC using machines of different capacities, as shown in Figure 6.21.

Specimen size   The effects of size have been examined in some detail by Bindiganavile [63]. He tested three different sizes of geometrically similar steel FRC

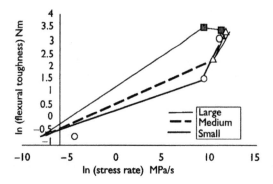

Figure 6.21 Influence of machine capacity on the impact response of an FRC [64].

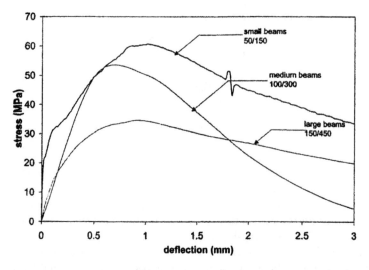

Figure 6.22 Stress vs. deflection responses of beams of different sizes, with beam depth:span ratios of 50:150, 100:300 and 150:450 [63].

beams using an instrumented drop-weight machine. As may be seen in Figure 6.22, not only the strengths of the beams, but also the shapes of the stress vs. deflection curves depend on the specimen size. As well, he found that the well-known size effects for concrete were more pronounced at higher loading rates. Similarly, from drop-weight impact tests on plain concrete cylinders of various sizes, experience has shown that one of the most important factors affecting impact tests is the relationship between the size and weight of the impacting hammer, and the specimen size and strength. Unfortunately, we cannot yet quantify these relationships.

Contact between specimen and impacting device   Sukontasukkul [65] carried out impact tests on FRC plates using two different circular load tups, with diameters of 1/4 and 3/8 of the clear span of the simply supported plates. For the smaller tup, the failure mode was dominated by shear. For the larger one, mixed shear and flexure failure modes were found. This too changes the apparent resistance of the FRC to impact loading.

6.3.6   Which parameters best characterize the response of FRC to impact loading?

Different investigators have used different parameters to try to characterize the response of both plain concrete and FRC to impact loading. However, there is no agreement as to how best to do this. The most common parameters are:

Peak load (or stress)   Most impact tests include some measure of the peak load or maximum stress that the specimen can withstand. While this is often useful information, for FRC it is the *post-peak* behaviour that is most important. However, it is difficult to *quantify* the load vs. deflection curve for easy use in structural design. It is probably more useful to record the residual load-bearing capacity of the specimen (or structural member) at different deflections which reflect the particular service conditions. For specimens that do not fail completely under a particular impact event, it may also be useful to determine the residual strength of the specimen under static loading [66] as a measure of the damage.

Projectile impact tests and blast tests are more difficult to interpret than drop-weight or Split Hopkinson Pressure Bar tests in terms of specimen strength. For these tests, some arbitrary measure of damage may be necessary, but this again shows up the difficulties in comparing the results obtained from different test arrangements.

Fracture energy   In addition to the peak load, the fracture energy is also very commonly measured in impact tests. Most commonly, this is taken as the area either under the complete load vs. deflection curve, or under this curve out to some particular deflection. Unfortunately, as indicated earlier, the fracture energy (or *toughness*) determined in this way depends strongly upon the mass and velocity of the impact hammer (or projectile), and upon the relative masses of the specimen and the impact hammer. It is thus essentially impossible at the present time to compare different tests.

Crack velocity   There have been a few attempts to measure crack velocities in both plain and fibre reinforced concretes under impact loading by using high speed photography (e.g. [62,67–69]). While such measurements provide invaluable information regarding the nature of the fracture in FRC, and may be used to compare qualitatively the effectiveness of different fibre types, they do not provide any useful design information.

Degree of damage or fragmentation   The degree of damage, or the amount of fragmentation, are generally more qualitative measures of fibre effectiveness. These rather subjective measures may provide comparisons amongst different fibres, but do not provide values that can be used directly in design or analysis.

6.3.7 Static vs. impact tests

The various impact and static tests described earlier are all intended to characterize the same property of the material: its toughness. To evaluate these different tests, it is necessary to compare the data provided by each with respect to the enhancement of the toughness when fibres are incorporated in the matrix. Gopalaratnam and Shah [70] concluded that the enhanced toughness due to the addition of 1–2% steel fibres was of the same order of magnitude when estimated by the static toughness index test and by instrumented impact testing, both of which led to an increase by a factor of 16–30 over that for the unreinforced matrix. The static test would be expected to give more consistent data than the impact test, simply because of the experimental complexity of the latter. However, this may lead to improper assessment of the efficiency of the fibres when the load to be resisted in service is of the impact type. For example, Mindess *et al.* [71] found that the toughening effect of 0.5% by volume of polypropylene fibres in conventionally reinforced concrete was small under static loading, but was significantly higher under impact loading.

In summary, both static and impact tests have experimental complexities, though the static tests are more adaptable to routine testing. However, when deciding which type of test should be used to characterize the reinforcing effects of fibres, the choice should be linked closely to the expected service conditions.

6.3.8 Fibre pull-out

Various test methods have been developed to determine the pull-out resistance of fibres embedded in a cementitious matrix. In *indirect* tests, the properties of the composite are determined and the shear bond strength is calculated on the basis of a theoretical approach in which the composite property is described as a function of the average interfacial bond. This may be done by measuring the crack spacing and basing the calculation on the ACK model (Section 4.5), or by measuring the composite strength as a function of $V_f l/d$, and determining the average interfacial bond using the composite materials approach (Section 4.3). *Direct* measurements are based on the pull-out of a single fibre (or an array of parallel fibres), from which the load vs. displacement curve is obtained. Some of the various experimental techniques used to determine pull-out characteristics are described later, based largely on the critical review by Gray [72].

Schematic descriptions of the various configurations that may be (or have been) used for pull-out tests are shown in Figure 6.23 [72]. The tests are classified on

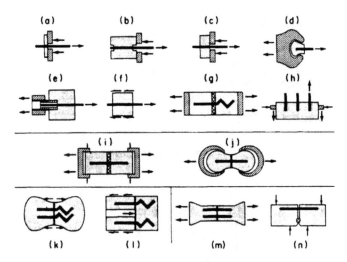

Figure 6.23 Configurations used for determining direct pull-out resistance, and their classification (after Gray [72]). (a)–(h) Single fibre, single sided; (i), (j) single fibre, double sided; (k), (l) multiple fibre, single sided; (m), (n) multiple fibre, double sided.

the basis of the number of fibres being pulled out simultaneously (single fibre or multiple fibres) and on the specimen configuration (single-sided or double-sided). In the single-sided configuration, fibre debonding and pull-out occur only on one side of the artificially induced crack, whereas in the double-sided test it may occur on either, or on both sides. In most of the test configurations shown in Figure 6.23 the pull-out is axial, and thus the results lend themselves more readily to analysis on the basis of the models described in Chapter 3, from which the average shear bond strength or the elastic and frictional interfacial bond components can be calculated. The only exception is the configuration in Figure 6.23(n) in which the pull-out is induced by flexural loading. Such a test was used by Mangat *et al.* [73], with the rationale that it better simulates pull-out and debonding in the actual composite, which is usually loaded in flexure. Another test configuration (not shown in Figure 6.23) is one in which the effect of fibre orientation on pull-out is determined, that is, the fibre, or array of parallel fibres, is inclined with respect to the direction of the pull-out load (e.g. [74]).

   Gray [72] evaluated these experimental techniques in terms of several criteria:

1   The stress–strain conditions in the tests specimen should simulate those which occur in the actual composite, that is, mainly tensile stresses in the matrix. Changes in the matrix stresses can affect the interfacial response due to various influences, such as the Poisson effect. This condition is not satisfied in

(a)

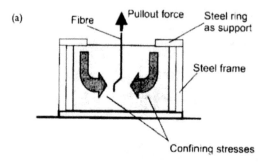

(b)

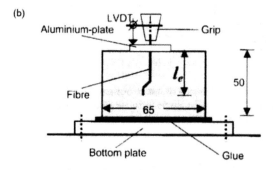

Figure 6.24 (a) Confining stresses due to supporting steel ring; (b) glued block produces very low lateral stresses (after Markovic et al. [75]).

techniques in which the load is applied to the surface of the matrix block from which the fibre is pulled out ((a)–(c) in Figure 6.23) leading to compression in the matrix. For instance, Markovic *et al.* [75] found, from a linear elastic finite element analysis, that the common test arrangement shown in Figure 6.24(a) leads to the presence of considerable arch-like compressive stresses, with a maximum value around the fibre hook. They therefore adopted the test set-up shown in Figure 6.24(b), in which the gluing of the concrete block to the bottom plate generates only very low lateral stresses. In the double-sided test configurations this criterion is usually satisfied, except where the gripping is of the briquette type (j).

2   The pull-out response is highly sensitive to the normal stresses developed across the interface, in particular those induced by drying shrinkage. To obtain reproducible results, these conditions must be kept constant, and they should represent the service conditions in the actual composite (Section 3.2.4).

3   It is advantageous to have a specimen shape which will allow testing of fibres or arrays of fibres at different orientations, as well as permit the application of radial loads to enable the simulation of the effect of normal stresses across the interface. A test configuration which allows an adjustment

to the radial loading during the pull-out test was described by Pinchin and Tabor [76].

4   The testing apparatus and techniques should be such that they minimize the variability. Although the properties measured are inherently variable, the testing arrangement, and in particular the gripping and alignment, may lead to considerable additional variations. Coefficients of variation in the range of 20–55% have been reported by Gray and Johnston [77].

As an addition to the usual load and displacement measuring devices that are normally used in pull-out tests, Weiler *et al.* [78] also used acoustic emission to monitor the failure of the fibre. It was found that this helped to clarify the interaction of their hooked end fibre with the matrix.

None of the pull-out tests that have so far been developed are ideal in terms of meeting all these criteria, though some come closer than others. It should also be noted that at this time, there are no *standard tests* available in any of the codes or specifications for fibre pull-out characteristics. Thus, the data reported by the various investigators are often not truly comparable.

A special problem arises with fibres composed of bundles of filaments which do not disperse in the actual composite, such as glass fibres. Here the reinforcing unit is not a single fibre surrounded by the cementitious matrix, but rather a bundle of filaments surrounded (and sometimes partially infiltrated) by the cementitious matrix, where the outer filaments are in direct contact with the matrix, while the inner filaments are less well bonded. Pull-out tests for such bundles have been developed by several investigators [79,80]. However, the interpretation of the test results in terms of interfacial bond stresses is impossible, because of the variable bonding in the outer and inner filaments. The approach usually taken in such cases is to express bond in terms of load per unit length rather than load per unit area. More recently, Trtik and Bartos [81] developed a microindentation method to carry out both push-in and push-out tests on composites reinforced with bundled fibres. With this technique, the bond characteristics of single filaments within a bundle could be assessed. Their experimental set-up is shown in Figure 6.25.

There is also a need to test the effects of high rates of loading on fibre pull-out. A testing procedure for that purpose, which enables the application of both high strain rates and impact loading, was developed by Gokoz and Naaman [52]. The specimen consists of two cylindrically shaped cement matrices, one hollow and one solid, bonded together by fibres. The mode of loading is shown in Figure 6.26; it can be applied by a testing machine or a drop-weight impact hammer.

Another approach to high loading rate fibre pull-out tests has been developed [82], in which an air gun is linked to the bottom of the specimen to create a high speed pull-out rate; the upper half of the specimen is suspended from a dynamic load cell. Pull-out rates of greater than 2 m/s can be achieved. The apparatus is shown in Figure 6.27 [83].

A different approach to pull-out and bond testing was developed by Wang *et al.* [84], in which the critical fibre length was determined, rather than pull-out

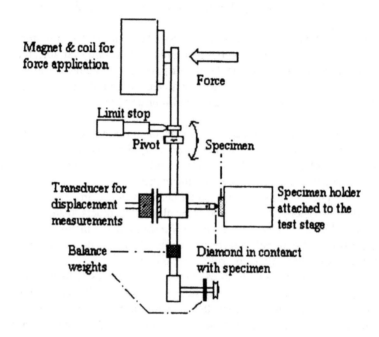

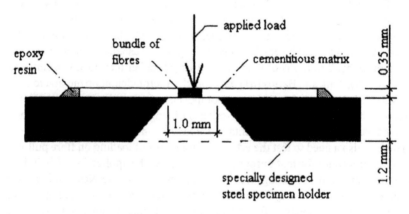

Figure 6.25 (a) Microindentation apparatus; (b) schematic diagram of push-out test (after Trtik and Bartos [81]).

resistance. The critical length test is based on the pulling out of an array of parallel fibres, varying in length, and determining the shortest length, ℓ_c, at which fibre fracture is observed (Figure 6.28). Assuming a uniform interfacial shear stress distribution, and knowing the fibre strength, σ_f, the average shear bond strength,

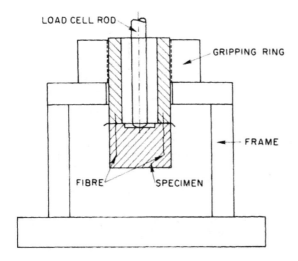

Figure 6.26 Pull-out testing arrangement for high rate loading (after Gokoz and Naaman [52]).

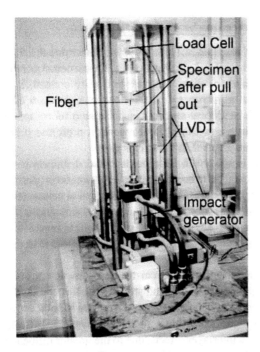

Figure 6.27 Dynamic fibre pull-out test set-up, and specimen geometry [83].

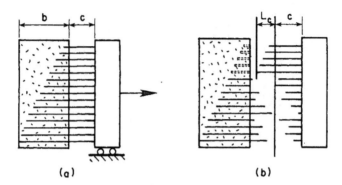

Figure 6.28 Bond strength determination by the critical length (L_c) method (a) before pull-out; (b) after pull-out (after Wang *et al.* [84]).

τ, can also be calculated, where d is the fibre diameter:

$$\tau = d\sigma_f/4\ell_c \qquad\qquad (6.8)$$

6.3.9   Determination of fibre content and distribution

Frequently, both in practice and in research, there is a need to determine the fibre content and fibre distribution in the FRC. Unfortunately, for hardened concrete, there are no standard test methods available. For routine quality control, it is possible to measure the average fibre *content* of the fresh mix (Section 6.2.2). However, the fibre *distribution* is seldom uniform, and because of compaction techniques and geometrical constraints within the formwork, a preferred (non-random) fibre orientation is the rule rather than the exception.

Many of the methods used to study the fibre content and distribution were developed for steel fibre reinforcement. In this case, the electromagnetic properties of the fibres can be used to develop test methods based on electrical measurement, such as that previously mentioned (Section 6.2.2) of Uomoto and Kobayashi [8]. In the course of their studies, they also found that the induced current is sensitive to fibre orientation, and there is thus the potential for developing a method to detect the extent of the anisotropy.

Electromagnetic measurements of fibre content using Pachometers (cover-meters) were reported by Malmberg and Skarendahl [85]. In this method, the measuring device is simply placed on the concrete surface, so there is no need for an external ring as in the method suggested by Uomoto and Kobayashi [8]. When using a Pachometer, one encounters interlayer effects, due to the greater influence on the readings of fibres closer to the surface, and boundary effects induced when measuring close to the specimen edge. By using interlayers of different thicknesses between the concrete surface and the measuring apparatus, the contribution of fibres at different distances from the surface could be made either weaker or stronger. However, to obtain quantitative fibre volume values, there is a need to calibrate the system against specimens with known fibre contents.

Another, more direct, approach to the measurement of fibre content and distribution is based on X-ray techniques. For instance, Ashley [86] described the use of X-ray fluorescence for the measure of glass fibre content, though this method also requires calibration against composites of known fibre volume. X-ray radiography has been used by a number of investigators to obtain a photographic representation of the projection on a plane of fibres in sections cut from the composite [87–89]. The thicknesses used were in the range of 10–50 mm. The clarity of the radiograph depends on the thickness of the specimen, the amount of input energy and the exposure time. The images obtained can then be quantified by various means, such as image analysis.

Stereological concepts can also be applied to the interpretation of these data (or of data obtained by a direct counting of fibre intersections with a sawn surface) to account for the 3D morphological characteristics. This development has been based on the extensive work of Stroeven [e.g. 88,90–92]. However, a detailed description of the stereological approach is beyond the scope of this book.

None of the methods mentioned earlier lend themselves well to synthetic fibres, which do not show up well on X-rays, do not affect the electromagnetic properties of the FRC significantly, and are difficult at best to count on a sawn surface. In addition, none of the methods lend themselves to routine quality control in the field; they remain confined primarily to laboratory research.

6.3.10 Shrinkage and cracking of hardened FRC

The problem of evaluating the shrinkage and cracking potential in hardened FRC involves the same concepts discussed earlier for the plastic shrinkage of fresh FRC. Here too, the evaluation of free shrinkage by methods such as ASTM C157 or ASTM C341 is not sufficient. The cracking sensitivity is a function of both the shrinkage strain and the improved toughness provided by the fibres. Thus, data which suggest that the presence of fibres may lead to reductions in free shrinkage (e.g. [9,10,93]) alone are not sufficient to establish the full potential of the fibres in reducing shrinkage cracking sensitivity in the hardened FRC. This can only be evaluated in test procedures in which restrained shrinkage is measured (as for plastic shrinkage).

A ring test, similar to that described in Section 6.2.3, has also been used for hardened concrete. In this case, however, the concrete is allowed to cure, and the external ring and the base are removed before drying is initiated. When the upper surface is sealed, the drying is only from the outer surface, in the radial direction. As the concrete shrinks, compressive stresses are induced in the steel ring, while tensile stresses develop in the concrete. By monitoring the strains in the steel ring, the average tensile stress in the shrinking concrete, σ_t, can be calculated from equilibrium considerations [93–95]:

$$\sigma_\theta = (r_e^2/r^2 + 1)p/(r_e^2/r_i^2 - 1) \tag{6.9}$$

$$\sigma_R = -(r_e^2/r^2 - 1)p/(r_e^2/r_i^2 - 1) \tag{6.10}$$

$$p = E\varepsilon_{0s}/[(r_e^2 + r_i^2)/(r_e^2 - r_i^2) + \mu] \tag{6.11}$$

where r, θ is the radius and angle of the point were stress is calculated; p, the internal pressure in the ring; σ_θ, the circumferential stress at a radius r; σ_r, the radial stress at a radius r; r_e, the external radius; r_i, the internal radius; ε_{0s}, the shrinkage strain; E, the modulus of elasticity of concrete; μ, the Poisson ratio.

Kovler et al. [94] have taken into consideration the modulus of elasticity of the restraining internal ring, in addition to that of the concrete:

$$\sigma_r/\varepsilon_{0s}E = -[r_i^2/r_e^2 - r_i^2/r^2)k/2] \tag{6.12}$$

$$\sigma_\theta/\varepsilon_{0s}E = -[r_i^2/r_e^2 + r_i^2/r^2)k/2] \tag{6.13}$$

where

$$k = [e(1 + \mu_0) + 1 - \mu_0]/[e(1 + \mu_0) + (1 - \mu_0)(r_i^2/r_e^2)]$$

$$e = \{1 + [(E_0/E_1)(1 - \mu_1)/(1 + \mu_0)]\}/\{1 - [(E_0/E_1)(1 - \mu_1)/(1 - \mu_0)]\}$$

where r is the radius of the point where stress is calculated; σ_θ, the circumferential stress at a radius r; σ_r, the radial stress at a radius r; r_i, the internal radius; r_e, the external radius; ε_{sh}, the shrinkage strain; E_0, the modulus of elasticity of the concrete; E_1, the modulus of elasticity of the core material; μ_0, the Poisson ratio of the concrete and μ_1, the Poisson ratio of the core ring material.

In view of the non-uniformity of the stress field in such tests, their common use has for a long time been limited to comparative purposes, to evaluate, for example, the efficiency of different methods applied to reduce cracking sensitivity.

However, in recent years the ring test has been developed into a quantitative test which can provide output having physical significance. Gryzbowski and Shah [96] calculated the stresses in hardened concrete which would be expected to develop when it is tested in a ring apparatus with concrete dimensions of internal and external diameters of 152 and 187 mm, respectively, and a steel ring core which is of a tube shape, with external and internal diameters of 152 and 127 mm, respectively. They concluded that the difference between the tangential radial stress at the inner and outer surfaces of the concrete ring is 10% and the radial compressive stress is less than 20% of the tensile tangential stress. Therefore they suggested that a geometry of this kind practically provides a uniform state of stress. Based on this concept, the ring test was developed into a quantitative tool (ASTM C-1581). This test can be used to determine the stress build-up in the concrete during restrained shrinkage as well as the time to cracking, with the possibility of considering the viscoelastic behaviour of the concrete which acts to relieve the stress development [97,98]. An overview of additional tests for restrained shrinkage is provided in Bentur [99].

Swamy and Stavrides [93] found that at the onset of visible cracking, the gauge readings were influenced by the location of the crack and its width. In the FRCs, the appearance of additional cracks caused some stress relaxation and the gauge readings gradually reduce. In the concretes without fibres, the formation of the first crack led immediately to total failure, that is, the gauge reading dropped suddenly to zero.

It is interesting to note the differences in the data obtained from free and restrained shrinkage tests. For example, from the results reported by Malmberg and Skarendahl [10] it can be seen that the addition of 1% by volume of steel fibres led to a reduction of only about 10% in the free shrinkage, while the total crack width in a ring test was reduced by about 80%. Thus, the information provided by the two types of tests with respect to the effect of fibres can be quite different.

Nevertheless, a free shrinkage test is sometimes useful, especially when different matrix formulations are expected to have a profound effect on performance. This is particularly the case with thin sheet components, such as glass fibre reinforced cements, in which the cement content is quite high. Here, testing of free shrinkage may provide information regarding the effect of changing the sand:cement ratio on potential cracking problems. RILEM Committee 49TFR [100], for instance, recommended not only a drying shrinkage test, but also the behaviour in wetting/drying cycles. It should be noted that with such materials, because of the complex shapes in which they may be cast, and due to restraining effects at the connections, it is often necessary to evaluate the shrinkage cracking sensitivity by testing the component itself, or to use a specialized test to simulate the special restraining effects (e.g. [101]).

6.3.11 Durability testing

Durability evaluation is of prime interest for FRC, as it is for plain concrete. The service life of concrete is expected to be many decades at least, and therefore there is a need to evaluate the long-term performance of the various FRC composites on the basis of accelerated tests; it is not possible to rely on data from natural exposure sites for a material which only came into widespread use in the 1970s. Since the causes for potential degradation may be different in each fibre reinforced cementitious composite, it is impossible to develop a single accelerated procedure that is universally applicable (see Chapter 5). Therefore, durability testing will be discussed separately in each of the chapters dealing with the different FRC systems. In this section only the general concepts in choosing or developing accelerated ageing tests will be presented. With the development of new types of fibres and their applications in composites in different climates and exposure conditions, a need often arises to devise a test for a particular material and application, and such tests are seldom detailed in the standards or specifications of the various agencies.

In developing accelerated ageing tests, two stages must be considered. First, the potential ageing mechanisms should be identified, before one can choose an appropriate means of accelerating them, such as temperature, moisture or radiation conditions. Second, the duration or number of cycles in the accelerated test should be translated into time in natural exposure conditions. The correlation between time in accelerated and natural ageing is not a unique function, since it depends on the climatic conditions in different zones, and even within the same zone there may be differences in the microclimate, for instance, the direction in which the component faces. Establishing correlations of this kind requires at least some limited time data from behaviour in exposure sites, which could be compared with the results of accelerated tests. However, even without this information, it is very useful to carry out accelerated tests, since they can provide an indication of whether there is an ageing problem, and how severe it might be. Also, accelerated tests can provide data for comparing the durability performance of different commercial products. A general outline for assessing long-term performance is provided in ASTM E632, *Standard Practice for Developing Accelerated Tests to Aid Prediction of the Service Life of Building Components and Materials.*

There are three classes of ageing processes to be considered for accelerated durability tests of FRC composites: ageing effects associated with (i) the matrix; (ii) the fibre and (iii) changes at the fibre–matrix interface.

Matrix problems, such as sulphate attack or freeze–thaw cycling are not unique to FRC composites, and they can be tested by the procedures developed for conventional mortars and concretes. Generally, the presence of fibres will have little effect on these chemical and physical processes.

The most common fibre problem is its sensitivity to alkali attack. This is a particular concern for glass, and for some natural and synthetic polymeric (organic) fibres. The most common and direct test is to expose the fibres to cement extract solutions at different temperatures, and to test the strength of the fibres before and after ageing. Since there is uncertainty as to how well the extract solution represents the actual pore solution in the cementitious matrix, various alternatives have been developed, in which the fibres are cast in a block of cement, and the whole assembly is tested before and after exposure to hot water [102].

Immunity of the fibres to alkali attack is not by itself sufficient to ensure adequate durability. Ageing effects which lead to densification of the matrix at the interface and to improved bond may cause a reduction in toughness (i.e. embrittlement), and under certain conditions may also result in reduced tensile or flexural strengths. This can be most efficiently evaluated by testing the properties of the composite before and after ageing. This is the most suitable way to proceed with durability testing, that is to obtain the overall effects associated with the combination of changes in the fibre itself and in the interfacial properties.

The change in properties by which the durability performance should be assessed must include strength as well as toughness. The choice of accelerated ageing conditions is important, and depends on the particular composite. For example, ageing in hot water was found to be appropriate for glass fibre reinforced cements

[103], while drying and wetting cycles were shown to simulate the ageing of natural fibre reinforced cements [104]. Carbonating conditions during ageing are thought to be necessary to simulate the ageing of cellulose–cement composites [105]. These conditions are not interchangeable, and an accelerating environment that is useful to simulate ageing in one composite may not be suitable for another. For example, Bergstrom and Gram [104] showed that the conditions developed for accelerated ageing of glass fibre reinforced cement (immersion in 50°C water) did not cause an ageing effect in a natural fibre-reinforced cementitious composite, in spite of the fact that the latter tended to undergo embrittlement after natural ageing. This embrittlement, however, could be simulated in an accelerated test involving wetting–drying cycles.

References

1. C.D. Johnston, 'Measures of the workability of steel fiber reinforced concrete and their precision', *Cem. Concr. Agg.* 6, 1984, 74–83.
2. ACI Committee 544, 'Measurement of properties of fiber reinforced concrete', *ACI Mater. J.* 85, 1988, 583–593.
3. E.P. Koehler and D.W. Fowler, *Summary of Workability Test Methods*, ICAR Report 105.1, The University of Texas at Austin, International Center for Aggregates Research, 2003, 92pp.
4. P.J.M. Bartos (ed.), *Fresh Concrete: Properties and Tests,* Elsevier Science Publishers, Amsterdam, 1992.
5. P.J.M. Bartos, M. Sonebi and A.K. Tamimi (eds), *Workability and Rheology of Fresh Concrete: Compendium of Tests,* RILEM Publications, Bagneux, 2002.
6. P. Balaguru and V. Ramakrishnan, 'Comparison of slump cone and V-B tests as measures of workability for fibre-reinforced and plain concrete', *Cem. Concr. Agg.* 9, 1987, 3–11.
7. PCI Committee on Glass Fiber Reinforced Concrete Panels, *Recommended Practice for Glass Fibre Reinforced Concrete Panels*, Prestressed Concrete Institute, Chicago, 1987.
8. T. Uomoto and K. Kobayashi, 'Measurement of fibre content of steel fiber reinforced concrete by electro-magnetic method', in G.C. Hoff (ed.) *Fiber Reinforced Concrete,* SP-81, American Concrete Institute, Farmington Hills, MI, 1984, pp. 233–246.
9. R.F. Zollo, J.A. Ilter and G.B. Bouchacourt, 'Plastic and drying shrinkage in concrete containing collated fibrillated polypropylene fibre', in R.N. Swamy, R.L. Wagstaffe and D.R. Oakley (eds) *Developments in Fibre Reinforced Cement and Concrete*, Proc. RILEM Symp., Sheffield, 1986, paper 4.5.
10. B. Malmberg and A. Skarendahl, 'Method of studying the cracking of fibre concrete under restrained shrinkage', in R.N. Swamy (ed.) *Testing and Test Methods of Fibre Cement Composites*, Proc. RILEM Symp., The Construction Press, UK, 1978, pp. 173–179.
11. H. Krenchel and S.P. Shah, 'Restrained shrinkage tests with pp-fiber reinforced concrete', in S.P. Shah and G.B. Batson (eds) *Fiber Reinforced Concrete: Properties*

and Applications, SP-105, American Concrete Institute, Farmington Hills, MI, 1987, pp. 141–158.

12. P.A. Dahl, 'Influence of fibre reinforcement on plastic shrinkage and cracking', in A.M. Brandt and I.H. Marshall (eds) *Brittle Matrix Composites – 1*, Proc. European Mechanical Colloquium 204, Elsevier Applied Science Publishers, London and New York, 1986, pp. 435–441.

13. P.P. Kraii, 'A proposed test to determine the cracking potential due to drying shrinkage of concrete', *Concrete Construction*, Sept., 1985, 775–778.

14. N. Banthia, C. Yan and S. Mindess, 'Restrained shrinkage cracking in fibre reinforced concrete: A novel test technique', *Cem. Concr. Res.* 26, 1996, 9–14.

15. N.S. Berke and M.P. Dallaire, 'The effect of low addition rates of polypropylene fibers on plastic shrinkage cracking and mechanical properties of concrete', in J.I. Daniel and S.P. Shah (eds) *Fiber Reinforced Concrete: Developments and Innovations*, SP-142, American Concrete Institute, Farmington Hills, MI, 1994, pp. 19–42.

16. C. Qi, J. Weiss and J. Olek 'Characterization of plastic shrinkage cracking in fiber reinforced concrete using semi-automated image analysis', *Mater. Struct. (RILEM).* 36, 2003, 386–395.

17. L. Zhang and S. Mindess, 'The compressive toughness of high strength fiber reinforced concrete', in N. Banthia, T. Uomoto, A. Bentur and S.P. Shah (eds) *Construction Materials*, Proc. ConMat '05 and Mindess Symp., University of British Columbia, Vancouver, Canada, 2005, CD-ROM.

18. RILEM TC 148-SCC, 'Test method for the measurement of the strain-softening behaviour of concrete under uniaxial compression', *Mater. Struct. (RILEM).* 33, 2000, 347–351.

19. S.P. Shah, P. Stroeven, D. Dahlusien and P. Van Stekelenburg, 'Complete stress-strain curves for steel fibre reinforced concrete in uniaxial tension and compression', in R.N. Swamy (ed.) *Testing and Test Methods of Fibre Cement Composites*, Proc. RILEM Symp., The Construction Press, UK, 1978, pp. 399–408.

20. C.D. Johnston and R.J. Gray, 'Uniaxial tensile testing of steel fibre reinforced cementitious composites', in R.N. Swamy (ed.) *Testing and Test Methods of Fibre Cement Composites*, Proc. RILEM Symp., The Construction Press, UK, 1978, pp. 451–461.

21. RILEM TC 162-TDF, 'Test and design methods for steel fibre reinforced concrete: Uni-axial tension test for steel fibre reinforced concrete', *Mater. Struct. (RILEM).* 34, 2001, 3–6.

22. S. Bar Shlomo, 'Bending of thin flexible fibre reinforced sheets', *Int. J. Cem. Comp. & Ltwt. Concr.* 9, 1987, 243–248.

23. C.D. Johnston, 'Definition and measurement of flexural toughness parameters for fiber reinforced concrete', *Cem. Concr. Agg.* 4, 1982, 53–60.

24. S. Mindess, J.F. Young and D. Darwin, *Concrete*, second edn., Prentice Hall, Upper Saddle River, NJ, 2003.

25. L. Chen, S. Mindess, D.R. Morgan, S.P. Shah, C.D. Johnston and M. Pigeon, 'Comparative toughness testing of fiber reinforced concrete', in D. Stevens, N. Banthia, V.S. Gopalaratnam and P.C. Tatnall (eds) *Testing of Fiber Reinforced Concrete*, SP-155, American Concrete Institute, Farmington Hills, MI, 1995 , pp. 41–75.

26. N. Banthia and S. Mindess, 'Toughness characterization of fiber-reinforced concrete: Which standard to use?', *J. Test. Eval.* 32, 2004, 138–142.

27. S. Mindess, L. Chen and D.R. Morgan, 'Determination of the first-crack strength and flexural toughness of steel fibre reinforced concrete', *J. Advanced Cement Based Materials.* 1, 1994, 201–208.

28. H. Xu, S. Mindess and N. Banthia, N. 'Toughness of polymer modified, fiber reinforced high strength concrete: beam tests vs. round panel tests', in *Advances in Concrete through Science and Engineering*, 2004, Hybrid-Fiber Session, Paper 20, CD-ROM.

29. P. Balaguru, R. Narahari and M. Patel, 'Flexural toughness of steel fiber reinforced concrete', *ACI Mater. J.* 89, 1992, 541–546.

30. N. Banthia and A. Dubey, 'Measurement of flexural toughness of fiber reinforced concrete using a novel technique, Part 1: assessment and calibration', *ACI Mater. J.* 96, 1999, 651–656.

31. N. Banthia and A. Dubey, 'Measurement of flexural toughness of fiber reinforced concrete using a novel technique, Part 2: performance of various composites', *ACI Mater. J.* 97, 2000, 3–11.

32. E.S. Bernard, 'Behaviour of round steel fibre reinforced concrete panels under point loads', *Mater. Struct. (RILEM).* 33, 2000, 181–188.

33. E.S. Bernard, 'Correlations in the behaviour of fibre reinforced shotcrete beam and panel specimens', *Mater. Struct. (RILEM).* 35, 2002, 156–164.

34. D. Dupont and L. Vandewalle, 'Comparison between the round plate test and the RILEM 3-point bending test', in M. di Prisco, R. Felicetti and G.A. Plizzari (eds) *Fibre-Reinforced Concretes, BEFIB 2004*, RILEM Proceedings PRO 39, RILEM Publications, Bagneux, 2004, Vol. 1, pp. 101–110.

35. RILEM TC 162-TDF, 'Test and design methods for steel fibre reinforced concrete – bending test', *Mater. Struct. (RILEM).* 35, 2002, 579–582.

36. J.A.O. Barros, V.M.C.F. Cunha, A.F. Ribeiro and J.A.B. Antunes, 'Post-cracking behaviour of steel fibre reinforced concrete', *Mater. Struct. (RILEM).* 38, 2005, 47–56.

37. L. Vandewalle, 'Test and design method for steel fibre reinforced concrete based on the σ-ε relation', in A.M. Brandt (ed.) *Some Aspects of Design and Application of High Performance Cement Based Materials*, Lecture Notes 18, Institute of Fundamental Technological Research, Warsaw, 2004, pp. 135–190.

38. EFNARC, *European Specification for Sprayed Concrete*, 1996.

39. E.S. Bernard, 'Correlations in the performance of fibre reinforced shotcrete beams and panels', *Engineering Reports No. CE9 and CE15*, School of Civil Engineering and Environment, University of Western Sydney, Australia, 1999, 2000.

40. D.R. Morgan, S. Mindess and L. Chen, 'Testing and specifying toughness for fibre reinforced concrete and shotcrete', in N. Banthia and S. Mindess (eds) *Fiber Reinforced Concrete: Modern Developments*, University of British Columbia, Vancouver, Canada, 1995, pp. 29–50.

41. J.-F. Trottier, D. Forgeron and M. Mahoney, 'Influence of construction joints in wet-mix shotcrete panels', *Shotcrete*, Fall 2002, 26–30.

42. D.L. Grote, S.W. Park and M. Zhou, 'Dynamic behavior of concrete at high strain rates and pressures : I. experimental characterization', *Int. J. Impact Engineering.* 25, 2001, 869–886.

43. L.J. Malvar and C.A. Ross, 'Review of strain rate effects for concrete in tension', *ACI Mater. J.* 95, 1998, 735–739.

44. A.J. Zielinski and H.W. Reinhardt, 'Stress-strain behavior of concrete and mortar at high rates of tensile loading', *Cem. Concr. Res.* 12, 1982, 309–319.

45. N. Banthia, S. Mindess, A. Bentur and M. Pigeon, 'Impact testing of concrete using a drop weight impact machine', *Experimental Mechanics.* 29, 1989, 63–69.

46. W. Suaris and S.P. Shah, 'Strain rate effects in fibre reinforced concrete subjected to impact and impulsive loading', *Composites.* 13, 1982, 153–159.

47. Comité Euro-International du Béton, *Concrete Structures under Impact and Impulsive Loading – Synthesis Report*, CEB, Bulletin 187, 1988, 3.6.

48. N. Jones and S. Mindess, 'Experience and capabilities in precision impact testing', in T. Krauthammer, A. Jenssen and M. Langseth (eds) *Precision Testing in Support of Computer Code Validation and Verification*, Workshop Report, Fortifikatorisk Notat Nr 234/96, Norwegian Defence Construction Service, 1996, pp. 35–51.

49. P.H. Bischoff, 'Role of physical testing in impact analysis of concrete structures', in W. Bounds (ed.) *Concrete and Blast Effects*, SP-175, American Concrete Institute, Farmington Hills, MI, 1998, pp. 241–259.

50. N. Banthia, V. Bindiganavile and S. Mindess, 'Impact resistance of fiber reinforced concrete: a progress report', in A.E. Naaman and H.W. Reinhardt (eds) *High Performance Fiber Reinforced Cement Composites (HPFRCC4)*, RILEM Proceedings PRO 30, RILEM Publications, Bagneux, 2003, pp. 117–131.

51. A. Bentur, S. Mindess and N. Banthia, 'The behaviour of concrete under impact loading: experimental procedures and method of analysis', *Mater. Struct. (RILEM).* 19, 1986, 371–378.

52. U.N. Gokoz and A.E. Naaman, 'Effect of strain rate on the pull out behaviour of fibres in mortar', *Int. J. Cem. Comp.* 3, 1981, 187–202.

53. T.S. Lok, P.J. Zhao and G. Lu, 'Using the split Hopkinson pressure bar to investigate the dynamic behaviour of SFRC', *Mag. Concr. Res.* 55, 2005, 183–191.

54. A. Kustermann and M. Keuser, 'High strength fibre reinforced concrete (HSFRC) under high dynamic impact loading', in M. di Prisco, R. Felicetti and G.A. Plizzari (eds) *Fibre-Reinforced Concretes, BEFIB 2004*, RILEM Proceedings PRO 39, RILEM Publications, Bagneux, 2004, Vol. 2, pp. 1217–1226.

55. A.N. Dancygier and D.Z. Yankelevsky, 'Effects of reinforced concrete properties on resistance to hard projectile impact', *ACI Struct. J.* 96, 1999, 259–267.

56. W.L. Server, 'Impact three point bend testing for notched precracked specimens', *J. Test. Eval.* 6, 1978, 29–34.

57. W. Suaris and S.P. Shah, 'Inertial effects in the instrumented impact testing of cementitious composites', *Cem. Concr. Agg.* 3, 1981, 77–83.

58. N. Banthia, S. Mindess and A. Bentur, 'Impact behaviour of concrete beams', *Mater. Struct. (RILEM).* 20, 1987, 293–302.

59. P. Gupta, N. Banthia and C. Yan, 'Fiber reinforced wet-mix shotcrete under impact', *ASCE J. Materials in Civil Eng.* 12, 2000, 102–111.

60. E.P. Chen and G.C. Sih, 'Transient response of cracks to impact loads', in *Mechanics of Fracture, Vol. 4: Elasto-Dynamic Crack Problems*, Nordhoff, Groningen, The Netherlands, 1977, pp. 1–58.

61. V.S. Gopalaratnam, S.P. Shah and R. John, 'A modified instrumented Charpy test for cement based composites', *Experimental Mechanics.* 24, 1984, 102–111.

62. S. Mindess, 'Crack velocities in concrete subjected to impact loading', *Can. J. Phys.* 73, 1995, 310–314.

63. V.S. Bindiganavile, *Dynamic Fracture Toughness of Fiber Reinforced Concrete.* PhD Thesis, University of British Columbia, Vancouver, Canada, 2003.

64. V.S. Bindiganavile and N. Banthia, 'Machine effect in the drop-weight impact testing of plain concrete beams', in N. Banthia, K. Sakai and O.E. Gjorv (eds), *Concrete under Severe Conditions*, (CONSEC'01), University of British Columbia, Vancouver, Canada, 2001, Vol. 1, pp. 589–596.

65. P. Sukontasukkul, *Impact Behaviour of Concrete under Multiaxial Loading*, PhD Thesis, University of British Columbia, Vancouver, Canada, 2001.

66. S. Mindess and K.-A. Rieder, 'Fracture behavior of biaxially confined concrete due to impact loading', in M. Langseth and T. Krauthammer (eds) *Transient Loading and Response of Structures*, Fortifikatorisk Notat Nr 257/98, Norwegian Defence Construction Service, pp. 151–160.

67. S. Mindess and A. Bentur, 'A preliminary study of the fracture of concrete beams under impact loading using high speed photography', *Cem. Concr. Res.* 15, 1985, 474–484.

68. S. Mindess, N. Banthia, J.P. Skalny and A. Ritter, 'Crack development incementitious materials under impact loading', in S. Mindess and S.P. Shah (eds) *Cement-Based Composites: Strain Rate Effects on Fracture*, Materials Research Society Symposia, Pittsburgh, PA, Vol. 64, 1986, pp. 217–224.

69. N.P. Banthia, *Impact Resistance of Concrete*, PhD Thesis, University of British Columbia, Vancouver, Canada, 1987.

70. V.S. Gopalaratnam and S.P. Shah, 'Strength, deformation and fracture toughness of fiber cement composites at different rates of flexural loading', in S.P. Shah and A. Skarendahl (eds) *Steel Fiber Concrete*, Elsevier Applied Science Publishers, London and New York, 1986, pp. 299–331.

71. S. Mindess, N. Banthia and A. Bentur, 'The response of reinforced concrete beams with a fibre concrete matrix to impact loading', *Int. J. Cem. Comp. and Ltwt. Concr.* 8, 1986, 165–170.

72. R.J. Gray, 'Experimental techniques for measuring fibre/matrix interfacial bond shear strength', *Int. J. Adhesion and Adhesives.* 3, 1983, 197–202.

73. P.S. Mangat, M. Motamedi-Azari and B.B. Shakar-Ramat, 'Steel fibre-cement matrix interface bond characteristics under flexure', *Int. J. Cem. Comp. and Ltwt. Concr.* 6, 29–37.

74. A.E. Naaman and S.P. Shah, 'Pull out mechanisms in steel-fibre reinforced concrete', *ASCE J. Struct. Div.* 102, 1976, 1537–1548.

75. I. Markovic, J.C. Walraven and J.G.M. van Mier, 'Experimental evaluation of fibre pullout from plain and fibre reinforced concrete', in A.E. Naaman and H.W. Reinhardt (eds) *High Performance Fiber-Reinforced Cement Composites (HPFRCC4)*, RILEM Proceedings PRO 30, RILEM Publications, Bagneux, 2003, pp. 419–436.

76. D.J. Pinchin and D. Tabor, 'Inelastic behaviour in steel wire pull out from portland cement mortar', *J. Mater. Sci.* 13, 1978, 1261–1266.

77. R.J. Gray and C.D. Johnston, 'The measurement of fibre-matrix interfacial bond strength in steel fibre reinforced cementitious composites', in R.N. Swamy (ed.) *Testing and Test Methods of Fibre Cement Composites*, Proc. RILEM Symp., The Construction Press, London, 1978, pp. 317–328.

78. B. Weiler, C. Grosse and H.W. Reinhardt, 'Debonding behavior of steel fibres with hooked ends', in H.W. Reinhardt and A.E. Naaman (eds) *High Performance Fiber Reinforced Cement Composites (HPFRCC 3)*, RILEM Proceedings PRO 6, RILEM Publications, Bagneux, 1999, pp. 423–433.

79. P. Bartos, 'Bond in glass reinforced cements', in P. Bartos (ed.) *Bond in Concrete*, Proc. Int. Conf., London: Applied Science Publishers, 1982, pp. 60–72.

80. V. Laws, A.A. Langley and J.M. West, 'The glass fibre/cement bond', *J. Mater. Sci.* 21, 1986, 289–296.

81. P. Trtik and P.J.M. Bartos, 'Assessment of micromechanical properties of cementitious composites by microindentation', in H.W. Reinhardt and A.E. Naaman (eds) *High Performance Fiber Reinforced Cement Composites (HPFRCC 3)*, RILEM Proceedings PRO 6, RILEM Publications, Bagneux, 1999, pp. 477–486.

82. V.S. Bindiganavile and N. Banthia, 'Polymer and steel fiber-reinforced cementitious composites under impact loading—Part 1: bond-slip response', *ACI Mater. J.* 98, 2001, 10–16.

83. H. Xu and S. Mindess, 'Dynamic pullout performance of steel fibers in polymer modified mortar', in N. Banthia, T. Uomoto, A. Bentur and S.P. Shah (eds) *Construction Materials*, Proc. ConMat '05 and Mindess Symp., University of British Columbia, Vancouver, 2005, CD-ROM.

84. Y. Wang, S. Backer and V.C. Li, 'An experimental study of synthetic fibre reinforced cementitious composites', *J. Mater. Sci.* 22, 1987, 4281–4291.

85. B. Malmberg and A. Skarendahl, 'Determination of fibre content, distribution and orientation in steel fibre concrete by electromagnetic technique', in R.N. Swamy (ed.) *Testing and Test Methods of Fibre Cement Composites*, Proc. RILEM Symp., The Construction Press, UK, 1978, pp. 289–296.

86. D.G. Ashley, 'Measurement of glass content in fibre cement composites by X-ray fluorescence analysis', in R.N. Swamy (ed.) *Testing and Test Methods of Fibre Cement Composites*, Proc. RILEM Symp, The Construction Press, UK, 1978, pp. 265–274.

87. P. Stroeven and R. Babut, 'Wire distribution in steel wire reinforced concrete', *Acta Stereol.* 5, 1986, 383–388.

88. P. Stroeven and S.P. Shah, 'Use of radiography-image analysis for steel fibre reinforced concrete', in S.P. Shah and G.B. Batson (eds) *Fiber Reinforced Concrete Properties and Application*, SP-105, American Concrete Institute, Farmington Hills, MI, 1987, pp. 275–288.

89. J. Kasperkiewicz, B. Malmberg and A. Skarendahl, 'Determination of fibre content, distribution and orientation in steel fibre concrete by X-ray technique', in S.P. Shah and G.B. Batson (eds) *Fiber Reinforced Concrete Properties and Application*, SP-105, American Concrete Institute, Farmington Hills, MI, 1987, pp. 297–306.

90. P. Stroeven, 'Morphometry of fibre reinforced cementitious materials', *Mater. Struct.* 12, 1979, 9–20.

91. P. Stroeven and Y.M. de Haan, 'Structural investigations on steel fibre concrete by stereological methods', in H.W. Reinhardt and A.E. Naaman (eds) *High Performance Fibre Reinforced Cement Composites,* London: E&FN SPON, 1992, pp. 407–418.

92. P. Stroeven, 'Steel fibre reinforcement at boundaries in concrete elements', in H.W. Reinhardt and A.E. Naaman (eds) *High Performance Fiber Reinforced Cement Composites (HPFRCC 3)*, RILEM Proceedings PRO 6, RILEM Publications, Bagneux, 1999, pp. 413–422.

93. R.N. Swamy and H. Stavrides, 'Influence of fibre reinforcement on restrained shrinkage and cracking', *J. Amer. Conc. Inst.* 73, 1979, 443–460.

94. K. Kovler, J. Sikuler and A. Bentur, 'Restrained shrinkage tests of fibre reinforced concrete ring specimens: effect of core thermal expansion', *Mater. Struct.* 26. 1993, 231–237.

95. S.P. Shah, C. Ouyang, S. Marikunte, W. Yang and E.A. Becq-Giraution, 'Method to predict shrinkage cracking of concrete', *ACI Mater. J.* 94, 1998, 339–346.

96. M. Gryzbowski and S.P. Shah, 'Shrinkage cracking of fiber reinforced concrete', *ACI Mater. J.* 87, 1990, 395–404.

97. A.B. Hossain and W.J. Weiss, 'Assessing restrained stress development and stress relaxation in restrained concrete ring specimens', *Cem. Concr. Compos.* 26, 2004, 531–540.

98. H.T. See, E.K. Attiogbe and M.A. Miltenberger, 'Shrinkage cracking characteristics of concrete using ring specimens', *ACI Mater. J.* 100, 2003, 239–245.

99. A. Bentur, 'Early age cracking tests', in A. Bentur (ed.) *Early Age Cracking in Cementitious Systems: State of the Art*, RILEM Report 25, RILEM Publications, Bagneux, 2003, pp. 275–294.

100. RILEM TC 49FTR, 'Testing and test methods for fiber reinforced cement based composites', *Mater. Struct. (RILEM).* 17, 1984, 441–456.

101. E. Tesfaye, L.L. Clarke and E.B. Cohen, 'Test method of measuring moisture movements in fibre concrete panels', in R.N. Swamy (ed.) *Testing and Test Methods of Fibre Cement Composites*, Proc. RILEM Symp., The Construction Press, UK, 1978, pp. 159–172.

102. K.L. Litherland, P. Maguire and B.A. Proctor, 'A test method for the strength of glass fibres in cement', *Int. J. Cem. Comp. and Ltwt. Concr.* 6, 1987, 39–45.

103. K.L. Litherland, D.R. Oakley and B.A. Proctor, 'The use of accelerated ageing procedures to predict the long term strength of GRC composites', *Cem. Concr. Res.* 11, 1981, 455–466.

104. S.G. Bergstrom and H.-E. Gram, 'Durability of alkali-sensitive fibres in concrete', *Int. J. Cem. Comp. and Ltwt. Concr.* 6, 1984, 75–80.

105. S.A.S. Akers and J.B. Studinka, 'Ageing behaviour of cellulose fibre cement composites in natural weathering and accelerated tests', *Int. J. Cem. Comp. and Ltwt. Concr.* 11, 1989, 93–97.

Part II

Cementitious composites with different fibres

Chapter 7

Steel fibres

7.1 Introduction

The early theoretical studies of FRC in the 1960s [e.g. 1,2] dealt primarily with the behaviour of steel fibre reinforced concrete (SFRC). Since then, SFRC has become the most commonly-used fibre concrete, though it is fast being over-taken by synthetic fibre reinforced concrete (see Chapter 10). Steel fibres greatly increase the toughness of cements and concretes. Originally, they were used primarily for *crack control*, to replace the secondary reinforcement often used for that purpose in flat slabs, pavements and tunnel linings, as well as in various repair applications. Today, while steel fibres are still used extensively for those purposes, they are also used increasingly in truly structural applications, either to replace conventional steel reinforcement, or to act in a complementary fashion with it (see Chapter 14).

The increase in toughness can prevent, or at least minimize, cracking due to changes in temperature or relative humidity, and can increase the resistance to dynamic loading (due to fatigue, impact, blast or seismic events). It should be noted, however, that improvements in strength due to fibre additions are very modest, except for high fibre volumes; the principal effect of fibres is to improve the post-peak load carrying capacity of the concrete (i.e. toughness), as shown in Figure 7.1 [3].

The requirements for steel fibres for SFRC have been specified in ASTM A820 [4]. Five types of fibres are defined, all of which should be small enough to be dispersed randomly in a concrete mixture:

1   Pieces of smooth cold-drawn wire;
2   Pieces of deformed cold-drawn wire;
3   Smooth or deformed cut sheet;
4   Melt-extracted fibres;
5   Mill-cut or modified cold-drawn wire.

The steels used for making fibres are generally carbon steels or alloy (stainless) steels; the latter are used primarily for corrosion-resistant fibres, in refractory applications and marine structures. Steel fibres may be manufactured in a number

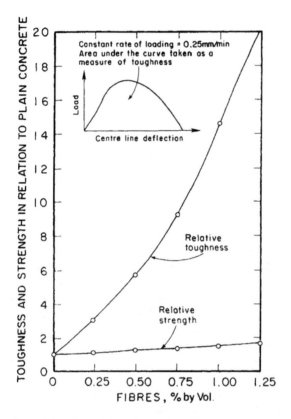

Figure 7.1 Effect of the volume of steel fibres on the strength and toughness of SFRC (after Shah and Rangan [3]).

of ways, depending on the fibre geometry desired. Depending on the type of steel and the particular production process, they may have tensile strengths in the range of 345–2100 MPa, and ultimate elongations of 0.5–35%. However, it has been found from pull-out experiments [5] that higher fibre yield strengths result in more severe matrix spalling around the fibre exit point, thus limiting the improvements in reinforcing efficiency found with higher fibre strengths.

It should be noted that straight, smooth fibres such as those used originally are now rarely seen, as they do not develop sufficient bond with the matrix; modern fibres have either rough surfaces, hooked or enlarged ends, or are crimped or deformed along their lengths, all of which are intended to improve the bond. *Round fibres* are produced by cutting or chopping wires, with diameters typically in the range of 0.25–1.0 mm. *Flat fibres* may be produced either by shearing sheets or flattening wire; cross-sectional dimensions are typically in the range of 0.15–0.40 mm thick, 0.25–0.90 mm wide. Crimped and deformed fibres of a

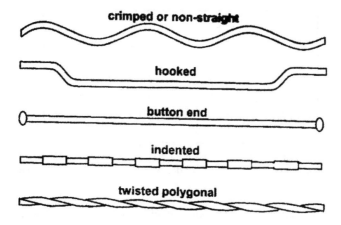

Figure 7.2 Commonly available deformed steel fibres [6].

number of different geometries have also been produced; the deformations may extend along the full length of the fibre, or be restricted to the end portions. Some examples of deformed fibres are shown in Figure 7.2 [6]. Fibres are also produced by the *hot melt extraction* process, in which a rotating wheel is brought in contact with the molten steel surface, lifting off some liquid metal which then solidifies (freezes) and is thrown off in the form of fibres. In order to make handling and mixing easier, some fibres are collated into bundles of 10–30 fibres using a water-soluble glue, which dissolves during the mixing process.

7.2 Technologies for producing SFRC

SFRC can, in general, be produced using conventional concrete practice, though there are obviously some important differences. The current technology for mixing, placing and finishing SFRC is described in detail in ACI 544.3R [7], which emphasizes these differences. ASTM C 1116 [8] also provides valuable information on the methods of specifying and manufacturing SFRC. The basic problem is to introduce a sufficient volume of *uniformly dispersed* fibres to achieve the desired improvements in mechanical behaviour, while retaining sufficient workability in the fresh mix to permit proper mixing, placing and finishing. As described earlier (Chapter 4), the performance of the hardened concrete is enhanced more by fibres with a higher aspect ratio, since this improves the fibre–matrix bond. On the other hand, a high aspect ratio adversely affects the workability of the fresh mix. In general, the problems of both workability and uniform distribution increase with increasing fibre length and volume. This is shown in Figure 7.3 [9].

It was these contradictory requirements that led to the development of the deformed fibres that are currently in use, with which bonding is achieved

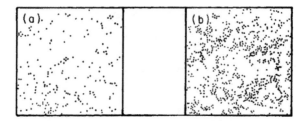

Figure 7.3 Section image of SFRC specimens containing (a) 0.5%; (b) 2.0% of steel wires. Wire density is lowest in the boundary layer, while wires have also moved somewhat towards the bottom side. The segregation is less pronounced in the 0.5% specimen [9].

largely through mechanical anchoring, which is much more efficient than the frictional shear bond stress mechanism associated with straight fibres. For instance, Mangat and Azari [10] have shown that the apparent coefficient of friction is about 0.09 for hooked end fibres, compared with only about 0.04 for straight fibres. Nonetheless, even with modern deformed fibres in the practical range of lengths (12–60 mm), and with modern superplasticizers, only up to about 2% of fibres can be incorporated using conventional concrete practice.

When steel fibres are introduced into a concrete mix, they generally have a detrimental effect on the packing density of the aggregates; this effect limits the maximum fibre content [11]. For plain concretes, the maximum packing density is obtained with about a 40% volume of fine aggregate. For low fibre content mixes (< about 0.5%), it is not really necessary to change the concrete mixture design from that used for plain concrete. However, for higher fibre volumes, the maximum packing density can be achieved only with a higher fines content. For instance, for 2% fibres, about a 60% fines content is required to achieve the maximum packing density. This effect is shown in Figure 7.4 [11]. Different fibres have different geometries, and therefore different packing densities of their own; this will affect the slope of the line in Figure 7.4. It should be noted that it is not really practical or possible to use the highest fibre contents (approaching 10%) since the fibres would interlock during mixing.

One of the chief difficulties in obtaining a uniform fibre distribution is the tendency for steel fibres to ball or clump together. *Clumping* may be caused by a number of factors:

- The fibres may already be clumped together before they are added to the mix; normal mixing action will not break these clumps down.
- Fibres may be added too quickly to allow them to disperse in the mixer.
- Too high a volume of fibres may be added.
- The mixer itself may be too worn or inefficient to disperse the fibres.
- Introducing the fibres to the mixer before the other ingredients will cause them to clump together.

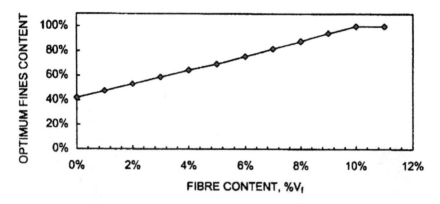

Figure 7.4 Fines content vs. fibre content for determination of optimum packing density (Harex fibres, 32 mm long, 0.9 mm diameter), from Hoy and Bartos [11].

In view of this, care must be taken in the mixing procedures. Most commonly, when using a transit mix truck or revolving drum mixer, the fibres should be added last to the wet concrete. The concrete alone, typically, should have a slump of 50–75 mm greater than the desired slump of the SFRC. The fibres should be added free of clumps, often by first passing them through an appropriate screen. (The use of collated fibres held together by water-soluble sizing which dissolves during mixing largely eliminates the problem of clumping.) Once the fibres are all in the mixer, about 30–40 revolutions at mixing speed should properly disperse the fibres [7]. Alternatively, the fibres may be added to the fine aggregate on a conveyor belt during the addition of aggregates to the concrete mix.

It should be noted that fibre balling or clumping usually occurs *before* the fibres are added to the mix. If they enter the mix free of balls, they tend to remain that way; if they enter in clumps, they will remain as clumps throughout the mixing process. Thus, fibre balling most commonly occurs either because the fibres are added too quickly to the mix, adding too many fibres, adding them to the mixer first before the other ingredients, or using worn out mixing equipment. Mixing for too long (overmixing) may also cause balls to form in the mixer.

SFRC can be placed adequately using normal concrete equipment. While it may appear to be very stiff because the fibres tend to inhibit flow, when *vibrated* the material will flow readily into the forms. It should be noted that water should be added to SFRC mixes to improve the workability only with great care, since above a w/c ratio of about 0.5, additional water may increase the slump of the SFRC without increasing its workability and placeability under vibration [7]. The finishing operations with SFRC are essentially the same as for ordinary concrete, as shown in Figure 7.5, though perhaps more care must be taken regarding workmanship.

Figure 7.5 Finishing steel fibre reinforced concrete. Photograph courtesy of the Bekaert Corporation.

A number of more specialized technologies for placing SFRC (and other FRCs), such as pumping, extrusion and fibre shotcreting, are described in some detail in Chapter 14. Slurry infiltrated fibre reinforced concrete (SIFCON) and slurry infiltrated mat concrete (SIMCOM) are described in Chapter 12. Though SIFCON was developed in 1984 [12], and SIMCON in 1992 [13,14], neither material has found much commercial use, perhaps because of the high costs involved in their production, and the lack of uniformity in the fibre orientation [15].

As an interesting extension of the SIFCON process, Bentur and Cree [16] have prepared SFRC by impregnating steel wool by slurry infiltration. Even with relatively modest fibre volumes, up to about 2.4%, the flexural strength was doubled compared with plain mortars, and the flexural toughness increased by a factor of ~7. They suggested that this material might be suitable for thin sheet applications.

7.3 Mix design of SFRC

As with any other type of concrete, the mix proportions for SFRC depend upon the requirements for any particular project, in terms of workability, strength, durability and so on. For relatively small fibre volumes (less than ~0.5%), the conventional mix designs used for plain concrete, based on the normal strength and durability considerations, may be used without modification. However, for larger fibre volumes, mix design procedures which emphasize the workability of the SFRC should be used. A number of such procedures are available (e.g. 17–20). However, there are several considerations that are particular to SFRC.

Edgington et al. [21] showed that for a particular fibre type, the workability of the mix decreased as the size and quantity of the aggregate particles larger than 5 mm increased; the presence of aggregate particles less than 5 mm in size had little effect on the compacting properties of the mix. They proposed an equation with which to estimate the critical percentage of fibres which would just make the SFRC unworkable:

$$PWc_{crit} = 75 \cdot \frac{\pi \cdot SG_f}{SG_c} \cdot \frac{d}{l} \cdot K \tag{7.1}$$

where PWc_{crit} is the critical percentage of fibres (by weight of mix); SG_f, the specific gravity of fibres; SG_c, the specific gravity of concrete matrix; d/l, the inverse of fibre aspect ratio; K, the $W_m/(W_m + W_a)$ and W_m, is the weight of mortar fraction (particle size < 5 mm); W_a, the weight of aggregate fraction (particle size > 5 mm).

They recommended that, to permit proper compaction, the fibre content should not exceed 0.75 PWc_{crit}. Figure 7.6 shows the effect of maximum aggregate size on workability.

The second factor which has a major effect on workability is the aspect ratio (l/d) of the fibres. The workability decreases with increasing aspect ratio, as shown in Figure 7.7 [21]. In practice, it is very difficult to achieve a uniform mix if the aspect ratio is greater than about 100.

In general, to provide better workability, SFRC mixes contain higher cement contents and higher ratios of fine to coarse aggregate than do ordinary concretes. Thus, the mix design procedures that apply to conventional concrete may not be entirely applicable to SFRC. Commonly, both to improve workability and to reduce the quantity of cement, up to 35% of the cement may be replaced with fly ash. In addition, to improve the workability of high fibre volume mixes, water reducing admixtures and, in particular, superplasticizers are often used, in conjunction with

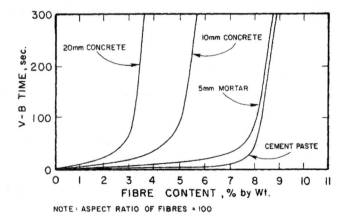

Figure 7.6 Workability versus fibre content for matrices with different maximum aggregate sizes [21]. Aspect ratio of fibres = 100.

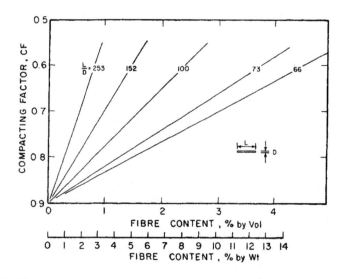

Figure 7.7 Effect of fibre aspect ratio on the workability of concrete, as measured by the compacting factor [21].

air entrainment. The typical range of proportions for normal weight SFRC is shown in Table 7.1 [7].

It must be emphasized that the values given in Table 7.1 are only 'typical'. Any given fibre must be evaluated on its own merits to find the optimum mix designs.

Table 7.1 Range of proportions for normal weight fibre reinforced concrete, adapted from ACI Committee 544 [7]

	9.5 mm maximum aggregate size	19 mm maximum aggregate size	38 mm maximum aggregate size
Cement (kg/m³)	355–600	300–535	280–415
w/c	0.35–0.45	0.35–0.50	0.35–0.55
Fine/coarse aggregate (%)	45–60	45–55	40–55
Entrained air (%)	4–8	4–6	4–5
Fibre content (%) by volume			
Smooth fibres	0.8–2.0	0.6–1.6	0.4–1.4
Deformed fibres	0.4–1.0	0.3–0.8	0.2–0.7

Note
1.0% steel fibres by volume have a weight of 78.5 kg/m³.

Table 7.2 Typical steel fibre reinforced shotcrete mixes [22]

Property	Fine aggregate mixture (kg/m³)	9.5 mm aggregate mixture (kg/m³)
Cement	446–558	445
Blended sand (< 6.35 mm)[a]	1483–1679	697–880
9.5 mm aggregate	–	700–875
Steel fibres[b,c]	35–157	39–150
Accelerator	Varies	Varies
w/c ratio	0.40–0.45	0.40–0.45

Notes
a The sand contained about 5% moisture.
b 1% steel fibres by volume = 78.6 kg/m³.
c Since fibre rebound is generally greater than aggregate rebound, there is usually a smaller percentage of fibres in the shotcrete in place.

For steel fibre reinforced shotcrete (see Chapter 14), different considerations apply, and most mix designs are still arrived at empirically. Typical mix designs for steel fibre shotcrete are given in Table 7.2 [22].

7.4 Mechanical properties of SFRC composites

The mechanical properties of SFRC are influenced by a number of factors:

1 Fibres: type
geometry
aspect ratio
volume
orientation
distribution

2   Matrix: strength
                maximum aggregate size
3   Specimen: size
                geometry
                method of preparation
                loading rate

As was described in detail in Chapter 3, fibres act through stress transfer from the matrix to the fibre by some combination of interfacial shear and mechanical interlock between the deformed fibre and the matrix. Up to the point of matrix cracking, the load is carried by both the matrix and the fibres; once cracking has occurred, the fibres carry the entire stress by bridging across the cracked portions of the matrix until they pull out completely, as shown in Figure 7.8. Thus, while fibres influence the properties of cementitious composites under all types of loading, they are particularly effective under direct tensile stresses, and in flexure, shear, impact and under fatigue loading; they are less effective under compressive loading.

Figure 7.8 Fracture surface of steel fibre reinforced concrete pipe; note that failure involves primarily pull-out of the fibres. Photograph courtesy of the Bekaert Corporation.

7.4.1   Fibre orientation and distribution

It must be remembered that, when examining or describing the effects of fibre additions, changes in SFRC properties are always expressed in terms of *average* fibre content. It is implicitly assumed that the fibres are uniformly distributed throughout the matrix, and, moreover, that they are randomly oriented. Unfortunately, neither assumption is likely to be correct after the SFRC has been placed and compacted by vibration [23], and this leads not only to a considerable amount of scatter in the test data, but also to a considerable variability in measured values due to the direction of loading (in relation to the direction of casting).

When using table vibration, fibres tend to become aligned in horizontal planes [24] as shown schematically in Figure 7.9 [25]. Internal vibration exerts a smaller influence on fibre alignment, but excessive vibration can also lead to preferred horizontal orientation of the fibres, and also segregation [24]. Fibres also generally show preferential alignment close to the bottom and sides of the forms, parallel to the forms. The 'wall effect' also leads to higher fibre concentrations near the faces of the specimens. (This suggests that in preparing small test specimens for research or quality control, it may be preferable to saw the specimens from larger samples, to eliminate these wall effects.) Insertion of a vibrator can also reduce the local fibre concentrations [24]. The type of vibration and the direction of casting can have a considerable effect on flexural strength, as shown in Figure 7.10 [26]. When the testing direction is perpendicular to the casting direction, specimens exhibit reductions in both flexural strength and toughness, compared with the case where casting and testing directions are parallel [27]. They found that this effect was stronger for more flowable mixes, which led to more fibre settlement during casting. However, little effect of fibre orientation is found on the dynamic modulus of elasticity, since elastic properties are measured at low strains, before any substantial cracking has taken place [26].

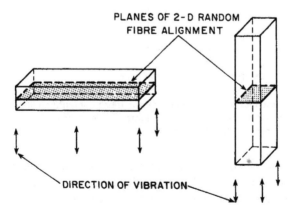

Figure 7.9 Effect of table vibration on fibre alignment (after Edgington and Hannant [25]).

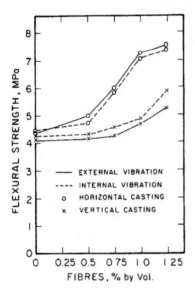

Figure 7.10 Influence of direction of casting and type of vibration on the flexural strength of SFRC (after Swamy and Stavrides [26]).

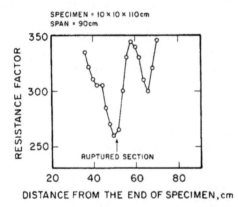

Figure 7.11 Distribution of resistance factor (fibre content/bending moment ratio) along the length of a beam. Note that fracture occurred at the point of minimum fibre content [28].

The fibres will generally show not only some preferential alignment, but also a non-uniform distribution along the length of a beam. This has been demonstrated clearly by electromagnetic measurements of fibre content carried out by Uomoto and Kobayashi [28], as shown in Figure 7.11.

Finally, on any cross section of a beam, the fibres are unlikely to be truly uniformly distributed, and this, too, will depend on the orientation of the cross

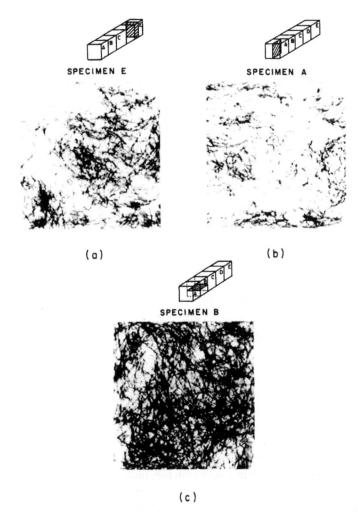

Figure 7.12   X-ray photographs of thin plates cut from specimens on planes perpendicular to the vibration plane and parallel to the failure plane [29].

section with respect to the direction of casting. This is shown clearly in X-ray photographs of fibre distribution, such as those in Figure 7.12 [29]. Figure 7.13 [29] shows the effect of the differences in distribution on the splitting tensile strengths, which could be related to the number of fibres intersecting the fractured section. This suggests that if fibres can be aligned uniaxially in some way, then the mechanical behaviour of the SFRC can be considerably enhanced, as long as the stress acts in the appropriate direction.

As has been shown by Vodicka *et al.* [30], a non-homogeneous fibre distribution becomes increasingly likely for very low fibre contents (less than about 25 kg/m^3

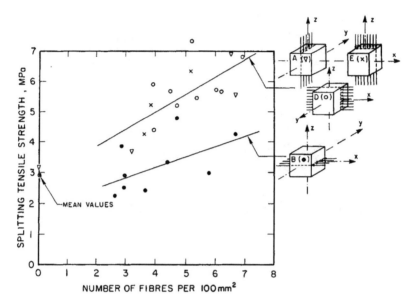

Figure 7.13 Splitting tensile strength as a function of the number of fibres intersecting the fractured section [29].

of fibres, corresponding to a volume percentage of about 0.3%). They suggested that for any particular fibre, there is a minimum fibre content required if a homogeneous distribution is to be obtained, which they estimated to be in the range of 30–40 kg/m³ (about 0.4–0.5%).

There have also been some attempts to model the effects of a non-uniform fibre distribution. For instance, based on a rigid-body-spring network of FRC, Bolander [31] showed that fibre distribution non-uniformity could lead to significant variations in the matrix stress of a specimen in tension, as shown in Figure 7.14. Bolander [32] also showed that, according to his model, the drying shrinkage of SFRC for aligned fibres would be about 10% less than for randomly distributed fibres.

Despite the above qualifications, it *is* quite possible to produce SFRC with an acceptably low variability in fibre distribution, as long as proper mixing, handling, placing and finishing procedures are followed, as shown, for example, by Taerwe *et al.* [33]. Indeed, Benson and Karihaloo [34] were able to achieve uniform fibre distributions of short steel fibres at volume fractions of up to 8%.

7.4.2 Fibre efficiency

It has been shown (Chapter 4) that, theoretically, the fibre efficiency (i.e. the resistance of the fibres to pull-out) increases with increasing aspect ratio. The bond

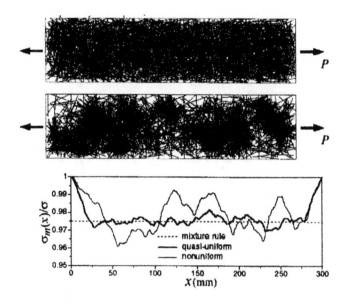

Figure 7.14 Matrix stress as affected by fibre distribution non-uniformity, for $V_f = 1\%$ [31].

Table 7.3 Pull-out bond stresses for fibres with different diameters [39]

Fibre type	Diameter (μm)	Length (mm)	Fibre strength (MPa)	Mean bond stress (MPa)
Plain straight	0.30	Various	1205	4.17
Indentions, straight	0.50	30	955	8.10
Plain, hooked ends	0.40	40	1355	4.93
Plain, weak crimped	0.35	30	1295	5.25
Plain, heavy crimped	0.40	25	1615	13.40
Plain, enlarged ends	0.3 × 0.4	14.5	510	7.27

strength also increases with increasing strength of the matrix (35–37). However, for smooth, straight fibres, beyond an aspect ratio of about 100, it is not generally possible to prepare a mix with both sufficient workability and uniform fibre distribution. Therefore, in a typical properly proportioned SFRC, failure is primarily by fibre pull-out, even with the deformed fibres shown in Figure 7.2.

Nonetheless, the post-cracking behaviour of SFRC depends heavily on the fibre geometry, since it is the fibre geometry that governs the pull-out force vs. displacement relationships of the fibres. This effect is shown in Figures 7.15 [35], Figure 7.16 [38] and in Table 7.3 [39].

Moreover, even the same type of fibre may exhibit different pull-out behaviours, depending on the matrix and the method of casting, as shown in Figure 7.17 [40].

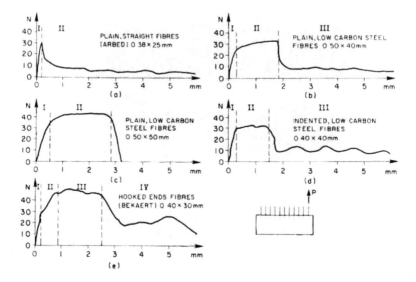

Figure 7.15 Typical load vs. slip curves for different types of fibres [35].

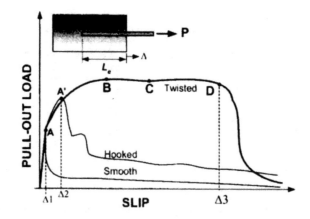

Figure 7.16 Typical pull-out vs. slip responses of smooth, hooked and twisted steel fibres [38].

Thus, different fibre geometries and fibre–matrix interactions can affect the flexural behaviour of SFRC. This is demonstrated in Figure 7.18 [41], which shows the effects of different fibre shapes on the flexural behaviour of steel fibre shotcrete. These data suggest that the aspect ratio (l/d) concept, which was developed for smooth, straight fibres, is not really useful when applied to deformed fibres.

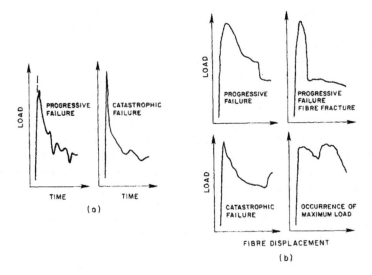

Figure 7.17 (a) Typical pull-out load vs. time patterns for plain, undeformed fibres; (b) typical pull-out load vs. fibre displacement curves for 'hooked' fibres [40].

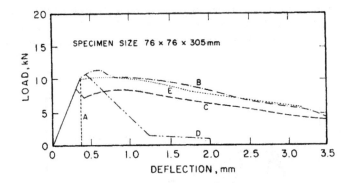

Figure 7.18 Flexural behaviour of steel fibre shotcrete prepared with fibres of different geometries [41]. A, No fibre; B, hooked ends, $L = 30$ mm, $d = 0.5$ mm, $\sigma_f = 1172$ MPa; C, hooked ends, $L = 30$ mm, $d = 0.4$ mm, $\sigma_f = 1172$ MPa; D, low carbon steel, 0.25 mm $\times$ 0.56 mm $\times$ 25.4 mm, $\sigma_f = 379$ MPa; E, brass-plated steel wire cord, $L = 18$ mm, $d = 0.25$ mm, $\sigma_f = 2958$ MPa.

7.5 Static mechanical properties

7.5.1 Compressive strength

Fibres do little to enhance the static compressive strength of concrete (e.g. [42–46]), with increases in strength ranging from essentially nil to perhaps 25%

for the normal range of fibre contents (<2%). Even in members which contain conventional reinforcement in conjunction with the steel fibres, the fibres have little effect on compressive strength [47, 48]. However, the fibres do substantially increase the post-cracking ductility or energy absorption of the material. These effects are shown graphically in the stress–strain curves of SFRC in Figure 7.19 [44]. Increasing the aspect ratio of the fibres also increases the compressive toughness, but again has little effect on strength, as shown in Figure 7.20 [44].

The effect of fibres on the compressive strength of ultra high strength concretes, such as the reactive powder concretes (RPC) described in Chapter 12, is less clear. While the fibres are added primarily to provide considerable ductility to these materials, they can also provide some significant strengthening at high enough fibre volumes. For instance, Karihaloo and de Vriese [49] found an increase in compressive strength from about 120 MPa to about 145 MPa (21%) on going from no fibres to 4% fibres by volume; similarly, Sun *et al.* [50] found an increase from about 150 MPa to about 200 MPa (33%) on going from no fibres to 4% fibres by volume.

Even extremely high fibre contents do not increase the compressive strength by as much as might be expected. For instance, Sun *et al.* [51] found that SIF-CON with fibre contents of up to 10% by volume had compressive strengths only in the range of about 25–50% higher than the plain concrete matrix; different fibre geometries led to quite different increases in strength. Similarly, using SIM-CON, Krstulovic-Opara [52] found an increase of about 30% in going from a fibre volume of 2.16% to 5.39%. It may be that for these very high fibre volume

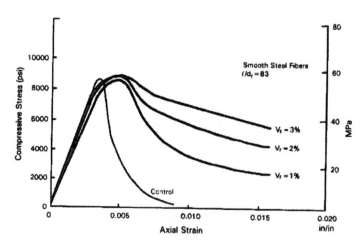

Figure 7.19 Effect of volume fraction of steel fibres on the stress–strain behaviour of concrete [44].

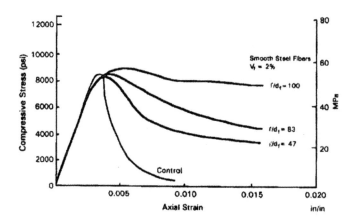

Figure 7.20 Effect of aspect ratio on stress–strain behaviour of SFRC [44].

materials, difficulties in achieving full compaction may lead to increased matrix porosities.

7.5.2 Tensile strength

Fibres *aligned* in the direction of the tensile stress may bring about very large increases in direct tensile strength, as high as 133% for 5% by volume of smooth, straight steel fibres [3]. However, for more or less randomly distributed fibres, the increase in strength is much smaller, ranging from as little as no increase in some instances [53] to perhaps 60%, with many investigations indicating intermediate values, as shown in Figure 7.21 [54]. Splitting tension tests of SFRC by Nanni [55] showed similar results, but Wafa and Ashour [45] found an increase in splitting tensile strength of 160% with 1.5% by volume of hooked steel fibres. Thus, adding fibres merely to increase tensile strength is probably not cost-effective. However, as in compression, steel fibres do lead to major increases in the post-cracking behaviour or toughness of the composites, by 1–2 orders of magnitude [56,57].

Very high fibre volumes, however, appear to be more effective in tension than in compression. Krstulovic-Opara [52] found about a 150% increase in the tensile strength of SIMCON on increasing the steel fibre content from 2.16% to 5.25%. With a fibre volume fraction of 5.29%, Krstulovic-Opara and Malak [14] and Krstulovic-Opara [52] were able to achieve SIMCON tensile strengths of 17 MPa. Similarly, Naaman and Homrich [58] were able to achieve tensile strengths of up to 28 MPa with SIFCON. Similarly high tensile strengths have also been obtained with the various reactive powder concretes (see Chapter 12)

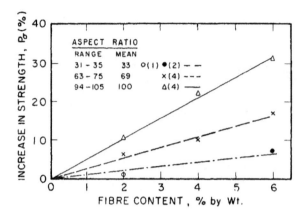

Figure 7.21 Influence of fibre concentration on tensile strength [54].

that have been developed, again due to the presence of high volume fractions of fibres.

7.5.3 Flexural strength

Steel fibres are often found to have a much greater effect on the flexural strength of SFRC than on either the compressive or tensile strengths, with increases of more than 100% having been reported. The increase in flexural strength is particularly sensitive not only to the fibre volume, but also to the aspect ratio of the fibres, with higher aspect ratios leading to larger strength increases. Thus, the flexural strength is often related to the term Wl/d, where l/d is the aspect ratio and W is the weight per cent of fibres. As would be expected, deformed fibres are more effective than straight, smooth fibres. It should be noted that for $Wl/d > 600$, the mix characteristics tend to become unsatisfactory due to inadequate workability or non-uniform fibre distribution [59].

For very high volume steel fibre concretes, the effect of the fibres on the flexural strength can be quite dramatic. For some reactive powder concretes (RPC), flexural strengths can reach 60 MPa with about 2.4% fibres by volume, and 102 MPa with 8% fibres [60]. Ductal®, another RPC, can achieve a flexural strength of about 45 MPa, with 2% by volume of 13–15 mm long fibres with diameters of 0.2 mm [61]; similarly, Sun *et al.* [62] obtained strengths of about 60 MPa with 4% by volume of 13 mm long by 0.175 mm diameter fibres.

7.5.4 Flexural toughness

As has been stated previously, fibres are added to concrete primarily to improve the toughness, or energy absorption capacity; any improvements in strength are of secondary importance. There are a number of different ways of defining and measuring the toughness of SFRC, and these have been described in detail in Chapter 6. Most commonly, the flexural toughness is defined as the area under the complete load–deflection (or stress–strain) curve in flexure; this is sometimes referred to as the total energy to fracture. Alternatively, the toughness may be defined as the area under the load–deflection curve out to some particular deflection, or out to the point at which the load has fallen back to some fixed percentage of the peal load. As is the case with flexural strength, flexural toughness also increases as the parameter Wl/d increases.

The load–deflection curves for different types and volumes of steel fibres can vary enormously, as was shown previously in Figure 6.8. For all empirical measures of toughness, fibres with better bond characteristics (i.e. deformed fibres or fibres with greater aspect ratios) give higher toughness than do smooth, straight fibres at the same volume concentrations. As well, many of the modern very high performance steel fibre composites, such as the reactive powder concretes, have a tensile stress–strain response that exhibits strain hardening, accompanied by multiple cracking; these will have a much higher toughness than the conventional, low fibre volume SFRCs which exhibit strain-softening behaviour beyond the peak load. These two types of behaviour are compared schematically in Figure 7.22 [63].

7.5.5 Shear and torsion

Steel fibres are widely recommended for use in conjunction with conventional steel reinforcement, in order to improve the shear resistance of structural elements (as discussed in some detail in Chapter 14). The fibres bridge across the cracks that form in the concrete matrix, and help to maintain the integrity of the material. This improvement in the residual strength of the matrix leads to an increase in the shear capacity (e.g. [64]).

However, the data are somewhat mixed on what happens to the shear strength of plain concrete when fibres are added. For instance, Barr [65] found that the shear strength of SFRC was independent of fibre content, though the toughness increased with increasing fibre content. On the other hand, Valle and Buyukozturk [66] found that steel fibres significantly increased the shear strength and ductility of concrete. This was particularly true for high strength concrete, due to the improved bond characteristics of the fibres in a high strength matrix. Similarly, Mirsayah and Banthia [67] found that steel fibres significantly improved both the shear strength and shear toughness. As well, Sun *et al.* [51] found a fourfold increase in shear strength (from about 4 MPa to 16.6 MPa) on going from plain concrete to a fibre volume of 2.5%.

The data are similarly scattered for torsion, with different studies showing anything from no increase in torsional strength [68] to a 100% increase [69], with other studies showing intermediate values.

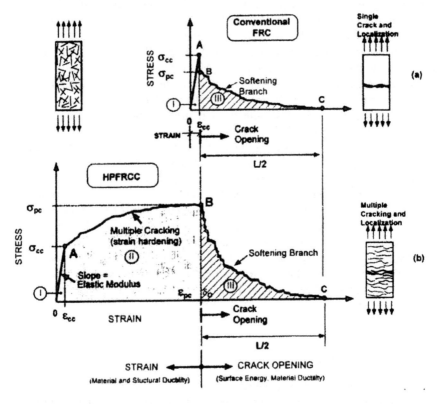

Figure 7.22 Comparison of typical stress–strain responses in tension of high performance fibre reinforced concrete and conventional FRC [63].

In both cases (shear and torsion), these different results are probably due to significant differences in specimen geometry and test procedures, since there are no 'standard' tests for these properties.

7.5.6 Multiaxial loading

Not a great deal of work has been carried out on the properties of SFRC under multiaxial (or confined) loading. Interestingly, Kosaka *et al.* [70] showed that, in uniaxial compression the confining effect of the steel fibres themselves was similar to that of a lateral confining pressure of slightly less than 1 MPa.

More generally, Brandt *et al.* [71] found that SFRC was stronger than plain concrete under triaxial loading, particularly when tensile stresses were involved.

Berthet *et al.* [72] found that the ultimate compressive strengths and strains of SFRC increased significantly with increasing lateral confinement, though the effect was somewhat less with higher strength concretes. Mander *et al.* [73] had observed a similar effect for plain concrete). The effects of confinement on SFRC under impact loading will be discussed in Section 7.7.

7.5.7 Abrasion, erosion, cavitation

SFRC is more resistant to abrasion and erosion than plain concrete [74], though the difference is not large. For scour by water containing debris at *low* velocities, SFRC is no better than plain concrete [75]; the rate of scour is determined by the quality of the aggregates and the hardness of the concrete. However, with *high* velocity flow which induces either cavitation or the impact of large debris, SFRC has been found to be significantly more effective than plain concrete [76–78].

Examples of the effectiveness of SFRC in this regard are the repairs carried out to the Dworshak, Libby and Tarbella dams [76,78]. All three structures were originally constructed with good, conventional concrete; they were all designed for discharge velocities as high as 30 m/s. These dams quickly exhibited severe damage due to cavitation and erosion. After repair with SFRC, the dams have performed satisfactorily. More recently, a 10-year old SFRC concrete lining on the Barr Lake Dam embankment in Colorado was examined [79]. No excessive wear or erosion was found, and the thin SFRC lining remained watertight.

7.5.8 Drying Shrinkage and creep

There is still not a great deal of shrinkage or creep data on SFRC. The conventional view, as expressed by Nawy [80] is that 'No appreciable improvement in the shrinkage and creep performance of concrete results from the addition of fibers'. This is, perhaps, a somewhat simplistic view. Mangat and Azari [10, 81] found that deformed fibres could reduce the *free shrinkage* of SFRC by up to 40%, with the reduction increasing with increasing fibre volume, as shown in Figure 7.23 [81]. The restraint depended on the fibre geometry, with deformed fibres being more effective than straight, smooth fibres. They developed an expression to predict the shrinkage of SFRC, which they related to the shrinkage of plain concrete as follows:

$$\varepsilon_{\mathrm{fs}} = \varepsilon_{\mathrm{os}}(1 - 2.45\mu V_{\mathrm{f}} l/d) \tag{7.2}$$

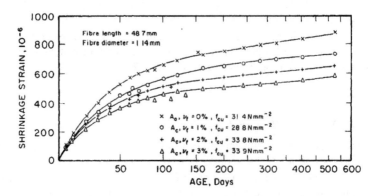

Figure 7.23 The influence of crimped steel fibres on the free shrinkage of concrete [81].

where ε_{fs}, is the free shrinkage of the SFRC; ε_{os}, the free shrinkage of the plain concrete; μ, the coefficient of friction between the fibres and the concrete (ranging from 0.04 for smooth, straight fibres to 0.12 for deformed fibres); V_f, the volume % of fibres; l/d, the aspect ratio.

Free shrinkage, however, is not really a useful indication of the effectiveness of fibres in reducing shrinkage problems. It is not the free shrinkage strain which needs to be reduced, but the cracking associated with *restrained shrinkage*. [Tests for restrained shrinkage have been described in Chapter 6.] For restrained shrinkage, steel fibres reduce the amount of cracking and the crack widths [82]; they found that drying shrinkage was reduced by about 15–20% by the presence of 1% of steel fibres. The beneficial effects of steel fibres on both the maximum crack widths and total crack widths are shown clearly in Figure 7.24 [83]. Indeed, Shah *et al.* [84] found that 0.5% by volume of steel fibres reduced the maximum crack widths by 80% and the average crack width by 90%. The fibre geometry also appears to be important; Chern and Young [85] found that higher aspect ratios led to reduced drying shrinkage. Chern and Chang [86] also showed that the combined use of fibres and silica fume was beneficial in reducing drying shrinkage.[1]

Mangat and Azari [87] also showed that steel fibres would be expected to have only a small effect on the creep of concrete. This is due to the fact that creep does not generally involve any macrocracking. However, Balaguru and Ramakrishnan [88] found that steel fibres slightly increased the creep of concrete, while Houde *et al.* [89] found that fibres increased the creep of concrete by 20–40%. On the other hand, Chern and Chang [86] found that, with the addition of silica fume, steel fibres reduced the creep of concrete. There is no obvious explanation for these contradictory results, except perhaps differences in the test procedures.

---

1 The issue of plastic shrinkage of fresh concrete will be addressed in detail in Chapter 13.

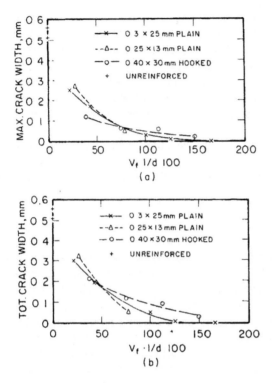

Figure 7.24 Effect of steel fibres on (a) maximum crack width; (b) total crack width, on the outer surface of ring specimens after 6 weeks of drying at 50% RH [83].

7.6 Dynamic mechanical properties

It has been found that steel fibres are particularly effective in concretes exposed to dynamic loading of any kind, such as fatigue, impact and blast.

7.6.1 Fatigue

There has been a considerable amount of work done on the fatigue properties of SFRC. Just as fibres lead to no significant improvements in static compressive strength, they also generally lead to no improvements in fatigue strength under compressive loading [42,90,91]. Contrary results, however, were presented by Paskova and Meyer [92], who found that steel fibres did improve the compressive fatigue behaviour. In direct tensile fatigue loading [91,93], considerable improvements in fatigue behaviour have been found, again mirroring the behaviour of SFRC under static tensile loading.

The bulk of the research has been carried out on flexural fatigue, and all studies agree that steel fibres do increase the fatigue properties of the concrete, in terms of higher endurance limits, finer cracks and much more energy absorption to failure. Higher fibre volumes, and deformed fibres (rather than straight, smooth ones), lead to greater improvements, as do higher aspect ratios [94]. For instance, Ramakrishnan et al. [95] reported that, with deformed fibres, the flexural endurance limit at 2×10^6 cycles was increased to 90–95% of the static strength (compared with about 55% for plain concrete). Similar results were obtained by Jun and Stang [96].

Steel fibres have been found to improve the fatigue performance of pavements [97], particularly when high performance steel fibre concretes are used [98]. They also appear to improve the fatigue properties of conventionally reinforced structural elements [99–101].

7.6.2 High strain rates

SFRC has much better properties under impact loading than does plain concrete, in terms of both strength and fracture energy. The role of the fibres is, essentially, the control of cracking by bridging across the cracks as they develop in the matrix. A variety of high strain rate tests have been used to evaluate the properties of SFRC, including instrumented drop weight tests, missile impact and, explosive loading, as described in some detail in Chapter 6. However, as indicated there, it is difficult to correlate the results of these various tests, as the high strain rate characteristics of SFRC appear to be strongly dependent on the details of the specimen geometry and the test arrangement. For instance, the mass of the drop hammer and its drop height [102], or the size and shape of an explosive charge, the specimen size and the specimen support conditions all affect the numerical data. Thus only a general description of the effects of high strain rates on SFRC can be presented here.

Shah and his colleagues, in an extensive set of experiments using an instrumented Charpy machine [103–107], showed that the fracture energy values of SFRC under impact loading were about 70–80% higher than those under static loading. Increases in fracture energy of SFRC were of the order of 40–100 times the value for plain concrete. In addition, impact strengths of SFRC were perhaps 50–100% higher than static strengths. Substantially similar results were obtained in a series of studies using a drop-weight impact machine [108,109]. The dynamic behaviour of plain concrete and SFRC in such a machine is shown in Figure 7.25 [108].

While the results of these earlier studies are still valid, more recent work has provided much more detail about the impact behaviour of SFRC. Pacios and Shah [110] showed that at higher rates of loading, the fibres developed more pull-out resistance and slip at the peak load. However, if the impact velocity is too great, the mode of failure of the fibres may change from pull-out to fracture, leading to a considerable drop in fracture energy [111,112]. This suggests that the fibre

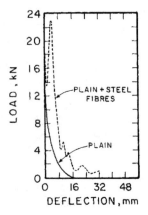

Figure 7.25 Behaviour of plain concrete and SFRC under impact loading, using an instrumented drop-weight impact machine [108].

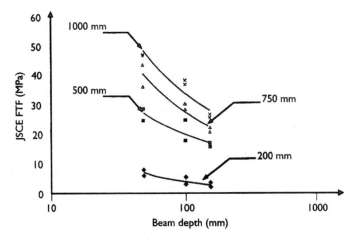

Figure 7.26 The JSCE SF4 toughness factor vs. beam depth for four different hammer drop heights [114].

geometry (which governs its pull-out characteristics) should be chosen carefully for any specific impact event.

Solomos and Berra [113] found that under compressive impact, SFRC exhibited a distinct size effect: the impact strength of 40 mm cubes was up to 25% higher than that of 60 mm cubes. Similar effects are found in flexure. Figure 7.26 [114] shows the effect of specimen size on the JSCE SF4 toughness factor (see Chapter 6) for four different drop heights of the impact hammer. The size effect appears to become more pronounced at higher drop heights (or impact velocities).

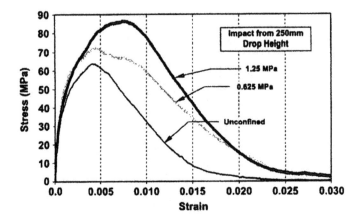

Figure 7.27 Effect of confinement on the compressive impact response of SFRC [116].

The capacity of the testing machine also has a very large effect on the impact response of SFRC. This has been seen in Figure 6.21 [114], in which beams were tested in three geometrically similar impact machines of very different capacities, with the impact velocity kept constant.

As one would expect, the impact strength and toughness generally increase with increasing fibre volume. If one goes to very high fibre volumes, as in compact-reinforced composites (CRC), further significant increases in strength and toughness are found [115].

If the SFRC is laterally confined, the strength and toughness also appear to increase significantly, as shown in Figure 7.27 [116] for compressive impact. Similar effects are found in confined flexural tests [117,118]. The mode of failure may also change under sufficient confining pressure, as shown in Figure 7.28 [116] for compressive impact, from the usual diagonal shear failure to vertical splitting.

Turning to more structural applications, Yan and Mindess [119–121] showed that steel fibres could considerably improve the bond between the matrix and conventional reinforcing bars under impact, primarily by inhibiting the growth of cracks emanating from the deformations (lugs) on the reinforcing bars. Also, Mindess *et al.* [122] showed that steel fibres could considerably improve the impact performance of prestressed concrete railroad ties.

These effects can be explained largely in terms of the more general rate-of-loading effects, since both plain concrete and SFRC are highly strain rate sensitive materials. For instance, Naaman [123] found that the compressive strength of SFRC increases by up to 40% with an increase in strain rate from 32 to 30,000 microstrain/s, while Rostasy and Hartwick [124] found that the ultimate compressive strength and strain in dynamic loading were about 20% higher than the corresponding static values.

(a)   (b)

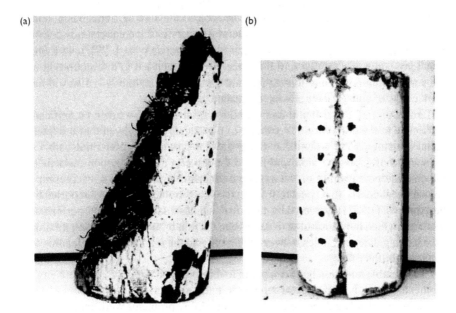

Figure 7.28 Failure of SFRC prisms under compressive impact loading: (a) unconfined;
(b) laterally confined [116].

Under tensile loading, as the strain rate was increased from 1.25×10^{-6}/s to 20/s, the tensile strength of SFRC was found to increase by about 70%, the maximum strain by about 25%, and the fracture energy by about 60% [125]. Under flexural loading, Naaman and Gopalaratnam [126] found a threefold increase in flexural strength and energy absorption with an increase in strain rate of about 5 orders of magnitude. Other investigators have obtained similar results, though the exact magnitudes of the increases in mechanical properties depend on the details of the experimental techniques employed.

7.7   Durability of SFRC

The discussion to this point has dealt only with the mechanical properties of SFRC. However, the durability of SFRC is at least as important, and there have been a number of studies in this area. At first glance, it would appear that steel fibres would be susceptible to severe corrosion, particularly near the surface of the concrete, where the cover is quite small. Since the diameter of the fibre is effectively reduced by corrosion, any substantial corrosion of the relatively thin fibre would lead to a considerable decrease in both the strength and the toughness of the SFRC, as has been confirmed experimentally by Kosa and Naaman [127]. They also found that the mode of failure could change from fibre pull-out to fibre fracture, which explains the decrease in toughness. However, Hoff [128] indicated that in practice this has not been the case. Generally, corrosion of the surface fibres has neither a significant adverse effect on the structural integrity of the concrete, nor does it lead to surface spalling. This was confirmed by Sustersic *et al.* [129], who found that corrosion was confined to the surface fibres, which led to discolouration of the surface, but not to any loss of flexural strength or toughness. It also did not lead to loss of underwater abrasion resistance.

Teruzzi *et al.* [130] found that the addition of steel fibres made no significant difference to the durability of concrete. In particular, they showed that the interfacial transition zone around the fibres did not provide a preferential path for the ingress of gases or fluids, and thus did not affect the carbonation resistance, the chloride permeability or the oxygen permeability; it also did not represent a zone of weakness with respect to frost resistance. Similar results were reported by Ferrara *et al.* [131]. However, they did find that brass-coated steel fibres performed less well, which they attributed to damage to the thin brass coating during mixing. This led to galvanic couples between the brass and the (exposed) steel, leading to corrosion of the fibres.

Considerable attention has been given to the corrosion of SFRC in severe environments, which is of particular interest in marine structures. According to Hoff [128], in completely submerged SFRC, no durability problems were encountered. In SFRC exposed in the splash zone, the corrosion was largely dependent on the extent of surface cracking. In sound, uncracked components, no adverse effects of the marine environment were observed. However, when cracks developed at the surface, ordinary carbon steel fibres tended to corrode [132,133]. Presumably, this could be prevented if stainless steel fibres were used.

The critical crack width below which corrosion of regular carbon steel fibres can be prevented has been estimated to be in the range of 0.10–0.25 mm [128,134–136]. It appeared that the extent of corrosion was a function of the initial crack width, with only limited corrosion occurring for hairline cracks. For instance, Corinaldesi and Moriconi [137] found that for a self-compacting SFRC, significant corrosion of steel fibres exposed to Cl^- ions did not occur, probably because the steel fibres also reduced the amount of drying shrinkage, and hence the width of the shrinkage cracks. Granju and Balouch [136] reported that the corrosion of fibres exposed to the surface did not extend more than 2–3 mm from the surface into the fibre.

Mangat and Gurusamy [138] showed that neither the generally accepted activation level of 0.41% Cl^- (by weight of cement), nor the value of 0.61 for the Cl^-/OH^- ratio for initiation of corrosion, apply to SFRC. These values can be greatly exceeded in SFRC before significant corrosion occurs. This is in agreement with the observations of Schupack [139] that penetration of chlorides into SFRC was not accompanied by corrosion, or that the corrosion was limited only to the surface fibres.

The much smaller sensitivity to corrosion of the steel fibres in SFRC compared with reinforcing bars in concrete can be accounted for by several mechanisms which were described in a review by Bentur [140]: (i) lower tendency to cracking and increased tendency for self-healing of cracks in FRC composites; (ii) small diameter of fibre which does not lead to the formation of sufficient rust for scaling of the concrete; (iii) lack of electrical conductivity because of the discrete nature of the steel fibre; (iv) improved matrix microstructure and denser interfacial transition zone in the fibres compared with conventional steel. An additional mechanism was suggested by Mangat and Molloy [141], that the enhanced durability performance of steel fibres relative to reinforcing bars may have to do with their structure: the drawing operation for producing wires and fibres result in an alignment of ferrite grains which provide surface homogeneity and therefore corrosion resistance.

7.8 Temperature extremes

At elevated temperatures, Purkiss [142,143] showed that below 600°C, SFRC performs better than plain concrete. Interestingly, this appeared to be largely independent of both the fibre type and the fibre content. In particular, SFRC exhibits a somewhat better *retention* of strength at elevated temperatures than plain concrete, as shown in Figure 7.29 [143]. There seems also to be an effect of the strength of the SFRC. Felicetti *et al.* [144] found that very high strength concretes, such as compact reinforced composites (CRC) and reactive powder concretes (RPC), with compressive strengths of about 160 MPa were much less affected by exposure to temperatures of 600°C than were 'ordinary' high strength concretes, with compressive strengths of about 95 MPa, and not containing fibres.

Tests on SFRC specimens exposed to even higher temperatures (800°C) and fire conditions showed that fire resistance is greatly enhanced by the presence of steel fibres [145]. On the other hand, other work [146] has suggested that steel fibres do *not* improve the behaviour of high strength concrete under the thermal strains associated with fires, though they appear to work well when combined with polypropylene fibres. Similar conclusions were reached by Horiguchi *et al.* [147]. However, as Gambarova [148] has pointed out, 'Trying to draw general conclusions (on the fire resistance of SFRC) is not easy, since several aspects of FRC behavior under fire and at high temperature are still open to investigation, like the combination of concrete long-term residual behavior with fiber residual behavior. To cite a specific problem, after a fire the fibers may get oxidized and lose most of their stiffness and strength, as it has been recently observed in some brass-coated steel microfibers.'

At very low temperatures, approaching those of liquefied natural gas, Rostasy and Sprenger [149] found that the addition of steel fibres greatly reduced the loss of both compressive and tensile strengths under repeated thermal cycles down to −170°C. Reactive powder concretes were found to have excellent freeze–thaw resistance, even in the absence of air entrainment [150].

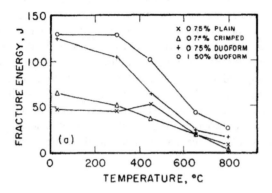

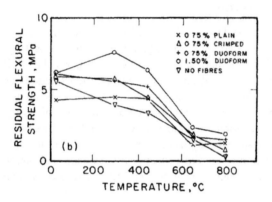

Figure 7.29 Variation of (a) total fracture energy, and (b) residual flexural strength of SFRC with temperature [143].

References

1. J.P. Romualdi and G. Batson, 'Mechanics of crack arrest in concrete', *ASCE, J. Eng. Mech. Div.* 89, 1963, 147–168.
2. J.P. Romualdi and J.A. Mandel, 'Tensile strength of concrete affected by uniformly distributed closely spaced short lengths of wire reinforcement', *J. American Concrete Institute.* 61, 1964, 657–671.
3. S.P. Shah and V.B. Rangan, 'Fiber reinforced concrete properties', *J. American Concrete Institute.* 68, 1971, 126–135.
4. ASTM A 820, *Standard Specification for Steel Fibers for Fiber-Reinforced Concrete*, ASTM International, West Conshohocken, NJ, 2004.
5. C.K.Y. Leung and N. Shapiro, 'Optimal steel fiber strength for reinforcement of cementitious materials', *ASCE Journal of Materials in Civil Engineering.* 11, 1999, 116–123.
6. C. Sujivorakul and A.E. Naaman, 'Modeling bond components of deformed steel fibers in FRC composites', in A.E. Naaman and H.W. Reinhardt (eds) *High*

Performance Fiber Reinforced Cement Composites, RILEM Proceedings PRO 30, RILEM Publications, Bagneux, 2003, pp. 35–48.

7. ACI Committee 544, *Guide for Specifying, Proportioning, Mixing, Placing and Finishing Steel Fiber Reinforced Concrete*, ACI 544.3R-93, American Concrete Institute, Farmington Hills, MI, 1993.

8. ASTM C 1116, *Standard Specification for Fiber-Reinforced Concrete and Shotcrete*, ASTM International, West Conshohocken, NJ, 2003.

9. P. Stroeven and R. Babut, 'Wire distribution in steel wire reinforced concrete', *ACTA Stereologica*. 5/2, 1986, 363–388.

10. P.S. Mangat and M.M. Azari, 'A theory of free shrinkage of steel fibrereinforced cement matrices', *J. Mater. Sci*. 14, 1984, 2183–2194.

11. C.W. Hoy and P.J.M. Bartos, 'Interaction and packing of fibres: Effects on the mixing process', in H.W. Reinhardt and A.E. Naaman (eds) *High Performance Fiber Reinforced Cement Composites (HPFRCC 3)*, RILEM Proceedings PRO 6, RILEM Publications, Bagneux, 1999, pp. 181–191.

12. D.R. Lankard, 'Slurry infiltrated fiber concrete (SIFCON)', *Concr. Int*. 6 (12), 1984, 44–47.

13. L.E. Hackman, M.B. Farrell and O.O. Dunham, 'Slurry infiltrated mat concrete (SIMCON)', *Concr. Int*. 14 (12), 1992, 53–56.

14. N. Krstulovic-Opara and S. Malak, 'Tensile behavior of SIMCON', *ACI Mater. J*. 94, 1997, 39–46.

15. N. Krstulovic-Opara and M.J. Al-Shannag, 'Slurry infiltrated mat concrete (SIMCON)-based shear retrofit of reinforced concrete members', *ACI Struct. J*. 96, 1999, 105–114.

16. A. Bentur and R. Cree, 'Cement reinforced with steel wool', *Int. J. Cement Composites and Lightweight Concrete*. 9, 1987, 217–223.

17. D.W. Ouanian and C.E. Kesler, '*Design of Fibre Reinforced Concrete for Pumping'*, Report No. DOT-TST76T-17, Federal Railroad Administration, Washington, DC, 1976, 53 pp.

18. E.K. Schrader and A.V. Munch, 'Deck slab repaired by fibrous concrete overlay', *ASCE Journal of the Construction Division*. 102, 1976, 179–196.

19. E.K. Schrader, 'Fiber reinforced concrete pavements and slabs: A state-of-the-art report', in S.P. Shah and A. Skarendahl (eds) *Steel Fibre Concrete*, US–Sweden Joint Seminar, Stockholm, 1985, pp. 109–131.

20. ACI Committee 544, *State-of-the-Art Report on Fiber Reinforced Concrete*. ACI 544.1R-96, American Concrete Institute, Farmington Hills, MI, 1996.

21. J. Edgington, D.J. Hannant and R.I.T. Williams, 'Steel fibre reinforced concrete', Current Paper CP69074, Building Research Establishment, Garston, Watford, 1974, 17 pp.

22. C.H. Henager, 'Steel fibrous shotcrete: a summary of the state-of-the-art', *Concr. Int.: Design and Construction*. 251–2583 (1), 1981, 50–58.

23. P. Soroushian and C.-D. Lee, 'Distribution and orientation of fibers in steel fiber reinforced concrete', *ACI Mater. J*. 87, 1990, 433–439.

24. R. Gettu, D.R. Gardner, H. Saldivar and B.E. Barragan, 'Study of the distribution and orientation of fibres in SFRC specimens', *Mater. Struct. (RILEM)*. 38, 2005, 31–37.

25. J. Edgington and D.J. Hannant, 'Steel fibre reinforced concrete. The effect on fibre orientation of compaction by vibration', *Mater. Struct. (RILEM)*. 5, 1972, 41–44.

26. R.N. Swamy and H. Stavrides, 'Some properties of high workability steel fibre concrete', in A. Neville (ed.) *Fibre-Reinforced Cement and Concrete*, RILEM Symp., The Construction Press, Lancaster, England, 1975, pp. 197–208.

27. H. Toutanjii and Z. Bayasi, 'Effects of manufacturing techniques on the flexural behavior of steel fiber-reinforced concrete', *Cem. Concr. Res.* 28, 1998, 115–124.

28. T. Uomoto and K. Kobayashi, 'Measurement of fiber content of steel fiber reinforced concrete by electro-magnetic method', in G.C. Hoff (ed.) *Fiber Reinforced Concrete: International Symposium*, SP-81, American Concrete Institute, Farmington Hill, MI, 1984, pp. 233–246.

29. J. Potrebowski, 'The splitting test applied to steel fibre reinforced concrete', *Int. J. Cement Composites and Lightweight Concrete.* 5, 1983, 49–53.

30. J. Vodicka, D. Spura and J. Kratky, 'Homogeneity of steel fiber reinforced concrete (SFRC)', in M. di Prisco, R. Felicetti and G.A. Plizzari (eds) *Fibre-Reinforced Concretes, BEFIB 2004*, RILEM Proceedings PRO 39, RILEM Publications, Bagneux, 2004, Vol. 1, pp. 537–544.

31. J.E. Bolander, 'Spring network model of fiber-reinforced cement composites', in H.W. Reinhardt and A.E. Naaman (eds) *High Performance Fiber Reinforced Cement Composites (HPFRCC 3)*, RILEM Proceedings PRO 6, RILEM Publications, Bagneux,1999, pp. 341–350.

32. J.E. Bolander, 'Numerical modeling of fiber reinforced cement composites: linking material scales', in M. di Prisco, R. Felicetti and G.A. Plizzari (eds) *Fibre-Reinforced Concretes, BEFIB 2004*, RILEM Proceedings PRO 39, RILEM Publications, Bagneux, 2004, Vol. 1, pp. 45–60.

33. L. Taerwe, A. van Gysel, G. de Schutter, J. Vyncke and S. Schaerlaekens, 'Quantification of variations in the steel fibre content of fresh and hardened concrete', in H.W. Reinhardt and A.E. Naaman (eds) *High Performance Fiber Reinforced Cement Composites (HPFRCC 3)*, RILEM Proceedings PRO 6, RILEM Publications, Bagneux, 1999, pp. 213–222.

34. S.D.P. Benson and B.L. Karihaloo, 'CARDIFRC® - manufacture and constitutive behaviour', in A.E. Naaman and H.W. Reinhardt (eds) *High Performance Fiber Reinforced Cement Composites (HPFRCC 4)*, RILEM Proceedings PRO 30, RILEM Publications, Bagneux, 2003, pp. 65–79.

35. A. Burakiewicz, 'Testing of fibre bond strength in cement matrix', in R.N. Swamy (ed.) *Testing and Test Methods of Fibre Cement Composites*, The Construction Press, Lancaster, 1978, pp. 355–365.

36. A. van Gysel, 'A pullout model for hooked end steel fibres', in H.W. Reinhardt and A.E. Naaman (eds) *High Performance Fiber Reinforced Cement Composites (HPFRCC 3)*, RILEM Proceedings PRO 6, RILEM Publications, Bagneux, 1999, 351–359.

37. I. Markovic, J.C. Walraven and J.G.M. van Mier, 'Experimental evaluation of fibre pullout from plain and fibre reinforced concrete', in A.E. Naaman and H.W. Reinhardt (eds) *High Performance Fiber Reinforced Cement Composites (HPFRCC 4)*, RILEM Proceedings PRO 30, RILEM Publications, Bagneux, 2003, pp. 419–436.

38. A.E. Naaman, 'Evaluation of steel fibers for applications in structural concrete', in M. di Prisco, R. Felicetti and G.A. Plizzari (eds) *Fibre-Reinforced Concretes, BEFIB 2004*, RILEM Proceedings PRO 39, RILEM Publications, Bagneux, 2004, Vol. 1, pp. 389–400.

39. M. Maage, 'Fibre bond and friction in cement and concrete', in R.N. Swamy (ed.) *Testing and Test Methods of Fibre Cement Composites*, The Construction Press, Lancaster, 1978, pp. 329–336.

40. R.J. Gray, 'Fiber-matrix bonding in steel fiber-reinforced cement based composites', in *Fracture Mechanics of Ceramics*, Vol. 7, Plenum Press, New York, 1986, pp. 143–155.

41. V. Ramakrishnan, W.V. Coyle, L.F. Dahl and E.K. Schrader, 'A comparative evaluation of fibre shotcretes', *Concr. Int.: Design and Construction.* 3 (1), 1981, 59–69.

42. A.D. Morris and G.G. Garrett, 'A comparative study of the static and fatigue behaviour of plain and steel fibre reinforced mortar in compression and direct tension', *Int. J. Cement Composites and Lightweight Concrete.* 3, 1981, 73–91.

43. P.S. Mangat and M.M. Azari, 'Influence of steel fibre reinforcement on the fracture behaviour of concrete in compression', *Int. J. Cement Composites and Lightweight Concrete.* 6, 1984, 219–232.

44. D. Fanella and A.E. Naaman, 'Stress-strain properties of fiber reinforced concrete in compression', *J. American Concrete Institute.* 82, 1985, 475–483.

45. F.F. Wafa and S.A. Ashour, 'Mechanical properties of high-strength fiber reinforced concrete', *ACI Mater. J.* 89, 1992, 449–455.

46. P. Balaguru and H.S. Franklin, 'High performance user-friendly fiber reinforced composite under cyclic loading', in H.W. Reinhardt and A.E. Naaman (eds) *High Performance Fiber Reinforced Cement Composites (HPFRCC 3)*, RILEM Proceedings PRO 6, RILEM Publications, Bagneux, 1999, pp. 225–238.

47. D. Atepegba and P.E. Regan, (1981), 'Performance of steel fibre reinforced concrete in axially loaded short columns', *Int. J. Cement Composites and Lightweight Concrete.* 3, 1981, 255–259.

48. P.S. Mangat and M.M. Azari, 'Influence of steel fibre and stirrup reinforcement on the properties of concrete in compression members', *Int. J. Cement Composites and Lightweight Concrete.* 7, 1985, 183–192.

49. B.L. Karihaloo and K.M.B. de Vriese, 'Short-fibre reinforced reactive powder concrete', in H.W. Reinhardt and A.E. Naaman (eds) *High Performance Fiber Reinforced Cement Composites (HPFRCC 3)*, RILEM Proceedings PRO 6, RILEM Publications, Bagneux, 1999, pp. 53–63.

50. W. Sun, S. Liu and J. Lai, 'Study on the properties and mechanism of ultra-high performance ecological reactive powder concrete', in A.E. Naaman and H.W. Reinhardt (eds) *High Performance Fiber Reinforced Cement Composites*, RILEM Proceedings PRO 30, RILEM Publications, Bagneux, 2003, pp. 409–417.

51. W. Sun, G. Pan, H. Yan, C. Qi and H. Chen, 'Study on the anti-exploding characteristics of fiber reinforced cement based composite', in H.W. Reinhardt and A.E. Naaman (eds) *High Performance Fiber Reinforced Cement Composites (HPFRCC 3)*, RILEM Proceedings PRO 6, RILEM Publications, Bagneux, 1999, pp. 565–574.

52. N. Krstulovic-Opara, 'Use of SIMCON in seismic retrofit and new construction', in H.W. Reinhardt and A.E. Naaman (eds) *High Performance Fiber Reinforced Cement Composites (HPFRCC 3)*, RILEM Proceedings PRO 6, RILEM Publications, Bagneux, 1999, pp. 629–649.

53. B.P. Hughes, 'Experimental test results for flexure and direct tension of fibre cement composites', *Int. J. Cement Composites and Lightweight Concrete.* 3, 1981, 13–18.

54. C.D. Johnston and R.A. Coleman, 'Strength and deformation of steel fiber reinforced mortar in uniaxial tension', iIn *Fiber Reinforced Concrete*, ACI SP-44, American Concrete Institute, Farmington Hills, MI, 1974, pp. 177–193.

55. A. Nanni, 'Spitting-tension test for fiber reinforced concrete', *ACI Mater. J.* 85, 1988, 229–233.

56. S.P. Shah, 'Complete stress-strain curves for steel fibre reinforced concrete in uni-axial tension and compression', in R.N. Swamy (ed.) *Testing and Test Methods of Fibre Cement Composites*, RILEM Symp., The Construction Press, Lancaster, 1978, pp. 399–408.

57. K. Visalvanich and A.E. Naaman, 'Fracture model for fiber reinforced concrete', *J. American Concrete Institute*. 80, 1983, 128–138.

58. A.E. Naaman and J.R. Homrich, 'Tensile stress-strain properties of SIFCON', *ACI Mater. J.* 86, 1989, 244–251.

59. C.D. Johnston, 'Steel fiber reinforced mortar and concrete: A review of mechanical properties', in *Fiber Reinforced Concrete*, ACI SP-44, American Concrete Institute, Farmington Hills, MI, 1974, pp. 127–142.

60. P. Richard and M.H. Cheyrezy, 'Reactive powder concrete with high ductility and 200–800 MPa compressive strength', in P.K. Mehta (ed.) *Concrete Technology Past, Present and Future*, SP-144, American Concrete Institute, Farmington Hills, MI, 1994, pp. 507–518.

61. G. Chanvillard and S. Rigaud, 'Complete characterization of tensile properties of DUCTAL® UHPFRC according to the French recommendations', in A.E. Naaman and H.W. Reinhardt (eds) *High Performance Fiber Reinforced Cement Composites (HPFRCC 4)*, RILEM Proceedings PRO 30, RILEM Publications, Bagneux, 2003, pp. 21–34.

62. W. Sun, J. Lai and C. Jiao, 'Study of the mechanical behavior of ECO-RPC under static and dynamic loads', in M. di Prisco, R. Felicetti and G.A. Plizzari (eds) *Fibre-Reinforced Concretes, BEFIB 2004*, ed. RILEM Proceedings PRO 39, RILEM Publications, Bagneux, 2004, Vol. 2, pp. 1411–1420.

63. A.E. Naaman and H.W. Reinhardt, 'Setting the stage: Toward performance based classification of FRC composites', in A.E. Naaman and H.W. Reinhardt (eds) *High Performance Fiber Reinforced Cement Composites (HPFRCC 4)*, RILEM Proceedings PRO 30, RILEM Publications, Bagneux, 2003, pp. 1–4.

64. A.K. Sharma, 'Shear strength of steel fiber reinforced concrete beams', *ACI Struct. J.* 83, 1986, 624–628.

65. B. Barr, 'The fracture characteristics of FRC materials in shear', in S.P. Shah and G.B. Batson (eds) *Fiber Reinforced Concrete – Properties and Application*, ACI SP-105, American Concrete Institute, Farmington Hills, MI, 1987, pp. 27–54.

66. M. Valle and O. Buyukozturk, 'Behavior of fiber reinforced high-strength concrete under direct shear', *ACI Mater. J.* 90, 1993, 122–133.

67. A.A. Mirsayah and N. Banthia, 'Shear strength of steel fiber-reinforced concrete', *ACI Mater. J.* 99, 2002, 473–479.

68. S. Mindess, 'Torsion tests of steel-fibre reinforced concrete', *Int. J. Cement Composites and Lightweight Concrete*. 2, 1980, 85–89.

69. R. Narayanan and A.S. Kareem-Palanjian, 'Steel fibre reinforced concrete beams in torsion', *Int. J. Cement Composites and Lightweight Concrete*. 5, 1983, 235–246.

70. Y. Kosaka, Y. Tanigawa and S. Hatanaka, 'Lateral confining stresses due to steel fibres in concrete under compression', *Int. J. Cement Composites and Lightweight Concrete*. 7, 1985, 81–92.

71. A.M. Brandt, J. Kasperkiewicz, M.D. Kotsovos and J.B. Newman, 'Preliminary tests of SFRC under triaxial loading', *Int. J. Cement Composites and Lightweight Concrete.* 3, 1981, 261–266.

72. J.F. Berthet, E. Ferrier and P. Hamelin, 'Effect of confinement on high performance fiber reinforced cement composite', in A.E. Naaman and H.W. Reinhardt (eds) *High Performance Fiber Reinforced Cement Composites (HPFRCC 4)*, RILEM Proceedings PRO 30, RILEM Publications, Bagneux, 2003, pp. 133–144.

73. J.B. Mander, M.J.N. Priestley and R. Park, 'Observed stress-strain behaviour of confined concrete', *ASCE Journal of Structural Engineering.* 114, 1988, 1827–1849.

74. J. Sustersic, E. Mali and S. Urvancic, 'Erosion-abrasion resistance of steel fiber reinforced concrete', in V.M. Malhotra (ed.) *Durability of Concrete: Second International Conference*, SP-126, American Concrete Institute, Farmington Hills, MI, 1991, pp. 729–744.

75. T.C. Lim, 'Maintenance and preservation of civil works structures; abrasion-erosion resistance of concrete', Technical Report No. C-78-74, Report No. 3, US Army Engineers Waterways Experiment Station, Vicksburg, MS, 1980, 129pp.

76. E.K. Schrader and A.V. Munch, 'Fibrous concrete repair of cavitation damage', *ASCE Journal of the Construction Division.* 102, 1976, 385–399.

77. D.L. Houghton, O.E. Borge and J.A. Paxton, 'Cavitation resistance of some special concretes', *J. American Concrete Institute.* 75, 1978, 664–667.

78. ICOLD 'Fiber reinforced concrete', Bulletin No. 40, International Commission on Large Dams (ICOLD), 1982.

79. G.R. Mass, 'SFRC lining for an embankment dam', *Concr. Int.* 19 (6), 1997, 24–27.

80. E.G. Nawy, 'Part A: Fiber-reinforced concrete (FRC)', in E.G. Nawy (ed.) *Concrete Construction Engineering Handbook*, Chapter 22, CRC Press, Boca Raton, FL, 1997.

81. P.S. Mangat and M.M. Azari, 'Shrinkage of steel fibre reinforced cement composites', *Mater. Struct. (RILEM).* 21, 1988, 163–171.

82. R.N. Swamy and H. Stavrides, 'Influence of fiber reinforcement in restrained shrinkage and cracking', *J. American Concrete Institute.* 76, 1979, 443–460.

83. B. Malmberg, and A. Skarendahl, 'Method of studying the cracking of fibre concrete under restrained shrinkage', in R.N. Swamy (ed.) *Testing and Test methods of Fibre Cement Composites*, The Construction Press, Lancaster, 1978, pp. 173–179.

84. S.P. Shah, M. Sarigaphuti and M.E. Karaguler, 'Comparison of shrinkage cracking performance of different types of fibers and wiremesh', in J.I. Daniels and S.P. Shah (eds) *Fiber Reinforced Concrete – Developments and Innovation*, SP-142, American Concrete Institute, Farmington Hills, MI, 1994, pp. 1–18.

85. J.-C. Chern and C.-H. Young, 'Factors influencing the drying shrinkage of steel fiber reinforced concrete', *ACI Mater. J.* 87, 1990, 123–139.

86. J.-C. Chern and C.Y. Chang, 'Effect of silica fume on creep and shrinkage of steel fiber reinforced concrete', in V.M. Malhotra (ed.) *High Performance Concrete*, SP-149, American Concrete Institute, Farmington Hills, MI, 1994, pp. 561–574.

87. P.S. Mangat and M.M. Azari, 'A theory for the creep of steel fibre reinforced cement matrices under compression', *J. Mater. Sci.* 20, 1985, 1119–1133.

88. P. Balaguru, and V. Ramakrishnan, 'Properties of fiber reinforced concrete: workability, behavior under long-term loading, and air-void characteristics', *ACI Mater. J.* 85, 1988, 189–196.

89. J. Houde, A. Prezeau and R. Roux, 'Creep of concrete containing fibres and silica fume', in S.P. Shah and G.B. Batson (eds) *Fiber Reinforced Concrete – Properties*

and Application, SP-105, American Concrete Institute, Farmington Hills, MI, 1987, pp. 101–118.

90. W. Yin and T.T.C. Hsu, 'Fatigue behavior of steel fiber reinforced concrete in uniaxial and biaxial compression', ACI Mater. J. 92, 1995, 71–81.

91. G.A. Plizzari, S. Cangiano and N. Cere, 'Postpeak behavior of fiber-reinforced concrete under cyclic tensile loads', ACI Mater. J. 97, 2000, 182–192.

92. T. Paskova and C. Meyer, 'Low-cycle fatigue of plain and fiber-reinforced concrete', ACI Mater. J. 94, 1997, 273–286.

93. N.L. Lovata and P.B. Morrill, 'Dynamic tension fatigue of fibrous concrete composites', in J.I. Daniels and S.P. Shah (eds) Fiber Reinforced Concrete – Developments and Innovation, SP-142, American Concrete Institute, Farmington Hills, MI, 1994, pp. 295–314.

94. C.D. Johnston and R.W. Zemp, 'Flexural fatigue performance of steel fiber reinforced concrete – influence of fiber content, aspect ratio and type', ACI Mater. J. 88, 1991, 374–383.

95. V. Ramakrishnan, G. Oberling and P. Tatnall, 'Flexural fatigue strength of steel fiber reinforced concrete', in S.P. Shah and G.B. Batson (eds) Fibre Reinforced Concrete – Properties and Applications, SP-105, American Concrete Institute, Farmington Hill, MI, 1987, pp. 225–245.

96. Z. Jun and H. Stang, 'Fatigue performance in flexure of fiber reinforced concrete', ACI Mater. J. 95, 1998, 58–67.

97. V.S. Gopalaratnam and T. Cherian, 'Fatigue characteristics of fiber reinforced concrete for pavement applications', in P. Balaguru, A. Naaman and W. Weiss (eds) Concrete: Materials Science to Application – A Tribute to Surendra P. Shah, SP-206, American Concrete Institute, Farmington Hill, MI, 2002, pp. 91–108.

98. N. Krstulovic-Opara, A.R. Haghayeghi, M. Haidar and P.D. Krauss, 'Use of conventional and high-performance steel-fiber reinforced concrete for bridge deck overlays', ACI Mater. J. 92, 1995, 669–677.

99. H.A. Kormeling, H.W. Reinhardt and S.P. Shah, 'Static and fatigue properties of concrete beams reinforced with continuous bars and with fibers', J. American Concrete Society. 77, 1980, 36–43.

100. K.-H. Kwak, J. Suh and C.-T.T. Hsu, 'Shear-fatigue behavior of steel fiber reinforced Concrete', ACI Struct. J. 88, 1991, 155–160.

101. R.L. Jindal and K.A. Hassan, 'Behavior of steel fiber reinforced concrete beam-column connections', in G.C. Hoff (ed.) Fiber Reinforced Concrete: International Symposium SP-81, American Concrete Institute, Farmington Hill, MI, 1984, pp. 107–124.

102. N. Banthia and V. Bindiganavile, 'Fiber reinforced cement based composites under drop weight impact loading: Test equipment and material influences', in P. Balaguru, A. Naaman and W. Weiss (eds) Concrete: Material Science to Applications – A Tribute to Surendra P. Shah, SP-206, American Concrete Institute, Farmington Hill, MI, 2002, pp. 411–428.

103. W. Suaris and S.P. Shah, 'Inertial effects in the instrumented impact testing of cementitious composites', Cement Concrete and Aggregates (ASTM) 3, 1981, 77–83.

104. W. Suaris and S.P. Shah, 'Strain-rate effects in fibre reinforced concrete subjected to impact and impulsive loading', Composites. 13, 1982, 153–159.

105. W. Suaris and S.P. Shah, 'Properties of concrete subjected to impact', ASCE J. Structural Engineering. 109, 1982, 1727–1741.

106. S.P. Shah, 'Concrete and fiber reinforced concrete subjected to impact loading', in S. Mindess and S.P. Shah (eds) *Cement-Based Composites: Strain Rate Effects on Fracture*, Proc. Materials Research Society Symp., Vol. 64, Materials Research Society, Pittsburgh, PA, 1986, pp. 181–202.

107. V.S. Gopalaratnam and S.P. Shah, 'Properties of steel fiber reinforced concrete subjected to impact loading', *J. American Concrete Institute*. 83, 1986, 117–126.

108. N.P. Banthia, *Impact Resistance of Concrete*, PhD Thesis, University of British Columbia, Vancouver, Canada, 1987.

109. N.P. Banthia, S. Mindess and A. Bentur, 'Impact behaviour of concrete beams', *Mater. Struct. (RILEM)*. 20, 1987, 293–302.

110. A. Pacios, and S.P. Shah, 'Measurement of the pull-out force at different rates of loading', in D.J. Stevens (ed.) *Testing of Fiber Reinforced Concrete*, SP-155, American Concrete Institute, Farmington Hill, MI, 1995, pp. 189–216.

111. V. Bindiganavile and N. Banthia, 'Polymer and steel fiber-reinforced cementitious composites under impact loading – Part 1: Bond-slip response'. *ACI Mater. J.* 98, 2001, 10–16.

112. H. Xu, S. Mindess and I.J. Duca, 'Performance of plain and fiber reinforced concrete panels subjected to low velocity impact loading', in M. di Prisco, R. Felicetti and G.A. Plizzari (eds) *Fibre-Reinforced Concretes, BEFIB 2004*, RILEM Proceedings PRO 39, RILEM Publications, Bagneux, 2004, Vol. 2, pp. 1257–1266.

113. G. Solomos and M. Berra, 'Compressive behaviour of high performance concrete at dynamic strain-rates', in M. di Prisco, R. Felicetti and G.A. Plizzari (eds) *Fibre-Reinforced Concretes, BEFIB 2004*, RILEM Proceedings PRO 39, RILEM Publications, Bagneux, 2004, Vol. 1, pp. 421–430.

114. N. Banthia, V. Bindiganavile and S. Mindess, 'Impact resistance of fiber reinforced concrete: A progress report', in A.E. Naaman and H.W. Reinhardt (eds) *High Performance Fiber Reinforced Cement Composites (HPFRCC 4)*, RILEM Proceedings PRO 30, RILEM Publications, Bagneux, 2003, pp. 117–131.

115. V. Bindiganavile, N. Banthia and B. Aarup, 'Impact response of ultra-high-strength fiber-reinforced cement composite', *ACI Mater. J.* 99, 2002, 543–548.

116. P. Sukontasukkul, S. Mindess and N. Banthia, 'Properties of confined fibre-reinforced concrete under uniaxial compressive impact', *Cem. Concr. Res.* 35, 2005, 11–18.

117. P. Sukontasukkul, S. Mindess, N. Banthia and T. Mikami, 'Impact resistance of laterally confined fiber reinforced concrete plates', *Mater. Struct. (RILEM)*. 34, 2001, 612–618.

118. P. Sukontasukkul and S. Mindess, 'The shear fracture of concrete under impact using end confined beams', *Mater. Struct. (RILEM)*. 36, 2003, 372–378.

119. C. Yan and S. Mindess, 'Bond between epoxy coated reinforcing bars and concrete under impact loading', *Can. J. Civ. Eng.* 21, 1994, 131–141.

120. C. Yan and S. Mindess, 'Effect of concrete strength on bond behavior under impact loading', in V.M. Malhotra (ed.) *High-Performance Concrete*, Proc. Int. Conf., Singapore, SP-149, American Concrete Institute, Farmington Hill, MI, 1994, pp. 679–700.

121. C. Yan and S. Mindess, 'Fracture mechanics analysis of bond behavior under dynamic loading', in O. Buyukozturk and M. Wecharatana (eds) *Interface Fracture and Bond*, SP-156, American Concrete Institute, Farmington Hill, MI, 1995, pp. 107–124.

122. S. Mindess, C. Yan and W.J. Venuti, 'Impact resistance of fibre reinforced prestressed concrete railroad ties', in V.M. Malhotra (ed.) *Evaluation and Rehabilitation*

of Concrete Structures and Innovations in Design, Int. Conf., Hong Kong, SP-128, American Concrete Institute, Farmington Hill, MI, 1991, Vol. 1, pp. 183–199.

123. A.E. Naaman, 'Fiber reinforced concrete under dynamic loading', in G.C. Hoff (ed.) *Fiber Reinforced Concrete: International Symposium*, SP-81, American Concrete Institute, Farmington Hill, MI, 1984, pp. 169–186.

124. F.S. Rostasy and K. Hartwick, 'Compressive strength and deformation of steel fibre reinforced concrete under high rate of strain', *Int. J. Cement Composites and Lightweight Concrete.* 7, 1985, 21–28.

125. H.A. Kormeling and H.W. Reinhardt, 'Strain rate effects on steel fibre concrete in uniaxial tension', *Int. J. Cement Composites and Lightweight Concrete.* 9, 1987, 197–204.

126. A.E. Naaman and V.S. Gopalaratnam, 'Impact properties of steel fibre reinforced concrete in bending', *Int. J. Cement Composites and Lightweight Concrete.* 5, 1983, 225–233.

127. K. Kosa and A.E. Naaman, 'Corrosion of steel fiber reinforced concrete', *ACI Mater. J.* 87, 1990, 27–37.

128. G.C. Hoff, 'Durability of fiber reinforced concrete in a severe marine environment', in J.M. Scanlon (ed.) *Katherine and Bryant Mather International Symposium on Concrete Durability*, ACI SP-100, American Concrete Institute, Farmington Hill, MI, 1987, Vol. I, pp. 997–1041.

129. J. Sustersic, A. Zajc, I. Leslovar and R. Ercegovic, 'Study of corrosion resistance of steel fibres in SFRC', in B.H. Oh, K. Sakai, O.E. Gjorv and N. Banthia (eds) *Concrete Under Severe Conditions: Environment and Loading (CONSEC '04)*, Seoul National University and Korea Concrete Institute, Vol. 2, pp. 1540–1547.

130. T. Teruzzi, E. Cadoni, G. Frigideri, S. Cangiano and G.A. Plizzari, 'Durability aspects of steel fibre reinforced concrete', in M. di Prisco, R. Felicetti and G.A. Plizzari (eds) *Fibre-Reinforced Concretes, BEFIB 2004*, RILEM Proceedings PRO 39, RILEM Publications, Bagneux, 2003, Vol. 1, pp. 625–634.

131. L. Ferrara, R. Fratesi, S. Signorini and F. Sonzogni, 'Durability of steel fibre-reinforced concrete precast elements : experiments and proposal of design recommendations', in M. di Prisco, R. Felicetti and G.A. Plizzari (eds) *Fibre-Reinforced Concretes, BEFIB 2004*, RILEM Proceedings PRO 39, RILEM Publications, Bagneux, 2004, Vol. 1, pp. 565–574.

132. P.S. Mangat and K. Gurusamy, 'Steel fibre reinforced concrete for marine applications', in J.A. Battjes (ed.) Proc. 4th Int. Conf. Behaviour of Offshore Structures, Elsevier Science Publishers, Delft, 1985, pp. 867–879.

133. N. Banthia and C. Foy, 'Marine curing of steel fibre composites'. *ASCE J. Materials in Civil Engineering.* 1, 1989, 86–96.

134. D.C. Morse and G.R. Williamson, 'Corrosion behaviour of steel fibrous concrete', Technical Report M-217, US Army Corps of Engineers Construction Engineering Research Laboratory, Champaign, IL, 1977, 36pp.

135. D.J. Hannant, 'Additional data on fibre corrosion in cracked beams and theoretical treatment of the effect of fibre corrosion on beam load capacity', in A. Neville (ed.) *Fibre Reinforced Cement and Concrete*, Proceedings RILEM Symposium, The Construction Press, Lancaster, England, 1975, pp. 533–538.

136. J.-L. Granju and S.U. Baluch, 'Corrosion of steel fibre reinforced concrete from the cracks', *Cem. Concr. Res.* 35, 2005, 572–577.

137. V. Corinaldesi and G. Moriconi, 'Some durability aspects of fiber reinforced self-compacting concrete', in M. di Prisco, R. Felicetti and G.A. Plizzari (eds) *Fibre-Reinforced Concretes, BEFIB 2004*, RILEM Proceedings PRO 39, RILEM Publications, Bagneux, 2004, Vol. 1, pp. 555–564.

138. P.S. Mangat and K. Gurusamy, 'Corrosion resistance of steel fibres in concrete under marine exposure', *Cem. Concr. Res.* 18, 1988, 44–54.

139. M. Schupack, 'Durability of SFRC exposed o severe environments', in S.P. Shah and A. Skarendahl (eds) *Steel Fiber Concrete*, Stockholm: Proceedings US–Sweden Joint Seminar, 1985, pp. 479–496.

140. A. Bentur, 'Durability of fiber reinforced cementitious composites, in J.P. Skalny and S. Mindess (eds), *Materials Science of Concrete-V*, The American Ceramic Society, Westerville, OH, 1998, pp. 513–536.

141. P.S. Mangat and B.T. Molloy, 'Size effect of reinforcement on corrosion initiation, in P. Rossi and G. Chanvillard (eds) *Fibre-Reinforced Concretes (FRC) BEFIB' 2000*, RILEM Proceedings PRO 15, RILEM Publications, Bagneux, 2000, pp. 691–701.

142. J.A. Purkiss, 'Steel fibre reinforced concrete at elevated temperatures', *Int. J. Cement Composites and Lightweight Concrete.* 6, 1984, 179–184.

143. J.A. Purkiss, 'Toughness measurements on steel fibre concrete at elevated temperatures', *Int. J. Cement Composites and Lightweight Concrete.* 10, 1988, 39–47.

144. R. Felicetti, P.G. Gambarova, M.P. Natali Sora, F. Corsi and G. Giannuzzi, 'On tension and fracture in thermally-damaged high-performance concrete: VHSC versus HSC, in H.W. Reinhardt and A.E. Naaman (eds) *High Performance Fiber Reinforced Cement Composites (HPFRCC 3)*, RILEM Proceedings PRO 6, RILEM Publications, Bagneux, 1999, pp. 437–448.

145. M. di Prisco, R. Felicetti, P.G. Gambarova and C. Failla, 'On the fire behavior of SFRC and SFRC structures in tension and bending', in A.E. Naaman and H.W. Reinhardt (eds) *High Performance Fiber Reinforced Cement Composites (HPFRCC 4)*, RILEM Proceedings PRO 30, RILEM Publications, Bagneux, 2003, pp. 205–220.

146. F. Dehn and G. Konig, 'Fire resistance of different fibre reinforced high-performance concretes', in A.E. Naaman and H.W. Reinhardt (eds) *High Performance Fiber Reinforced Cement Composites (HPFRCC 4)*, RILEM Proceedings PRO 30, RILEM Publications, Bagneux, 2003, pp. 189–204.

147. T. Horiguchi, T. Sugawara and N. Saeki, 'Fire resistance of hybrid fiber reinforced high strength concrete', in M. di Prisco, R. Felicetti and G.A. Plizzari (eds) *Fibre-Reinforced Concretes, BEFIB 2004*, RILEM Proceedings PRO 39, RILEM Publications, Bagneux, 2004, Vol. 1, pp. 669–678.

148. P.G. Gambarova, 'Overview of recent advancements in FRC knowledge and applications, with specific reference to high temperature', in M. di Prisco, R. Felicetti and G.A. Plizzari (eds) *Fibre-Reinforced Concretes, BEFIB 2004*, RILEM Proceedings PRO 39, RILEM Publications, Bagneux, 2004, Vol. 1, pp. 125–140.

149. F.S. Rostasy and K.H. Sprenger, 'Strength and deformation of steel fibre reinforced concrete at very low temperature', *Int. J. Cement Composites and Lightweight Concrete.* 6, 1984, 47–51.

150. J. Tanesi, B. Graybeal and M. Simon, 'Effects of curing procedure on freeze-thaw durability of ultra-high performance concrete', in M. di Prisco, R. Felicetti and G.A. Plizzari (eds) *Fibre-Reinforced Concretes, BEFIB 2004*, RILEM Proceedings PRO 39, RILEM Publications, Bagneux, 2004, Vol. 1, pp. 603–613.

Chapter 8

Glass fibres

8.1 Introduction

Glass fibre reinforced cementitious composites have been developed mainly for the production of thin sheet components, with a paste or mortar matrix, and ~5% fibre content [1]. Other applications have been considered, either by making reinforcing bars with continuous glass fibres joined together and impregnated with plastics [2], or by making similar short, rigid units, impregnated with epoxy, to be dispersed in the concrete during mixing [3]. In practice, however, the main application of the glass fibre reinforcement is in thin sheets, and this will be the focus of this chapter. Such systems are referred to as GRC (glass fibre reinforced cement) or GFRC (glass fibre reinforced concrete).

Glass fibres are produced in a process in which molten glass is drawn in the form of filaments, through the bottom of a heated platinum tank or bushing. Usually, 204 filaments are drawn simultaneously and they solidify while cooling outside the heated tank [4,5]; they are then collected on a drum into a strand consisting of the 204 filaments. The structure of such a strand was shown in Chapter 2, Figure 2.3. Prior to winding, the filaments are coated with a sizing which protects the filaments against weather and abrasion effects, as well as binding them together in the strand. Several strands are wound together to form a roving. This stage of production is controlled so that the roving can be separated back into the individual strands when it is combined with the matrix in the production of the GRC without breaking up into the individual filaments. The fibres can be marketed as either continuous or chopped roving, or they can be formed into a mat (Figure 8.1). For cement reinforcement the mat is usually woven, with sufficiently large openings to allow penetration of cement grains.

The structure of the reinforcing glass fibres has required the development of special technologies to incorporate the fibres into the matrix. A particular problem that also had to be dealt with was the low alkali resistivity of the E-glass fibres, which is the glass type commonly used for reinforcing plastics, and which was originally used for GRC. This type of fibre deteriorates quite rapidly in the highly alkaline environment of the cementitious matrix. To overcome this problem, special alkali-resistant glass formulations (AR glass fibres) had to be developed.

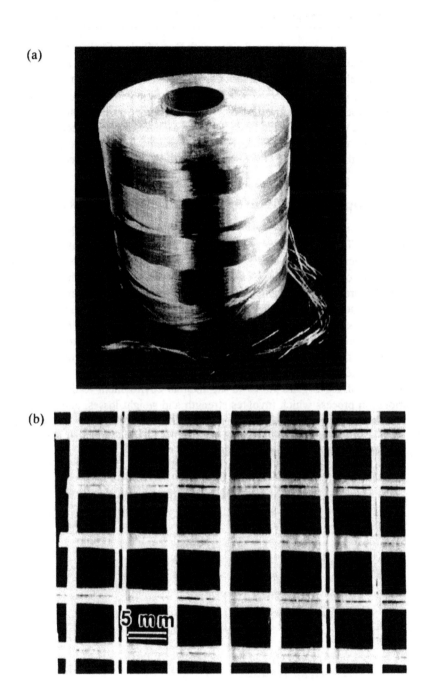

(a)

(b)

Figure 8.1 Various forms of glass fibre reinforcement: (a) continuous roving; (b) Woven mat.

An alternative approach, to be used either with E or AR glass fibres, was to modify the matrix in order to improve the resistance of the fibres, either by lowering the alkalinity level (e.g. silica fume additions; high alumina cement) or by sealing the matrix (e.g. polymer modified cements). The long-term performance of the GRC composite is still the major criterion by which its quality is assessed, and considerable effort is continuously being made to improve the service life of such materials. These topics will be discussed in detail in Sections 8.4–8.6. The production methods and the properties of the composites will be dealt with in Section 8.3.

8.2   Composition and properties of fibres for GRC

A typical inorganic glass consists of an amorphous silicon–oxygen network (Figure 8.2). The drawn fibres can be quite long, in addition to having a modulus of elasticity which is greater than that of the matrix. The properties and compositions of E and AR glasses are given in Tables 8.1 and 8.2. The properties in Table 8.1 are for single filaments. However, it should be noted that the properties of the bundled, multifilament strand can be different; in particular the tensile strength can be lower (e.g. strand strength of about 1500 MPa for the AR glass strand, compared with 2500 MPa for the single filament). Such differences can be accounted for theoretically by statistical considerations of the behaviour of a fibre bundle (e.g. Ref. [6]).

Exposure of the E-glass fibres to an alkaline environment leads to a rapid deterioration process which involves strength and weight losses, and reduction in the filament diameter [7,8]. This process can be attributed to breaking of the $Si-O-Si$ bonds in the glass network, by the $OH-$ ions which are highly concentrated in the alkaline pore solution:

$$- \overset{|}{\underset{|}{Si}} - O - \overset{|}{\underset{|}{Si}} \quad + O\bar{H} \rightarrow \overset{|}{\underset{|}{Si}} - OH + SiO^- \text{(in-solution)}$$

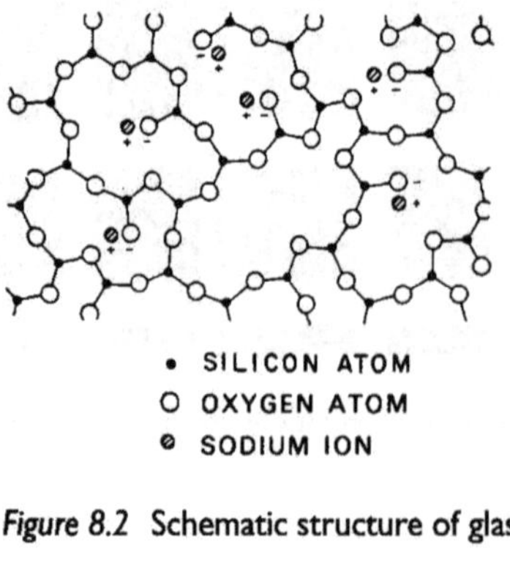

- • SILICON ATOM
- O OXYGEN ATOM
- ⊘ SODIUM ION

Figure 8.2 Schematic structure of glass (after Hull [5]).

Table 8.1 Properties of single filaments of glass (after Majumdar and Nurse [1])

	E glass	AR glass
Density (kg/m^3)	2540	2780
Tensile strength (MPa)	3500	2500
Modulus of elasticity (GPa)	72.5	70.0
Elongation at break (%)	4.8	3.6

Table 8.2 Chemical composition of E and AR glass (after Majumdar and Nurse [1])

	E glass	AR glass
SiO_2	52.4%	71%
$K_2O + Na_2O$	0.8	11
B_2O_3	10.4	—
Al_2O_3	14.4	18
MgO	5.2	—
CaO	16.6	—
ZrO_2	—	16
Li_2O	—	1

The network breakdown leads to surface damage of the glass (e.g. etch pits) and the corrosion reaction products may either dissolve and leach out or accumulate on the surface of the glass [9–11]. Such characteristics are seen in Figure 8.3(a) for E-glass fibre removed from a cement paste matrix. Early work by Biryukovich *et al.* [12] attempted to overcome this problem by using the E-glass fibres in combination with a matrix of low alkali cement, or cement with polymer.

A different approach was taken by Majumdar [7,8] to develop glass fibres which are resistant to alkali attack. In developing AR glass fibres attention was given to two main aspects:

1   The chemical resistivity of the glass, to enhance its performance in an alkaline medium.
2   The physical properties of the melt glass, to enable its fabrication in a commercial process.

Proctor and Yale [13] discussed the significance of treating these two aspects simultaneously, in order to develop AR glass fibres of adequate quality, that could be produced commercially. The most efficient means currently used for enhancing the alkali resistivity is the incorporation of ~16% ZrO_2 in the glass composition

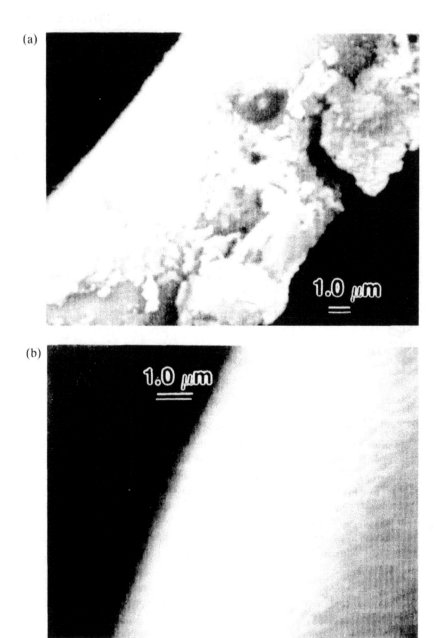

Figure 8.3 The surface of glass fibres removed from aged composites (after Bentur [11]):
(a) E glass removed from a composite after ageing for 2 months in water at 20°C,
showing severe corrosion; (b) AR glass (CemFIL-I) removed from a composite
after ageing in water at 20°C for 1/2 year, showing only mild damage.

(AR glass composition in Table 8.2). Such compositions were developed by Majumdar [7,8], based on observations that fibres produced from compositions in the $Na_2O-SiO_2-ZrO_2$ system are more chemically stable in alkaline solutions, as shown in Figure 8.3(b).

The alkali resistivity can be evaluated by chemical means (i.e. the portion of the glass constituents which break down and are extracted into the alkaline solution), physical means (change in diameter or weight) or mechanical means (changes in strength), after exposure to alkaline solutions which are supposed to represent the pore solution of the cementitious matrix. Solution studies of this kind have some limitations:

1   The pore solution in the actual composite is not readily simulated by synthetic solutions. The pH level can be 13.5 or even greater in the actual matrix, due to the influence of the alkalis in the cement [14]; this is quite different from saturated lime solutions (pH of about 12.5) or even cement extract solutions.
2   The microstructural constraints of the matrix, in relation to the multifilament structure of the strand (Chapter 2, Figures 2.10 and 2.11) may enhance effects which are not simulated properly in the study of the strength of single filaments exposed to alkaline solutions.

In view of such limitations the trend is to evaluate the performance of the fibres by studying their behaviour under realistic conditions (i.e. in the real cementitious matrix) and relying on strength retention measurements as the performance criterion. Special methods have been developed for this purpose, most notably the strand in cement (SIC) test [15] which is shown schematically in Figure 8.4. Whether tested in solutions or in the cementitious matrix, the ageing effects are frequently accelerated by carrying out the tests at elevated temperatures, in the range of 40–60°C.

The improved performance of the AR glasses containing ZrO_2 is demonstrated in Figures 8.5 and 8.6, where they are compared with E-glass. Such compositions were the basis for the development of the first commercial AR glass fibres for cement reinforcement, called CemFIL-1 (produced by Pilkington Brothers, UK).

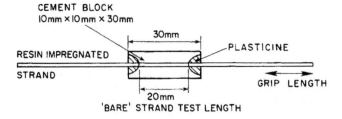

Figure 8.4 Schematic description of the strand in cement (SIC) test (after Litherland et al.[15]).

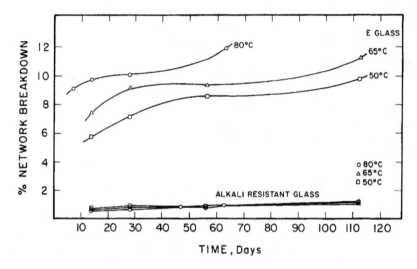

Figure 8.5 Effect of glass composition on the network breakdown in glass fibres exposed to cement extract solutions at various temperatures. Network breakdown was calculated by summing up the network formers removed from the glass, as reported by Larner et al. [16] (after Majumdar [17]).

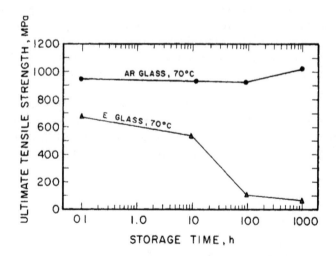

Figure 8.6 The effect of glass composition on the tensile strength of fibres after exposure to Portland cement extract solution (pH = 13.4) at 70°C (after Franke and Overbeck [18]).

The mechanisms by which the modified composition, and in particular the presence of ZrO_2, impart alkali resistivity have been studied extensively [7,8,13,16–24]. The Zr–O bonds, in contrast to Si–O bonds, are only slightly attacked by the OH⁻ ions, and thus the incorporation of ZrO_2 as part of the network imparts stability to the glass structure in the alkaline environment [16], that is, considerably reduces the network breakdown (Figure 8.5).

This can be either the result of an overall improvement in the stability of the glass network when ZrO_2 is present, or the formation of a ZrO_2-rich protective surface layer as some SiO_2 is broken down and extracted. The undissolved ZrO_2-rich layer thus formed remains on the glass and may serve as a diffusion barrier to reduce the rate of further attack. Microanalysis of the AR glass surface has confirmed the formation of a ZrO_2-rich layer after ageing in a cement paste [19] while bulk analysis did not show any appreciable change in the ZrO_2/SiO_2 ratio [20]. More complex characteristics of the surface layer were reported in several references [21, 23, 24], which involves interactions with the Ca in the pore solution: the Ca/Si and Zr/Si ratios of the surface layer increase with time [23] and there is some evidence for the formation of calcium silicate hydrogel which may bear some similarity to the hydration product of cement, as might be seen in Figure 8.3(a).

It should be noted that although strength retention tests always show the superior behaviour of AR-glass compared with E-glass, there is contradictory evidence regarding the magnitude of the strength reduction with AR-glass. Sometimes, there is a decline in strength as high as 50–60% [22,25] while in other studies [18,26] (Figure 8.7) only a small, or even no reduction has been observed. Of particular interest are the data of West and Majumdar [26] (Figure 8.7) which show only a mild reduction in the strength of glass fibres removed from a 10 year naturally aged composites, but a greater reduction in a water aged composite. Yet, most of the strength reduction occurred within the first year, and even in the water aged composite the tensile strength of the fibres after 10 years of ageing remained as high as ~ 1000 MPa. This is in agreement with SEM observations showing only mild damage to the glass surface after ageing in cementitious composites [11,20,27] (Figure 8.3).

Orlowsky et al. [22] studied the strength loss of glass filaments stored in 50°C solution having a pH of 13.5 and analysed the results in terms of the Weibull statistical model. The strength loss over time (60% reduction after 28 days in accelerated conditions) could be interpreted in terms of the growth of flaws, since strength loss was not accompanied by reduction in filament diameter. The calculated flaw size of filaments with sizing was 24–58 nm before ageing, and 75–396 nm after ageing, showing not only an increase in size but also in scatter. The flaw population represents weaker spots on the surface, which are unevenly distributed. These flaws are more sensitive to leaching and network breakdown. It should be noted that this study evaluated the durability of AR glass fibre, and the mechanisms here (dominated by flaws) may be different than in E-glass fibres (dominated by heavy corrosion of the external layers, in addition to flaw generation). Orlowsky et al.'s model suggests that the growth of flaws slows down and is arrested as

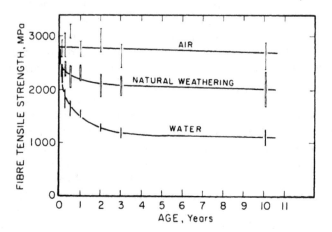

Figure 8.7 The strength of AR glass fibres removed from GRC composites after ageing in different environments (after West and Majumdar [26]).

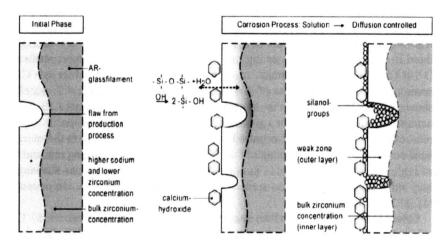

Figure 8.8 Formation of flaws in AR glass filaments and their growth and arrest in alkaline environment (after Orlowsky *et al.* [22]).

they approach the zirconia-rich layer (Figure 8.8). This mechanism may explain the apparent contradictory results in the literature on loss of strength in AR glass, recorded by some, and stability, reported by others [18,22,25,26]. Stability may be taken to imply that loss in strength is limited.

Improved strength retention over the currently available AR glass fibres has been reported by Fyles *et al.* [28], using new glass compositions. A surface coating for improvement in the glass fibre durability [29] has been used to enhance the performance in the second generation of AR glass fibres (CemFIL-2). Hayashi *et al.* [30] have demonstrated the improved performance of AR glass fibres which were treated with various coatings such as phenol, furon and polyvinyl alcohol (PVA).

8.3   Production of GRC composites

Various methods have been developed for the production of GRC components. These methods have mostly been adapted from the glass fibre reinforced plastics industry, with proper modifications to adjust for the special nature of the cementitious matrix. To obtain a product of an adequate quality the mix composition should be carefully controlled, to be compatible with the production process, while at the same time providing the needed physical and mechanical properties in the hardened composite. Thus, the properties of GRC composites vary over a wide range, and are a function of a complex combination of the production process and the mix composition. A detailed discussion of the design and production of GRC components is beyond the scope of this chapter. These topics are covered in various publications and guidelines [31–37] and only some essential points will be discussed in Section 8.7 and Chapter 14.

GRC is used mainly for the production of thin sheet components, which can assume a relatively simple shape such as flat (Figure 8.9) or corrugated sheets [38,39], or a complex shaped panel to be used in precast construction (Chapter 1, Figure 1.2). Complex shapes can readily be achieved due to the plasticity and thixotropy induced in the fresh mix by the fibres. Sufficiently high tensile strength and toughness are required in order to maintain a thin component in the hardened state that will be resistant to cracking. Cracks may occur either during the early life of the component (due to transient loads imposed during transportation and erection) or later on, when it is fixed in the structure. The production process must permit the incorporation and uniform dispersion of a sufficiently large content of fibres, to achieve an adequate reinforcing effect; at the same time it should be adjusted to the method by which the mix is applied and shaped in the mould. These two aspects are quite different from those of conventional concrete production, and will be detailed here. In dealing with GRC production, attention should be given to the special nature of the mix:

1   It should contain a relatively large volume of fibres ($\sim$5% by volume) since this is the primary reinforcement in the thin component.
2   The matrix is a cement–sand mixture (i.e. mortar), with a relatively small sand content (1:0.5–1:1 cement/sand ratio), to enable both the production of a thin component and the incorporation of a large fibre content.

Five different production processes have been developed for making GRC components: premixing, spray-up, extrusion, winding and lay-up of mats. Some of

(a)

(b)

Figure 8.9 GRC precast cladding panels: (a) As cast; (b) assembled in a structure. Photographs courtesy of Pilkington Brothers PLC, UK.

Table 8.3 Comparison of the 28 days properties of GRC composites of paste matrix produced by different methods (after Majumdar and Nurse [1])

Method of manufacture	Spray-suction	Hand-moulded premix	Extruded premix	Pressed premix
Flexural strength, MPa	21.2	9.8	18.0 [a]	13.5
Impact resistance (Charpy) kJ/m^2	8.1	8.6	9.0	9.2
Density (oven dry) kg/m^3	2180	1600	1740	1800
Glass fibre (wt, % of dry materials [b]	3.25	2.5	2.5	2.5
Water/cement ratio	0.30	0.40	0.26	0.10

Notes
a Tested parallel to the extrusion direction.
b 20 mm long.

these processes can be further modified by applying pressure and/or vacuum-dewatering immediately after casting. Premixing, spray-up and winding are the main methods currently used in the production of GRC components, and only they will be discussed further.

8.3.1 Premixing

The premixing process is based on mixing the fibres with the matrix followed by moulding. It has some inherent difficulties: uniform dispersion of the fibres is difficult to achieve, and usually a higher w/c ratio is required. Increasing the intensity of the mixing to overcome such difficulties may lead to other adverse effects, such as fibre damage and separation of the strand into smaller units (filamentization), thus reducing the workability of the fresh mix. These limitations usually lead to a GRC of lower quality than that produced by the spray-up method (Table 8.3). This is due in part to the higher w/c ratio matrix, and in part to the 3D fibre distribution, rather than the 2D distribution in the spray-up process. Special mixers and mixing procedures have been developed to minimize such difficulties, and to enable an increase in the fibre content. The use of water reducing admixtures is essential for these formulations.

8.3.2 Spray-up

The spray-up process is based on the direct application of a GRC mix onto the surface of a mould, by using a special spray head, shown schematically in Figure 8.10.

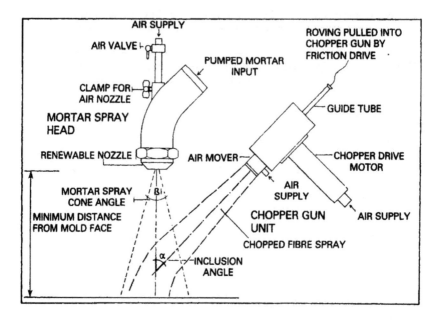

Figure 8.10 Schematic description of a GRC spray head (after True [31]).

The head consists of two units, one to chop and spray the fibres and the other to spray the mortar. The two streams meet at the mould surface and mix into a well-dispersed composite. The glass fibres are fed into the chopping unit from a continuous roving, while the mortar mix is pushed into the mortar spray head by a pumping unit.

This technique produces a high quality material, with a relatively large volume of fibres which are dispersed relatively uniformly in two dimensions. These characteristics show up clearly when comparing the properties of various GRC composites (Table 8.3). The glass fibre content can be controlled by changing the rate of flow through each of the nozzles. The spray head can be operated manually, or used in an automated process. The whole process is continuous and can easily be adapted to produce components of both simple and complex shapes. This versatility and the simplicity of the production process, is one of the main advantages of this technique.

The spray-up product, as well as the one produced by premixing, can be made in layers, with each being compacted by a roller and fitted to the geometry of the mould. Further densification can be obtained by pressing and vacuum-dewatering. The effects of such further treatments are demonstrated in Table 8.3, showing the increase in flexural strength of the premix composite with further pressure processing.

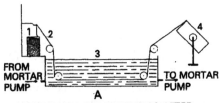

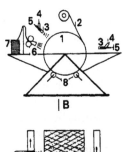

A WINDING OF GLASS ROVINGS AFTER
  IMPREGNATION IN BATH OF CEMENT PASTE

  1. Bobbin
  2. Roving
  3. Cement Paste
  4. Winding Frame

B WINDING OF GLASS ROVINGS WITH
  ADDITIONAL SPRAYING OF CHOPPED GLASS
  FIBRE AND CEMENT PASTE

  1. Winding Cylinder
  2. Glassfibre Roving
  3. Cement Paste Spray
  4. Compressed Air
  5. Cement Paste
  6. Chopped Roving
  7. Roving
  8. Compaction Rollers

Figure 8.11 Schematic description of production of GRC composite by winding (after True [31]).

8.3.3 Continuous reinforcement

Use of continuous reinforcement in the form of mats, fabrics and filament winding has been reported in several references [31,40–45].

The winding process, shown schematically in Figure 8.11 consists of drawing the reinforcing glass fibres through a cement slurry bath, and then winding them around a rotating mould. This can be combined with additional spraying of the paste matrix and chopped fibres, as well as compaction by compressive rollers (Figure 8.11(b)). With this process, it is possible to incorporate a higher fibre content, with better control of its dispersion, orientation and mechanical properties [42,43]. Continuous glass rovings were used for production of corrugated sheets of hybrid reinforcement of the rovings with polypropylene network [42].

Mats have been used to prepare special cladding components, such as a composite with lightweight core with external faces of cement reinforced with woven mat [41] (see also Chapters 13 and 14).

8.3.4 Curing

The curing stage is not essentially different from that in normal concrete technology, except that the GRC product is much more sensitive to the deleterious effects of improper water curing. The higher surface area and the low thickness of the GRC component can lead to greater rates of drying, and the resulting shrinkage can lead to distortion and warping of the component, as well as to reductions in its strength. It is interesting to note that the effect of poor water curing can have quite different influences on the first crack strength and the flexural strength. The former decreases with poor curing while the latter may sometimes increase [31]. This difference may be related to the fact that the first crack is a matrix controlled property (and poor curing results in weaker matrix), while the flexural strength is also influenced by the fibre–matrix bond which may increase as the matrix shrinks during poor water curing, resulting in a higher frictional resistance.

Special methods have been suggested to reduce the sensitivity to poor and non-uniform water curing. The addition of a polymer latex has been reported to be effective in eliminating the adverse effects of lack of water curing. It has been suggested that for AR-GRC, the addition of 5% polymer solids by volume, without any moist curing, may replace the recommended practice of seven days curing in a composite without the polymer [45].

8.4   Properties and behaviour of GRC composites

The properties of GRC composites are a function of the production process (Table 8.3), fibre content and length. When considering the long-term performance, the environmental conditions and the type of glass fibres (E vs. AR glass fibres) must also be considered. Even for fibres that are classified as AR fibres, differences in the ageing characteristics are observed. In addition, the mechanical behaviour of the GRC composite can be analysed on the basis of some of the concepts discussed in Chapters 3 and 4. However, to be applied to this composite, the special microstructure of the GRC must be considered (see Section 2.3.2 and Chapter 5).

8.4.1   Mechanical properties of unaged GRC composites

The mechanical properties of GRC composites have been studied mostly in composites produced by the spray-up system or its modification by vacuum dewatering. In these thin sheet components the fibres are essentially dispersed in two dimensions, and their content is typically about 5% by volume. With this range of reinforcement, the fibres are effective in enhancing tensile strength and ultimate strain, as shown by the stress–strain curves in Figure 8.12. The increase in fibre content leads to an increase in first crack stress, tensile strength and ultimate strain (Figure 8.12(a)) as well as in flexural strength (Figure 8.13(a)) and impact resistance (Figure 8.13(b)). An increase in fibre length leads to an improvement in properties, but not to the same extent as an increase in the fibre content (Figures 8.12 and 8.13). It should be noted from Figure 8.12(a) that the rate of increase in strength with increasing fibre content is rapid initially, but tends to slow down or even decrease as the fibre content exceeds ~ 6% by volume. This

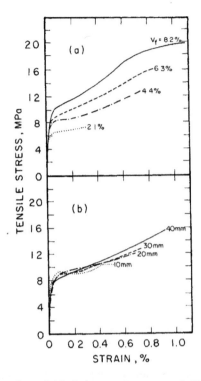

Figure 8.12 Stress–strain curves of GRC composites produced by the spray-up vacuum-dewatering method (after Ali *et al.* [46]). (a) Effect of fibre content (fibre length 30 mm); (b) effect of fibre length (fibre content 4% by volume).

is apparently the result of the reduction in density as the fibre content increases, in particular over 6% (Figure 8.14). This reflects the difficulties in compaction at higher fibre contents, and suggests that the optimal fibre content is about 6%. Various specifications (e.g. [33]) recommend fibre contents in this range. The impact resistance does not, however, show a trend to levelling off at 6% fibre content.

The data in Figures 8.12 and 8.13 indicate that tensile and flexural strength values of about 15 and 40 MPa, respectively, can be obtained with fibre contents of ~5% and lengths of ~20 mm. This is considerably higher than the matrix strength, and is accompanied by an increase of more than an order of magnitude in the ultimate strain in tension, or the impact resistance. The values reported in Figures 8.12 and 8.13 are for air-stored composites, and they tend to be somewhat lower for water stored composites [46]. This is true for the ultimate strength properties (tensile or flexural strength and impact resistance), though first crack strength is higher in the water stored composites.

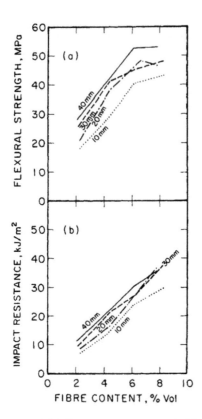

Figure 8.13 Flexural properties of GRC composites produced by the spray-up vacuum-dewatering method (after Ali *et al.* [46]). (a) Flexural strength; (b) Impact resistance (Izod).

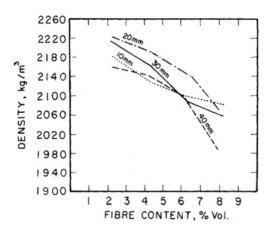

Figure 8.14 Effect of fibre content on the density of GRC composite produced by the spray-up vacuum-dewatering method (after Ali *et al.* [46]).

8.4.2   The mechanics and microstructure of FRC composites

8.4.2.1   Internal structure

Most of the GRC production processes lead to dispersion of the roving into the individual strand units. However, the multifilament strand is not broken up and maintains its bundled structure (see Section 2.3.2). This special microstructure is shown schematically in Figure 8.15, [47] where it is represented in terms of an idealized unit cell. The spaces between the bundled filaments are mostly vacant after the first few weeks, as they are too small to be penetrated by cement grains (Figure 8.15 and Figure 2.3 in Chapter 2). However, when kept in a moist environment for a sufficiently long time they eventually become filled with reaction products [11,20,27,48,49]. The nature of these products is dependent on the type of glass fibre; it tends to be massive CH in the case of the first generation of AR glass fibres (CemFIL-1, Figure 2.11(a)), but more porous material in the case of the second generation of AR glass fibres (CemFIL-2, Figure 2.11(b)). This difference

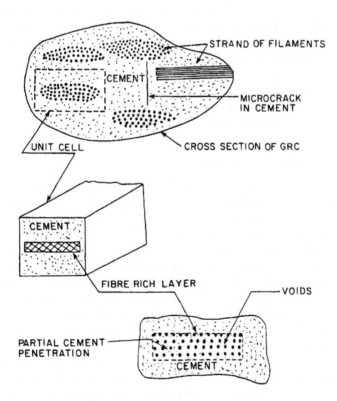

Figure 8.15 Schematic description of the internal structure of GRC composite, presenting an idealized unit cell (after Nair [47]).

may be the result of the modified surface of the CemFIL-2 fibres, which are treated to improve their durability; if the deposition of the hydration products between the filaments is by a nucleation on the fibre surface followed by a growth stage, then the modification in the fibre surface can change the composition and microstructure of the products deposited around it.

The composition of the matrix may also affect the nature of hydration products developing and depositing around the strand and within it. The presence of supplementary cementing materials, and in non-Portland cement matrices, the growth into the strand may be reduced and the nature of products deposited there may be much more porous, affecting the mechanical properties of the composite (see Section 8.6.2.2).

The significance of the strand microstructure and the deposition of hydration products in it led to several studies in which indentation and micro-indentation techniques were applied to characterize the strand and the nature of the matrix around individual filaments within the strand [50–52]. Zhu and Bartos [51] applied micro-indentation techniques to push in individual filaments to determine the resistance to their slip and through that assess the development of the microstructure within the strand. They demonstrated that in young composites the microstructure and interfacial bond of individual filaments was weaker within the fibre bundle than at the bundle–matrix interface. Upon ageing the inner microstructure within the strand became denser and stronger, and the resistance to the push-in force of the internal filaments increased to become similar to that of the external filaments. Purnell *et al.* [52] resolved by thin section petrographic techniques similar characteristics of strand filling upon ageing, but commented that the deposits of hydration products within the strand were not uniform and monolithic.

8.4.2.2 Mechanics of the composite

The shape of the stress–strain curves of GRC composites, such as those shown in Figure 8.12, can be predicted by the ACK model (Chapter 4), as it clearly shows the three zones predicted by this model: elastic, multiple cracking and post-multiple cracking. Thus, the concepts of this model, and their various modifications, have been used to analyse the mechanics of GRC when loaded in tension [46,53–56]. The tensile stress–stress relation was used to predict the flexural behaviour, based upon the concepts discussed in Section 4.7 [57]. An essential element in an analysis of this kind is the determination of the fibre–matrix interfacial bond strength.

The special bundled structure of the reinforcing strand poses a particular problem, since the fibre surface in contact with the matrix is not well defined; the external filaments in the strand are expected to be bonded to a greater extent than the inner ones. Thus, there is a difficulty in interpreting the results of pull-out tests of strands in terms of bond stresses, as well as in calculating the actual effective bond in the composite, even if the bond of a single filament surrounded by the matrix is known. To deal with this problem it has been suggested that the value of the shear flow strength, q_{fu}, be used, expressed in units of force per length

(Section 3.2.5). Bartos [58], Laws *et al.* [59] and Oakley and Proctor [55] estimated by microscopy techniques the perimeter, p, of the filaments in the strand in real contact with the matrix, and determined the actual bond strength, τ_{fu}. Oakley and Proctor [55] reported a value of 1.1 MPa in the unaged composite. The estimated critical length from such data is about 20–30 mm. De Vekey and Majumdar [60] reported a higher value of about 5–10 MPa in pull-out tests of single glass filaments of ~ 1 mm diameter. Oakley and Proctor [55] also estimated the bond strength on the basis of the crack spacing in the loaded composite (Eq. 4.65, Section 4.5) and found it to be in the range of 0.2–1.1 MPa, with the lower value being in a high sand content composite (sand/cement ratio of 1.5 and water/cement ratio of 0.44) and the higher one in a composite with a pure paste matrix (water/cement ratio of 0.32). The efficiency factors for tensile strength can be determined on the basis of Eq. 4.28, and they were found to be about 0.27 for the longitudinal direction [55,61] and 0.17 for the transverse direction [55].

In order to better understand the mechanics of strand reinforcement, which is characteristic of GRC, several studies in recent years combined micromechanical testing with advanced microscopical observation of the strand before and after loading, as well as during loading [51,52,62,63–65]. Zhu and Bartos [51] applied micro-indentation techniques to characterize the micromechanics of strand reinforcement (Figure 8.16). Their observation that in an unaged Portland cement composite the resistance of the outer filaments to push-in was higher than that of the interior filaments is consistent with a recent report by Banholzer [65] in which a novel technique of pull-out combined simultaneously with microscopical observation using light transmission was developed. Comparison of the pull-out behaviour with image analysis of the strand in the course of pull-out indicated the presence of sleeve filaments (i.e. external filaments) which break during the loading, whereas the internal filaments pull out (Figure 8.17). These characteristics support an earlier concept of Bartos [58] who proposed a model of the unaged bundle in which the pull-out of a strand can be described in terms of a telescopic mechanism, which is induced by the distribution of bond within the bundle. Observations obtained by *in situ* loading of strands within the SEM are consistent with this model [62].

Particular attention was given to the enhancement in the first crack stress due to the glass fibre addition (Figures 8.12 and 8.18). This could be predicted on the basis of the crack suppression equation (Eq. 4.49) developed by Aveston *et al.* [66] by using bond strength values in the range of 1–3 MPa [46,55], the perimeter of the strand in direct contact with the matrix, and an efficiency coefficient in the range of 1/2–3/8.

8.4.3   Long-term mechanical properties and behaviour

Ageing of GRC composites in a wet environment leads to a change in properties, as seen from the stress–strain curves in Figure 8.19. The most significant change is the marked reduction in the ultimate strain in natural weathering and water storage, accompanied by an increase in the first crack stress and a reduction in the tensile

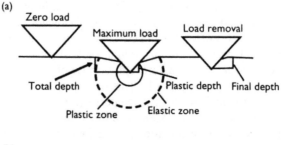

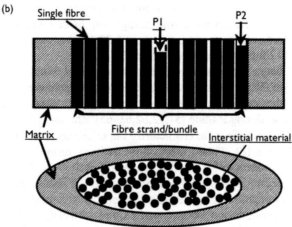

Figure 8.16 Schematic diagram of (a) the microstrength test, and (b) the fibre push-in test (after Zhu and Bartos [51]).

strength [67,68]. The composite kept in dry air maintained its initial properties. Similar trends were observed for flexural properties, showing stability in dry air but a gradual decline in flexural strength (Figure 8.20(a)) [69] and a more rapid decline in toughness (estimated by impact resistance), as seen in Figure 8.20(b) and Table 8.4 for water stored specimens. The rate of decline in properties is smaller for natural weathering (in the UK environment) compared with 20°C water storage [70] and higher residual properties are maintained at higher fibre contents [68]. It should be noted that the decline in flexural strength tends to level off (Figure 8.20), while the impact resistance is reduced by about an order of magnitude (Table 8.4), and the composite is not much tougher than the matrix itself.

The better preservation of the strength reflects the fact that the composite cannot become weaker than the matrix cracking stress. At this point, as the strength is reduced to its lower limit, the properties of the composite are not much different from those of the matrix, which is characterized by a very low toughness, but with a

(a)

(b)

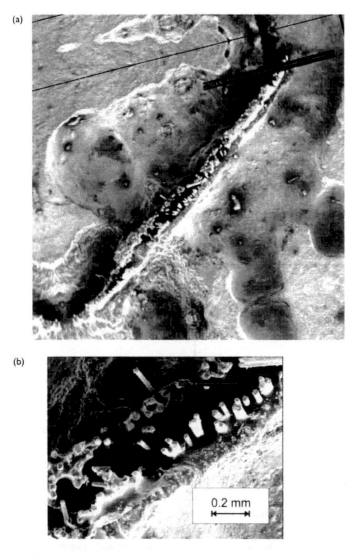

0.2 mm

Figure 8.17 SEM micrograph of a strand after pull-out test showing the external (sleeve) filaments which have been fractured during the loading, and the empty space in the core from which filaments pulled out without fracturing (after Banholzer [65]).

sufficiently high tensile or flexural strength. Thus, judging the ageing performance of the composite on the basis of tensile or flexural strength only may be misleading; it should also be assessed by some measure of toughness. The toughness is much more sensitive to ageing, since the main effect of the fibres is to increase this

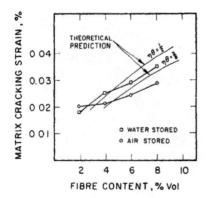

Figure 8.18 The effect of glass fibre content on the matrix cracking strain (after Ali *et al.* [46]).

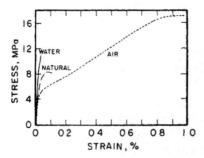

Figure 8.19 Effect of ageing for 5 years on the stress–strain curves of AR-GRC composites (after Singh *et al.* [68]).

property (by more than an order of magnitude compared with the brittle matrix), with only a modest effect on increasing the strength.

The rate of ageing is also a function of the type of glass fibre. With E glass it is much faster [1], while with the newer generation of AR glass fibres (e.g. CemFIL-2) it is much slower than for the earlier products (such as CemFIL-1) [29,71] which were described in Figures 8.19 and 8.20. Such differences are demonstrated in Figure 8.21. Improved AR glass fibre formulations [28, 30] are expected to lead to even better preservation of properties.

The ageing performance of the composite is also sensitive to the weathering conditions. The rate of reduction in strength increases with the relative humidity [69] and the temperature [25]. Litherland *et al.* [25] found correlations between the rates of loss of flexural strength at various exposure sites and the average yearly temperature at those sites. This is an indication that the reduction in mechanical

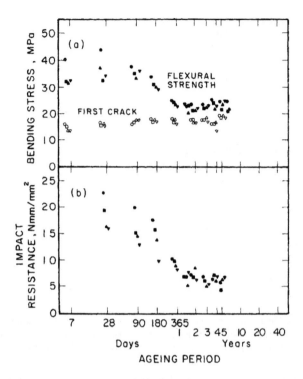

Figure 8.20 Changes in the properties of AR-GRC after water storage at 20°C (from [69]).
(a) Flexural strength and matrix cracking stress (LOP, limit of proportionality);
(b) Impact resistance.

Table 8.4 Estimated mean properties of spray-up vacuum dewatered
AR-GRC composites (5% wt. Glass fibres) after 20 years
(from [70])

Property	Unaged	Aged		
		Air	Water	Weather
Bending				
Strength (MPa)	35–50	26–34	15–20	13–17
First Crack (MPa)	14–17	14–16	15–18	13–16
Tensile				
Strength (MPa)	14–17	11–15	6–8	5–7
First Crack (MPa)	9–10	7–8	6–8	5–7
Modulus of elasticity (GPa)	20–25	25–33	28–34	25–32
Impact resistance (Nm/mm^2)	17–31	14–20	2–3	2–4

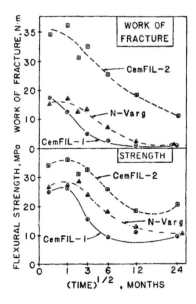

Figure 8.21 Different ageing trends observed with various commercial AR glass fibres estimated by changes in toughness (area under the load-deflection curve) and flexural strength, after water storage at 20°C (after Bentur *et al.* [71]).

properties is the result of processes which are temperature controlled. Yet, results such as those reported in reference [69] may suggest that the influence of humidity should not be ruled out completely.

Other long-term properties which were evaluated are creep and fatigue endurance. With regard to creep and fatigue [72,73] the performance was largely a function of the applied stress, whether below or above the first crack. The fatigue life was considered to be adequate when the applied stress was kept below the first crack stress [72]. Creep strains are very large when the applied stress is close to the first crack stress [74,75]. Allen [74] suggested that at higher stresses there is also a risk of sudden premature failure. Humidity variations were found to have a significant effect on creep as might be expected from a composite with a cementitious matrix.

Freeze–thaw durability was evaluated, showing adequate long-term performance. This was attributed to the reinforcing effect of the fibres [76] as well as to the lower absorption of the matrix [77]. The latter effect was reported for a polymer latex modified matrix.

8.4.4 Accelerated ageing

The necessity of finding methods for the rapid assessment of the long-term performance of GRC composites, in order to predict the service life in various climates, and the potential improvements by using new glass compositions and modified matrices, has led to extensive study and development of accelerated test methods. The most common accelerating environment is immersion in hot water for various periods of time, and determining the change in the mechanical properties induced by such exposure [25,78–80]. The properties evaluated are those of the composite (usually flexural behaviour) and of the fibres, by the SIC test described in Section 8.2 (Figure 8.4). The temperature most commonly used is 50°C [25] which is sufficiently high to accelerate the ageing process, but is believed to be low enough not to change the nature of the processes taking place at lower temperatures.

Good correlations were found between the behaviour in natural and accelerated ageing, and quantitative relations could be established between changes in the fibre strength (SIC test) and the flexural strength of the composite in accelerated and natural environments [25,78–80]. Typical ageing curves of these two properties are shown in Figure 8.22, demonstrating a decline with a tendency for levelling off. The data were used to construct normalized Arrhenius curves, by plotting the logarithm of the time (for a given strength loss) at some temperature, T, relative to the time at some chosen standard temperature (50°C), against $1/T$ (Figure 8.23). The natural weathering data were analysed in terms of the average annual temperature at the site. All of the data (SIC and composite flexural strength obtained in accelerated and natural weathering) could be described by a single linear relationship.

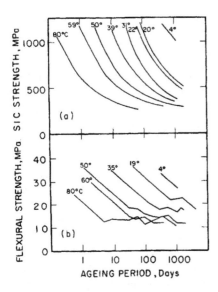

Figure 8.22 Effect of ageing on the strength retention (after Litherland et al. [25]). (a) SIC strength retention in water at various temperatures; (b) Flexural strength retention of the GRC composite in water at different temperatures and at natural weathering.

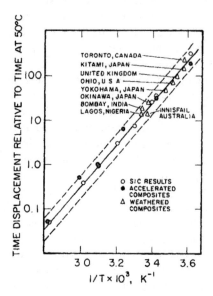

Figure 8.23 Normalized Arrhenius plots of strength retention (SIC test and composite flexural strength) for specimens in accelerated and natural weathering (after Litherland et al. [25]).

The excellent correlation shown in Figure 8.23 enables the prediction of the changes in the flexural strength of the composite in natural weathering from the accelerated test data, by considering the average annual temperature on the site and determining an 'acceleration factor' from Figure 8.23. This factor is the ratio between the life time at a given climate (described by the single parameter of average annual temperature) and the time in accelerated ageing. This procedure was demonstrated by Aindow et al. [78] for the UK weather, and good agreement was observed between the flexural strength predicted from the accelerated tests and the values actually observed on site (Figure 8.24). This procedure was also used to compare the ageing curves in various climates, in particular hot ones, by treating the data at the hot site as if they were accelerated ageing data in relation to the cooler UK climate. For this purpose 'acceleration factors' relative to the UK mean annual temperature were calculated from Figure 8.23 (Table 8.5). The resulting curve is shown in Figure 8.25, demonstrating again that all the data can be described by a single curve. The levelling off at prolonged ageing, which is confirmed in this figure, is particularly important for design purposes which must account for long-term performance (Section 8.7).

It is interesting to note that curves of this kind imply that the over-riding parameter in controlling ageing is temperature, and the influence of humidity can be ignored. This was confirmed particularly for the two Australian sites (Figure 8.25)

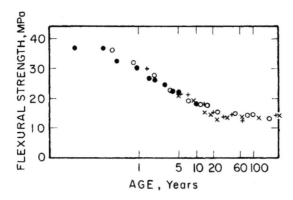

Figure 8.24 Flexural strength retention of AR-GRC in UK weather (•) compared with predictions from accelerated ageing tests at 50°C(O), 60°C(+) and 80°C(×) (after Aindow *et al.* [78]).

Table 8.5 Acceleration time factor of ageing of GRC in hot climates relative to the cooler UK weather (after Aindow *et al.* [78])

Site	Mean annual temperature, °C	Time equivalent to UK
United Kingdom	10.4	1
Arizona, USA	21.7	4.2
Innisfail, Australia	23.6	5.5
Cloncurry, Australia	26.2	7.3
Lagos, Nigeria	26.7	7.8
Bombay, India	27.6	8.4

where one is hot and dry (Cloncurry) while the other is hot and wet (Innisfail). This is somewhat contradictory to laboratory results which indicate excellent retention of properties when the relative humidity is about 50%.

This method of interpreting accelerated ageing tests was also found to be applicable for predicting the improvement in long-term performance when using glass fibres of different composition [71,79]. Therefore, it is a valuable tool for development purposes as well as for the assessment of the quality of different commercial products.

It has been suggested [25,79] that the excellent agreement between the Arrhenius plots of the SIC test and the flexural strength of the actual composite (Figure 8.23) is an indication that changes in the strength of the AR glass fibre due to its insufficient alkali resistivity is the process which controls the ageing of the composite. However, it has been shown [79] that such an agreement cannot be obtained when other properties of the composite, in particular those related to toughness (e.g. ultimate strain in tension and impact resistance) are analysed in a similar way.

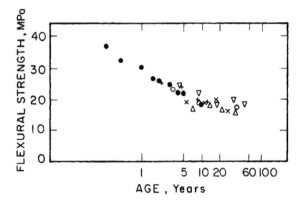

Figure 8.25 Flexural strength retention of AR-GRC in UK weather (•) compared with predictions from real weather results obtained over several years in warmer climates, Arizona (+), Innisfil (×), Cloncurry (Δ), Lagos (O) and Bombay (∇) (after Aindow *et al.* [78]).

This implies that other ageing mechanisms are also involved. Thus, for the prediction of toughness retention, these procedures may be adequate for the comparison between different products, but fall short of providing quantitative service life prediction as they do for flexural strength.

8.4.5 Dimensional changes

GRC composites can undergo large dimensional changes induced by drying shrinkage. Values in the range of 0.05–0.25% have been reported, depending on the sand content of the matrix [34,81–83] (Figure 8.26). These values are much higher than those typical for normal concretes (~0.05% or less). In these composites, shrinkage is largely a matrix-controlled property. Thus, the higher shrinkage of GRC composites compared with concrete is due to the low aggregate (i.e. sand) content in the GRC and the high cement content. The high shrinkage is a source of potential cracking problems especially in elements which are restrained (see Section 8.7). One of the most efficient means of reducing the shrinkage is by increasing the sand content.

However, a high sand content may lead to a reduction in the strength of the composite (Figure 8.27). Usually, a sand:cement ratio of 0.5:1 is recommended, which is sufficient to reduce considerably the shrinkage of a paste matrix composite (Figure 8.26) without being accompanied by a marked reduction in mechanical properties.

Volume changes due to thermal effects are also matrix dependent. The thermal expansion coefficient of GRC is in the range of 11×10^{-6} to 16×10^{-5} °C^{-1} [33], similar to that of the mortar matrix.

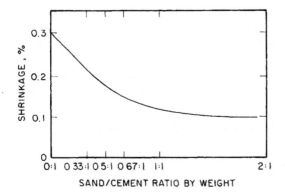

Figure 8.26 The effect of sand/cement ratio on the shrinkage of GRC composites (from [34]).

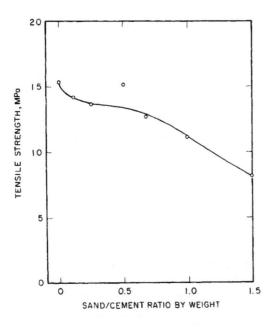

Figure 8.27 The effect of sand/cement ratio on the tensile strength of GRC composites, derived from data presented by Oakley and Proctor [55]).

8.4.6   Fire resistance

The GRC composite is made of non-combustible ingredients and therefore it can provide resistance to the passage of flames. However, from the point of view of fire endurance, the performance is dependent on the configuration of the panel and the nature of the insulation or core material backing up the GRC skin. Some recommendations and test results of the fire endurance time as a function of panel configuration are compiled in [33].

Majumdar and Nurse [1] pointed out the risk of making a matrix which is too dense and does not allow the escape of water during a fire, which can lead to explosive failure. They indicated, however, that the fibres can play an important positive role by holding the matrix together even if it is completely dehydrated.

8.5 Ageing mechanisms of GRC composites

The changes in the properties of the GRC composite during ageing, and in particular the loss in toughness (i.e. embrittlement) are an extremely important characteristic of this material. This problem is a central one for the application of GRC in the construction industry, and it is dealt with at two different levels:

1   Development of composites of improved durability performance using modified glass and matrix formulations.
2   Design of GRC components taking into account the changes in properties during ageing.

These two topics will be further discussed in Sections 8.6 and 8.7. However, a comprehensive treatment of these two aspects requires a more detailed analysis of the processes which lead to this change in properties.

It is now well established that the ageing of GRC cannot be simply attributed to alkali attack of the glass fibres by the highly alkaline cementitious matrix. Two mechanisms must be considered to account for the reduction in strength and loss in toughness of this composite:

1   Chemical attack of the glass fibres – chemical attack mechanism.
2   Growth of hydration products between the glass filaments – microstructural mechanism.

There is some controversy regarding the relative importance of these two mechanisms. Resolving these aspects of the ageing process is not merely of academic interest, it is essential to clarify this question if composites of improved long-term performance are to be developed and used. In particular, should the effort be directed towards improvement of the fibres, or modification of the interface through the matrix composition? In this section a critical review of this problem will be presented.

8.5.1   Chemical attack mechanism

The nature of the chemical attack, its causes, and the methods used to alleviate its adverse effect by changing the glass composition have been discussed in

Section 8.2. The extent of this problem can be evaluated most readily by determining the change in the strength of glass fibre strands exposed to an alkaline environment, either in solutions or in a cementitious matrix. Tests of this kind clearly reveal the rapid strength loss due to chemical degradation in the case of E glass (Figure 8.6). Even with commercial AR glass fibres, a reduction in strength of glass filaments removed from GRC aged in a wet environment could be observed [26] (Figure 8.7). A greater strength loss was reported in the SIC tests [25]. Thus, even the commercial AR glass fibres are not completely immune to chemical attack. However, data such as those shown in Figure 8.7 indicate that the reduction in strength takes place mainly during the first 2–3 years, and then the strength becomes stable, for at least up to 10 years. Also, the residual strength of the fibres is quite high, being 1000 MPa in the data in Figure 8.7 of West and Majumdar [26] and more than 500 MPa in the SIC tests reported by Proctor *et al.* [29]. This is sufficiently high to provide useful reinforcing effects.

8.5.2 Microstructural mechanisms

Marked microstructural changes can take place at the fibre–matrix interface as the composite ages in a humid environment (Sections 2.3.2 and 8.4.2.1). This is the result of the special bundled structure of the reinforcing strand (Figure 2.3 in Chapter 2), where the spaces between the filaments are initially largely empty (Figure 2.10 in Chapter 2), and then gradually become filled with hydration products that in the case of a Portland cement matrix are in the form of dense CH crystals [20,27,48,49] (Figure 2.11(a) in Chapter 2). It has been suggested that the densification of the matrix micro-structure at the glass interface can lead to embrittlement [20,71,84]. This can be explained in terms of three mechanisms, or a combinations of the three: (i) effective increase in the fibre–matrix bond [85]; (ii) generation of local flexural stresses [85]; (iii) development of stress concentrations resulting in notching or surface flaw enlargement which lead to static fatigue effects [21,52,86,87]. A general treatment of these concepts is provided in Chapter 5. The discussion here will highlight their application to the specific nature of GRC.

(i) Fibre–matrix bond   The improvement in bond in fibrous composites is known to be a possible cause for the reduction in toughness. This was discussed in Section 5.2 and presented graphically in Figure 5.2, on the basis of a calculation which takes into account the change of the failure mode from fibre pull-out to fibre fracture, as the bond increases. The results of such a calculation, assuming typical GRC parameters (fibre length – 37.5 mm, fibre diameter – 10 μm, fibre strength – 1500 MPa) are shown graphically in Figure 8.28, as work of fracture (toughness) vs. interfacial bond strength. It can be seen that increasing the bond strength beyond ~0.2 MPa results in a loss in toughness. An increase of this kind may be expected due to the microstructural changes discussed earlier. Investigators have shown that the interfacial bond strength of glass filaments surrounded completely by cement matrix ranges from about 1 MPa [55] to 5 MPa [51]; these are well

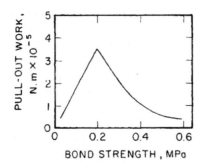

Figure 8.28 Pull-out work vs. bond strength calculated for GRC composite (after Bentur [11]).

within the range in which the work to fracture becomes small due to the high bond. Results of the effect of ageing on pull-out tests indicated an increase in the bond strength with ageing when single filaments were studied [60]. Zhu and Bartos [51] demonstrated that in Portland cement matrices, the resistance to push-in of the inner filaments in the strand, relative to the external ones, increased during ageing, from about 30% to 100%, after 40 days of accelerated ageing at 60°C. This correlated with the reduction in flexural strength and toughness of the composite. In a 25% metakaolin matrix, the resistance ratio of the inner filaments relative to the external ones remained low during the ageing period, below 20%, in agreement with the preservation of flexural strength and toughness of the composite made from this matrix. This was consistent with *in situ* SEM testing [62] showing pull-out of filaments in unaged composite and their fracture in an aged composite.

Mobasher and Li [88] demonstrated that the changes in the pull-out of glass fibres upon ageing could be accounted for by an increase in bond as well as the stiffness of the interfacial matrix. Using a fracture mechanics model they calculated the increases in adhesional resistance, frictional resistance and stiffness at the interface from 0.62 N/mm, 0.59 N/mm and 0.04 mm^{-1} to 1.23 N/mm, 0.88 N/mm and 0.15 mm^{-1}, respectively (for 3 days of accelerated ageing). These changes could account for a change in the pull-out curve, from one which exhibits elastic–plastic behaviour to one which shows a sharp decline after the peak.

There are however other reports in which inconclusive data regarding relations between bond enhancement and composite performance are presented. Laws *et al.* [59] observed an increase in bond during the ageing of CemFIL-2 in water at 20°C and CemFIL-1 in water at 50°C, but no change in bond when CemFIL-2 was aged in 50°C water. Purnell *et al.* [64] observed an increase in bond upon ageing of GRC composites with different matrices, but this increase was not necessarily associated with degradation in mechanical properties: the bond increase was similar in Portland cement and in a 25% metakaolin matrix, but the Portland cement-matrix composite showed a much more rapid decline in strength and toughness.

(ii) Local flexural stresses It was shown in Section 3.3 that when fibres oriented at an angle bridge across a crack, local flexural stresses can develop. If the matrix around the fibre is dense, it will resist the bending, and local flexural stress will develop. For high modulus and brittle fibres such as glass, this may lead to premature failure and decline in strength and toughness, as outlined in Section 5.2. Aveston *et al.* [89] showed that in the case of a glass fibre in a cementitious matrix, this local flexural stress can be higher by a factor of 7 than the tensile stress which is calculated by assuming pure tension in the fibre (Section 3.3, Figure 3.21). This bending stress will be reduced if the matrix around the fibre is weak and porous, and can crumble; such effect relieves the build-up of flexural stresses. Stucke and Majumdar [20] showed by a simple calculation that the densification of the matrix microstructure around the glass filaments in aged GRC could lead to local flexural stresses in the glass fibre which may exceed its tensile strength. Thus fibre fracture and embrittlement can occur even if the glass fibre does not lose strength due to a chemical attack. Bentur and Diamond [90] have shown by *in situ* testing in an SEM that such local flexure can occur even if the glass fibre strand is oriented perpendicular to the crack (Figure 5.5 in Chapter 5). The interaction of the crack with the glass fibre strand is associated with some shift in the crack path and the development of local bending. In order to bridge across the crack, the glass filaments must be able to accommodate large flexural deformations (Figure 5.5). If the spaces between the filaments are filled with hydration products, the strand becomes rigid since the freedom of movement of one filament relative to the other is lost or reduced; as a result, it cannot accommodate the large local flexural deformations, and brittle failure will occur.

(iii) Flaw enlargement and notching The bonding and flexural mechanisms to account for reduction in mechanical properties due to microstructural changes have been recently challenged by Purnell and co-workers [52,64,86,87], based on the observation of ageing of GRC with different types of matrices: Portland cement with 20% metakaolin and calcium sulphoaluminate cement.

Ageing curves of strength changes at 60°C of GRC with different matrices indicated that in modified matrices (Portland cement with 20% metakaolin and calcium sulphoaluminate cement) a two-stage process is obtained, with strength increasing at the beginning and reducing later on. This behaviour cannot fit the linear relation between strength and inverse of log time, which was used by Litherland *et al.* to develop plots of rate of strength loss v. *1/T*, to construct Arrhenius type relations (Figure 8.23). Since this did not hold true for the modified matrices, Purnell *et al.* concluded that the ageing cannot be simply described in terms of a single mechanism of growth of hydration products in between the filaments, with accelerated ageing being the result of a simple acceleration of the hydration process [86]. They added that this conclusion is supported by SEM observations that in some of the modified matrices there was growth of hydration products, and yet, the ageing in mechanical properties took a different course than that of a Portland cement matrix [52]. They proposed an alternative mechanism [87] based on the concept of enlargement of flaws, which is essentially a static fatigue type mechanism of a stress corrosion nature. They discussed several sub-failure driving stresses for such processes: (i) thermal stresses developed at the higher temperature ageing, or during temperature cycling in service, due to mismatch in the thermal coefficient of expansion of the fibres and the matrix; (ii) precipitation of CH and its nucleation at pre-existing flaws; (iii) preferential leaching of components from the glass surface. Based on the relation between the rate of growth of a critical flaw, which is dependent on the induced stress, OH$^-$ concentration and the temperature, and invoking a fracture mechanics approach, they determined the ageing curves in terms of a characteristic parameter for each matrix, k, which was obtained by curve fitting. The k parameter was shown to fit Arrhenius type equations for ageing at different temperatures, for the various matrices studied.

The notching of the glass filaments as an ageing mechanism was also suggested by Yilmaz and Glasser [21], based on the observation of growth of CH crystals at the glass surface, which seem to be inducing notches into it. This may be consistent with the observations by Purnell et al. [52].

(iv) Combined mechanisms The three mechanisms identified above may be acting simultaneously, or in sequence, and therefore it is difficult perhaps to distinguish between them, since they are all associated with deposition of hydration products between the filaments in the strand.

Leonard and Bentur [84] observed that when the hydration products grown are porous in nature the rate of degradation in mechanical properties is reduced. This was explained in terms of lower level of bond increase induced by such products, relative to that obtained when CH is deposited, which is characterized by a dense crystalline structure and intimate contact with the filaments. However, one can also make the argument that such porous hydration products are not inducing stress concentrations and notches, consistent with the model of Purnell et al. [87]. Therefore, one cannot rule out that the increase in bond and notching/crack growth mechanisms are occurring simultaneously, with both being dependent on the nature of hydration products deposition.

In an extreme case, when the spaces between the filaments were filled with minute silica fume particles that did not react with the surrounding matrix [91], no ageing occurred, although the spaces were filled with relatively dense material. This is an indication that indeed, it is not just filling of the space between the filaments which leads to ageing, but rather the nature of the material deposited, and its interaction with the glass. A weak interaction will not lead to ageing, and this can be predicted in terms of the various mechanisms suggested, whether bonding or crack growth.

In conjunction with the two stage process observed by Purnell et al. [86] in the modified matrices, one should note that this has been quantified by considering the bonding and flexural mechanism simultaneously, the first one leading to an increase in strength at the early stage of ageing and the second one becoming more effective later on, leading to strength reduction (Section 5.2, Figure 5.3). Such an approach was used to model the ageing of carbon fibres, and the same concept might be applied to glass fibres, to explain an increase in strength upon ageing, followed by a decrease.

8.5.3 The effectiveness of the various ageing mechanisms in controlling the long-term performance

The question of the relative importance of the two major ageing mechanisms described above is not clear, since both processes, fibre degradation and microstructural changes, occur at the same time, and it is difficult to devise a critical experiment to distinguish between the two. However, a critical analysis of the data published in the literature enables us to reach some conclusions at least regarding the qualitative trends of the effects of the two mechanisms.

GRC with E-glass fibres Observations of GRC composites with E-glass fibres indicate that, after several weeks of storage in water at 20°C, the glass fibres begin to corrode and etch pits are clearly evident on their surface (Figure 8.29(a)). But, at the same time, hydration products are seen to be deposited between the filaments (Figure 8.29(b)). Since the defects caused by the chemical attack are quite severe (Figure 8.29(a)), it is likely that the chemical corrosion is the overriding process leading to embrittlement.

GRC with AR glass fibres In the case of GRC with AR glass fibres, one would expect that if chemical attack was a major factor in controlling the long-term performance, the durability would be sensitive to the pH of the matrix. Proctor *et al.* [29] reported that there is such a correlation (Table 8.6), while Leonard and Bentur [84] showed that differences in durability characteristics could not be correlated with the pH of the matrix (Table 8.7). When considering such apparent contradictions, it should be borne in mind that the changes in pH were obtained by variations in the composition of the matrix, which were accompanied also by changes in the microstructure. Therefore, one cannot rule out the possibility that the improved ageing performance of the lower pH matrices in Table 8.6 is the result of differences in the microstructure of the matrix around the glass filaments. The differences in the performance of the GRC composites whose results are shown in Table 8.7 were explained on the basis of differences in microstructure [84]. The marked improvement in the durability of composites made using the Japanese low alkali CGC cement was attributed to microstructural changes at the interface rather than to its low alkalinity [30,92].

It is evident that some loss in strength of the AR glass fibres in aged GRC is taking place ([26], Figure 8.7). It is interesting to note that the curve appears to be levelling off, which is similar to the trend observed for the loss of the flexural strength of the composite. On the basis of extensive analysis, Litherland *et al.* [25] demonstrated that there is a quantitative relation between the strength retention of the glass fibres (estimated by the SIC test) and the flexural strength retention of the GRC composite (Section 8.4.4). This may indicate that chemical degradation is a dominant factor controlling the durability of GRC.

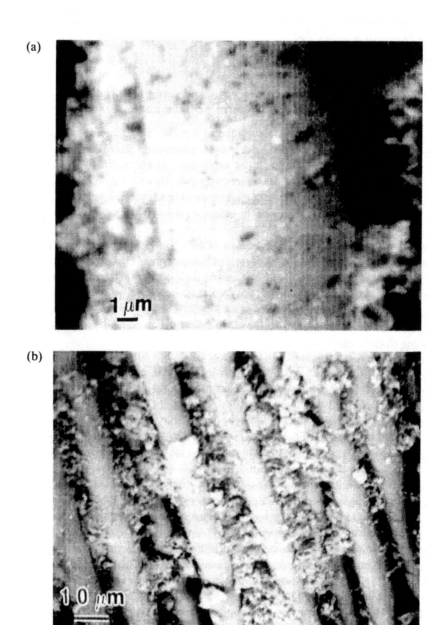

Figure 8.29 Microstructure of GRC with E glass fibres after ageing in water at 20°C for 2 months showing (a) degradation of the glass surface; (b) growth of hydration products between the filaments (after Bentur [11]).

Table 8.6 Effect of cement slurry pH on strength retention of CemFIL-1 fibre (after Proctor, Oakley and Litherland [29])

Cement	pH	SIC strength (MPa)[a]
RHPC[b]	13.0	500
RHPC[b] + 20% Danish Diatomite	12.6	700
Supersulphated cement	12.0	800
RHPC[b] + 40% precipitated Silica	11.0	950

Notes
a Strand in cement strength after accelerated ageing for 2 months in 50°C water. Similar trends were observed in the performance of the GRC composite.
b Rapid hardening Portland cement.

Table 8.7 Effect of matrix pH on retention of flexural strength and toughness of GFRC composites with AR fibres (after Leonard and Bentur [84])

Cement	pH	Retention of mechanical properties[a] % of unaged composite	
		MOR	Toughness
Portland cement	13.02	44%	5%
Portland cement + 35% Fly Ash B	12.91	48%	29%
Portland cement + 35% Fly Ash A	12.93	92%	76%
	12.93	92%	76%

Note
a After ageing in water at 20°C for 5 months.

However, if the toughness of the composite is considered, the GRC becomes brittle at about the point at which the decline in its modulus of rupture has levelled off [69,71]. These trends are demonstrated in Figure 8.30, which shows that the composite loses practically all its toughness, but only part of its flexural strength. Thus, although a correlation can be established between the flexural strength of the composite and the strength of the glass fibres [25], no such correlation is evident between the glass fibre strength and the toughness of the composite. In spite of the fact that the glass fibres retain a considerable portion of their strength even after prolonged ageing (Figure 8.7), the composite becomes brittle and loses more than 95% of its original toughness. The embrittlement observed at this stage can be accounted for by microstructural changes. In the specific example in Figure 8.30, at the time the composite became brittle (1/2 year in water at 20°C), the filaments were engulfed with massive CH [71]. In several other studies [30,71,84] it was also shown that improvement in the durability of GRC composites made with

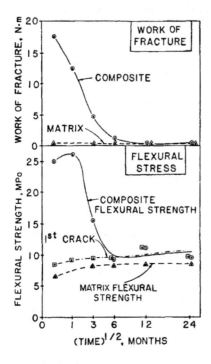

Figure 8.30 Effect of ageing in water at 20°C on the flexural strength and work of fracture (area under the load-deflection curve) of GRC composite with AR glass fibre (after Bentur *et al.* [71]).

different types of AR glass fibres or blended cement matrices, could be correlated with microstructural changes, but not with chemical attack on the glass fibre.

Examination of the glass surface of filaments removed from composites after 1/2 year of ageing in water at 20°C, when some of them became brittle (CemFIL-1) while others remained ductile (CemFIL-2), showed that in all of them hardly any surface damage was visible [71]. In a few filaments, some roughening of the surface was seen, but such roughening was shown to result only in a limited strength loss of the fibres [20]. On the other hand, in the composites which became brittle after 1/2 year in water at 20°C, dense CH had engulfed the filaments (Figure 8.31(a)) and the failure mode was brittle (Figure 8.31(b)). In the other composites which, at this stage of ageing, were still maintaining most of their toughness, hardly any hydration products were deposited around the filaments (Figure 8.32(a)) and the failure mode was ductile (Figure 8.32(b)).

It should be noted that, after prolonged ageing (5 months at 50°C water) signs of more severe chemical attack could be observed in all of the GRC composites. Few of the filaments showed signs of severe surface damage (Figure 8.33). However,

(a)

(b)

Figure 8.31 Microstructure of GRC with AR glass fibre (CemFIL-I) after 1/2 year of ageing in water at 20°C, showing (a) dense CH deposited around the glass filaments; (b) a brittle mode of failure (after Bentur et al [71]).

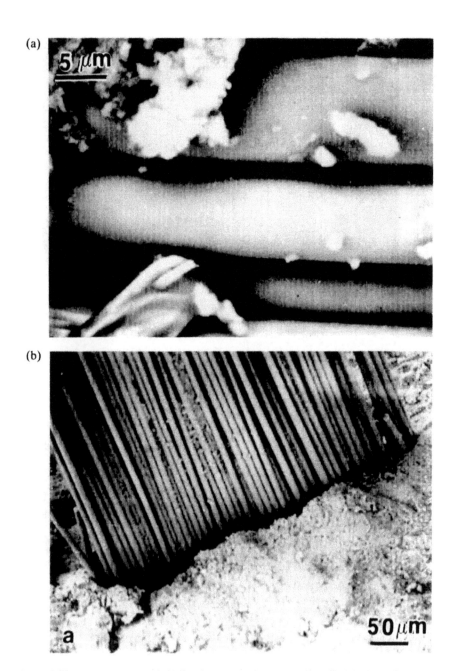

Figure 8.32 Microstructure of GRC with the AR glass fibres (CemFIL-2) after 1/2 year of ageing in water at 20°C, showing (a) empty spaces between the filaments; (b) a ductile mode of failure with pulled out fibres (after Bentur et al. [71]).

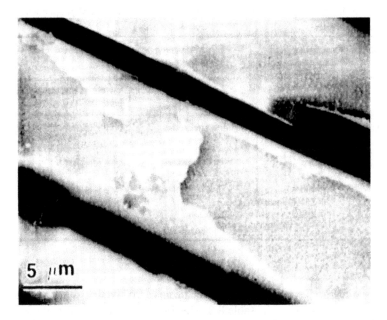

5 μm

Figure 8.33 Extensive surface damage due to chemical attack observed in an AR glass
filament removed from GFC composite aged in 50°C water for 5 months
(after Bentur [11]).

the embrittlement of the composites occurred at a much earlier stage, when no such
damage was seen, but when the matrix around the glass filaments became dense.

This discussion suggests that the major cause of the embrittlement of GRC reinforced with AR glass fibres is the microstructural changes resulting in deposition
of CH around the glass filaments. At the stage when the composite becomes brittle
($\sim$ 5–30 years of natural weathering or $\sim$ 1 year in water at 20°C) chemical attack
of the glass fibres is very mild. There is evidence that prolonged exposure to a
wet environment may result in a marked increase in the damage to the glass fibres
due to the chemical attack. However, the embrittlement occurs at an earlier stage,
when the chemical attack is very mild, if it occurs at all.

In a study of the effect of non-Portland cement matrices on the durability of
E and AR glass fibres [93] it was shown that in the case of E glass a markedly
improved performance was obtained when the alkalinity of the matrix was sufficiently low to delay or prevent alkali attack and the spaces between the filaments
were not filled with hydration products, or if filled with hydration products they
tended to be very porous and friable. In the case of AR glass fibres the condition for
marked durability improvement was associated mainly with the latter condition,
either (i) prevention of filling of the spaces between the filaments, or (ii) deposition
of hydration products which are porous and friable. Based on the studies reviewed

Table 8.8 Effect of ageing mechanisms on reduction in mechanical properties of
GRC

Type of fibre	Ageing period	Effect of ageing mechanism on reduction in mechanical properties	
		Chemical degradation of fibres	Growth of dense hydration products
E glass	Short (< 1 year)	Very effective	Mildly effective
AR glass	Short (< 1 year)	Not effective	Not effective
	Medium (5–40 years)	Mildly effective	Very effective
	Long (> 30–50 years)	Effective	Very effective

above the effectiveness of the various mechanisms involved in loss in strength and
toughness are summarized in Table 8.8. The Table provides a qualitative assessment of the influence of these processes, for GRC with E and AR glass fibres,
at different periods of natural ageing. These ageing periods should be viewed as
rough estimates, giving an order of magnitude only. The filling of spaces referred
to in this Table is for dense hydration products.

8.6   GRC Systems of improved durability

The discussion in Section 8.5 suggests that there is still room for improvement in
the long-term performance of GRC composites. If E glass fibres are to be used,
attempts should be made to seal the fibres completely from the matrix, or alternatively to use a very low alkali cementitious material. These means are essential,
since alkali attack is the main degradation mechanism in E glass. On the other
hand, in order to improve the durability of GRC with AR glass fibres, the main
effort should be directed towards modifying the microstructure of the matrix in the
vicinity of the glass filaments. This could be achieved by treatment of the glass
surface to modify the nucleation and growth of the hydration products in its vicinity,
or by changing the composition of the matrix. However, although improvements
achieved by these means could considerably increase the life expectancy of GRC
with AR glass fibres, it may not solve the durability problem completely, since
at later ages (Table 8.8) chemical attack may still become significant. Therefore, while there is the need to develop glass fibres of better alkali resistivity, the
more immediate problem which causes embrittlement at intermediate ages is a
microstructural one.

This section deals with various systems of improved durability, and discusses their performance in view of these mechanisms. A review of this topic covering developments until 1985 was presented by Bijen [94].

8.6.1   Composites with E glass fibres

Some of the early uses of GRC composites were based on E glass fibres in combination with high alumina cement, which is a relatively low alkali matrix (pH = 12) compared with Portland cement (pH ~ 13.5), and could thus lead to a reduced rate of chemical attack. This was one of the formulations developed by Biryukovich *et al.* [12] and a product of this type was used in the UK in the 1960s. Although this formulation led to an improvement in durability, the composite lost strength and toughness to a greater extent than a GRC with AR glass fibres made with the same matrix [1]. However, the main drawback was the limited durability of the high alumina cement itself (due to the conversion reaction), and in some countries this cement is excluded from use in the construction industry [17].

A variety of non-Portland cement matrices was evaluated for GRC composites using E glass [93], which provided low alkalinity of about 11–11.4 and reduced CH content (glass cement, calcium aluminate phosphate, magnesia phosphate and ettringite cement). The matrices which eliminated the chemical attack as well as microstructural densening (magnesia phosphate and ettringite cements) resulted in a composite which maintained its strength and toughness over 56 days of accelerated ageing at 56°C. Cuypers *et al.* [95] reported the use of phosphate cement with E glass fibres, and its performance under cyclic ageing of 2 h water spray, followed by drying at 40°C and 60°C. They found that under these conditions there was no degradation in properties in the multiple cracking part of the stress–strain curve. These ageing conditions are much less severe than those of water immersion, and they may represent the environment in some applications.

Artificial carbonation to reduce the alkalinity of the matrix and thus improve the durability of E glass composite was studied by Bentur [96]. Although a marked improvement could be obtained, the performance achieved was still no better than that of AR-GRC composites with a conventional Portland cement matrix. Also, to achieve complete carbonation, an elaborate vacuum drying process was involved, which is difficult to apply in practice. An alternative method of supercritical carbonation of GRC was studied by Purnell *et al.* [97,98], using AR fibres. This study evaluated only the properties of unaged composites.

The use of a polymer modified cement matrix reinforced with E glass fibres was developed by Bijen and Jacobs [99–104], using a special acrylic polymer (Forton) which was developed for that purpose. In the production of this composite, the tiny polymer latex particles (~0.1 μm in diameter) infiltrate and fill the spaces between the filaments in the strand and eventually coalesce into a film. Thus they can provide a protective effect, both to reduce the chemical attack, and to reduce the extent of the microstructural mechanism, by eliminating the growth of dense and rigid hydration products around the filaments. The coalescence of

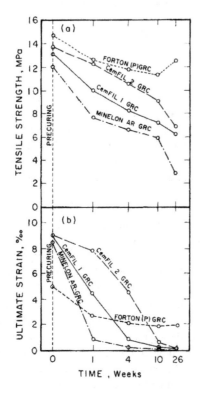

Figure 8.34 The effect of accelerated ageing in water at 50°C on the properties of E glass composites with a polymer modified matrix, (P)GRC, and AR glass composites with Portland cement matrix AR-GRC (a) Tensile strength development; (b) strain at ultimate stress (after Bijen [101]).

the film requires dry curing in air which is usually preceded by a short initial wet curing, to promote cement hydration. The formation of the film in the matrix itself can provide an additional sealing effect to slow down moisture movement, thus sometimes eliminating the need for water curing [35]. The strength and strain retention of this composite was as good as that of AR-GRC (Figure 8.34). However, this composite still tends to lose toughness on ageing, and the improvement in property retention may be due, at least in part, to the higher quality of the polymer modified matrix, particularly with respect to strain capacity. It was also noted [105,106] that the polymer film formed over the glass filaments was not always perfect and sometimes contained openings and perforations (Figure 8.35). This may account for the observations [29,105] that in these systems the E glass can still be attacked, and this may happen in the unprotected zones where the surface is not covered by the film.

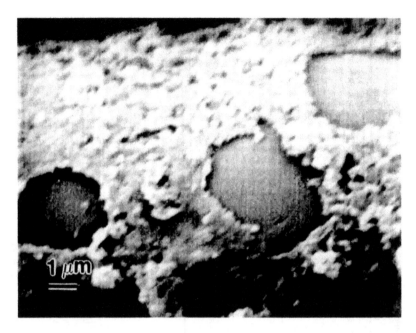

Figure 8.35   Perforated polymer film covering glass filament in GRC with polymer modified Portland cement matrix (after Bentur [106]).

8.6.2   Composites with AR glass fibres

8.6.2.1   Fibres of improved composition

The potential for improving the performance by modifying the glass composition was discussed briefly in Section 8.2. The two approaches that have been used are changing the glass composition and surface treatments [28,30]. Both can lead to new fibres of improved performance, as shown by the examples in Figures 8.36 and 8.37. The improved performance of the CemFIL-2 fibres over the CemFIL-1 fibres, which is demonstrated in Figure 8.21, was also achieved by means of surface treatment [29]. Such treatments can lead to a more stable surface which will suppress alkali attack, and can also change the nucleation and growth of the hydration products to delay or prevent the formation of a dense interfacial matrix. The latter effect was apparently effective in the case of the CemFIL-2 fibres [71]. Bijen suggested that such effects can be brought about by changes in the zeta potential of the surface. He suggested that in CemFIL-2 the organic compounds in the coating may interfere with the precipitation of CH.

There has also been a growing interest in AR glass compositions produced by the sol–gel method. Here, the glass is made through hydrolysis and gelling rather than by conventional melting techniques. This is potentially advantageous

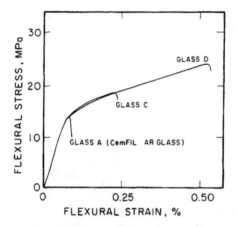

Figure 8.36 The effect of glass composition on the stress–strain curve of a GRC composite after ageing in water at 60°C for 35 days (after Fyles *et al.* [28]).

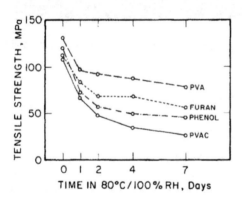

Figure 8.37 The effect of surface treatments of AR glass fibres on the strength retention of a strand in a Portland cement matrix (after Hayashi *et al.* [30]).

for AR formulations, which require high process temperatures in the conventional technique [107]. This method has also been applied to coat E glass with an alkali resistant SiO_2–ZrO_2 film [108,109].

8.6.2.2 Matrix modification

The modification of the matrix can have a profound effect on the durability of AR-GRC composites, as demonstrated in the data compiled by Majumdar [17] in

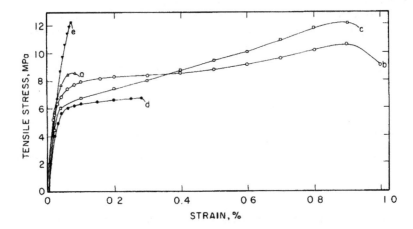

Figure 8.38 The effect of matrix composition on the stress–strain curves of AR-GRC composites containing ∼ 5% by weight fibres, after ageing in natural environment (after Majumdar [17]). (a) Portland cement; (b) alumina cement; (c) supersulfated cement; (d) Portland + 40% flyash; (e) polymer modified cement.

Figure 8.38. Thus, this too may be an effective means of improving the long-term behaviour, even when using the currently available AR glass fibres.

MODIFICATION OF THE PORTLAND CEMENT MATRIX WITH POZZOLANIC FILLERS

The use of pozzolanic fillers, such as the more 'conventional' natural pozzolans and fly ash and the 'advanced' reactive silica fume and metakaolin can reduce the alkalinity of the matrix as well as the content of CH, and thus slow down the two processes which lead to the degradation in the properties of the composite: chemical attack and microstructural changes at the interface (in particular deposition of CH).

The use of 'conventional' pozzolanic materials has been studied extensively [29,84,110–112], in particular the use of fly ash. These additives did not prevent a loss in strength and toughness, but they did slow down the rate of loss. Singh and Majumdar [112] reported that the best results, in terms of strength and toughness retention, were obtained with a 40% fly ash content. They pointed out, however, that these high level additions were accompanied by some reduction in the initial properties. The extent of the improvement in durability varied with different fly ashes and natural pozzolans [84,110]. Leonard and Bentur [84] reported that such differences could not be correlated with the pozzolanic activity, but rather with the effect of the fly ash on the interfacial microstructure developed.

The improvements in durability with the addition of 'conventional' pozzolans such as fly ash were relatively modest. Thus, special attention has more recently

Table 8.9 Effect of 25% replacement of cement with silica fume and metakaolin on the toughness retention of AR glass fibre composite after 50°C accelerated ageing, adopted from Marikunte *et al.* [116] (the values in the table are relative to unaged composite)

Matrix composition	Toughness retention after accelerated ageing at 50°C(%)	
	28 days	84 days
Portland cement	37.0	12.9
25% silica fume replacement	43.6	20.9
25% metakaolin replacement	78.7	71.0

been given to the use of more reactive supplementary cementing materials, in particular blast furnace slag, silica fume and metakaolin [30,51,52,86,91, 113–120]. Replacement of part of the Portland cement with 10–25% silica fume brought about a modest slowing down of the degradation process, [30,91,112–115,117], Figure 8.39 and Table 8.9. Combination of blast furnace slag and dimension stabilizing admixture was found to be much more effective [119], and its influence was attributed to the elimination of CH and its deposition between the filaments in the strand. Metakaolin was consistently reported to be extremely effective in drastically reducing the rates of loss [51,52,86,91,116), Table 8.9. The difference between the two could be explained by the difference in their influence on the changes in the bonding between the filaments, especially as shown from the data of Zhu and Bartos [51] based on the microindentation of the filaments in the strand, after ageing: the bonding of the inner filaments increased and became as high as that of the outer filament in the case of the Portland cement and silica fume matrices, but remained less than 20% of the bond of the outer filaments in the case of the metakaolin matrix. This may reflect differences in the nature of hydration products and their deposition within the filaments, along the line of the mechanisms discussed in Section 8.5.2.

The combined influence of strand size (i.e. number of filaments per strand) and matrix modification as a means for improved durability was reported by Brandt and Glinicki [120], using metakalonin and diatomite. Increase in the size of the bundle resulted in enhanced toughness retention (Figure 8.40) in the systems in which calcium hydroxide was formed (control, and systems with lower content of replacements, 20–30%). The influence of the bundle size could be accounted for by the bonding and flexural mechanisms outlined in Section 8.5.2: with larger number of filaments the relative influence of the calcium hydroxide, which preferentially deposits at the external filaments, becomes smaller. With larger replacement contents (40–50%), where the calcium hydroxide is completely eliminated, the toughness retention is much higher, and the system is practically insensitive to the bundle size. This behaviour can be interpreted in terms of the elimination of

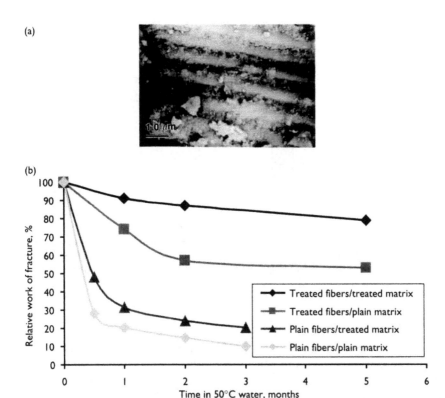

Figure 8.39 The treatment of glass fibre strands in silica fume slurry (a) Filling of the spaces between the filaments by the silica fume particles, (b) Effect of matrix modification with silica fume (treated matrix), slurry treatment of the strands (treated fibres) and the combination of the two on the toughness retention in accelerated ageing at 50°C (after Bentur [91]).

the mechanism of deposition of crystalline material and other hydration products which may bind the filaments in the bundle.

It should be noted that if the silica fume was applied in a different way, by immersing the glass fibre strands in a silica fume slurry prior to their incorporation in the matrix, much greater improvements were observed. Here, the tiny ($\sim 0.1 \ \mu$m) silica fume particles penetrated into the spaces between the filaments (Figure 8.39(a)), leading to a limited reduction in strength and toughness, and yielding a tough composite after prolonged accelerated ageing (Figure 8.39(b)). This improvement was attributed to the ability to place the silica fume at the interface, where it is most needed to delay or prevent the development of CH [91]. At present the limitation of this process is that it is completely manual, and has not been developed for industrial applications.

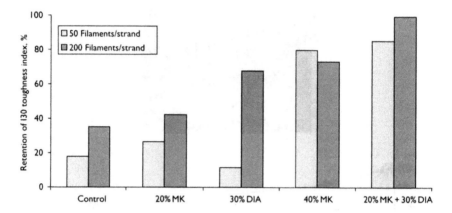

Figure 8.40 Effect of size of bundle (number of filaments) and supplementary material (metakaolin – MK; Diatomite – DIA) on the retention of the I_{30} toughness index after accelerated ageing at 50°C for 84 days (plotted from data of Brandt and Glinicki [120]).

NON-PORTLAND CEMENT MATRICES

Aluminate cements Majumdar *et al.* [121] studied the ageing of AR glass fibres in a high alumina cement matrix. They reported that the composite retained a substantial proportion of its strength and impact resistance after 10 years in both air and water storage. This is a much better performance than that of a similar composite with a Portland cement matrix. However, in natural weathering there was a larger decline in impact resistance, and here the performance of the high alumina cement composite was no better than that of the Portland cement composite. This may be due in part to the conversion reaction which leads to matrix weakening. This was confirmed when testing composites at temperatures of 30°C–40°C, where the conversion reaction is more likely to occur. As indicated in Section 8.6.1, there is a need for caution when using high alumina cement because of the matrix durability problem, even though it has a favourable effect on the durability of the GRC composite.

Another type of aluminate cement in which hydrogarnet and gehlenite hydrate are the major hydration products was reported by Bentur *et al.* [93], and compared with other non-Portland cement matrices. It consisted of $CaO–Al_2O_3–SiO_2$ glass which was finely ground to exhibit cementitious properties. It gave excellent ageing performance in accelerated ageing, showing an initial increase in toughness and later a decline from a peak, but at 56 days of accelerated ageing its toughness was similar to that of the unaged composite.

Supersulphated cement Supersulphated cement is a mixture of granulated blast furnace slag, calcium sulphate and a small content of activator (Portland cement or lime). It is a low pH material without CH, thus making it a potentially advantageous matrix for GRC composites. Indeed, long-term tests have shown that the strength and toughness of such composites are much better preserved in humid environments than GRC with a Portland cement matrix [122,123], as demonstrated in Figure 8.38. However, in a dry environment, whether indoor or in natural weathering, the composite showed a marked reduction in first crack stress and modulus of elasticity, and these values were much lower than for GRC with a Portland cement matrix. This was usually accompanied by the formation of a friable layer on the surface of the composite. These deleterious effects are due to the weakening of the supersulphated cement matrix by carbonation [122–124]. Carbonation takes place more readily in a dry or partially wet environment, but is limited when the material is kept wet. Thus, although the use of this matrix has the potential for limiting the ageing effect in GRC when kept wet, it may not offer a practical solution since in drier conditions its use will lead to a reduction in matrix controlled properties (first crack stress and modulus of elasticity).

The carbonation of this matrix leads to the decomposition of ettringite into gypsum and $CaCO_3$. It was suggested [124] that the ettringite in this system bonds together the other components of the matrix (CSH and unreacted slag) and therefore this phase, even when present in small quantities, exerts a considerable influence on the mechanical properties. Various methods have been considered to overcome the carbonation problem in the supersulphated cement GRC, including surface compaction and the application of a protective layer [123].

Portland blast furnace slag cement   The use of a Portland blast furnace slag matrix was not found to be particularly advantageous, in spite of its lower alkalinity and CH content [125,126]. Majumdar and Singh [125] suggested this to be the result of the high Portland cement content used in their work ($\sim 80\%$). However, Mills [126] showed that even with matrices of higher blast furnace slag content (50%) no significant improvement was observed. Mills attributed this behaviour to the large CH content that was still available, in spite of the presence of 50% blast furnace slag, and to the stronger affinity of the CH for the AR glass surface than for the blast furnace slag.

Calcium sulpho-aluminate cements   Calcium sulpho-aluminate cements were developed in Japan for special use as a matrix for GRC (labelled as low alkali cements). They are based on a mixture of calcium silicate – $C_4A_3\bar{S}$ – $C\bar{S}$ – slag [30,92,127]. This cement, known also as CGC cement, is free of CH, and its main hydration products are CSH and ettringite. Its pH is about 11.0 [92]. With this cement it was possible to preserve the strength and toughness (ultimate strain) of an AR-GRC composite after extended accelerated ageing (Figure 8.41). No improvement was observed when this cement was used in combination with E glass (Figure 8.41). SEM observations suggested that the main effect of the CGC cement in the AR composite is in eliminating the growth of CH around the glass filaments [92]. The properties of this matrix are at least as good as those of Portland cement, and in some respects even better. For example, it exhibits less shrinkage on drying [92,127].

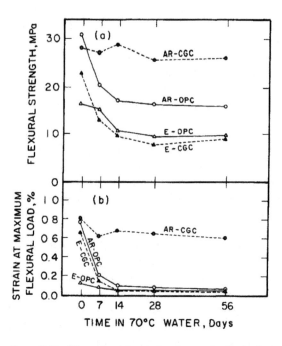

Figure 8.41 Effect of accelerated ageing on the flexural strength (a) and ultimate strain, (b) of GRC composites (E and AR glass) with Portland cement or CGC cement matrix (after Hayashi et al. [30]).

Considerably improved durability performance was reported by Purnell *et al.* [52,86] for a matrix of shrinkage compensating cement with 10% metakaolin (MK). The main components of the blended cement were C_3S, C_2S, $C_4A_3\bar{S}$, $C\bar{S}$, MK and gypsum, at contents of 34%, 8%, 14%, 8%, 9% and 4% by weight, respectively. Major hydration products, in addition to CSH were ettringite and hydrogarnet, with the pH being 13.4, compared with 13.7 for a Portland cement matrix, and all the CH was eliminated. The toughness retention was much better than that of a Portland cement matrix, with 'half life time' in accelerated ageing at 65°C being about 200 days, about an order of magnitude bigger than with Portland cement. Purnell *et al.* [52] observed bundle filling even in the unaged composite, and attributed the improved durability performance to slower flaw growth, as outlined in Section 8.5.2. Pera and Ambroise [128] evaluated a blend of Portland cement and calcium sulpho-aluminate with phosphogypsum. They reported the properties of the unaged composite.

Another type of sulphoaluminate cement in which ettringite is a major hydration product was reported by Bentur *et al.* [93], based on a composition consisting of a $CaO–Al_2O_3–SiO_2$ glass (80%) and gypsum (20%). In this system the ettringite formed through solution and therefore the material did not experience expansion.

Accelerated ageing for a period up to 56 days in 50°C water did not result in any toughness loss, and this was attributed to the prevention of formation of dense hydration products.

POLYMER–PORTLAND CEMENT MATRICES

Most of the developments of polymer–cement matrices for GRC were based on the use of polymer latex modified GRC. The concepts behind this approach are essentially the same as those described for E glass in Section 8.6.1: providing improved chemical resistance, and the prevention of the growth of dense and rigid hydration products around the glass filaments. The influence of various types of polymers was evaluated at the Building Research Establishment in the UK [129–131], while the effect of an acrylic polymer latex (Forton) which was developed specifically for E-GRC (Section 8.6.1) was studied in the Netherlands with AR-GRC [101,104].

The incorporation of the Forton polymer latex was found to exert a considerable positive effect, eliminating any strength loss during 4 years of water storage, compared with a marked strength loss in the composite without polymer modification (Figure 8.42). Both composites showed a reduction in ultimate strain, but at a slower rate in the modified system (Figure 8.42(b)).

West et al. [129] also reported an improved durability with acrylic polymer modified GRC. This was observed only in natural weathering, and no improvement was seen in water storage. A very limited improvement in accelerated ageing in water was also reported by Soroushian et al. [132]. This is different from the data reported by Jacobs and Bijen [104], in which improvement was observed in natural weathering as well as in water storage. They suggested that such differences may be due to the nature of the acrylic latex dispersions studied; in their case the poor performance in water may have been due to the loss of the integrity of the polymer film when kept continuously in water. Similar improvements in performance were also reported using vinyl proprionate–vinyl chloride copolymer latex and styrene butadiene rubber latex [130,131]. Here too, improvement in durability was observed in natural weathering but not in water storage. Polyvinidene dichloride polymer latex provided no improvement in durability [130].

There are clearly benefits in combining polymer latex with the GRC composites. However, the influence of this modification may depend on the type of polymer used. Also, assessing its performance by the standard tests, based on immersion in water, may underestimate the effect of the polymer, because of possible instability of the polymer film when kept continuously immersed. It should be noted that in the work reviewed here the polymer content was about 10–15% solids by weight of cement. Lower contents may not necessarily give this enhancement in durability.

Another approach to the use of polymers in a cementitious matrix was based on polymer impregnation. This was studied in conjunction with a light-weight GRC system, in which fly ash cenospheres of low density were used [133]. Impregnation with methyl methacrylate was effective in improving the initial properties as well as the long-term performance.

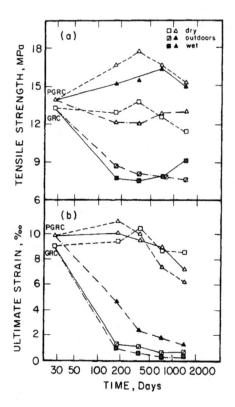

Figure 8.42 Effect of ageing in water on the tensile strength (a) and ultimate strain; (b) of composites with CemFIL-I AR glass fibres with Portland cement, GRC, or polymer latex modified cement matrices, PGRC (after Jacobs and Bijen [104]).

8.7   Design and application considerations

GRC has been used in a variety of applications, the most common of which is cladding panels. Other uses that have been developed or considered include shell structures, prefabricated windows, pipes, channels, permanent formwork, floor slabs and the rendering of masonry construction to enhance its strength and stability. In all these applications the GRC is particularly attractive since it can readily be produced in various complex shapes (due to the strength and flexibility of the green product, induced by the fibres), and be made as a thin component (~10 mm thick) that in its hardened state is sufficiently strong and tough, in particular prior to ageing. It can thus provide a basis for making various lightweight precast units. The range of current applications and potential new ones has been discussed and reviewed in several publications [31–37,134–140].

For successful application of these components, special attention must be given to the long-term performance, in particular with respect to the deterioration in the mechanical properties on weathering, and volume changes induced by environmental effects. These two properties must be taken into account in the design of the component in order to reduce or eliminate sensitivity to cracking. At present, because of the changes in mechanical properties during ageing, there is a tendency to limit the use of GRC to non-structural or semi-structural components, such as cladding panels. Various design procedures have been developed, particularly with respect to cladding panels, to ensure adequate long-term performance [31–37,134].

The reduction in mechanical properties is dealt with on the basis of the observation that even after prolonged ageing, the flexural strength does not decrease below the first crack strength of the unaged composite (Figure 8.20). Thus, for example, the PCI [33] recommendation is that this value be used for design, and hence the significance of characterizing the load–deflection curve of the unaged composite for quality control. Various factors and coefficients are also considered, and the first crack stress (or limit of proportionality) is further adjusted (i.e. reduced), to take into account effects such as loading conditions, variability in quality and the shape of the component. In the structural calculation it is essential to take into account also stresses induced by thermal and moisture movements, because these could be as high or even higher than those induced by loads. The typical range of GRC properties is provided in Table 8.10 for a composite produced by the spray-up method with about 5% fibres.

Experience has indicated that the problem of dimensional changes is the one which has caused most of the cracking that has been experienced with GRC panels. This problem can be aggravated with time, since the composite tends to lose its toughness. In addition, the differential strains induced by thermal and moisture movement can lead to stress generation if the component is restrained; if these stresses are sufficiently high, or if the material is not sufficiently ductile to relieve them, cracking will occur [32,34,141]. Such problems are usually more acute in sandwich type elements with a rigid core, as a result of the restraint in movement of the external skin relative to the core, and the generation of differential strains between the external face of the skin, and the inner face, in contact with the core.

Cracking can also occur more readily in panels which are connected in a rigid manner to a frame, or around connections which do not allow freedom of movement. Cracking may also be induced more readily in panels of complex configuration, where internal restraint may be generated by the shape of the panel.

The most important means of overcoming such cracking risks is to design the component properly, taking these effects into account and in particular providing flexible connections. For this purpose a variety of different connections have been developed (see Chapter 14).

Another approach which has become common in recent years is to use a system with a single skin GRC panel mounted on a studded steel frame. The connections

Table 8.10 Typical range of GRC properties[a] (from [33])

Property	28 days	Aged[b]
Density	1900–2500 kg/m³	1900–2500 kg/m³
Impact strength (Charpy)	9.5–25.4 Nmm/mm²	3.5–5.0 Nmm/mm²
Compressive strength	48–83 MPa	70–83 MPa
Flexural		
First Crack[c]	6–10 MPa	7–11 MPa
Ultimate strength	17–28 MPa	9–14 MPa
Modulus of elasticity	10–20 GPa	17–24 GPa
First Crack[c]	4.8–7 MPa	4.8–8 MPa
Tensile		
Ultimate strength	7–11 MPa	5–8 MPa
Modulus of elasticity	0.6–1.2%	0.03–0.06%
Shear		
Interlaminar	2.8–5.5 MPa	2.8–5.5 MPa
In plane	7–11 MPa	5–7.6 MPa
Coefficient of thermal expansion	$11–16 \times 10^{-6}$°C^{-1}	$11–16 \times 10^{-6}$°C^{-1}

Notes
a These values are typical but should not be used for design purposes. The values achieved in practice depend on mix design, quality control, production process and curing.
b Developed from accelerated test results.
c First crack values are obtained from the deviation from linearity in the stress–strain curve in tension or load–deflection curve in flexure, and they are sometimes referred to as yield, limit of proportionality (LOP) and bend over point (BOP).

as well as the steel frame are sufficiently flexible to minimize the risk of cracking, and a survey by Williamson [142] has indicated favourable experience with such systems (see Chapter 14).

References

1. A.J. Majumdar and R.W. Nurse, *'Glass fibre reinforced cement'*, Building Research Establishment Current Paper, CP79/74, Building Research Establishment, England, 1974.
2. E.G. Nawy, G.E. Neuwerth and C.J. Phillips, 'Behaviour of fibre glass reinforced concrete beams', *ASCE, J. Struct. Div.* 97, 1971, 2203–2215.
3. S. Klink, 'Fycrete-glass fibre reinforced plastics to strengthen concrete structures', Presented at the *Fibre-Cement Composites Expert Working Group Meeting*, United Nationals Industrial Development Organization, Vienna, 20–24 October, 1969.
4. M.S. Aslanova, 'Glass fibres', in W. Watt and B.V. Perov (eds) *Handbook of Composites, Part I – Strong Solids*, Chapter I, Elsevier Science Publishers, London, 1985, pp. 4–85.
5. D. Hull, *An Introduction to Composite Materials*, Cambridge Solid State Science Series, Cambridge University Press, England, 1981.
6. Z. Chi, T.W. Chou and G. Shen, 'Determination of single fibre strength distribution from fibre bundle testing', *J. Mat. Sci.* 19, 1984, 3319–3324.

7. A.J. Majumdar and J.F. Ryder, 'Glass fibre reinforcement for cement products', *Glass Technol.* 9, 1968, 78–84.

8. A.J. Majumdar, 'Glass fibre reinforced cement and gypsum products', *Philos. Trans. R. Soc. London, A* 319, 1970, pp. 69–78.

9. A. Al Cheikh and M. Murat, 'Kinetics of non-congruent dissolution of E-glass fiber in saturated calcium hydroxide solution', *Cem. Concr. Res.* 18, 1988, 943–950.

10. M. Murat and A. Al Cheikh, 'Behavior of E-glass fiber in basic aqueous medium resulting from the dissolution of mineral binders containing metkaolinite', *Cem. Concr. Res.* 19, 1989, 16–24.

11. A. Bentur, 'Mechanisms of potential embrittlement and strength loss of glass fibre reinforced cement composites', in S. Diamond (ed.) Proc. Durability of Glass Fiber Reinforced Concrete Symp., Prestressed Concrete Institute, Chicago, IL, 1985, pp. 109–123.

12. K.L. Biryukovich, L. Yu and D.L. Biryukovich, *Glass fibre reinforced cement*, Budivelnik, Kiev, 1964, Translation No. 12, Civil Engineering Research Association, London, 1966.

13. B.A. Proctor and B. Yale, 'Glass fibres for cement reinforcement', *Philos. Trans. R. Soc. London A* 294, 1980, 427–436.

14. R.S. Barneyback and S. Diamond, 'Expression and analysis of pore fluids from hardened cement pastes and mortars', *Cem. Concr. Res.* 11, 1981, 279–285.

15. K.L. Litherland, P. Maguire and B.A. Proctor, 'A test method for the strength of glass fibres in cement', *Int. J. Cem. Comp. & Ltwt. Concr.* 6, 1984, 39–45.

16. L.J. Larner, K. Speakman and A.J. Majumdar, 'Chemical interactions between glass fibres and cement', *J. Non-Cryst. Solids* 20, 1976, 43–74.

17. A.J. Majumdar, 'Properties of GRC', in *Fibrous Concrete*, Proc. Symp., The Concrete Society, London, 1980, pp. 48–68.

18. L. Franke and E. Overbeck, 'Loss in strength and damage to glass fibres in alkaline solutions and cement extracts', *Durab. Bldg. Mat.*, 4, 1987, 73–79.

19. M. Chakraborty, D. Das, S. Basu and A. Paul, 'Corrosion behaviour of a ZrO-containing glass in aqueous acid and alkaline media and in a hydrating cement paste', *Int. J. Cem. Comp.* 1, 1979, 103–109.

20. M.S. Stucke and A.J. Majumdar, 'Microstructure of glass fibre reinforced cement composites', *J. Mat. Sci.* 11, 1976, 1019–1030.

21. V.T. Yilmaz and F.P. Glasser, 'Reaction of alkali-resistant glass fibres with cement: part 1, review, assessment and microscopy', *Glass Technol.* 32, 1991, 91–98.

22. J. Orlowsky, M. Raupach, H. Cuypers and J. Wastiels, 'Durability modelling of glass fibre reinforcement in cementitious environment', *Mater. Struct.* 38, 2005, 155–162.

23. A. Makashima, M. Tsutsumi, T. Shimohira and T. Nagata, 'Characterization of insoluble layers formed by NaOH attack on the surface of ZrO_2-containing silicate glass', *J. Amer. Ceram. Soc.* 66, 1983, 139–140.

24. Y. Oka, K.S. Ricker and M. Tomozawa, 'Calcium deposition on glass surface as an inhibitor to alkaline attack', *J. Amer. Ceram. Soc.* 62, 1979, 631–632.

25. K.L. Litherland, D.R. Oakley and B.A. Proctor, 'The use of accelerated aging procedures to predict the long term strength of GRC composites', *Cem. Concr. Res.* 1981, 455–466.

26. J.M. West and A.J. Majumdar, 'Strength of glass fibres in cement environments', *J. Mat. Sci. Letters.* 1, 1982, 214–216.

27. A. Bentur and S. Diamond, 'Aging and microstructure of glass fibre cement composites reinforced with different types of glass fibres', *Durab. Bldg. Mat.* 4, 1987, 201–226.

28. K. Fyles, K.L. Litherland and B.A. Proctor, 'The effect of glass fibre compositions on the strength retention of GRC', in R.N. Swamy, R.L. Wagstaffe and D.R. Oakley (eds) *Developments in Fibre Reinforced Cement and Concrete*, Proc. RILEM Symp., Sheffield, The Construction Press, Lancaster, 1986, Paper 7.5.

29. B.A. Proctor, D.R. Oakley and K.L. Litherland, 'Development in the assessment and performance of GRC over 10 years', *Composites.* 13, 1982, 173–179.

30. M. Hayashi, S. Sato and H. Fujii, 'Some ways to improve durability of GFRC', in S. Diamond (ed.) Proc. Durability of Glass Fibre Reinforced Concrete Symp., Prestressed Concrete Institute, Chicago, IL, 1985, pp. 270–284.

31. G. True, *GRC Production and Use*, A Viewpoint Publications, Palladium Publications, Ltd. London, 1986.

32. M.W. Fordyce and R.G. Wodehouse, *GRC and Buildings*, Butterworths, England, 1983.

33. PCI Committee on Glass Fiber Reinforced Concrete Panels, *Recommended Practice for Glass Fiber Reinforced Concrete panels*, fourth edition, Prestressed Concrete Institute, Chicago, IL, 2001.

34. Anon, *Design Guide: Glass Fibre Reinforced Cement*, Pilkington Brothers, L.P.O., CemFIL GRC Technical Data Manual, UK, 1984.

35. D.M. Oesterle, D.M. Schultz and J.D. Glikin, 'Design considerations for GFRC facades', in J.I. Daniel and S.P. Shah (eds) *Thin-section Fiber Reinforced Concrete and Ferrocement*, ACI SP-124, American Concrete Institute, Detroit, MI, 1990, pp. 157–182.

36. N.W. Hanson, J.J. Roller, J.I. Daniel and T.L. Weinmann, 'Manufacture and installation of GFRC facades', in J.I. Daniel and S.P. Shah (eds) *Thin-section Fiber Reinforced Concrete and Ferrocement*, ACI SP-124, American Concrete Institute, Detroit, MI, 1990, pp 183–213.

37. A. Bentur and S.A.S. Akers, 'Thin sheet cementitious composites', in N. Banthia, A. Bentur and A. Mufti (eds) *Fiber Reinforced Concrete: Present and Future*, Canadian Society for Civil Engineers, Montreal, Canada, 1999, pp. 20–45.

38. J.M. West, 'Durability of non-asbestos fibre reinforced cement', in J.M. Baker, P.J. Nixon, A.J. Majumdar and H. Davies (eds) *Durability of Building Materials and Components*, Durability of Building Materials and Components, Proc. 5th Int. Conf., E&FN SPON, 1991, pp. 709–714.

39. D. Cherubin-Grillo, A.Vautrin, F. Pierron and P. Soukatchoff, 'Mechanical behaviour of glass reinforced cement corrugated plates', in A.M. Brandt, V.C. Li and I.H. Marshall (eds) *Brittle Matric Composites*, Proc. Int. Conf., ZTUREK RSI and Woodhead Publications, Warsaw, 2000, pp. 137–146.

40. B. Mobasher, 'Micromechanical modeling of angle ply cement based composites', in A.M. Brandt, V.C. Li and I.H. Marshall (eds) *Brittle Matrix Composites*, Proc. Int. Conf., ZTUREK RSI and Woodhead Publications, Warsaw, 2000, pp. 62–72.

41. M. Perez-Pena and B. Mobasher, 'Mechanical properties of fiber reinforced lightweight concrete composites', *Cem. concr. Res.* 24, 1994, 1121–1132.

42. G. Xu, S. Magnani and D.J. Hannant, 'Durability of hybrid polypropylene-glass fibre cement corrugated sheets', *Cem. Concr. Compos.* 20, 1998, 79–84.

43. A. Pivacek, G.J. Haupt and B. Mobasher, 'Mechanical response of angle ply cement based composites', in A.M. Brandt, V.C. Li and I.H. Marshall (eds) *Brittle Matrix Composites*, Proc. Int. Conf., BIGRAF and Woodhead Publications, Warsaw, 1997, pp. 187–196.

44. M. Schupack, 'Thin sheet glass and synthetic fabric reinforced concrete 60–120 pfc density', in J.I. Daniel and S.P. Shah (eds) *Thin-section Fiber Reinforced Concrete and Ferrocement*, ACI SP-124, American Concrete Institute, Detroit, MI, 1990, pp. 521–436.

45. J.I. Daniel and M.E. Pecoraro, *Effect of Forton Polymer on Curing Requirements of AR-Glass Fiber Reinforced Cement Composites*, Research Report (Sponsored by Forton Inc., Sewickley, Pennsylvania), Construction Technology Laboratories, Division of Portland Cement Association, Skokie, IL, October, 1982.

46. M.A. Ali, A.J. Majumdar and B. Singh, 'Properties of glass fibre cement – the effect of fibre length and content', *J. Mat. Sci.* 10, 1975, 1732–1740.

47. N.G. Nair, 'Mechanics of glass fibre reinforced cement', in A. Neville (ed.) *Fibre Reinforced Cement and Concrete*, Proc. RILEM Symp., The Construction Press, England, 1975, pp. 81–93.

48. R.H. Mills, 'Preferential precipitation of calcium hydroxide on alkali resistant glas fibres', *Cem. Concr. Res.* 11, 1981, 689–697.

49. A.C. Jaras and K.L. Litherland, 'Microstructural Freatures in glass fibre reinforced cement composites', in A. Neville (ed.) Fibre Reinforced Cement and Concrete, Proc. RILEM Symp., The Construction Press, England, 1975, pp.327–334.

50. S. Igarashi and M. Kawamura, 'Effects of a size in bundled fibers on the interfacial zone between the fibers and the cement paste matrix', *Cem. Concr. Res.* 24, 1994, 695–703.

51. W. Zhu and P.J.M. Bartos, 'Assessment of the interfacial microstructure and bond properties in aged GRC using a novel microindentation method', *Cem. Concr. Res.* 27, 1997, 1701–1711.

52. P. Purnell, N.R. Short, C.L. Page and A.J. Majumdar, 'Microstructural observations in new matrix glass fibre reinforced cement', *Cem. Concr. Res.* 30, 2000, 1747–1753.

53. V. Laws and M.A. Ali, 'The tensile stress/strain curve of brittle matrices reinforced with glass fibre', in *Fibre Reinforced Materials*, Institution of Civil Engineers, London, 1977, pp. 115–123.

54. V. Laws, P. Lawrence and R.W. Nurse, 'Reinforcement of brittle matrices by glass fibres', *J. Phys. D: Appl. Phys.* 6, 1973, 523–537.

55. D.R. Oakely and B.A. Proctor, 'Tensile stress–strain behaviour of glass fibre reinforced cement composites', in A. Neville (ed.) *Fibre Reinforced Cement and Concrete*, Proc. RILEM Symp., The Construction Press, England, 1975, pp. 347–359.

56. B.A. Proctor, 'The stress–strain behaviour of glass-fibre reinforced cement composites', *J. Mat. Sci.* 21, 1986, 2441–2448.

57. V. Laws, 'On the mixture rule for strength of fibre reinforced cements', *J. Mat. Sci.* 2, 1983, 527–531.

58. P. Bartos, 'Bond in glass reinforced cements', in P. Bartos (ed.) *Bond in Concrete*, Proc. Int. Conf., Applied Science Publishers, London, 1982, pp. 60–72.

59. V. Laws, A.A. Langley and J.M. West, 'The glass fibre/cement bond', *J. Mat. Sci. Letters* 21, 1986, 289–296.

60. R.C. De Vekey and A.J. Majumdar, 'Interfacial bond strength of glass fibre reinforced cement composites', *J. Mat. Sci.* 5, 1970, 183–185.

61. H.G. Allen, 'Stiffness and strength of two glass-fibre reinforced cement laminates', *J. Compos. Mater.* 5, 1971, 194–207.
62. P. Trtik and P.J.M. Bartos, 'Assessment of glass fibre reinforced cement by in-situ SEM bending test', *Mater. Struct.* 32, 1999, 140–143.
63. P. Bartos, 'Brittle-matrix composites reinforced with bundles of fibres', in J.C. Maso (ed.) Vol. 2. Proc. 2nd Int. RILEM Congress from Materials Science to Construction Materials Engineering, Chapman and Hall, London and New York, 1984, pp. 539–554.
64. P. Purnell, A.J. Buchanan, N.R. Short, C.L. Page and A.R. Majumdra, 'Determination of bond strength in glass fibre reinforced cement using petrography and image analysis', *J. Material Science* 35, 2000, 4653–4659.
65. B. Banholzer, 'Bond Behaviour of a Multi-Filament Yarn Embedded in a Cementitious Matrix', in *Schriftenreihe Aachener Beiträge zur Bauforschung*, Institut für Bauforschung der RWTH Aachen, Diss., 2004.
66. A. Aveston, G.A. Cooper and A. Kelly, 'Single and multiple fracture', in *The Properties of Fibre Composites*, Proc. Conf. National Physical Laboratories, IPC, Science and Technology Press, England, 1971, pp. 15–24.
67. A.J. Majumdar, B. Singh, A.A. Langley and M.A. Ali, 'The durability of glass fibre cement – the effect of fibre length and content', *J. Mat. Sci.* 15, 1980, 1085–1096.
68. B. Singh and A.J. Majumdar, 'The effect of fibre length and content on the durability of glass reinforced cement – ten year results', *J. Mat. Sci. Letters* 4, 1985, 967–971.
69. Anon, *'A study of the properties of Cem-FIL/OPC composites'*, Building Research Establishment Current Paper CP 38/76, Building Research Establishment, England, 1976.
70. Anon, *Properties of GRC: ten year results*, Building Research Establishment Information Paper IP 38/79, Building Research Establishment, England, 1979.
71. A. Bentur, M. Ben Bassat and D. Schneider, 'Durability of glass fiber reinforced cements with different alkali resistant glass fibers', *J. Amer. Ceram. Soc.* 68, 1985, 203–208.
72. J.M. West and P.L. Walton, 'Fatigue endurance of aged glass fibre reinforced cement', *J. Mat. Sci.* 16, 1981, 2398–2400.
73. A.P. Hibbert and F.J. Grimer, *'Flexural fatigue of glass fibre reinforced cement'*, Building Research Establishment Current Paper CP 12/76, Building Research Establishment, England, 1976.
74. H.G. Allen, 'Creep of glass-fibre reinforced cement composites in tension', *Int. J. Cem. Comp.* 2, 1980, 185–191.
75. H.G. Allen and C.K. Jolly, 'The weather and its influence on the creep of glass fibre-reinforced cement composites', *J. Mat. Sci.* 17, 1982, 2037–2046.
76. J.I. Daniel and D.M. Schultz, 'Durability of glass fibre reinforced concrete system', in S. Diamond (ed.) Proc. Durability of Glass Fibre Reinforced Concrete Symp., Prestressed Concrete Institute, Chicago, IL, 1985, pp. 174–198.
77. M.J.N. Jacobs, *Forton PGRC – a many sided construction material*, Department of Material Application and Development, DSM – Central Laboratory, Galeen, The Netherlands, October 1981.
78. A.J. Aindow, D.R. Oakley and B.A. Proctor, 'Comparison of the weathering behaviour of GRC with predictions made from accelerated aging tests', *Cem. Concr. Res.* 14, 1984, 271–274.

79. K.L. Litherland, 'Test methods of evaluating the long term behaviour of GFRC', in S. Diamond (ed.) Proc. Durability of Glass Fibre Reinforced Concrete Symp., Prestressed Concrete Institute, Chicago, IL, 1985, pp. 210–221.

80. K.L. Litherland and B.A. Proctor, 'Predicting the long term strength of glass fibre cement composites', Paper 7.6 in R.N. Swamy, R.L. Wagstaffe and D.R. Oakley (eds) Developments in Fibre Reinforced Cement and Concrete, Proc. RILEM Symp., Sheffield, RILEM Technical Committee 49-FTR, 1986.

81. Anon, Shrinkage of GRC composites, CemFIL Bulletin No. 51, Pilkington Reinforcements, Ltd., St. Helens, England, May 1986.

82. J.A. Lee and T.R. West, 'Measurement of drying shrinkage of glass rienforced cement composites', in R.N. Swamy (ed.) Testing and Test Methods of Fibre Cement Composites, Proc. RILEM Symp., The Construction Press, England, 1978, pp. 149–157.

83. A.A. Langley, 'The dimensional stability of glass-fibre-reinforced cement', Mag. Concr. Res. 33, 1981, 221–226.

84. S. Leonard and A. Bentur, 'Improvement of the durability of glass fiber reinforced cement using blended cement matrix', Cem. Concr. Res. 14, 1984, 717–728.

85. A. Bentur, 'The role of interfaces in controlling the durability of fiber reinforced cements', ASCE J. of Materials in Civil Eng. 12, 2000, 2–7.

86. P. Purnell, N.R. Short, C.L. Page, A.J. Majumdar and P.L. Walton, 'Accelerated ageing characteristics of glass-fibre reinforced cement made with new cementitious matrices, Composites Part A. 30, 1999, 1073–1080.

87. P. Purnell, N.R. Short and C.L. Page, 'A static fatigue model for the durability of glass fibre reinforced cement', Journal of Materials Science 36, 2001, 5385–5390.

88. B. Mobasher and C.Y. Li, 'Modeling of stiffness degradation of the interfacial transition zone during fibre debonding', J. of Composite Engineering 5, 1995, 1349–1365.

89. J. Aveston, R.A. Mercer and J.M. Sillwood, 'Fibre reinforced cement-scientific foundation for specifications', in Composites, Standards Testing and Design, Proc. National Physical Laboratory Conference, England, 1974, pp. 93–103.

90. A. Bentur and S. Diamond, 'Effect of aging of glass fibre reinforced cement on the response of an advancing crack on intersecting a glass fibre strand', Int. J. Cem. Comp. & Ltwt. Concr. 8, 1986, 213–222.

91. A. Bentur, 'Silica fume treatments as means for improving durability of glass fiber reinforced cements', J. Mater. Civ. Eng. 1, 1989, 167–183.

92. M. Tanaka and L. Uchida, 'Durability of GFRC with calcium silicate $-C_4A_3\bar{S}-C\bar{S}$–Slag type low alkaline cement', in S. Diamond (ed.) Proc. Durability of Glass Fiber Reinforced Concrete Symp., Prestressed Concrete Institute, Chicago, IL, 1985, pp. 305–314.

93. A. Bentur, K. Kovler and I Odler, 'Durability of some glass fiber reinforced cementitious composites', in K.C.G. Ong, J.M. Lau and P. Paramasivam (eds) 5th International Conference on Structural Failure, Durability and Retrofitting, Singapore Concrete Institute, Singapore, 1997, pp. 190–199.

94. J. Bijen, 'A survey of new developments in glass composition, coatings and matrices to extend service lifetime of GFRC', in S. Diamond (ed.) Proc. Durability of Glass Fiber Reinforced Concrete Symp., Prestressed Concrete Institute, Chicago, IL, 1985, pp. 251–269.

95. H. Cuypers, J. Gu, K. Croes, S. Dumortier and J. Wastiels, 'Evaluation of fatigue and durability properties of E-Glass fibre reinforced phosphate cementitious composite', in A.M. Brandt, V.C. Li and I.H. Marshall (eds) Proc. Int. Symp. Brittle Matrix Composites 6, ZTUREK RSI and Woodhead, Warsaw, 2000, pp. 127–136.

96. A. Bentur, 'Durability of carbonated glass fibre reinforced cement composites', *Durab. Bldg. Mater.* 1, 1983, 313–326.

97. P. Purnell, N.R. Short and C.L. Page, 'Super-critical carbonation of glass-fibre reinforced cement. Part 1: mehanical testing and chemical analysis', *Composites Part A.* 32, 2001, 1777–1787.

98. P. Purnell, A.M.G. Seneviratne, N.R. Short and C.L. Page, 'Super-critical carbonation of glass-fibre reinforced cement. Part 2: microstructural observations', *Composites Part A.* 34, 2003, 1105–1112.

99. J. Bijen and M. Jacobs, 'Properties of glass fibre reinforced polymer modified cement', *Mater. Struct.* 15, 1982, 445–452.

100. J. Bijen, 'Glass fibre reinforced cement: improvements by polymer addition', in D.M. Roy, A.J. Majumdar, S.P. Shah and J.A. Manson (eds) *Advances in Cement-Matrix Composites*, Proc. Materials Research Society Meeting, Materials Research Society, Pittsburgh, PA, 1980, pp. 239–249.

101. J. Bijen, 'Durability of some glass fibre reinforced cement composites', *Amer. Concr. Inst. J.* 80, 1983, 305–311.

102. M.J.N. Jacobs, 'Durability of PGRC, design implications', *Betonwerk+Fertigteil-Technik.* 52, 1986, 228–233.

103. M.J.N.Jacobs, Durability of PGRC, design implications, part 2', *Betonwerk+Fertigteil-Technik.* 52, 1986, 756–761.

104. M.J.N. Jacobs and J. Bijen, 'Durability of Forton polymer modified GFRC', in S. Diamond (ed.) Proc. Durability of Glass Fiber Reinforced Concrete Symp., Prestressed Concrete Institute, Chicago, IL, 1985, pp. 293–304.

105. J.I. Daniels and D.M. Schultz, 'Long term strength durability of glass fibre reinforced concrete', in R.N. Swamy, R.L. Wagstaffe and D.R. Oakley (eds) *Developments in Fibre Reinforced Cement and Concrete*, Proc. RILEM Symp. Sheffield, RILEM Technical Committee 49-FTR, 1986, Paper 7.4.

106. A. Bentur, Unpublished results.

107. W.S. Shin, 'Recent development in alkali-resistant fibres at the Battelle Institute', in *International Congress on Glass Fibre Reinforced Cement*, Paris, 1981, Glass Fibre Reinforced Cement Association, 1981, pp. 359–374.

108. M. Guglielmi and A. Maddalena, 'Coating of glass fibres for cement composites by the sol-gel method', *J. Mat. Sci.* 4, 1985, 123–124.

109. A. Maddalena, M. Guglielmi, V. Gottardi and A. Raccanelli, 'Interactions with portland cement paste of glass fibres coated by the sol-gel method', *J. Non-Cryst. Solids.* 82, 1986, 356–365.

110. B. Singh and A.J. Majumdar, 'Properties of GRC containing inorganic fillers', *Int. J. Cem. Comp. & Ltwt. Concr.* 3, 1981, 93–102.

111. B. Singh, A.J. Majumdar and M.A. Ali, 'Properties of GRC containing PFA', *Int. J. Cem. Comp. & Ltwt. Concr.* 6, 1984, 65–74.

112. B. Singh and A.J. Majumdar, 'The effect of PFA addition on the properties of GRC', *Int. J. Cem. Comp. & Ltwt. Concr.* 7, 1985, 3–10.

113. S.G. Bergstrom and H.E. Gram, 'Durability of alkali-sensitive fibres in concrete', *Int. J. Cem. Comp. & Ltwt. Concr.* 6, 1984, 75–80.

114. A. Bentur and S. Diamond, 'Direct incorporation of silica fume into strands as a means for developing GFRC composites of improved durability', *Int. J. Cem. Comp. & Ltwt. Concr.* 9, 1987, 127–136.

115. A. Bentur and S. Diamond, 'Effects of direct incorporation of microsilica into GFRC composites on retention of mechanical properties after aging', in S. Diamond (ed.) Proc. Durability of Glass Fiber Reinforced Concrete Symp., Prestressed Concrete Institute, Chicago, IL, 1985, pp. 337–351.

116. S. Marikunte, C. Aldea and S.P. Shah, 'Durability of glass fiber cement composites: effect of silica fume and metakaolin', *Journal of Advance Cement Based Materials* 5, 1997, 100–108.

117. K. Rajczyk, E. Giergiczny and M.A. Glinicki, 'The influence of pozzolanic materials on the durability of glass fibre reinforced cement composites', in A.M. Brandt, V.C. Li and I.H. Marshal (eds) Proc. Int. Symp. Brittle Cement Composites 5, BIGRAF and Woodhead Publishers, Warsaw, 1997, pp. 103–112.

118. V.T. Yilmaz and F.P. Glasser, 'Reaction of alkali-resistant glass fibres with cement, Part 2: durability in cement matrices conditioned with silica fume', *Glass technol.* 32, 1991, 138–147.

119. A. Peled, J. Jones and S.P. Shah, 'Effect of matrix modification on durability of glass fiber reinforced cement composite', *Mater. Struct.* 38, 2005, 163–171.

120. A.M. Brandt and M.A. Glinicki, 'Effects of pozzolanic additives on long-term flexural toughness of HPGFRC', in A.E. Naaman and H.W. Reinhardt (eds) *Fourth International Workshop on High Performance Fiber Reinforced Cement Composites* (HPFRCC 4), RILEM Publications, Bagneux, France, 2003, pp. 399–408.

121. A.J. Majumdar, B. Singh and M.A. Ali, 'Properties of high-alumina cement reinforced with alkali resistant glass fibres', *J. Mat. Sci.* 16, 1981, 2597–2607.

122. A.J. Majumdar, B. Singh and T.J. Evans, 'Glass fibre-reinforced supersulphated cement', *Composites.* 12, 1981, 177–183.

123. B. Singh and A.J. Majumdar, 'GRC made from supersulphated cement: 10 year results', *Composites.* 18, 1987, 329–333.

124. A.J. Majumdar and M.S. Stucke, 'Microstructure of glass fibre reinforced supersulphated cement', *Cem. Concr. Res.* 11, 1981, 781–788.

125. A.J. Majumdar and B. Singh, Non portland cement GRC, Information Paper IP 7/82, Building Research Establishment, England, 1982.

126. R.H. Mills, 'Age embrittlement of glass-reinforced concrete containing blast furnace slag', in D.M. Roy, A.J. Majumdar, S.P. Shah and J.A. Manson (eds) *Advances in Cement-Matrix Composites*, Proc. Materials Research Society Symp., Materials Research Society, Bagneux, France, 1980, pp. 201–208.

127. S. Akihama, T. Suenaga, M. Tanaka and M. Hayashi, 'Properties of GRC with low alkaline cement', in S.P.Shah and G.B.Batson (eds) *Fiber Reinforced Concrete, Properties and Applications*, ACI SP-105, American Concrete Institute, Detroit, MI, 1987, pp. 189–209.

128. J. Pera and J. Ambroise, 'New applications of calcium sulfoaluminate cement', *Cem. Concr. Res.* 34, 2004, 671–676.

129. J.M. West, R.C. De Vekey and A.J. Majumdar, 'Acrylic-polymer modified GRC', *Composites.* 16, 1985, 33–38.

130. J.M. West, A.J. Majumdar and R.C. De Vekey, 'Properties of GRC modified by vinyl emulsion polymers', *Composites* 17, 1986, 56–62.

131. A.J. Majumdar, B. Singh and J.M. West, 'Properties of GRC modified by styrene-butadiene rubber latex', *Composites* 18, 1987, 61–64.
132. P. Soroushian, A. Tlili, M. Yohena and B.L. Tilsen, 'Durability characteristics of polymer-modified glass fiber reinforced concrete', *ACI materials Journal* 90, 40–49, 1993.
133. J.M. West, A.J. Majumdar and R.C De Vekey, 'Polymer impregnated lightweight GRC', *Composites* 11, 1980, 169–174.
134. K.N. Quinn, 'Design performance and durability of GRC panels', in *Design Life of Buildings*, Thomas Telford, London, 1985, pp. 139–156.
135. J.F.A. Moore, 'The use of glass-reinforced cement in cladding panels', Building Research Establishment Information Paper IP 5/84, Building Research Establishment, England, 1984.
136. E. Cziesielski, 'Possibilities for the application of glass fibre reinforced concrete in the building industry', *Betonwerk+Fertigteil-Technik.* 1978, 260–265.
137. E. Cziesielski, Possibilities for the application of glass fibre reinforced concrete in the building industry, *Betonwerk+Fertigteil-Technik.* 1978, 312–316.
138. H. Steinegger, *Building components of fibreglass-reinforced concrete*, Betonwerk+Fertigteil-Technik. 1984, 710–715.
139. F. Mayer, Double floor slab of glass fibre reinforced concrete, *Betonwerk+Fertigteil-Technik.* 1982, 39–40.
140. G.F. True, 'Glassfibre reinforced cement permanent formwork', *Concrete*, February 1985, 31–33.
141. E. Haeussler, Considerations on warping and cracking in sandwich panels, *Betonwerk+Fertigteil-Technik.* 1984, 774–780.
142. G.R. Williamson, 'Evaluation of glass fibre reinforced concrete panels for use in military construction', in S. Diamond (ed.) Proc. Durability of Glass Fiber Reinforced Concrete Symp., Prestressed Concrete Institute, Chicago, IL, 1985, 54–63.

Chapter 9

Asbestos fibres

Asbestos–cement was the first FRC composite in modern times, and was used extensively as a cladding material, for roofing and wall units as well as pipes. The asbestos fibres are made of natural crystalline fibrous minerals, consisting of bundles of filaments (sometimes known as fibrils), with individual filaments being as thin as 0.1 μm or less. In the manufacture of the actual composite, the bundles tend to be split up during the processing procedure, though considerable portions of the reinforcement remain in the form of small fibre bundles.

During the 1960s and 1970s it became evident that asbestos fibres pose considerable health hazards. Because of their small size they can be inhaled into the lungs, causing damage and disease. The incubation time for the illness to develop can be several decades, and therefore it took considerable time to realize the hazards involved. The risk is particularly high when the fibres are in a loose form and inhalable, which is the case during the production process. However, even when they are bound into the cement matrix in the cladding composite, risks may occur, especially during various operations during construction, maintenance and removal, when the composite can be affected by mechanical actions which expose the fibres. In view of these risks the use of asbestos for any application, including the production of asbestos–cement composites has been banned in most countries.

Although this composite is of no practical interest, the research and know-how developed over the years are of scientific and engineering value, as they may provide guidance to other composites, where similar production methods are used, and to other composite systems where microfibres of sizes similar to asbestos are being applied. The technical advantages of the asbestos fibres, facilitating efficient production of the composite as well as useful mechanical properties generated considerable effort to develop systems of similar performance using other fibres which pose no health hazards. For such systems, the understanding of the 'reference' of asbestos–cement composite is of considerable value. In view of these considerations, it was decided to keep this chapter in the revised book, to serve as a source of valuable information.

The technical and economic success of asbestos–cement in the past has been largely the result of the compatibility between the fibres and the cement matrix. This is due to the high modulus of elasticity and strength of the fibres, and to their affinity with the Portland cement, which permits effective dispersion of relatively large fibre volumes (10% and more) and enhances the fibre–matrix bond in the hardened composite. This affinity with the fresh mix is particularly useful in developing efficient production processes, some of them based on pulp industry principles such as the Hatschek process.

In reviewing asbestos–cement composites, one must present a treatment that combines the concepts involved in the production technology with those of the reinforcement of the hardened composite. These two aspects are a product of the special structure and properties of the asbestos fibres. In this chapter, the production technology will first be discussed, followed by a description of the fibre structure and properties, and of the fibre–matrix interaction in the hardened composite.

9.1   Production processes

The production technologies are based on the preparation of an asbestos–cement–water slurry, and the subsequent formation of a thin lamina of the dewatered slurry. The final product is obtained by the 'piling up' of laminae, one over the other while they are still wet, to obtain a laminate of the desired thickness. Frequently, the build-up of the layered laminate is obtained by continuous winding of the laminae onto a cylinder (mandrel). This cylindrical product can be kept in the form of a pipe, or it can be cut to form a flat sheet or a board of some other shape (e.g. corrugated sheet) by further shaping of the fresh flat sheet which has the green strength and flexibility required for such shaping. The final production stage is curing, which can be carried out at room temperature or by heat treatment.

The most common production method is the Hatschek process for sheet products, and its modification (Mazza process) for pressure pipe making. A schematic description of a Hatschek machine is presented in Figure 9.1 [1]. Five stages of production can be identified:

1   Slurry preparation (not shown in Figure 9.1).
2   Lamina formation on the sieve cylinder (I in Figure 9.1) and its continuous application on the running felt (3 and 12 in Figure 9.1).
3   Build-up of the laminae (layers) into a laminate by winding on the cylinder (mandrel), until the required thickness is obtained (7 in Figure 9.1).
4   Cutting and shaping if required (8 in Figure 9.1).
5   Curing (not shown in Figure 9.1).

The critical step in the process is the formation of the laminae; at this stage the asbestos fibres play a critical role. As the sieve cylinder rotates, a slurry layer is deposited on its surface, which turns into a coherent layer as water filters from it into the 'dry' side of the sieve cylinder, where some back-up water accumulates. The properties of the laminate depend upon the filtration rate. The control of filtration characteristics by the asbestos fibres is discussed in Section 9.2, in conjunction with the composition and surface properties of different kinds of fibres.

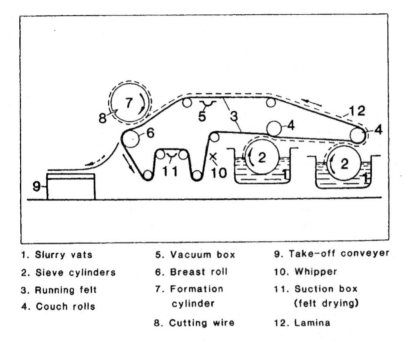

1. Slurry vats	5. Vacuum box	9. Take-off conveyer
2. Sieve cylinders	6. Breast roll	10. Whipper
3. Running felt	7. Formation	11. Suction box
4. Couch rolls	cylinder	(felt drying)
	8. Cutting wire	12. Lamina

Figure 9.1 Schematic description of the Hatschek process (after Williden [1]).

Attention must be given to the back-up water accumulated inside the sieve cylinder. Economical and ecological considerations require that this water be circulated back into the process. This can be most efficiently achieved if the water has a minimum content of solids. The presence of solid particles in the back-up water is associated with fine cement particles and short fibres which can most readily be filtered out. Asbestos fibres of the type which adsorb and retain cement particles in the fresh mix are thus the essential ingredient required for minimizing the solids content of the back water. The characteristics of the asbestos which can enhance this effect will be discussed in Section 9.2.

An additional important aspect of the deposition of the slurry on the sieve cylinder, and of the lamina formation, is the phenomenon of streaming of the deposited slurry. This may lead to a preferred orientation of the fibres. This effect, which is usually detrimental to the performance of the hardened composite, is enhanced by high speed and by the presence of longer fibres.

After adequate lamina formation and deposition on the running felt, there is a need to control the water content in the laminae to facilitate the successful completion of the subsequent production stages. An optimum water content (22–28%) is required to achieve the combination of high green strength and plasticity which enable handling and additional shaping of the green product (e.g. corrugation) without cracking or uncontrolled deformation (e.g. ovality in pipes). Thus, the water content of the newly formed laminae, which is about 55%, should be reduced with the aid of the vacuum boxes (5 in Figure 9.1) which are placed along the running felt. Here, too, the filtering characteristics of the green laminae are important. Poor filtration will require slowing down of the process, to extend the residence time over the vacuum boxes.

The final stage of the production process involves curing, which can be based on room temperature treatment in a tunnel in which moist conditions are maintained, or higher temperature steam curing to accelerate the hardening process. Autoclave curing is also common in the asbestos–cement industry, and in such instances part of the cement is replaced with finely ground silica. Replacement of part of the cement with low-cost inert fillers or fly ash is also sometimes used, for economic reasons.

Other types of production processes are also dependent on the special nature of the fresh mix, which is controlled by the special properties of the asbestos fibres. The Magnani process is based on direct deposition of slurry on a porous former, followed by dewatering and consolidation by vacuuming and pressure. The product can be produced by the build-up of several layers.

9.2   Structure and properties of asbestos fibres

Asbestos is the collective name given to a variety of naturally occurring fibrous silicates which are crystalline in structure. Depending on their mineralogical composition, they can be subdivided into two groups, serpentines and amphiboles. The fibrous form of the commercially used serpentine is called *chrysotile*, while in the amphibole groups there are several types of fibrous minerals, with *crocidolite* (blue asbestos) being the one most commonly used commercially. Some of the more common varieties of asbestos are classified in Figure 9.2 [2,3] , along with their chemical compositions. Amphibole has a layered structure, and its external surface is a silicate layer, of low surface potential (Figure 9.3(a)). Chrysotile consists of parallel sheets of brucite ($Mg(OH)_2$) and silicates (linked SiO_4^{4-}), with the brucite layer having slightly larger dimensions. As a result of this mismatch, the sheets tend to become curved, leading to the cylindrical structure shown in Figure 9.3(b), in contrast to the banded structure of amphibole. Since the brucite layer forms the external surface of the chrysotile, the fibre is basically hydrophylic, and develops a high surface potential when in contact with water. The differences in the surface properties of the fibres, with the chrysotile being much more reactive and hydrophylic compared with the amphibole, lead to considerable differences in the effect of each of these types of asbestos on the properties of the slurry in the fresh state.

The mean diameter of the elementary crystals (fibrils) is in the range of 0.02–0.2 μm, with chrysotile having the lower range. In practice, asbestos fibres consist of bundles of such fibrils (Figure 9.4) [4] and these bundles are split-up during processing, though not necessarily into individual fibrils. Splitting of the

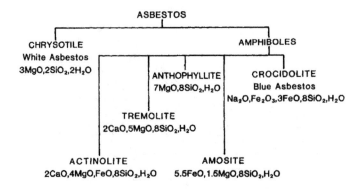

Figure 9.2 Classification of asbestos and some of its more common varieties [2,3].

Table 9.1 Strength and modulus of elasticity of asbestos fibres (after Aveston [5])

Type	Fibre	Tensile strength (MPa)	Modulus of elasticity (GPa)
Serpentine	Chrysotile	3060–4480	160
Amphibole	Crocidolite	1680–2790	169–178
	Amosite	1160–1710	171–172
	Termolite	310	152
	Actinolite	1890	146
	Anthophyllite	1350	154

fibre bundles and their opening (i.e. fiberization) is carried out prior to the slurry preparation stage by mechanical milling, followed by fiberization in a wet spinning process. The various kinds of asbestos are classified, treated and used, depending upon their length and degree of fiberization. These characteristics determine the effective aspect ratio, which is crucial in controlling the properties of the composite both in the fresh state and after hardening.

The crystalline nature of the various asbestos minerals, which is characterized by a stable silicon–oxygen structure, should theoretically lead to tensile strength values of more than 10,000 MPa [2]. In practice, the measured tensile strength is high (Table 9.1) (300–4000 MPa, depending on the type of asbestos, method of testing and specimen preparation), but is still considerably lower than the theoretical value. This is usually attributed to various defects in the structure and to the cation composition [2]. It is suggested that under strain, cation–oxygen bonds will break prior to failure of the silicate chains, and therefore the cation composition of amphiboles influences their strength. This may be one of the causes for the differences in strength of the various types of amphiboles (Table 9.1) [5]. The bundled structure of the fibres has also been considered when analysing the strength of asbestos.

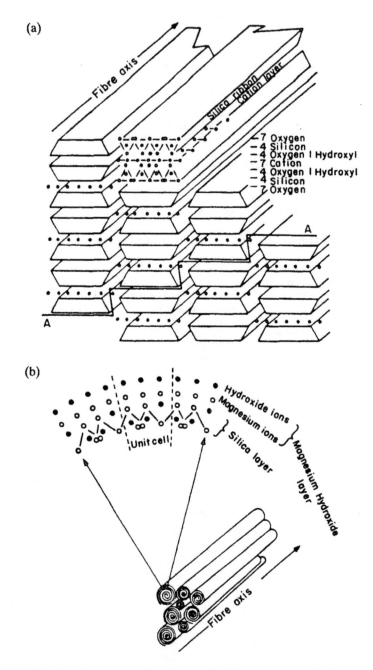

Figure 9.3 The crystalline structure of the two major forms of asbestos (after Hodgson [2]). (a) Amphibole and (b) chrysotile.

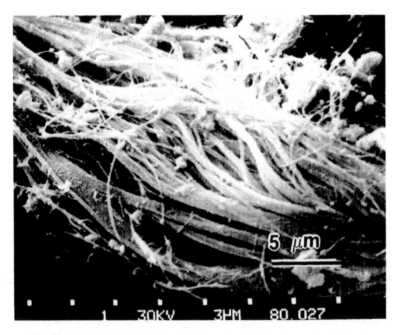

Figure 9.4 Micrograph of an asbestos fibre showing the structure of a bundle composed of numerous fibres (after Akers and Garret [4]).

Hodgson [3] has described this system as a series of overlapping crystallites (fibrils) of various lengths, held together in the fibre by hydrogen bonds. He argued that loss of strength during heating is associated with weakening of these bonds, and, as a result, the fibrils are pulled apart and broken more readily, thus leading to a reduction in the fibre strength. Aveston [5], on the other hand, suggested a model of continuous fibrils bundled together in a fibre in which the interfibrillar adhesion is low for chrysotile, somewhat higher for amphiboles, and much higher for the heat treated amphiboles. The strength of the fibre can be analysed in terms of weak link concepts; if the interfibrillar bond is too strong the stress concentration at the tip of a surface crack formed in a weak fibril cannot be relieved by a secondary crack propagating perpendicular to the fibril at the interface between fibrils. Thus, brittle fracture will be dominated by the strength of the weakest fibril in the bundle. Aveston [5] suggested that heating results in enhancement of interfibrillar bond, and when it exceeds a critical value, fibre weakening will take place. Therefore, the strength of the asbestos fibre is highly sensitive to the interfibrillar bond, and the average strength is a statistical function dependent upon, amongst other things, the length of fibre tested.

As indicated previously, the fibre properties depend also on the degree of fiberization obtained during their processing. This can be measured by surface

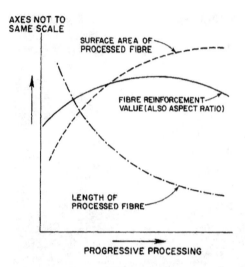

AXES NOT TO
SAME SCALE

SURFACE AREA OF
PROCESSED FIBRE

FIBRE REINFORCEMENT
VALUE (ALSO ASPECT RATIO)

LENGTH OF
PROCESSED FIBRE

PROGRESSIVE PROCESSING

Figure 9.5 Effect of processing on length, surface area and strengthening in the hardened composite (after Williden [1]).

area determination (or other related methods used in the industry) [1]. Typical surface areas after milling, as measured by N_2 adsorption, are approximately 3000 m^2/kg [3]. This is about 5 times the area prior to milling, and about 10 times the area of ordinary Portland cement.

The length of the fibres is an important parameter in controlling final strength, green strength and filterability. Longer fibres have a favourable effect on each of these properties. However, during the fiberization processing stage, the fibres undergo some degradation which leads to a reduction in length, and the generation of fines. Thus, from the point of view of strength, processing increases the surface area, which is a favourable effect, but at the same time reduces length, which is detrimental. As a result, optimum processing time and intensity are required for maximum strength (Figure 9.5). In addition, the filterability reduces continuously with progressive processing. The deterioration of the fibres depends on their composition, and when comparing fibres, their filterability should be assessed at the processing stage, at which the optimum level of strength is generated.

When considering the performance of the fibres in the wet stage, their surface properties must be considered. The filterability of amphiboles is usually better than that of chrysotiles. This can be attributed to the higher surface activity and zeta potential of the chrysotile and its more hydrophilic nature, which can lead to reaction with the hydrated material formed early on. The chrysotiles tend to pack into smaller sedimentation volumes, and react with the cement, thus leading to a low permeability to water. Zeta potential values which are significantly different from zero (which is the case with chrysotile) lead to repulsive forces between the fibres. This prevents flocculation, and causes them to pack more densely during sedimentation, which reduces their ability to form an efficient felt during filtration [6]. However, it should be pointed out that Schultz *et al.* [6] found that the surface charge is only a second-order effect, and parameters such as fibre content and diameter might be more crucial. Another aspect of the fibre performance in the fresh system is its solids retention, which is needed to prevent the fine portion of the raw materials from passing into the back water (see Section 9.1). Here, the amphiboles are superior. They are also more effective in dispersing the fibres and preventing their agglomeration and clumping. The stabilization of the slurry cannot be efficiently achieved by mechanical and chemical means only, and the presence of amphiboles is essential from this point of view.

In the asbestos–cement industry it is common to use a blend of different types of asbestos fibres, to optimize the characteristics of the mix in its green (processing) stage, as well as the strength in the hardened state. Some of the asbestos fibres are chosen for their processing characteristics, while others are present because of their efficiency in enhancing strength.

9.3 Mechanical properties of asbestos–cement composites

The processing parameters and the composition of the hardened composite control the mechanical properties of the composite [7,8]. From a simple theoretical point of view, one would expect strength and modulus of elasticity to increase with fibre content and length. However, this is not the case. The strength (tensile and flexural) shows a maximum at an intermediate fibre content, whereas the modulus of elasticity decreases with increasing fibre content (Figures 9.6, 9.7; Table 9.2).

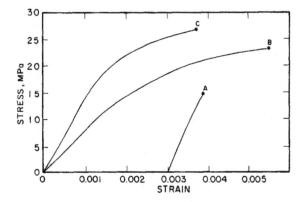

Figure 9.6 Stress–strain curves of asbestos–cement composites with different fibre contents (after Allen [7]). A – low fibre content, B – high fibre content and C – intermediate fibre content.

Table 9.2 Effect of fibre content on properties of abestos–cement composites (after Allen [7])

Fibre content (% vol.)	Matrix void content (% vol.)	Modulus of elasticity[a] (GPa)	Tensile strength[a] (MPa)	Ultimate elongation[a] (%)
2.91	14.3	17.29	14.6	0.95
5.10	26.7	16.09	20.3	2.32
7.32	32.3	14.73	25.4	3.70
14.85	60.4	8.45	21.3	5.06

Note
a Longitudinal direction.

A similar influence of the fibre length was reported by Allen [7], showing the maximum strength at an intermediate length (Table 9.3). These characteristics can readily be explained by the reduction in the density (increase in void content) of the composite with an increase in fibre content and length (Figure 9.7, Tables 9.2, 9.3), which reflects the less efficient consolidation of the green product. This can be partially eliminated by increasing the pressure applied to the green sheet, but even with this treatment there is still an adverse effect of high fibre content (Figure 9.7). The difference in the trends for strength and modulus of elasticity (i.e. the former showing a maximum at an intermediate fibre content, and the latter continuously decreasing) reflects the fact that the contribution of the fibre to the modulus of elasticity of the composite is small compared with the matrix effect, and therefore the reduction in the density of the matrix becomes the overriding parameter, leading to a continuous decrease in modulus of elasticity with increasing fibre content. In the case of strength, the contribution of the fibres is dominant; therefore, at low fibre contents the strength of the composite increases with fibre content and length. However, at a high fibre content or with long fibres, the effect of the reduction in density becomes more pronounced, and cannot be compensated for by the increase in fibre content or length.

It is interesting to note that the *toughness* of the composite, estimated by ultimate strain or impact strength, increases with fibre length (Table 9.3) and fibre content (Figures 9.6 and 9.7). Here, in contrast to strength, the decrease in density may not be detrimental when considering the energy involved in the pull-out of the fibres.

The fibre treatment is also important with respect to breaking the bundle structure in the mechanical milling process, and opening up and fiberization by techniques such as wet spinning. Allen [7] demonstrated that additional milling and fiberization were associated with considerable improvements in the properties of the composite, compared with the material produced with fibres which were in the same form as received from the mine (Table 9.4). He suggested that this is the result of the very poor bond in the coarse, unsplit crude fibre bundle. More complex relations between fibre processing and composite properties were resolved by Akers and Garrett [8], who studied the simultaneous effect of mechanical milling (collering) followed by fiberization (Figure 9.8). The best performance

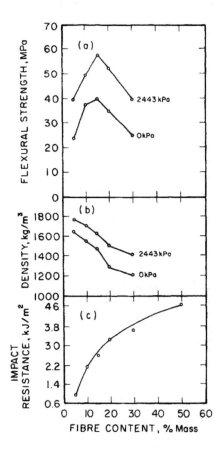

Figure 9.7 Effect of fibre content and compaction pressure on the properties of asbestos–cement composites (after Akers and Garrett [8]).

Table 9.3 Effect of fibre length on the properties of asbestos–cement composites (after Allen [7])

Fibre length	Fibre content (% vol.)	Matrix void content (% vol.)	Modulus of elasticity[a] (GPa)	Tensile strength[a] (MPa)	Ultimate elongation[a] (%)
Short	5.70	17.9	16.93	17.8	1.28
Medium	5.10	26.7	16.09	20.3	2.32
Long	4.72	32.6	13.05	16.1	2.54

Note
a Longitudinal direction.

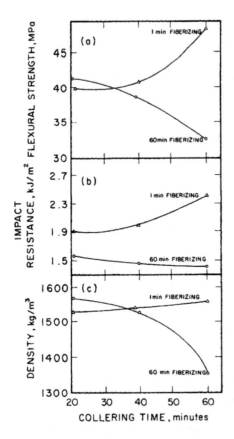

Figure 9.8 Effect of fibre treatment on the properties of asbestos–cement composites (after Akers and Garrett [8]).

was achieved by a combination of short fiberizing times and extended collering periods. It was suggested that the fiberization treatment leads to a greater water retention of the liquid suspension, compared with the milling process. Therefore, if the fiberization process is too long, the density of the composite will be reduced to such an extent that it cannot be compensated for by the additional opening up of the fibres.

When considering the properties of asbestos–cement composites in relation to the production processes, the orientation effect must also be taken into account. The data reported by Allen indicate that the strength in the transverse direction is only about 60–85% of that in the longitudinal direction, with similar ratios reported for the ultimate elongation. Zevin and Zevin [9,10] determined the orientation of the fibres by X-ray methods. They then calculated the strength ratio between the longitudinal and transverse directions, which was found to compare favourably with

Table 9.4 Effect of fibre processing on the properties of asbestos–cement composites (after Allen [7])

Processing	Fibre content (% vol.)	Matrix void content (% vol.)	Modulus of elasticity[a] (GPa)	Tensile strength[a] (MPa)	Ultimate elongation[a] (%)
Processed	5.10	26.7	16.09	20.3	2.32
Crude	6.02	12.8	20.41	14.5	0.86

Note
a Longitudinal direction.

Table 9.5 Longitudinal to transverse strength ratio in asbestos–cement composites produced by different processes (after Zevin and Zevin [10])

Process	Strength ratio
Uniaxial pressure	1
Rolling of one layer	1.07
Sheets on Hatschek machine	1.20
Pipes on Hatschek machine	1.50

experimental results. Some of their data for various asbestos–cement processing are given in Table 9.5.

9.4 Reinforcing mechanisms in asbestos–cement composites

The two main approaches for dealing with the mechanics of asbestos–cement composites are based on composite materials concepts (rule of mixtures) and fracture mechanics. Microstructural characterization of the composite has usually been carried out in conjunction with these studies, to determine the numerical parameters required for the modelling, such as the aspect ratio, and to resolve the pull-out and fracture processes during failure. The results of these microstructural studies will be reviewed first, since they provide the background required for the modelling of the processes which control the mechanical performance of the hardened composite.

9.4.1 Bonding and microstructure

SEM characterization of the fractured surface of asbestos–cement composites was reported by Akers and Garrett [4,11] and Mai [12]. It was shown that in the actual composite, a considerable portion of the fibres was in the form of bundles

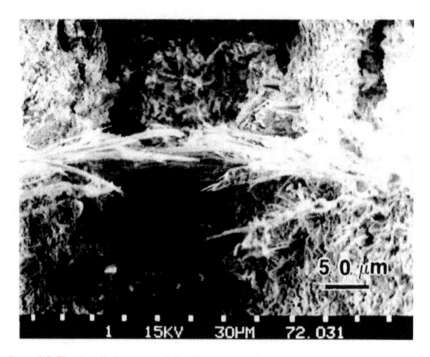

Figure 9.9 The bundled nature of the fibre in an asbestos–cement composite, as seen during *in situ* testing of a composite (after Akers and Garrett [11]).

(Figure 9.9), ranging in diameter from less than 1 μm to 300 μm. The length of the fibres ranged from a fraction of a millimetre to 5 mm, giving average aspect ratios of about 100 [11] and 160 [12]. Akers and Garrett [11] concluded that the pull-out and bond characteristics were the result of a complex combination of mechanisms associated with the special nature of the fibrillated, bundled structure of the asbestos fibre, and with the matrix interface, which is different from that developed in other fibre–cement composites.

In many cement composites there is evidence of the formation of a weak interfacial transition zone, rich in CH. In the asbestos–cement composite no such zone was found, and the C/S ratio of the matrix at the interface was the same as that of the bulk, showing no evidence of the accumulation of CH. This was correlated with observations after pull-out, which showed cement adhering to the surface of the pulled-out fibre, suggesting a strong interface, and debonding away from the interface in the matrix itself. This improved interfacial bonding is associated with the densification of the interfacial microstructure of the matrix. It may be related to the more hydrophilic surface of the asbestos fibres compared with other reinforcing inclusions, and to the processing of the asbestos–cement composites, which probably eliminates the tendency for formation of water-filled spaces in the

Figure 9.10 Interlocking of fibrils at the edge of a bundled fibre with adjacent hydration products (after Akers and Garrett [4]).

vicinity of the fibres in the fresh product. Akers and Garrett [4] observed an additional bonding mechanism, associated with the interlocking of the fine asbestos fibrils with hydration products (Figure 9.10). Such an interlocking can be particularly effective, since the size of the individual fibrils is of the same order of magnitude as that of the hydration products. Another mechanism which makes the bonding in asbestos–cement more complex is the sheath/core separation during fracture. A sheath of external fibrils will become separated from the pulled-out fibre bundle, leaving behind and revealing the core of the bundle (Figure 9.11). Thus, when considering the overall bond and pull-out resistance, the combination of interfacial debonding through matrix failure, and the influence of interfibre separation should be considered.

Single fibre pull-out tests have indicated that the pull-out bond strength is in the range of 0.8–2.4 MPa [4,12,13]. When considering a fibre strength in the range of 200–3000 MPa [5,14] the critical aspect ratio is greater than the aspect ratio of the reinforcement in the actual composite, which was found to be in the range of 100–160. Thus, the failure mechanism in the asbestos-cement composite is expected to be a combination of fibre fibrillation and fibre pull-out, as found by the *in situ* SEM testing of Akers and Garrett [11] (Figure 9.12).

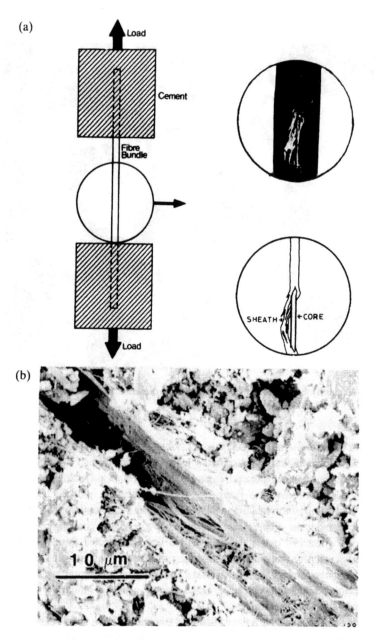

Figure 9.11 The complex mode of asbestos bundle pull out, showing a sheath separated from the core (after Akers and Garrett [4]). (a) Schematic description; (b) micrograph showing the outer sheath of chrysotile fibrils from a bundle remaining in a cement trough after separation and pull-out of the inner core.

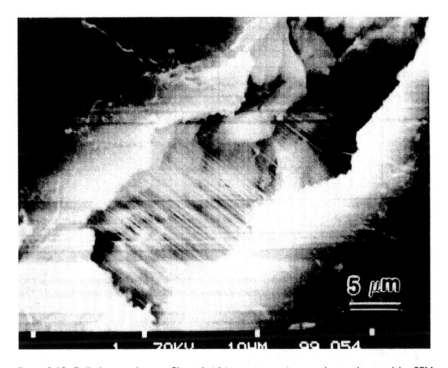

Figure 9.12 Pulled out asbestos fibres bridging over a microcrack, as observed by SEM *in situ* testing (after Akers and Garrett [11]).

9.4.2 Prediction of mechanical properties – composite materials approach

The composite materials approach, based on the rule of mixtures (reviewed in Section 4.3) has been applied by various investigators to account for the modulus of elasticity, tensile strength and flexural strength of asbestos–cement composites.

Allen [7] considered the modulus of elasticity in terms of the rule of mixtures (see Equation 4.26):

$$E_c = E_m(1 - V_f) + E_f V_f \tag{9.1}$$

However, he pointed out that since an increase in fibre content is associated with an increase in the matrix void content (i.e. the reduction in density discussed in Section 9.3), the matrix modulus of elasticity, E_m, in Eq. 9.1 is a variable, which changes with fibre content. He described it in terms of the hypothetical modulus of elasticity of a void-free matrix, E_{mo}, and the void content p:

$$E_m = E_{mo}(1 - p) \tag{9.2}$$

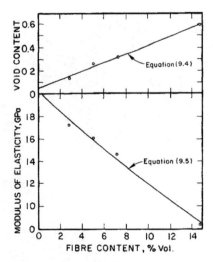

Figure 9.13 Relations between matrix void content (*p*), composite modulus of elasticity (E_c) and the fibre content of the composite (after Allen [7]).

Substituting E_m in Eq. 9.1 gives:

$$E_c = E_{mo}(1 - p)(1 - V_f) + E_f V_f \qquad (9.3)$$

On the basis of the results presented in Tables 9.2–9.4, Allen deduced a linear relationship between the matrix void content, *p*, and the fibre volume content (Figure 9.13(a)):

$$p = 0.0522 + 3.7407 V_f \qquad (9.4)$$

Substituting *p* (Eq. 9.4) into Eq. 9.3 leads to:

$$E = (0.9478 - 3.7407p)E_{mo}(1 - V_f) + E_f V_f \qquad (9.5)$$

The best fit of Eq. 9.5 to the experimental results (Figure 9.13(b)) was obtained for values of E_{mo} = 21.45 GPa and E_f = 10.30 GPa. With these values, Eq. 9.3 has general applicability. The value of E_{mo} is reasonable for a void-free cement paste. However, E_f is considerably lower than the measured modulus of elasticity of asbestos fibres by about an order of magnitude. This might be accounted for on the basis of their short length and random orientation. Thus, this value must take into account other factors which have not been considered explicitly.

Strength was also analysed [7,11,12] on the basis of the composite materials approach. Since it was demonstrated that the composite failure is associated with fibre pull-out, Eqs 4.31–4.33, 4.42 (Section 4.3) were applied [11,12]. In these studies, the fibre contribution is described by the term $\eta V_f \tau (\ell/d)$, in which η is the orientation efficiency factor. Akers and Garrett [11] found that the strength can be best described by assuming that there is no matrix contribution in tension, or in flexure, and that the efficiency factor for random orientation is the $2/\pi$ value derived by Aveston et al. [15]. Thus, for tension:

$$\sigma_{cu} = \frac{2}{\pi} \cdot V_f \tau (\ell/d) \tag{9.6}$$

For flexure, using the stress distribution block diagram of Hannant [16], which gives an increase in flexural strength by a factor of 2.44 compared with direct tension, the following relation was derived:

$$\sigma_b = 1.55 V_f \tau (\ell/d) \tag{9.7}$$

where σ_b is the flexural strength.

Substituting $V_f = 10\%$, $\ell/d = 100$ and $\tau = 2.4\,\text{MPa}$ (see Section 9.4.1), the values of tensile and flexural strength were calculated to be 15 and 37 MPa respectively, which compare well with experimental results.

Mai [12], on the other hand, used the approach developed by Swamy et al. [17] in which the matrix contribution to strength is also considered, and the fibre contribution is calculated as $2\tau V_f (\ell/d)$ for a longitudinal array, and 41% of that value for random 2D orientation:

$$\sigma_b = \sigma_m (1 - V_f) + 0.82\tau V_f (\ell/d) \tag{9.8}$$

If we define α as $\sigma_b \sigma_{cu}$ for the composite, and β as $\sigma_b \sigma_t$ for the matrix, the flexural strength of the composite can be derived by modification of Equation 9.8:

$$\sigma_b = \left(\frac{\alpha}{\beta}\right)(\sigma_b)_{\text{matrix}}(1 - V_f) + 0.82\tau V_f (\ell/d) \tag{9.9}$$

Mai determined the values of α and β from his experimental results to be 2.69 and 3.3, respectively. Using his estimated aspect ratio of 160 (Section 9.4.1) and τ of 0.83 MPa (after de Vekey and Majumdar [18]), a good correlation could be established between the experimental data and the predictions based on Eqs 9.8 and 9.9 (Figure 9.14).

It should be noted that in Mai's results, the strength levels off at high fibre volume contents (Figure 9.14), but does not decline as was observed by Akers and Garrett [8] (Figure 9.7) and Allen [7] (Table 9.2). This probably reflects the different production processes: Akers and Garrett [8] and Allen [7] tested composites produced by the Hatschek process, whereas Mai tested laboratory specimens prepared by casting in a vacuum filter box. The simple composite materials strength equations (Eqs 9.6–9.9) do not account for either the levelling off or reduction in strength at high fibre contents, which is associated with the less efficient compaction of the composite.

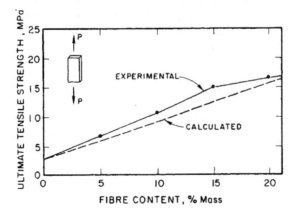

Figure 9.14 Calculated and analytical relations between tensile strength and fibre content of asbestos cement composites (after Mai [12]).

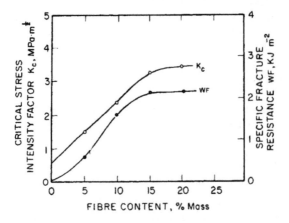

Figure 9.15 The effect of fibre content on fracture mechanics parameters (after Mai [12]).

9.4.3   Fracture mechanics

Fracture mechanics has been applied to model some of the fracture characteristics asbestos–cement composites, especially those related to crack bridging by the fibres. Mai [12] analysed the behaviour of both notched beams in flexural loading and double cantilever beams (DCB) to determine the critical stress intensity factor, K_c, and the specific work of fracture, WF, which measures the average fracture energy per apparent unit crack surface over the entire fracture process. The effect of fibre content on these parameters is shown in Figure 9.15. The trends resemble those observed for the effect of fibre content on strength (Figure 9.14).

The total specific work of fracture, WF_t, can be treated as the sum of three effects, the energies involved in:

1   matrix cracking, WF_m;
2   fibre–matrix debonding, WF_{db};
3   fibre pull-out, WF_{po}.

For a fibre orientation efficiency of 0.41, the following relationship was derived:

$$WF_t = WF_m + WF_{db} + WF_{po} = V_m \cdot G_m + 0.41 V_f \frac{4 \cdot \bar{\ell} \cdot G_{db}}{d} + \frac{0.41 V_f \bar{\ell}^2 \tau}{6d}$$
$$(9.10)$$

where $\bar{\ell} = l/4$; G_m, G_{db} are the fracture energies of the matrix and the fibre–matrix interface, respectively.

This equation is valid for a system in which fibres pull out without fracturing, and this has been shown to be the case in asbestos cement. By assuming $G_m = G_{db}$, Eq. 9.10 can be rewritten as:

$$WF_t = \left[V_m + 0.41 \frac{V_f \ell}{d} \right] G_m + \frac{0.41 V_f \ell^2 \tau}{6d} \qquad (9.11)$$

By substituting the appropriate values of the structural parameters in Eq. 9.11 (see Sections 9.4.1, 9.4.2), namely $\ell = 4$ mm, $d = 25$ μm, $\tau = 0.83$ MPa, $G_m = 20$ Jm^{-1}, the $(WF)_t$ values were found to be in good agreement with the experimental results of Figure 9.15. It was also concluded that the energies involved in creating new surfaces (the first term in Eq. 9.11) accounted for only ~5% of the total specific work of fracture, WF_t. Thus, pull-out can be considered as the main mechanism contributing to the toughness of these composites. It should be emphasized that, since the average pull-out length is so small, only a few millimetres, the work of fracture and toughness in asbestos–cement composites are much lower than in other fibre–cement composites, in which the fibres are much longer.

Lenain and Bunsell [19] characterized the fracture behaviour of the asbestos–cement composite by determining the R-curve. Fibre bridging over microcracks which formed ahead of the crack tip was shown to be a major mechanism in resisting crack propagation, and it thus played an important role in controlling the R-curve.

The fibres bridging the microcracks, as well as those across the main crack, induced a closing stress on the crack tips and thus increased the work of fracture of the material. Such effects could account quantitatively for the shape of the R-curve. Lenain and Bunsell [19] also tried to use acoustic emission to resolve some of the cracking processes. However, this technique has severe limitations for quantitative analysis as discussed by Akers and Garrett [20].

Mai et al. [12,21] also concluded that R-curves were most suitable for the characterization of the fracture behaviour of asbestos–cement and of asbestos–cellulose hybrid-reinforced cement. They applied the rule of mixtures to account for the fracture toughness, which included the sum of the contributions due to fibre pull-out and matrix fracture.

9.5   Long-term performance

Very little has been published in the open literature regarding the durability of asbestos–cement composites in natural weathering. However, many years of experience have suggested that this composite usually has an excellent long-term performance. For instances, Jones [22] has reported that no reduction of flexural strength was observed in weathered roofing sheets, the oldest of which was 26 years. One contributing factor could be the stable nature of the fibre, which does not degrade in the alkaline environment of the cement matrix. Majumdar *et al.* [14] extracted asbestos fibres from weathered boards and showed that no significant change in their strength occurred during seven years of exposure. The strength data showed considerable scatter but no real tendency for decline could be seen. This is consistent with accelerated ageing of the actual composite, which showed no reduction in flexural strength after one year of immersion in 50°C water [23].

In spite of these characteristics, the asbestos–cement composite is not completely immune to changes in properties. Te'eni and Scales [24] and Jones [22] reported a reduction of 50% in the impact resistance after 10–15 years of exposure, which was not accompanied by flexural strength reduction. Jones [22] suggested that the reduced impact resistance observed could be the result of carbonation. He showed that accelerated carbonation tests led to similar changes in mechanical properties as those observed in natural weathering, that is, a mild increase in flexural strength and about 50% reduction in impact resistance. The reduction in impact resistance might be associated with a somewhat greater tendency for cracking sometimes experienced in the aged composite. Majumdar [25] suggested that the reduced impact resistance, which is not accompanied by a reduction in static strength, may be associated with improved fibre–matrix bond in the asbestos bundle, similar to the mechanisms discussed in the case of glass fibre bundles in Chapter 8. Here, the bundled nature of the asbestos fibre may play an important role. Opoczky and Pentek [26] studied the microstructure of aged asbestos–cement sheets and concluded that fibres can undergo partial carbonation. In addition, the cement hydration products crystallize on the surface of the asbestos fibres and along the cleavage planes between fibrils, and they may react chemically with the fibres, especially in the presence of CO_2. Such effects may indeed lead to improved bond.

Sharman and Vautier [27] recognized the significant effect of carbonation on the mechanical properties of room temperature cured asbestos–cement. They found an increase in tensile and flexural strengths with increased carbonation, but this was not accompanied by a reduction in impact resistance as suggested by Jones [22]. Sharman and Vautier [27] suggested that this discrepancy may indicate that these effects depend on the type of asbestos–cement used, since even in Jones' [22] results not all of the carbonated composites exhibited reduced impact resistance. They suggested that the two stages of carbonation effects proposed by Opoczky and

Pentek [26] (partial carbonation of asbestos fibres and 'crystallization' of reaction products at the interface) may advance to different degrees in different asbestos–cement composites, resulting in different effects of carbonation on mechanical properties. Obviously, the ageing mechanisms of these composites are not clear, but field experience and the few studies reported in the open literature suggest that no strength reduction is expected in natural weathering, although a loss in toughness may sometimes occur.

The changes in toughness may affect the cracking sensitivity, especially when restrained components are exposed to volume changes due to temperature and humidity variations. Such cracking, which is not directly related to changes in the strength of the composite, has to be considered when discussing the long-term performance. However, this problem may not be critical, since it has been shown that restrained components exposed to wetting and drying do not necessarily crack even though the calculated drying induced stress exceeds their strength [28]. It was suggested that the asbestos–cement components can undergo relaxation which prevents the build-up of high stress.

There was particular interest in the performance of asbestos–cement in aggressive environments, particularly with respect to special components such as pipes. The problems that may be encountered here are usually matrix related, and they may be dealt with by changing the matrix compositions by means such as autoclaving or, in special circumstances, polymer impregnation [29]. For instance, ASTM 6500-74 specified the limiting free CH content in asbestos–cement pipes as a function of the aggressiveness of the water to be transported through the pipe. For highly aggressive water, the limit is so low that it can be met only by the autoclaving of a matrix consisting of cement and silica. Al-Adeeb and Matti [30] studied the performance of pipes in soft water, and concluded that the longitudinal cracking of failed pipes could not be initiated by mechanical forces, and hence were caused by chemical deterioration of the matrix, which was initiated by leaching of CH. Sarkar et al. [31] analysed the degradation of asbestos–cement sheets exposed to sulphate-rich water in cooling towers, and also concluded that the problem was due to chemical and microstructural changes in the matrix, while the fibres remained intact. Delamination of such sheets was induced by expansion of gypsum deposited at the fibre surface. Gypsum was formed by reaction between Ca^{++} ions released by the matrix and sulphate which was supplied by the water. Matti and Al-Adeeb [32] also attributed sulphate attack in asbestos–cement pipes to the sensitivity of the matrix, but they suggested that formation of ettringite was the main cause of the durability problem. They recommended the use of ASTM Type II and Type V cements to accommodate this problem, as is the practice with other Portland cement applications.

9.6 Asbestos–cement replacement

Some asbestos fibres, when inhaled, can constitute a health hazard leading to lung cancer. The health risks are greatest during the production process, but may be small during the use of the asbestos–cement component, as long as it does not degrade and fall apart due to environmental effects or mechanical abuse. As a result of such health risks, and strict environmental regulations, attempts were made to find replacements for the asbestos fibres, and to develop asbestos-free fibre reinforced cements. Some asbestos-free products are produced on a commercial basis, in particular cellulose pulp-reinforced cement (see Section 11.3). The challenges, difficulties, and the philosophy behind the development efforts of the industry to find such replacements were reviewed in [33] and [34]. The object was to develop a substitute of similar or better properties, while preserving the same efficient production process.

The asbestos fibre is unique in its properties. It permits control of the properties of the fresh mix, to make it processable by the efficient dewatering techniques, and, at the same time, it provides an excellent reinforcing efficiency in the hardened composite, due to its high strength, modulus of elasticity and bond. In addition, this fibre is stable in the alkaline cement environment, thus providing a composite of excellent durability. This combination of properties is difficult to match with any one type of fibre. There are fibres which can match the mechanical quality of asbestos, but they cannot, on the other hand, provide the processing characteristics.

The concept that seemed to be most actively pursued was to develop a substitute made up of a blend of different types of fibres (hybrid composites), some of them functioning as 'processing fibres', while the others are intended to reinforce the hardened composite [33,34]. The requirement of the processing fibres is to enable homogeneous dispersion in the slurry, and to provide it with good filtering characteristics, as well as solids retention. Such fibres should be of a high surface area, and with web-forming characteristics. Cellulose fibres are potentially suitable for this purpose. The reinforcing fibres should possess adequate mechanical

Table 9.6 Properties of some fibres considered for asbestos replacement in cementitious composites (after Studinka [34])

Fibre	Tensile strength (MPa)	Modulus of elasticity (GPa)	Ultimate strain (%)	Diameter (μm)	Durability in cement[a]
Asbestos	3600	150	0.1–0.3	0.1–1	+
Glass	1000–3500	70	2–5	10–50	−
Steel	2400–3800	200	1–2	2–150	−
Kevlar 29	2700	70	3–4	11	+ −
Polypropylene	200–550	0.5–5	10–15	10–50	+
Polyamide 6-6	700–1000	~60	~15	10–50	−
Polyester (PET)	800–1300	up to 15	8–15	10–50	−
Rayon	450–1100	up to 11	7–15	10–50	+ −
HM-Polyvinylalcohol	1600	30	6	14	+
HM-Polyacrylonitril	850	18	9	19	+

Note
a + Good durability; − poor durability.

properties (strength, modulus of elasticity, bond) as well as stability in an alkaline environment. Whenever high temperature or autoclave curing is considered, the fibre should be thermally stable. A list of some possible fibres, and their properties, is provided in Table 9.6.

In the development of successful replacements, consideration was also be given to matrix modifications [33], to adjust its properties in order to accommodate fibres that are not ideally compatible with Portland cement. For example, fibres with some sensitivity to an alkaline environment might potentially be considered adequate if means are taken to reduce the alkalinity of the matrix. The use of inert and active fillers, and control of the cement particle size distribution and the curing treatment provide additional means to adjust for the processing properties as well as the characteristics of the hardened composite.

References

1. J.E. Williden, *A Guide to the Art of Asbestos Cement*, J.E. Williden Publ., London, 1986.
2. A.A. Hodgson, 'Chemistry and physics of asbestos', in L. Michaels and S.S. Chissick (eds) *Asbestos*, Chapter 3, John Wiley, New York, 1979, pp. 67–114.
3. A.A. Hodgson, 'Fibrous Silicates', *Lecture Series No. 4*, The Royal Institute of Chemistry, London, 1965.
4. S.A.S. Akers and G.G. Garrett, 'Fibre-matrix interface effects in asbestos-cement composites', *J. Mater. Sci.* 18, 1983, 2200–2208.
5. J. Aveston, 'The mechanical properties of asbestos', *J. Mater. Sci.* 4, 1969, 625–633.
6. J. Schultz, E. Papirer and M. Nardin, 'Physicochemical aspects of the filtration of aqueous suspensions of fibers and cement', Parts 1–6, *Ind. Eng. Chem. Prod. Res. Dev.*, 22 (1983) 90–105; 23 (1984) 91–98.
7. H.G. Allen, 'Tensile properties of seven asbestos cements', *Composites.* 2, 1971, 98–103.
8. S.A.S Akers and G.G. Garrett, 'The influence of processing parameters on the strength and toughness of asbestos cement composites', *Int. J. Cem. Comp. Ltwt. Concr.* 8, 1986, 93–100.
9. L.S. Zevin and I.M. Zevin, 'Orientation of asbestos in asbestos cement', *Cem. Concr. Res.* 9, 1979, 599–606.
10. L.S. Zevin and I.M. Zevin, 'Asbestos cement: orientation of fibres and anisotropy of mechanical properties', *Int. J. Cem. Comp. Ltwt. Concr.* 4, 1982, 181–184.
11. S.A.S. Akers and G.G. Garrett, 'Observations and predictions of fracture in asbestos-cement composites', *J. Mater. Sci.* 18, 1983, 2209–2214.
12. Y.W. Mai, 'Strength and fracture properties of asbestos-cement mortar composites', *J. Mater. Sci.* 14, 1979, 2091–2102.
13. S.A.S. Akers and G.G. Garrett, 'The relevance of single fibre models to the industrial behavior of asbestos cement composites', *Int. J. Cem. Comp & Ltwt. Concr.* 5, 1983, 173–179.
14. A.J. Majumdar, J.M. West and L.J. Larner, 'Properties of glass fibres in cement environment', *J. Mater. Sci.* 12, 1977, 927–936.
15. J. Aveston, R.A. Mercer and J.M. Sillwood, 'Fibre reinforced cements – scientific foundations for specifications', in *Composites, Standards, Testing and Design*, Proc.

National Physical Laboratory Conference, England, IPC Science and Technology Press, Guildford, UK, 1974, pp. 499–508.

16. D.J. Hannant, 'The effect of post-cracking ductility on the flexural strength of fibre cement and fibre concrete', in A. Neville (ed.) *Fibre Reinforced Cement and Concrete*, Proc. RILEM Symp., The Construction Press, England, 1975, pp. 499–508.

17. R.N. Swamy, P.S. Mangat and C.V.S.K. Rao, 'The mechanics of fiber reinforcement of cement matrices', in *Fiber Reinforced Concrete. Amer. Conc. Inst. Special Publication SP-44*, Detroit, MI, 1974, pp. 1–28.

18. R.C. De Vekey and A.S. Majumdar, 'Determining bond strength in fibre reinforced composites', *Mag. Concr. Res.* 20, 1968, 229–239.

19. J.C. Lenain and A.R. Bunsell, 'The resistance to crack growth of asbestos cement', *J. Mater. Sci.* 14, 1979, 321–332.

20. S.A.S. Akers and G.G. Garrett, 'Acoustic emission monitoring of flexural failure of asbestos', *Int. J. Cem. Comp. Ltwt. Concr.* 5, 1983, 97–104.

21. Y.W. Mai, R.M.L. Foote and B. Cotterell, 'Size effects and scaling laws of fracture in asbestos cement', *Int. J. Cem. Comp.* 2, 1980, 23–39.

22. F.E. Jones, *Weathering tests on asbestos-cement roofing materials*, London HMSO, Building Research Technical Paper No. 29, 1947.

23. M.A. Ali, A.J. Majumdar and D.L. Rayment, 'Carbon fibre reinforcement of cement', *Cem. Concr. Res.* 2, 1972, 201–212.

24. M. Te'eni and R. Scales, *Fibre reinforced cement composites*, Technical Report 51–067, Materials Technology Division, Concrete Society, London, 1973.

25. A.J. Majumdar, 'Properties of fibre cement composites', in A. Neville (ed.) *Fibre Reinforced Cement and Concrete*, Proc. RILEM Symp., The Construction Press, England, 1975, pp. 279–313.

26. L. Opoczky and L. Pentek, 'Investigation of the corrosion of asbestos fibres in asbestos cement sheets weathered for long times', in A. Neville (ed.) *Fibre Reinforced Cement and Concrete*, Proc. RILEM Symp., The Construction Press, England, 1975, pp. 269–276.

27. W.R. Sharman and B.P. Vautier, 'Accelerated durability testing of autoclaved wood–fibre reinforced cement-sheet composites', *Durability of Building Mater.* 3, 1986, 255–275.

28. R. Becker and J. Laks, 'Cracking resistance of asbestos-cement panels subjected to drying', *Durab. Bldg. Mater.* 3, 1985, 35–49.

29. V. Kosi, 'Some properties of poly (methyl methacrylate) impregnated asbestos-cement materials', *Cem. Concr. Res.* 4, 1974, 57–68.

30. A.M. Al-Adeeb and M.D. Matti, 'Leaching corrosion of asbestos cement pipes', *Int. J. Cem. Comp. Ltwt. Concr.* 6, 1984, 233–240.

31. S.L. Sarkar, S. Jolicoeur and J. Khorami, 'Micromechanical and microstructural investigations of degradation in asbestos-cement sheet', *Cem. Concr. Res.* 17, 1987, 864–874.

32. M.A. Matti and A. Al Adeeb, 'Sulphate attack on asbestos cement pipes', *Int. J. Cem. Comp. Ltwt. Concr.* 17, 1985, 169–176.

33. J. Studinka, 'Replacement of asbestos in the fiber cement industry – state of substitution, experience up to now', Paper 25 in *International Man-Made Fibres Congress*, Dornbirn, Austria, Austrian Chemical Institute, 1986.

34. C. Bleiman, M. Bulens and P. Robin, 'Alternatives for substituting asbestos in fibre cement products', in Proc. Conf. on High Performance Roofing Systems, Plastic and Rubber Institute, London, 1984, pp. 8.1–8.12.

Chapter 10

Synthetic fibres

10.1 Introduction

Synthetic (polymer) fibres are increasingly being used for the reinforcement of cementitious materials. Some fibres, such as polypropylene, are used very extensively, and many fibres are available that have been formulated and produced specifically for reinforcement of mortars and concretes. The properties of synthetic fibres vary widely with respect to strength and modulus of elasticity, as shown for some common fibres in Table 10.1.

To increase the strength of the composites, the fibres must have a modulus of elasticity greater than that of the matrix. For cementitious materials, for which the modulus of elasticity ranges from about 15 to 40 GPa, this condition is difficult to

Table 10.1 Typical properties of synthetic fibres[a]

Fibre type	Diameter (μm)	Specific gravity	Tensile strength (GPa)	Elastic modulus (GPa)	Ultimate elongation (%)
Acrylic	20–350	1.16–1.18	0.2–1.0	14–19	10–50
Aramid (Kevlar)	10–12	1.44	2.3–3.5	63–120	2–4.5
Carbon (PAN)	8–9	1.6–1.7	2.5–4.0	230–380	0.5–1.5
Carbon (Pich)	9–18	1.6–1.21	0.5–3.1	30–480	0.5–2.4
Nylon	23–400	1.14	0.75–1.0	4.1–5.2	16–20
polyester	10–200	1.34–1.39	0.23–1.2	10–18	10–50
Polyethylene	25–1000	0.92–0.96	0.08–0.60	5	3–100
Polyolefin	150–635	0.91	275	2.7	15
Polypropylene	20–400	0.9–0.95	0.45–076	3.5–10	15–25
PVA	14–650	1.3	0.8–1.5	29–36	5.7
Steel (for comparison)	100–1000	7.84	0.5–2.6	210	0.5–3.5
Cement matrix	—	1.5–2.5	0.003–0.007	10–45	0.02

Note
a The values in the table are for fibres that are commercially available. They may vary considerably from manufacturer to manufacturer.

meet with most synthetic fibres. Therefore, some high tenacity fibres have been developed for concrete reinforcement, where the term 'high tenacity' refers to fibres with a high modulus of elasticity, accompanied by high strength. However, both theoretical and applied research have shown that, even with low modulus fibres, considerable improvements can be obtained with respect to the strain capacity, toughness, impact resistance and crack control of the FRC composites. In most applications, the enhancement of these properties is of much greater significance than a modest increase in tensile or flexural strength. In this chapter, a distinction will be made between low modulus and high modulus synthetic fibres, with the dividing value being the modulus of elasticity of the matrix. Within each group, specific fibres will be discussed separately.

The pioneering work on synthetic fibres has been described in detail by Zonsveld [1], Hannant [2] and Krenchel and Shah [3]. This early work emphasized the need to overcome disadvantages due to the low modulus of elasticity and poor bonding with the matrix. The latter was in particular a problem with many of the early synthetic fibres, due to their chemical composition and surface properties. Advances in this field, in particular with polypropylene fibres, only became feasible when it was recognized that fibres with special properties had to be developed for cement and concrete applications; this trend is still continuing. It must be noted, however, that many of the modern fibre modifications are proprietary, and information regarding the changes induced in the fibre structure is not readily available in the open literature. Thus, in the discussion later of the properties of each of the specific FRC systems, it should be kept in mind that differences in behaviour and properties may occur within the same 'family' of fibres. Characteristics such as modulus of elasticity, ease of dispersion, fibre geometry and alkali resistivity may not be the same for fibres classified under the same name.

10.2   Low modulus fibres

10.2.1   Polypropylene

10.2.1.1   Fibre structure and properties

Polypropylene fibres are produced from homopolymer polypropylene resin in a variety of shapes and sizes, and with differing properties. The main advantages of these fibres are their alkali resistance, relatively high melting point (165°C) and the low price of the material. Their disadvantages are poor fire resistance, sensitivity to sunlight and oxygen, a low modulus of elasticity (1–8 GPa) and poor bond with the matrix. These apparent disadvantages are not necessarily critical. Embedment in the matrix provides a protective cover, helping to minimize sensitivity to fire and other environmental effects. The mechanical properties, in particular modulus of elasticity and bond, can readily be enhanced. A number of different polypropylene fibres for use with cementitious matrices have been developed and are marketed commercially.

Polypropylene fibres are made of high molecular weight isotactic polypropylene. Because of the sterically regular atomic arrangement of the macro-molecule, it can be more readily produced in crystalline form, and then processed by stretching to achieve a high degree of orientation, which is necessary to obtain good fibre properties. Polypropylene fibres can be made in three different geome-tries, all of which have been used for the reinforcement of cementitious matrices: monofilaments [4,5], film [2] and extruded tape [6,7]. All three forms have been used successfully for mortar and concrete reinforcement.

Monofilament polypropylene fibres are produced by an extrusion process, in which the polypropylene resin is hot drawn through a die of circular cross section. A number of continuous filaments (tows) are produced at one time, and are then cut to the appropriate lengths.

Polypropylene film is also produced by an extrusion process, with the material being drawn through a rectangular die. The resulting film consists of a complex microstructure of amorphous material and crystalline microfibrils. The drawing of such films increases the orientation and improves the mechanical properties, but leaves the film weak in the lateral direction. The film is slit longitudinally into tapes of equal width. A lattice pattern of interconnecting fibres (Figure 10.1) can

Figure 10.1 The structure of fibrillated polypropylene fibre produced by splitting of a polypropylene film.

then be produced by mechanically drawing the tapes over a specially patterned pinwheel. Alternatively, the fibrillated tape may be twisted so that when it is cut to the desired fibre length, collated bundles of fibres result, as shown in Figure 10.2 [8,9].

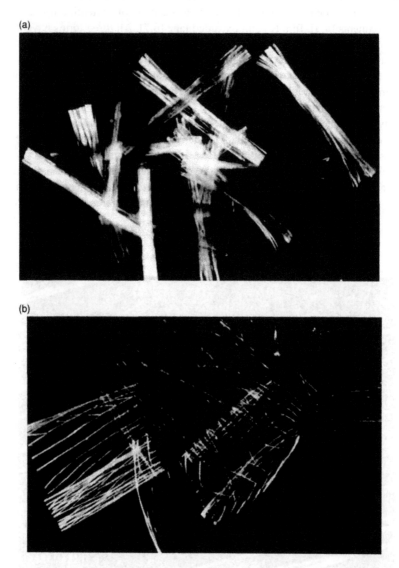

Figure 10.2 Collated fibrillated polypropylene fibres produced by Fibermesh before and after mixing (after Mindess and Vondran [8] and Bentur et al. [9]) (a) Collated fibrillated polypropylene fibres prior to mixing; (b) Open bundles of fibrillated polypropylene fibres.

(c)

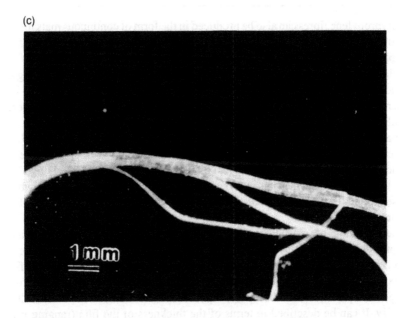

(d)

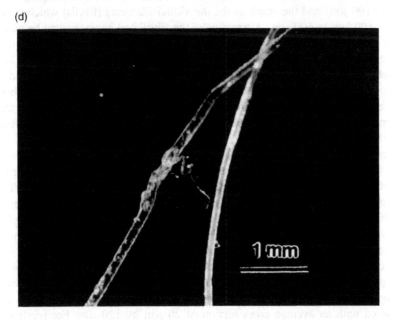

Figure 10.2 Continued. (c) Separation after mixing. (d) High magnification of a separated
unit showing the structure of a bundle consisting of two filaments 'cross linked'
together.

Polypropylene fibres can also be produced in the form of continuous mats for the production of thin sheet components [10–12], made by impregnating the fibrillated polypropylene mat with cement paste or mortar. The network-mesh structure of the fibrillated mat is intended to improve the bonding with the matrix by providing an interlocking effect. The fibrillated fibre, in its continuous mat (or mesh) form has been referred to by various names such as film, mat, mesh and fibrillated network. It will be referred to here as fibrillated polypropylene mat.

In applications in which the fibrillated polypropylene is mixed as short fibres in a cementitious matrix, the fibrillation plays an additional important role. It provides an initial structure of bundled, collated filaments, 'cross linked' together. This bundle can disperse uniformly in the matrix as filaments (fibrils) which are separated from the collated-fibrillated fibre during the mixing operation (Figure 10.2(c) and (d)).

The modulus of elasticity of both the monofilament and the fibrillated polypropylene (for fibres commercially available for FRC) is in the range of 3–5 GPa, and the tensile strength is about 140–690 MPa [13]. The monofilaments can be made in different diameters, ranging from about 50 μm to 0.5 mm. The geometry of the fibrillated polypropylene (Figure 10.2) is more difficult to quantify. It can be described in terms of the thickness of the film (ranging from ~15 to 100 μm) and the width of the individual filaments (fibrils) which range from ~100 μm to 600 μm. Alternatively, the fibrillated geometry can be quantified by the measurement of the specific surface area by adsorption techniques, where values in the range of ~80 to ~600 mm^2/mm^3 have been reported [14]. Sometimes the *denier unit* common to the textile industry is used, as defined in Section 2.2. While both the fibrillated and monofilament polypropylene fibres have essentially the same strength and elastic modulus, it has been suggested that in terms of their ability to arrest cracks, the monofilament fibres are more effective than the fibrillated fibres [15].

A high tenacity polypropylene fibre for cement reinforcement has been developed in Denmark, under the trade name Krenit [6]. This fibre is produced from an extruded tape which is split mechanically into single rectangular fibres. The splitting is carried out in such a way that the edges of the fibres become uneven and frayed, thus inducing a good mechanical bond with the matrix. The splitting operation can be controlled, to change the degree of fraying and the cross section of the fibres, to match them to the desired application. The width can vary considerably both from fibre to fibre, and also along the individual fibre, thus providing additional bond enhancement. For asbestos replacement, fibres of higher specific surface area are needed, and for that purpose the fibres are produced with an average cross section of 20 μm by 120 μm. For reinforcement of mortar and concrete matrices, the fibres have an average cross section of 30 μm by 200 μm. Krenit fibres have a relatively high modulus of elasticity (7–18 GPa) and tensile strength (500–1200 MPa), and can thus be considered as high tenacity polypropylene, even though the modulus of elasticity is still below that of the matrix. Trottier and Mahoney [16] developed a high tensile strength

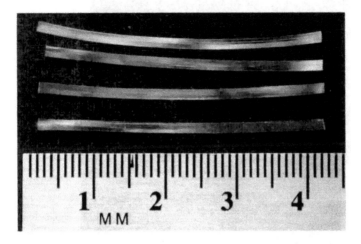

Figure 10.3 STRUX® fibre. Photo courtesy of Grace Construction Products.

fibre that partially fibrillates during mixing with concrete, thereby increasing the bonding capacity with the matrix. The fibre is produced by extruding a mixture of polypropylene and polyethylene. A variant of this technique is the STRUX® fibre (shown in Figure 10.3), which has an elastic modulus of 9.5 GPa and a tensile strength of 620 MPa. These higher tenacity fibres, when formulated to enhance bond with concrete have been particularly useful for toughening of hardened concrete (called sometimes 'structural fibres' [17]), Altoubat *et al.*, 2004, whereas the lower modulus discrete fibres find their application mainly in control of plastic shrinkage cracking. The lower modulus fibres, whether in the form of fibrillated mat [10] or technical fabrics [18], can be used as continuous reinforcement to produce thin sheet cementitious components, and they are dealt with in detail in Chapter 13.

The chemical structure of the polypropylene makes it *hydrophobic* with respect to the cementitious matrix, leading to reduced bonding with the cement, and negatively affecting its dispersion in the matrix. Thus, most of the polypropylene fibres developed for FRC undergo various proprietary surface treatments to improve the wetting of the fibres [9] in order to overcome these disadvantages. For instance, Zhang *et al.* [19] found that treating the fibres with low temperature cascade arc plasma was effective in improving the flexural performance and toughness of polypropylene FRC. Similarly, Tu *et al.* [20] found that treating the surface of the fibres with fluorination or oxyfluorination processes improved the performance of the FRC. SEM studies of the interface of such treated fibres indicate that they can develop an intimate contact with the cement matrix, as shown in Figure 10.4 [9].

(a)

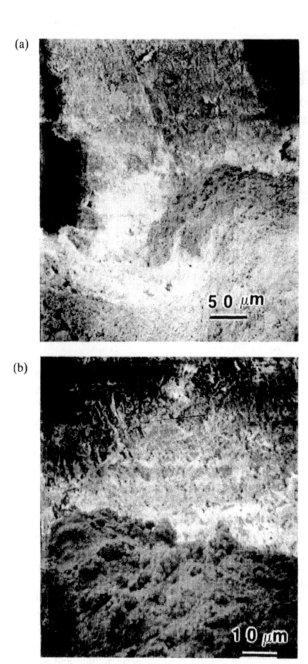

(b)

Figure 10.4 The intimate contact between a polypropylene fibre with proprietary surface treatment (Fibremesh) and the concrete matrix (after Bentur *et al.* [9]). (a) High magnification at the root of the fibre; (b) still higher magnification at the root of the fibre, showing the dense interface.

10.2.1.2 Mechanics of polypropylene FRC

The behaviour of polypropylene FRC has been analysed on the basis of the ACK model (see Sections 4.2 and 4.5), and attempts have been made to account for the shape of the stress–strain curve in terms of the fibre and matrix properties, the multiple cracking process and its dependency upon the fibre–matrix bond.

The nature of the pull-out vs. displacement curves and of the peak loads from which the average interfacial shear bond is calculated can be quite different, depending on the fibre geometry and on the various treatments used to improve the bond. Naaman *et al.* [21] demonstrated the influence of such parameters on the maximum load and the load-bearing capacity after debonding, as shown in Figure 10.5.

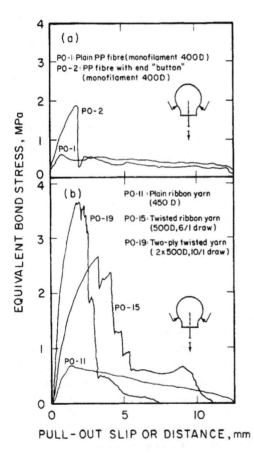

Figure 10.5 The effect of various treatments on the pull-out behaviour of polypropylene fibres embedded in a cement matrix (after Naaman *et al.* [21]). (a) Monofilament fibres; (b) ribbon yarn.

In monofilament fibres, the addition of 'buttons' at the ends of the fibres increased the maximum pull-out load (Figure 10.5(a)), while in ribbon type fibres a considerable increase in the maximum load and stress transfer in the debonded zone could be achieved by twisting the fibres (Figure 10.5(b)). With plain monofilament and bond was low (0.55 MPa), but for the treated fibres this was increased up to about 3 MPa. Walton and Majumdar [4] determined the average bond values for monofilament fibres in the range of 0.34–0.48 MPa after debonding, and 0.70–1.23 MPa at the maximum. These values were not sensitive to environmental conditions or to the w/c ratio of the matrix. It should be noted that with the polypropylene fibres developed specifically for FRC, where proprietary surface treatments are applied to enhance the wetting and the compatibility with the matrix (as demonstrated by the dense interfacial microstructure in Figure 10.4) the bond value reported in these studies should be considered as a lower bound. Also, when considering the dense interfacial microstructure in Figure 10.4) the bond value interpretation of the maximum pull-out load in terms of stress is of limited physical significance. The bond strength is not only the result of interfacial adhesion; the major contribution is due to mechanical anchoring and interlocking. This is certainly the case with the fibrillated mat which is hand laid in the composite [11,14]. Even with normally mixed FRC, such effects are evident; the fibrillated fibre does not separate into individual filaments (Figure 10.2(d)), and in the hardened concrete a complex fibre geometry of cross-linked multifilament strands can be seen (Figure 10.6). This can provide mechanical anchoring at different levels between the cross-linked filaments and by the interlocking of tiny fibrillations of the fibre surface [9,21,22].

The significance of the bond in generating multiple cracking of a desired crack spacing was analysed by Hannant et al. [11] using the ACK model. They derived the following equation for crack spacing with aligned fibre reinforcement:

$$x = \frac{V_m}{V_f} \frac{\sigma_{mu} \cdot A_f}{\tau_{fu} \cdot \psi} \tag{10.1}$$

where x is the minimum crack spacing; V_m, V_f, the matrix and fibre volume contents, respectively; A_f, ψ, the cross-sectional area and perimeter of the fibre, respectively; τ_{fu}, the interfacial shear bond strength.

In order to obtain a composite with adequate crack spacing (5 mm or less), they calculated the value of the interfacial shear bond strength, τ_{fu}, from Eq. (10.1), using typical values for the matrix and for the fibrillated polypropylene: $\sigma_{mu} = 4$ MPa, $V_f = 0.05$, $V_m = 0.95$, an equivalent fibre width of 2.5 mm and a thickness of 30 μm. The calculated τ_{fu} value was 0.23 MPa, which can readily be achieved even with monofilaments. However, it was pointed out by Kelly and Zweben [23] and Pinchin [24] that this level of bond may not be attained in practice once cracking is initiated, because of the Poisson effect (see Section 3.5). At the crack, the matrix strain is zero, while the bridging fibre is loaded; due to the Poisson effect the fibre will contract and may separate from the matrix, thus eliminating the normal

(a)

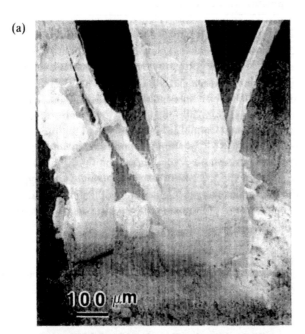

100 μm

(b)

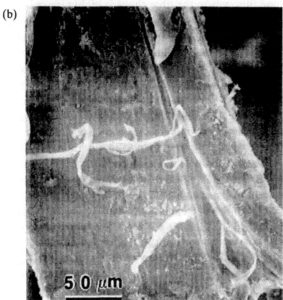

50 μm

Figure 10.6 SEM observation of a polypropylene fibre in concrete (after Bentur *et al.* [9]). (a) The structure of the fibre consisting of individual filaments 'cross-linked' together; (b) High magnification of the root of the branch at the left filament in (a), showing fibrils.

stresses which are required to generate frictional shear resistance. This problem is particularly acute for polymeric fibres. However, experimental studies [11,25,26] have shown that in practice, when the polypropylene fibre content exceeds the critical volume, multiple cracking does occur, with the sufficiently small spacing of 1–3 mm [11,26] and a calculated effective interfacial shear bond τ based on Eq. (10.1) of about 0.4 MPa. The apparent discrepancy between theory and practice was treated in detail by Baggott and Gandhi [25] who observed multiple cracking in their composites, containing up to 8.7% of aligned monofilament polypropylene fibres of 340 μm diameter. They also observed a linear relationship between the crack spacing, x, and $V_m V_f$, as expected from the ACK model. They suggested that effective bonding may be induced by mechanisms other than interfacial shear, which are not sensitive to the absence of intimate fibre–matrix contact. Three such mechanisms were considered:

1   The presence of asperities on the fibre surface;
2   Misalignment of the two surfaces of the crack, moving laterally as well as longitudinally relative to each other and to the fibre array;
3   Slight departure from parallelism of the reinforcing fibres.

In fibrillated polypropylene film, where the fibre shape is highly irregular, additional bonding effects of this kind can be induced, which are quite different from simple interfacial shear. Hannant [26] and Hughes [14] reached similar conclusions when analysing the behaviour of a composite with fibrillated polypropylene mat, and suggested that the contact area is not the controlling parameter in the bond of such fibres. They hypothesised that mechanical misfit between the fibrillated polypropylene and the matrix could occur, and that slip could be controlled by internal shearing within the film itself. The contribution of the variable profile of the fibre was considered to be very important for generating such effects.

Such conclusions are consistent with observations that the fibre–matrix bond strength is insensitive to the environmental conditions during the test. While a simple frictional bond would be a function of the normal stresses induced by volume changes of the matrix (see Section 3.2.4), bond induced by mechanical anchoring would be much less affected by volume changes in the matrix. Hannant [26] and Hannant and Hughes [27] found that for a fibrillated fibre the bond remained in the range of 0.3–0.4 MPa, regardless of the environmental conditions (water, air and natural weathering). Walton and Majumdar [4] found little effect of the w/c ratio and the curing conditions on the average bond strength of monofilaments tested in direct pull-out, which is again consistent with bonding mechanisms which are not sensitive to interfacial contact. Thus, it seems that, whatever the exact mechanism, there is sufficient bond between the polypropylene and the cement matrix to induce a composite action, for both fibrillated and monofilament fibres.

Hughes and Hannant [28] determined the stress–strain curves of composites prepared with up to 10% fibrillated polypropylene fibre mats, with their modulus of elasticity ranging from about 2.5 to 9 GPa (Figure 10.7); the multiple cracking can readily be observed. The strain at the end of the multiple cracking zone was calculated using the ACK model (Eq. (4.68)) and the known properties of the fibre and matrix. This was in agreement with the experimental curves (compare data in Table 10.2 and in Figure 10.7): the higher modulus fibre led, as expected, to a smaller multiple cracking strain. The curves in the post-multiple cracking zone were approximately linear, and their slopes were approximately equal to $E_f V_f$, as predicted by the ACK model (the slopes in curves S4 and S8 in Figure 10.7

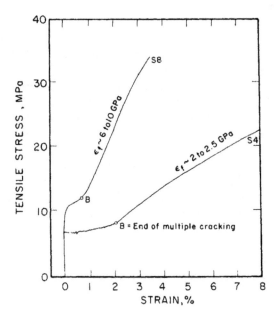

Figure 10.7 Effect of the modulus of elasticity of fibrillated polypropylene on the stress–strain curve of a composite prepared with 10% fibre by volume (after Hughes and Hannant [28]).

Table 10.2 Calculation of the strain at the termination of the multiple cracking zone, ε_{mc} for composites reinforced with fibrillated polypropylene mats of different moduli of elasticity (after Hughes and Hannant [28])

Polypropylene type	E_m (GPa)	E_f (GPa)	α	$\bar{\varepsilon}_{mu}$ (%)	ε_{mc} (%)
S4	32	2.5	115	0.0260	2.00
S8	32	9	32	0.0368	0.81

correspond to E_f of 2.29 and 8.36 GPa, respectively, which agrees well with the measured moduli of the fibres).

Although the fibre–matrix bond is important in determining the mechanics of the composite, the effect of the fibre content should also be considered. If the fibres are to act as *primary reinforcement*, that is to increase the strength as well as the toughness (e.g. polypropylene for asbestos fibre replacement), their content must exceed a critical volume, which is about 3% for typical polypropylene fibre and cement matrix properties (tensile strengths of 400–800 MPa and 2–6 MPa, respectively). The effect of the fibre content is shown in Figure 10.8. An ascending curve in the post-cracking zone can be achieved only at a relatively high fibre content; for an increase in strength beyond the first crack strength, the composite must first undergo considerable deformation.

The relatively low slope of the curve in the post-cracking zone is a consequence of the low modulus of elasticity of the polypropylene fibre. Indeed, the ACK model (Section 4.5) would predict that for such fibres the stress–strain curve would be almost horizontal in the post-cracking zone, as seen in Figure 10.8. However, although there is little improvement in the load-bearing capacity, the strain capacity is very high, leading to a tough composite. The difference between tensile and flexural loading should also be noted. Improved ductility in tension (i.e. an elastic–quasi–plastic stress–strain curve) will lead to strengthening in flexure, with an ascending load–deflection curve in the post-cracking zone. This can be the case with polypropylene reinforcement, as shown in Figure 10.9. While large strains might be needed to mobilize the strength of the polypropylene fibres in tension, relatively smaller deformations are needed to obtain strengthening in bending.

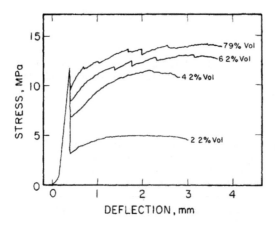

Figure 10.8 The effect of polypropylene fibre content (49 μm diameter monofilaments of 26 mm length) on the load–deflection curve of the composite (after Dave and Ellis [5]).

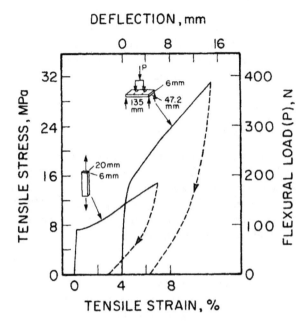

Figure 10.9 Tensile stress–strain curve and bending load–deflection curve of a composite containing 5.7% by volume of fibrillated polypropylene mat (after Hannant and Zonsveld [10]).

10.2.1.3   Properties and behaviour of the composite

Polypropylene fibres may be used in several different ways to reinforce cementitious matrices. One application is in thin sheet components, in which the polypropylene provides the *primary reinforcement*. Used in this way, the polypropylene may be considered as an alternative to asbestos reinforcement, and its volume content must be relatively high, exceeding 5% (i.e. above the critical volume). Such components can be made by hand lay-up of layers of continuous mats in a paste or mortar mix, or by industrial, mechanized processes (e.g. Vittone [12]). Composites with the same range of properties may also be produced using short fibres especially developed to be compatible with the Hatschek process [6,7]. Such materials are often referred to as *high performance* FRC. In a second application, the volume content of the polypropylene is low, usually less than 0.5% by volume, and it is intended to act mainly as *secondary reinforcement* for crack control, but not for structural load-bearing applications. In this use, the polypropylene is added as short fibres in a conventional concrete mixer. These materials are referred to as *conventional* FRC. The reinforcing effects of the fibres in these two types of materials are quite different, as may be seen in Figure 1.1, and they will be considered separately. High performance FRC composites are treated in Chapter 12.

More recently, a third application of polypropylene fibres has been in semi-structural applications such as fibre shotcrete and FRC slabs on grade, at dosages of 0.5% by volume and greater. While they appear to be less efficient than steel fibres at the same volume concentrations, they are gaining popularity in this application [17,29,30].

PRIMARY REINFORCEMENT IN THIN SHEET COMPONENTS

In this application, the fibre volume content must exceed the critical volume in order to obtain both strengthening and toughening. Composites with this level of fibre content can be produced in a number of ways. The early work on such materials involved hand lay-up of continuous fibrillated polypropylene mats [10,26]. However, since then a number of processes suitable for industrial applications have been developed using short, discontinuous fibres:

- Impregnating preplaced mats [21].
- Spraying (e.g. Walton and Majumdar [4]); most commonly now, continuous fibres are chopped and simultaneously air-sprayed with the cementitious slurry onto a mould. The thickness of each layer is typically about 4–6 mm. This method is particularly suited for complex architectural shapes, or to produce very high strength composites.
- Hatschek process [31]; see Section 9.1. Composites made using this process typically contain 7–15% fibres by volume [32].
- Mixing [5,6,21]. It should be noted that mixing with such high fibre contents is very difficult, and special techniques must be used to ensure proper dispersion. For instance, Dave and Ellis [5] used a high water/cement ratio [0.5–1.0] mix, with the excess water later extracted by suction and pressing. However, the highest fibre contents can more readily be incorporated by using continuous fibre mats.
- Extrusion processing [33–36]. A highly viscous mixture, with a dough-like consistency, is forced through a die of the desired shape by the pressure applied by either an auger extruder (Figure 10.10 [33]) or a piston extruder. The material must be plastic enough to flow through the die under pressure, but stiff enough to maintain its shape after exiting the die.
- Reinforcement by technical textiles (see Chapter 13).

The overall properties of the composites are dependent on the matrix properties, the fibre content (Figure 10.8), its modulus of elasticity, and the various means used to improve the fibre–matrix bond. A variety of different fibre types have been studied, including monofilaments of various diameters, fibres with 'buttons' on their ends [21], twisted tapes [21], fibrillated mats [26], textile fabrics [18] and high tenacity fibres with frayed ends [7]. Even with smooth monofilament

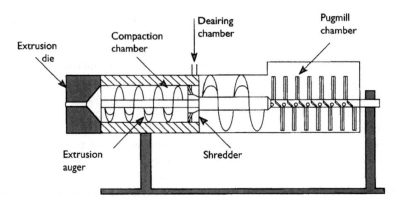

Figure 10.10 An auger extruder for manufacturing thin reinforced cementitious products [33].

fibres, which might be considered as the least attractive for cement reinforcement, a considerable improvement in performance can be obtained by reducing the fibre diameter and increasing its length. In spite of this, however, smooth monofilaments are not as good as the special fibres developed to achieve better bonding and compatibility with the cementitious matrix. With composites containing 6–10% by volume of fibres, the flexural strength achieved using monofilaments did not generally exceed ~15 MPa, while with fibrillated polypropylene mats or the frayed Krenit fibres, strength values greater than 25–30 MPa have been reported [10,37].

Also, for such composites, the significance of crack control in the post-cracking zone must be taken into account. Baggott [38] indicated that an ascending load–deflection curve in the post-cracking zone is not by itself a sufficient condition for proper composite action. Another important consideration is that the cracks should be *invisible* (to the naked eye). This can readily be achieved with continuous fibrillated polypropylene mats, but not always with discontinuous fibrillated fibres or monofilaments. For example, Baggott and Gandhi [25] observed crack spacings greater than 10 mm in the monofilament composites, due to an apparently low interfacial bond (~ 0.05 MPa), while with fibrillated fibres Hannant *et al.* [11] could obtain spacings of less than 1–3 mm. Baggott [38] also studied lightweight cementitious matrices, and concluded that with fibres less than 26 mm in length, the cracking criterion could be satisfied with fibrillated fibres if their content was 7.5% by volume, but could not be achieved with the short monofilament fibres, even though they provided a considerable strengthening effect and an ascending load–deflection curve in the post-cracking zone.

Composites reinforced with continuous fibres can be made by filament winding or by lay-up of several layers of mats have become in recent years a focus for new developments, and they are treated in detail in Chapter 13.

SECONDARY REINFORCEMENT IN CONCRETE

Polypropylene is most commonly used at contents much smaller than the critical volume (i.e. < 1–3%), and as such its function is mainly to act as secondary reinforcement, to control cracking due to environmental effects (e.g. temperature and moisture changes). Very low volume contents (~0.1%) are often recommended by various producers as adequate for controlling and reducing the plastic shrinkage cracking of the fresh concrete. An example of the use of polypropylene FRC in structures where crack control is important is shown in Figure 10.11. In such applications, the polypropylene is mixed with the concrete using conventional equipment. The polypropylene is added as discontinuous short fibres with lengths ranging from 6 to 60 mm. Newer generation of polypropylene fibres at moderate contents of 0.2–0.6% by volume are currently used to provide crack control in the hardened concrete. Such fibres (structural fibres) are specially formulated to achieve performance approaching that of steel fibres [17]. The performance and influence of the polypropylene fibres in the fresh and hardened concrete is quite different, and therefore several topics will be considered separately: fresh concrete, crack control in the plastic and hardened state and properties of hardened concrete.

Fresh concrete During the mixing of the fibrillated fibres in concrete they tend to separate into individual filaments from the parent fibrillated fibres (Figure 10.2). The dispersion of such fine units in the concrete matrix

Figure 10.11 Water tank in Australia construction with polypropylene FRC. Photograph courtesy of Fibermesh Co., Chattanooga, Tennessee.

has a considerable influence on the properties of the fresh mix, in particular on workability and plastic shrinkage cracking. As expected, an increase in fibre volume will result in a reduction in the consistency of the concrete [39,40]; this reduction is greater for longer fibres and higher fibre contents (Figures 10.12 and 10.13). Here, a distinction should be made between the finer fibrillated fibres, and the coarser rectangular ones such as those reported by Hasaba et al. [40], with a cross section of 2 × 0.6 mm. The former (Figure 10.12) lead to a marked reduction in consistency when added at a dosage of even a few tenths of a per cent

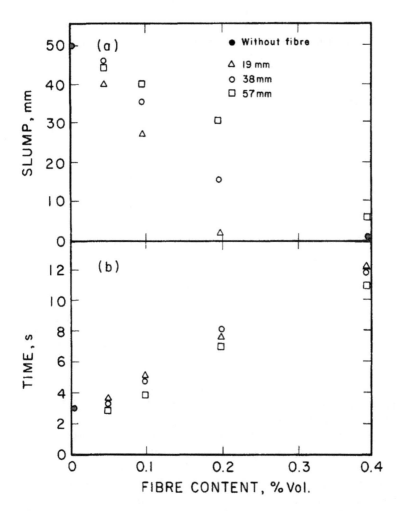

Figure 10.12 Effect of the content and length of fibrillated polypropylene fibres on the consistency of the concrete, measured by the slump (a) and VeBe; (b) tests (after Spadea and Frigione [39]).

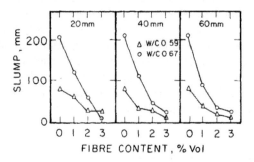

Figure 10.13 Effect of the content and length of rectangular polypropylene fibres (2 × 0.6 μm section) on the slump of concrete (after Hasaba *et al.* [40]).

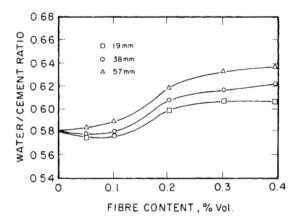

Figure 10.14 The effect of varying the volume content and length of fibrillated polypropylene fibres on the w/c ratio of the matrix which is required to maintain a constant slump of 50 pm (after Spadea and Frigione [39]).

by volume, while the latter can be added at contents of upto about 2% by volume before a drastic reduction in consistency is obtained (Figure 10.13). However, both of these fibre volumes are still below the critical fibre volume.

The reduction in the workability can be compensated for by using water reducing admixtures or increasing the water content. The latter method, however, leads to a higher w/c ratio matrix of lower quality. Still, as can be noted from Figure 10.14 [39], the change in w/c ratio required to maintain a constant slump is negligible when the fibre content does not exceed 0.1% by volume, and this is reflected in the recommendations of the various producers to use this level of reinforcement in the concrete. Even when considering the greater slump loss that occurs at 0.2% or 0.3% fibre addition, it should be remembered that the workability is not reduced to the extent indicated by the slump reduction. The limitations of the static slump test were discussed in Section 6.2.1. Under dynamic conditions, where the mix is being vibrated, the FRC can still exhibit good workability, and can be compacted readily, without excessive vibration. However, for fibre contents greater than a few tenths of a per cent, a different mix design is required, using superplasticizers in combination with fine mineral admixtures. For instance, Ramakrishnan *et al.* [41] found that satisfactory workability could be maintained at fibre volumes up to 2.0% by using appropriate amounts of superplasticizers, while maintaining equal strength and w/c ratio.

Crack control – plastic shrinkage cracking   Low modulus polypropylene fibres have been used for control of plastic shrinkage cracking. Fibrillated or monofilament ($\sim$50 μm and smaller) fibres at a small content of about 0.1% by volume are effective in suppressing most of the cracking, reducing its extent by an order of magnitude when evaluated in restrained tests such as those outlined in Section 6.2.3 [42–52], Figures 10.15 and 10.16. It should be noted that the performance of fibres for plastic shrinkage crack control is useful for compensating poor site practices, but where good finishing and curing practice is applied, their influence is small, as demonstrated in Figure 10.17.

The effect of the fibre on reduction of free shrinkage and weight loss on drying was shown to be small, and its influence stems from the crack bridging ability [49–51]. However, Zollo and Ilter [53] and Wang *et al.* [52] noted a considerable reduction ($\sim$25–40%) in plastic shrinkage with 0.1–0.3% fibres by volume. Zollo and Ilter [53] suggested that the fibres function as a network which stabilizes the matrix and prevents consolidation; this was also indicated by the elimination of bleeding in the fibre-reinforced mix. Yet, even in this case the reduction in cracking induced by the fibres was much greater than the reduction in free shrinkage. Indeed, one would expect that the reinforcing efficiency of low volume–low modulus fibres would be greatest in the fresh mix, when the concrete matrix is still weak and of lower stiffness than the fibres. At this stage, the critical fibre volume (Eq. 4.30) is

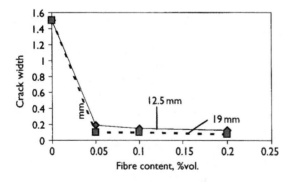

Figure 10.15   Effect of fibrillated polypropylene fibre content and length (fibrillated PP) on plastic shrinkage cracking (after Soroushian *et al.* [45]).

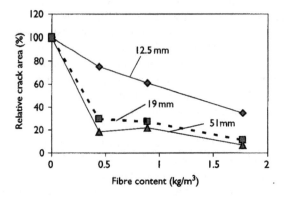

Figure 10.16 Effect of monofilament polypropylene fibre content and length (PP) on plastic shrinkage cracking relative to control (after Berke and Dallaire [46]).

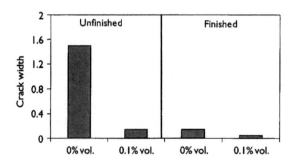

Figure 10.17 Effect of fibrillated polypropylene fibre and mode of finishing on plastic shrinkage cracking (after Soroushian *et al.* [45]).

smaller by about an order of magnitude than in the hardened concrete, and it is in the range of about 0.1% by volume.

Polypropylene fibres are not all equally effective in reducing plastic shrinkage cracking. Kraii [42] compared seven different types of fibres and found that some of them did not provide drastic improvement. In the case of monofilament fibres the effective bonding was shown to have a marked influence on the performance: increase in length, reduction in thickness and roughening of the surface (Tables 10.3 and 10.4), led to enhanced crack control performance.

It is well established that the addition of silica fume to concrete increases the tendency for plastic shrinkage cracking. However, Bayasi and McIntyre [54] showed that this could be compensated for by the addition of fibrillated polypropylene fibres; the use of 0.3% by volume of fibres could successfully combat the plastic shrinkage cracking of concrete containing 10% silica fume.

Table 10.3 Effect of fibre shape on plastic shrinkage cracking performance in a ring test
of 0.57 w/c ratio concrete, after Kovler
et al. [50]

Fibre shape		Time of plastic cracking (min)
Thickness	Surface	
40 μm	Smooth	107
20 μm	Smooth	0
40 μm	Smooth	0

Table 10.4 Effect of 0.1% of PP fibre reinforcement by volume on cracking of 0.57
w/c mortar in drying, after two days of water curing, after Kovler
et al. [50].

Fibre shape		Cracking		
Thickness	Surface	Time to cracking (min)	Crack width (mm)	Crack length (mm)
40 μm	Smooth	13	0.35	15[a]
20 μm	Smooth	13	0.35	13
40 μm	Rough	13	0.03	2[b]

Notes
a Through macrocrack across the ring.
b Microcracks on the surface of the concrete.

Crack control in set and hardened concrete   The low modulus polypropylene
fibres at the content of ~0.1% by volume applied for plastic shrinkage crack control are not effective for crack control of hardened concrete, as shown in Table 10.4
[50]. However, Sanjuan et al. [55] found that low volume propylene reinforcement could lead to somewhat decreased initial cracking of the concrete in the
plastic stage, and this resulted in somewhat lower corrosion rates of steel reinforcement in the hardened concrete. The data indicate that only the fibres which
received special surface treatment of roughening had an influence on reducing
cracking.

For the purpose of hardened concrete crack control, there is a need to use higher
modulus fibres, at a larger content, as was demonstrated by Krenchel and Shah
[56] and Swamy and Stavrides [57] using polypropylene fibres having 15–18 GPa
modulus and content of 2% by volume (Table 10.5). Swamy and Stavrides [57]
reported that the effect of fibres on reducing shrinkage was small (~10%) and
therefore their influence demonstrated in Table 10.5 is the result of a reinforcing
effect.

Table 10.5 Cracking characteristics of plan mortar and mortar reinforced with 2% of PP fibres (after Swamy and Stavrides [57])

Type of mortar	Fibre volume	Age at first crack (days)	Number of cracks	Max. crack width (mm)	Age at failure (days)
Unreinforced	0	8	1	1.35	10
Reinforced	2	14	3	0.15	> 21

Hardened concrete   The compressive and tensile strengths of concretes reinforced with low volumes of polypropylene fibres are not significantly different from those of the unreinforced matrix (e.g. [40,58,59], because the fibre content is below the critical volume. Indeed, the compressive strength at higher fibre contents may sometimes be reduced, probably because of the difficulty in fully compacting such mixes. However, polypropylene fibres have been reported to be effective in increasing the flexural strength [40, 58,60,61]; this may be due to their ability to enhance the load-bearing capacity in the post-cracking zone. It may also be due to the effect of the polypropylene in reducing the cracking, particularly in specimens tested air dry: such increases in flexural strength are not large, usually not exceeding 10–20%. It was reported that polypropylene fibres appear to have an adverse effect on the abrasion resistance of concrete [62], which must be considered when they are to be used in concrete pavements or floor systems.

A much more important effect of low volume fibre reinforcement is to enhance the energy absorption capacity of the composite in tension or flexure, as can be observed in both static testing (Figure 10.18) and impact (Figure 10.19). With a fibre content of only 0.1% by volume, only a small increase in impact resistance was reported [8]. With higher fibre contents of 0.3–0.5% by volume, somewhat greater improvements were found, in the range of 30–80% [8,63]. However, the enhancement in toughness by this low volume reinforcement seemed to be much greater when evaluated by static testing than by impact testing [63,64].

Improved impact resistance was also observed when 0.1–0.5% fibrillated polypropylene was incorporated in concrete reinforced with conventional steel bars, the increase being up to a factor of 3, relative to concrete with reinforcing bars but without polypropylene [65,66]. Also, in such concretes, small additions of polypropylene were found to help offset the brittleness associated with high strength concrete subjected to impact.

It has also been shown that polypropylene fibres are effective in enhancing concrete performance under blast loading [67], although for such purposes fibre contents much higher than 0.1% were used. Hibbert and Hannant [68] demonstrated that the impact resistance obtained with 1.2% of polypropylene was similar to that obtained using about the same steel fibre content. Finally, the addition of only ~0.3% of fibrillated polypropylene fibres led to a modest (~18%) increase in the flexural fatigue strength of concrete.

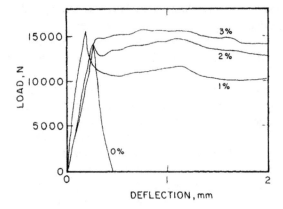

Figure 10.18 Effect of the content of rectangular polypropylene fibres (2 × 0.6 mm section) on the load–deflection curve (after Hasaba et al. [40]).

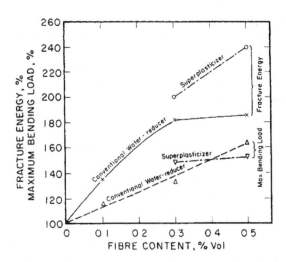

Figure 10.19 Effect of the content of fibrillated polypropylene fibres on the maximum bending load and fracture energy of concretes in impact, expressed as a percentage of the values for beams with no fibres (after Mindess and Vondran [8]).

10.2.1.4 Durability and performance at high temperatures

The durability of polypropylene FRC has mainly been evaluated in thin sheet components, where the fibres are the primary reinforcement. Polypropylene fibres are known for their high alkali resistivity, and therefore would be expected to retain their strength in the highly alkaline matrix. This has been confirmed by many investigators (e.g. Hannant and Hughes [27]). However, other problems might be expected, due to oxidation and softening effects at elevated temperature. The latter is a function of the melting temperature of about 165°C. The sensitivity to UV radiation is not expected to be critical since the polymer is protected by the matrix.

Krenchel [37] observed a marked reduction in the flexural strength of high fibre volume (10%) composites when the temperature exceeded $\sim 120°C$, which he attributed to the softening of the fibre. The modulus of elasticity of the composite was also reduced with increasing temperature, but much more gradually, reflecting the fact that this property is essentially matrix dependent. Bayasi and Dhaheri [69] also found that prolonged exposure to temperatures in the range of 100–200°C led to a decrease in flexural strength and post-peak load-bearing capacity of fibrillated polypropylene fibre concretes. However, there was no effect of exposure at temperatures below 100°C. These (and similar) results suggest that the composite will function properly at the high ambient temperatures that may be encountered in practice in hot climates (~ 40–50°C), but not when fire hazards are considered. The combustibility of the polypropylene is not a particular problem when the fibres are in the cement matrix; instead, they melt and volatilize during the fire, leaving behind empty channels and additional porosity. Indeed, this property of the fibres has been used to provide additional fire protection to the reinforcing steel, as the empty channels provide an 'escape route' for the steam that is formed during a fire, minimizing spalling of the concrete due to high steam pressures [70]. Kalifa et al. [71] showed the remarkable effect of the fibres in high performance concrete at temperatures above 200°C, and similar results were presented by Dehn and Konig [72] and Velasco et al. [73]. As well, Chan et al. [74] found that even after the polypropylene had evaporated off, this did not lead to a reduction in the mechanical properties of the concrete.

It should be noted that the polypropylene fibres themselves may shrink by 3–15% on heating to just below their melting point [75], but this in itself appears not to have any significant mechanical effects.

Resistance to oxidation of polypropylene fibres was studied by Mai et al. [76], to evaluate the performance of the composite in autoclave curing. This is particularly important in applications in which polypropylene FRCs are intended as asbestos–cement substitutes, which are often cured by autoclaving. They found that heat treatment of the polypropylene fibres at 140°C did not lead to a reduction in their strength or ductility. However, autoclaving at 140°C led to a loss in the ductility of the fibre without strength reduction, while additional extended heating at 116°C also led to a reduction in strength. They suggested that during the autoclaving, most of the antioxidant was deactivated or leached out. Beyond this stage, the usefulness of the residual antioxidants for preventing oxidation depends on the temperature and exposure time. Heat treatment (drying) at 116°C for 24 h caused the fibres to oxidize and lose strength, while maintaining the temperature after autoclaving at 65°C would prevent this strength loss. This procedure, of autoclaving followed by a low drying temperature, was suggested as a practical solution for the overall curing process. Obviously, these particular curing parameters depend upon the type of fibres, in particular the quality and content of the antioxidants.

Hannant and Hughes [27] found that polypropylene fibre-reinforced composites exposed to laboratory air or natural weathering did exhibit some changes in properties, but these were attributed primarily to changes in the matrix; the fibre strength and the fibre–matrix bond remained constant. However, the increase in modulus of elasticity of the matrix over time might lead to an increase in the critical fibre volume. For composites with a fibre volume only marginally greater than the critical volume, the ageing of the composite (i.e. the increase in E_c) could result in a situation in which the critical fibre volume content increases beyond the actual fibre content, thus changing the failure mode from a ductile (multiple cracking) to a brittle mode (single crack). Thus, changes in matrix properties can have a significant influence on the long-term performance of the FRC, even if the fibres remain intact. The extent of such effects will depend upon the fibre content, and therefore caution should be exercised when making general statements about the durability. On the whole, however, looking upon 18 years of data, Hannant [77] found that fibrillated polypropylene fibre cement sheets displayed excellent strength retention, and concluded that one could rely upon the long-term stability of such composites whether used inside buildings or in structures exposed to the weather.

Al-Tayyib and Al-Zahrani [78,79] showed that adding 0.2% polypropylene fibres to the concrete effectively enhanced the durability of the concrete skin when exposed to severe marine environments, and improved the concrete durability by retarding the corrosion of the reinforcing steel.

10.2.2 Other fibres

A number of other relatively low modulus synthetic fibres besides polypropylene have been developed for cement and concrete reinforcement. These include acrylic, nylon, polyester, polyethylene and polyolefin fibres.

10.2.2.1 Acrylic fibres

Acrylic fibres such as those used in the textile industry have tensile strengths typically in the range of 200–400 MPa. They contain at least 85% by weight of acrylonitrile units. They can form a strong bond with the cement matrix. However, since they have been found to lose strength in the highly alkaline cement environment [80] they are not suitable for use in FRC. More recently, however, high tensile strength acrylic fibres with tensile strengths up to 1000 MPa have been developed, which are being used in FRC. These will be discussed in some detail in Section 10.3.3.

10.2.2.2 Nylon fibres

Nylon fibres appear to be used increasingly in FRC, often as a substitute for polypropylene fibres. For FRC, the nylon fibres are generally produced as a high tenacity yarn that is heat and light stable, and is subsequently cut into appropriate lengths. These fibres typically have a tensile strength of about 800 MPa, and an elastic modulus of about 4 GPa. It should be noted that these fibres are *hydrophilic*, and can absorb about 4.5% of water; this must be considered if high volume contents of the fibre are being used, rather than the more usual 0.1–0.2%. Like polypropylene, nylon is chemically stable in the alkaline cement environment. It has been found to develop only a low bond strength with the cement matrix.

10.2.2.3 Polyester fibres

Polyester fibres for concrete are generally available in monofilament form, with tensile strength of 280–1200 MPa and elastic modulus of 10–18 GPa. However, they are not stable in the alkaline cement environment, exhibiting a strength and ductility loss over time [80–82]. Like the ordinary acrylic fibres described above, they are not suitable for use in FRC.

10.2.2.4 Polyethylene fibres

There is considerable interest in the use of polyethylene fibres in FRC [14,83–85]. These fibres can be readily mixed into the concrete using conventional batching and mixing techniques at fibre volumes of up to 4% [13]. The monofilaments have 'wart-like' surface deformations, which improve the mechanical bond. Polyethylene fibres currently used for concrete have tensile strengths in the range of 80–590 MPa, and an elastic modulus of about 5 GPa, but others have been reported to have elastic moduli in the range of about 15.4–31.5 GPa, similar to those of the concrete matrix [28].

Polyethylene fibres have been evaluated using either short, dispersed fibres mixed with concrete at volumes up to about 4% [84], or a continuous network of fibrillated fibres (polyethylene in pulp form) to produce a composite with about 10% by volume of fibres. In pulp form, they are intended for use as asbestos replacement. The effect of short, discontinuous fibres ($\sigma_t = 200$ MPa, $E = 5$ GPa) on the load–deflection of concrete is shown in Figure 10.20. At a volume content of 2% the fibres led to a marked post-cracking load-bearing capacity, while at 4% the maximum load in the post-cracking range exceeded the first crack stress. Thus, these fibres seem to be very effective for crack control. The fibrillated continuous polyethylene fibres appeared to be more effective than similar polypropylene fibres, probably because of the higher elastic modulus of the polyethylene.

It has recently been found that plasma treatment of polyethylene fibres can greatly increase the bond strength and the interfacial toughness [86]. They were also able to observe an interfacial chemical bond with the cement.

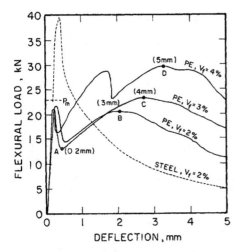

Figure 10.20 The effect of the content of short polyethylene fibres on the load–deflection curves of concretes (after Kobayashi and Cho [84]).

10.2.2.5 Polyolefin fibres

Polyolefin fibres, produced in monofilament form from a homopolymeric resin, appear to be well-suited for use in FRC. They have a tensile strength of about 275 MPa, and an elastic modulus of about 2.7 GPa. They exhibit a good bond with the concrete matrix, due to their surface roughness, consisting of bulges and grooves along the fibre surface [87]. They have much the same effects as polypropylene on the strength and toughness characteristics of FRC, and seem to be particularly effective under impact loading [88,89]. They also bring about a considerable improvement in fatigue properties; polyolefin fibre reinforced concretes reached an endurance limit at about 2 million cycles [90].

10.3   High modulus fibres

10.3.1   Carbon fibres

Carbon fibres consist of *tows*, each made up of numerous (~10, 000) filaments. The filaments are 7–15 μm in diameter, and consist of small crystallites of 'turbostratic' graphite, which is one of the allotropic forms of carbon. In the graphite crystal, carbon atoms are arranged in an hexagonal array in a plane (Figure 10.21(a)). The planes are stacked together, with covalent bonds acting within the planes and weaker Van der Waals forces holding the planes together [92]. To obtain a high modulus and strength, the layered planes of graphite must be aligned parallel to the fibre axis. However, the structure of the stacked planes is not

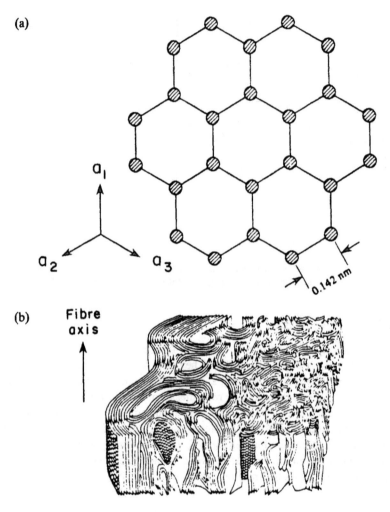

Figure 10.21 Structure of carbon fibres. (a) Arrangement of carbon atoms in graphite layer; (b) a schematic description of the structure of carbon fibres based on X-ray diffraction and electron microscopy [91].

ideally regular, as can be seen from the schematic representation in Figure 10.21(b). Therefore, the properties of carbon fibres can vary over a wide range, depending on the degree of perfection, which is a function of the production process. Carbon fibres are inert to most chemicals, and thus they are well-suited to perform in the alkaline cement environment.

The two main processes for making carbon fibres are based on different starting materials, either polyacrylonitrile (PAN carbon fibres) or petroleum and coal tar

Table 10.6 Properties of carbon fibres (after Hull [91], Nishioka et al. [93])

	PAN		Pitch
	Type I	Type II	
Diameter (μm)	7.0–9.7	7.6–8.6	18
Density (kg/m^3)	1950	1750	1600
Modulus of elasticity (GPa)	390	250	30–32
Tensile strength (MPa)	2200	2700	600–750
Elongation at break (%)	0.5	1.0	2.0–2.4
Coefficient of thermal expansion $\times 10^{-6}\,°C^{-1}$	−0.5 to −1.2 (parallel) 7–12 (radial)	−0.1 to −0.5 (parallel) 7–12 (radial)	—

pitch (pitch carbon fibres). Both processes involve heat treatments, and various grades of carbon fibres can be obtained with each, depending on the combination of heat treatment, stretching and oxidation. Typical properties are presented in Table 10.6[93]. The PAN carbon fibres are of higher quality (and higher cost), and are sometimes classified into two types, I and II, with type I having a higher modulus of elasticity and strength [91,93]. The pitch carbon fibres have a much lower modulus of elasticity and strength, but they are much less expensive than the PAN fibres. Nonetheless, the pitch fibres still have superior properties to most other synthetic fibres, and their modulus of elasticity is equal to or greater than that of the cement matrix. This, combined with their lower price, has made them more attractive for cement reinforcement. Mesophase (high-modulus) pitch-based carbon fibres are also now available, with properties generally intermediate between those of the other two types. The pitch fibres were developed in Japan [93], and much of the work on using these fibres in FRC has been carried out there.

10.3.1.1 Production of the composite

Carbon reinforced cement composites can be produced in a number of ways:

- Hand lay-up of continuous fibres or mats (e.g. [94]);
- Filament winding;
- Spraying;
- Conventional mixing [93,95].

The diameter of the filaments ($\approx$10 μm) is of the same order of magnitude as the size of a cement particle. For the cement particles to be able to penetrate between the individual filaments, that is, if the spacing between the filaments is about the same size as a cement grain, the maximum fibre content is about 12% by volume for aligned fibres, and less than 4–5% for randomly oriented fibres [96]. Briggs [94] reported that an increase of fibre content over 12% by volume resulted in fibre bundling and loss of strength. Thus Brown and Hufford [97] used carbon tow impregnated with polyvinyl acetate (PVA) to obtain a good fibre–matrix bond and to avoid the difficulty of opening up the spaces between filaments to allow penetration of cement particles. The PVA served as an intermediate medium for achieving effective stress transfer. Other attempts to improve bond and to compensate for potentially poor dispersion have also been based on polymer impregnation of the composite [98].

The uniform dispersion of short carbon fibres, particularly at higher fibre volumes, is difficult. The most that can be added using conventional mixing methods and materials is about 1% by volume. The dispersion of the fibres into individual filaments is controlled by the mixing technique. The methods used have included:

- fibre separation by a dispenser or a hammermill [93];
- mixing with an Omni mixer;
- mixing with an ordinary mortar mixer, but replacing 40% of the Portland cement with silica fume [99,100];
- using finer cements and large quantities of superplasticizers, which permitted the incorporation of up to 3% by volume of fibres [101];
- the use of suitable dispersing agents, such as carboxyl methyl cellulose [102], or slag [103].

In these applications, the mortar mix also contained various combinations of water reducing and air entraining agents. Using such methods, it was possible to incorporate up to 3–5% by volume of short fibres (3–10 mm) with effective dispersion. Although the flow properties decreased with an increase in fibre volume (Figure 10.22), mixes containing up to 5% short fibres were found to possess suitable flow properties, and their mouldability and finishability were similar to those of the unreinforced matrix.

Curing of carbon fibre composites can be carried out either by normal room temperature curing or by autoclaving. The fibres are sufficiently stable to endure the high pressure and temperature conditions in the autoclave.

10.3.1.2 Mechanical properties

The mechanical properties have generally been evaluated for two types of composites, prepared either by hand laying of continuous fibres and mats, or by mixing of short fibres. Most of the work with continuous, hand lay-up reinforcement has been carried out with the high quality PAN carbon fibres, while random short fibre reinforced composites have generally been studied using the lower quality pitch carbon fibres.

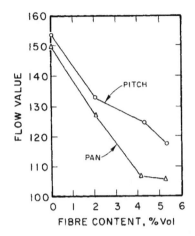

Figure 10.22   Effect of fibre content and type on the flow of mortar mixes with 3 mm long carbon fibres prepared in an Omni mixer (fibre diameters of 14.5 and 7.0 μm for the pitch and PAN type, respectively; aspect ratios of 207 and 429 for the pitch and PAN type fibre, respectively) (after Akihama *et al.* [98]).

Continuous hand laid reinforcement   Stress–strain curves of continuous and aligned carbon FRC are shown in Figure 10.23 [104]. They clearly demonstrate the effect of the increase in fibre content on the increase in modulus of elasticity, first crack stress and strain, and ultimate tensile strength. However, for random mat fibres, with 3% fibre content by volume, the composite failed at the limit of proportionality, without exhibiting the marked post-cracking behaviour typical of the continuous and aligned fibres. Others have found less dramatic strengthening effects, or less post-cracking ductility. These differences may be attributed to variations in the quality of the carbon fibres used in the different investigations, or differences in the dispersion of the filaments during specimen preparation. Thus, although a linear relationship between strength and fibre content may be established as expected on the basis of composite materials theory, the slopes of these curves can be quite different.

Similar effects have been found in the modulus of elasticity results, with different investigators all finding linear relationships between elastic modulus and fibre content, but with quite different slopes. In particular, the addition of pitch fibres does not lead to any significant increase in modulus of elasticity, reflecting the fact that their modulus of elasticity is similar to that of the matrix. The addition of PAN fibres does increase the elastic modulus, as their modulus of elasticity is much greater than that of the matrix.

Short and randomly mixed fibres   The effect of mixing short carbon fibres, mainly of the pitch type, has been studied extensively, mostly using fibres

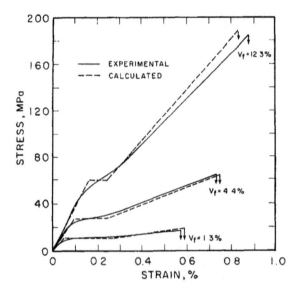

Figure 10.23 Effect of reinforcement with continuous carbon fibre on the stress–strain curve of cementitious composites (after Aveston *et al.* [104]).

less than 10 mm in length. It appears that the tow disperses into individual filaments and so the aspect ratio of such fibres is quite high, more than 100, since the diameter of the individual filaments is about 10 μm. Thus, the reinforcing effect of these short fibres is quite high, even with volume contents less than 4%.

Both pitch and PAN based fibres can greatly increase the tensile strength of the composite, as shown in Figure 10.24 [105]. Carbon fibres lead to a slight increase in compressive strength up to fibre volumes of about 3% as shown in Figure 10.25 [106], after which the strength begins to decline, probably because of the difficulty in fully compacting high fibre volume materials. The load–deflection curves in Figure 10.26 [107] clearly indicate the marked improvement in flexural strength and post-cracking behaviour achieved with carbon fibres. Similar effects are obtained for both autoclaved and room temperature cured composites.

It is interesting to note the effect of strengthening the matrix by reducing the w/c ratio (compare Figures 10.26(a) and (b)), which leads to a marked change in the shape of the load–deflection curve, reducing its post-cracking portion in the low w/c ratio matrix, without much changing its peak load. The matrix composition has a greater effect on toughness than on the flexural strength. It is worth noting, in the latter case, the particularly low toughness values for the low w/c ratio mix of 0.298. Similar effects have been reported by Delvasto *et al.* [108], who studied the influence of applying compaction pressure after casting, to expel surplus water. The influence of pressure on the load–deflection curve is shown in Figure 10.27, demonstrating again the adverse effect of densifying the matrix on

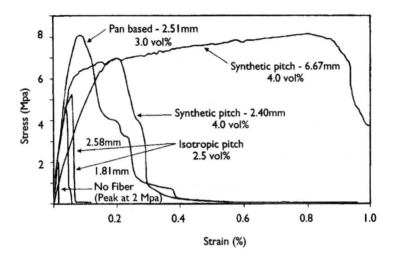

Figure 10.24 Uniaxial tensile testing of carbon fibre reinforced cement (after Li and Obla [105]).

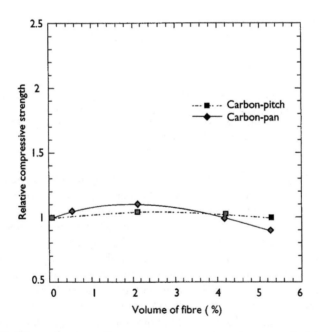

Figure 10.25 Effect of carbon fibre content on compressive strength (after Li and Mishra [106]).

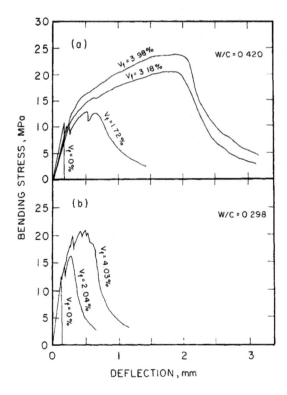

Figure 10.26 The effect of pitch carbon fibre content on the load–deflection curves in bending of mixes with paste matrix. The fibre is 10 mm long and 14.5 μm in diameter (after Akihama et al. [107]). (a) 0.420 w/c ratio paste matrix; (b) 0.298 w/c ratio paste matrix.

the post-cracking behaviour, though this is not accompanied by any reduction in the maximum load. As discussed earlier in Section 4.1, with these small diameter fibres a mild increase in bond (achieved by densifying the matrix) can result in an anchoring effect which is sufficiently high to lead to failure by fibre fracture than by pull-out, thus causing a marked reduction in the toughness.

For 3 mm long carbon fibres, the change from fibre failure by pull-out to failure by fracture (see Section 8.5, Figure 8.28 for a similar calculation for glass filaments) occurs at a frictional bond strength of about 0.6 MPa. Such a value can readily be achieved with the lower w/c matrices. Akihama et al. [107] and Nishioka et al. [93] noted that many of the fibres in their composites were broken when the composite was loaded to failure, rather than pulled out. Nishioka et al. [93] suggested that the critical length of the pitch carbon fibres is 0.8–1.4 mm. In the actual composite, many of the fibres would be in this length range, because the longer 10 mm fibres originally introduced into the mix tend to break upon mixing,

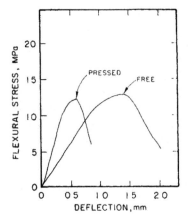

Figure 10.27 The effect of applying compaction pressure after casting on the load–deflection curve of carbon fibre reinforced paste with 1.7% fibres by volume (after Delvasto et al. [108]).

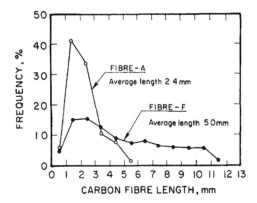

Figure 10.28 Length distribution of pitch carbon fibre after mixing of fibres having original length of 10 mm. Fibre A. 440 MPa tensile strength; Fibre F, 682 MPa tensile strength (after Nishioka et al. [93]).

and the resulting average fibre length is 1/2 to 1/4 of the original value. The length distribution of the fibres after mixing is thus quite different from the original uniform 10 mm length (Figure 10.28). The reduction in the average length and the resulting distribution apparently depends on the strength of the fibres, with the weaker ones breaking up more than the stronger ones. This change in fibre length results in a shortening of many of the fibres to the critical length of 0.8–1.4 mm, thus accounting for the mixed mode of fibre fracture and fibre pull-out.

The effect of the quality of the fibres on the properties of the composite was evaluated by Nishioka *et al.* [93] using pitch fibres with tensile strengths in the range of 440–764 MPa, and elastic moduli in the range of 26.6–32.4 GPa. An increase in fibre strength led to an increase in flexural strength and toughness (Figure 10.29). It was thus suggested that for effective reinforcement, the fibres should have a tensile strength greater than 640 MPa. Part of the poor performance of the lower strength fibres is associated with their breakdown into shorter lengths during mixing. For the data in Figure 10.29, a better correlation can be established when the composite property is plotted against the average fibre length *after* mixing (Figure 10.30). This again emphasizes the importance of accounting for the composite properties on the basis of the length of the fibres after mixing, rather than on the original length.

The modulus of elasticity of a pitch carbon composite produced with short fibres tended to decrease slightly with an increase in fibre content. This is because

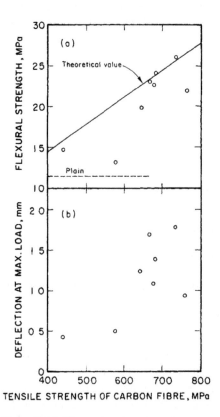

Figure 10.29 Effect of pitch carbon fibre strength on the flexural strength (a) and the deflection at maximum load (b) of a composite with 3% short fibres, 10 mm long and 18 μm in diameter (after Nishioska *et al.* [93]).

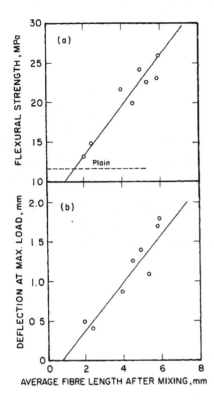

Figure 10.30 The effect of pitch carbon fibre length after mixing on the flexural strength (a) and the deflection at maximum load (b), of a composite with 3% fibres which were 10 mm long prior to mixing (after Nishioska *et al.* [93]).

the modulus of elasticity of this type of fibre is approximately the same as that of the matrix, combined with the reduced density of the matrix at higher fibre contents. The latter effect is the result of the greater difficulty in compacting the fresh composite containing fibres.

10.3.1.3   Dimensional stability

Carbon fibres have been found to be effective in decreasing strains due to swelling and shrinkage, both for PAN fibres [94] and pitch fibres [99]. The reduction was by about a factor of 2–3 for 3% of short fibres, and considerably greater for higher fibre contents. Briggs *et al.* [109] reported a decrease in shrinkage by a factor of 10 with a 5.6% fibre volume, and a reduction in creep by a factor of 6 with 2% fibres. The shrinkage values can be lowered further by autoclave curing of the composite [95]. As a result of the reduced shrinkage and improved toughness, this composite is both dimensionally stable and less sensitive to cracking.

10.3.1.4   Durability

The alkali resistivity of carbon FRC was tested using an accelerated procedure (immersion in water at 50°–75°C), similar to the test applied for GRC. Most of the strength and toughness were retained after prolonged testing, up to 3 months [95] as shown in Figure 10.31(a), or one yearc [110]. Exposure of the composite to freeze–thaw cycles and alternating wetting and drying cycles [94,95] also did not lead to any reduction in strength or modulus of elasticity (Figure 10.31(b)). However, in studies of high modulus PAN carbon fibres a reduction in strength was observed in hot water exposure, in a composite with a dense matrix (for more details see Chapter 5, Figure 5.3 [111]). In the discussion of the mechanisms leading to such changes it was suggested that the sensitivity to age effects would

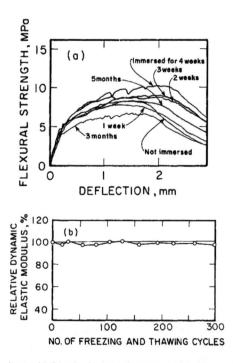

Figure 10.31   Durability of pitch carbon fibre reinforced cement produced with 2% fibres and autoclaved matrix (after Akihama et al. [95]). (a) After different periods of immersion in warm water; (b) as a function of the number of freeze-thaw cycles.

be greater in the high modulus PAN carbon fibre than in the lower modulus pitch carbon fibre.

10.3.1.5 Electrical resistivity

While this book is concerned primarily with the mechanical properties and durability of fibre concretes, it is worth mentioning here that carbon fibre reinforced cement or concrete has some very useful electrical properties, because the carbon fibres are electrically conductive. Chen and Chung [112] showed that carbon fibre reinforced concrete was intrinsically a 'smart' concrete that could sense elastic and inelastic deformations and fracture. The change in electrical resistance provides the signal for this. This property also allowed the composite to be used for traffic monitoring and weighing in motion [113]. The change in electrical resistivity with temperature also allowed the fibre reinforced cement to act as a thermistor [114].

10.3.1.6 Mechanics of the composite

A number of investigators have applied the ACK model (Chapter 4) to account for the stress–strain curves of continuous carbon fibre composites. Good agreement between the experimental and calculated curves was obtained, as shown in Figure 10.23. Aveston et al. [104] showed that the crack spacing developed in the post-cracking zone was proportional to $\varepsilon_{mu} V_m / V_f$, as predicted by Eq. (4.65) of their model, when substituting $E_m \varepsilon_{mu}$ for σ_{mu}. On the basis of this relationship, Sarkar and Bailey [115] and Akihama et al. [116] calculated the average bond strength τ to be ~ 0.8 MPa, with fibres of 9.2 μm diameter. A bond value of ~ 2.5 MPa might be calculated from the data and analysis of Aveston et al. [104] assuming a fibre diameter of 10 μm. Taking typical pitch carbon fibre strength values (600–750 MPa) and diameter (10–20 μm), and the bond strength in the range of 0.8–2.5 MPa, the critical fibre length can be calculated to be in the range of 0.6–2 mm (Eq. 4.2, Section 4.2). This is similar to the value of 0.8–1.4 mm reported by Nishioka et al. [93], and is in agreement with the observations that short (3–10 mm) randomly dispersed pitch carbon fibres are effective in reinforcing a cementitious matrix.

The rule of mixtures has been applied to account for the strength and modulus of elasticity of cementitious matrices reinforced with continuous or randomly dispersed carbon fibres. A reasonably good agreement has generally been reported between the calculated and experimental modulus of elasticity for the continuous reinforced composites [104,110,115,116]. The same researchers found that the tensile strength was 30–90% of the theoretical value of $\sigma_{fu} V_f$ for systems in which the fibre volume content exceeded the critical value ($\sim 0.5\%$). Akihama et al. [116] reported that the ratio between the experimental and calculated tensile strength decreased with an increase in fibre content. The lower strength in the actual composite was suggested [115] to be the result of the fibres failing progressively (i.e. without attaining their ultimate strain simultaneously). Similarly, Ali et al.

[110] argued that since the reinforcing fibres are in the form of bundles, from statistical considerations the strength of the bundle is likely to be significantly lower than the average strength of the individual filaments.

The rule of mixtures can also be used to account for the tensile strength of the composite with short and randomly dispersed pitch carbon fibres, based on Eq. (4.28) in Section 4.3. For a fibre length of 10 mm, which exceeds the critical length, the overall efficiency coefficient was found to be in the range of 20–40 %, with the value decreasing with an increase in fibre volume [107]. Since, during mixing, some of the fibres break and become shorter, this should also be taken into account. Nishioka *et al.* [93] considered the actual length distribution of the fibres after mixing (Figure 10.30), and calculated the strength by summing the contributions of the fibres with lengths shorter than the critical length and those with lengths greater than the critical length:

$$\sigma_{cu} = \eta_1 \left(1 - \frac{l_c}{2l_1}\right) \sigma_f V_{f1} + \eta_1 \left(\frac{l_2}{2l_c}\right) \sigma_f V_{f2} \tag{10.2}$$

The critical length was in the range of 0.8–1.4 mm. The same equation was used to calculate the flexural strength, by multiplying the value of σ_{cu} by 2.44 (see Section 5.6). Good agreement was reported between experimental and calculated values.

10.3.2 Aramid (Kevlar) fibres

10.3.2.1 Introduction

Aramid was one of the first commercial polymeric fibres in which high strength and elastic modulus were achieved by chain alignment. It is an aromatic polyamide called poly(*para*-phenylene terephthalamide), with the chemical formula:

It is also known by the trade name Kevlar, given by the DuPont Company to its aramid fibre. The aromatic rings provide the molecule with rigidity, and in the production process these stiff molecules are aligned parallel to the fibre axis, thus leading to a modulus of elasticity which can be as high as 130 GPa. A fibre consists of planar sheets of molecules linked together by hydrogen bonding (Figure 10.32(a)) [117]. The sheets are stacked together radially to form the fibre (Figure 10.32(b)). The bonds between the sheets are weak, and therefore the fibre has a low longitudinal shear modulus and poor transverse properties. The properties of the fibre, and in particular the modulus of elasticity, depend on the degree of alignment achieved during production, and therefore aramid fibres can

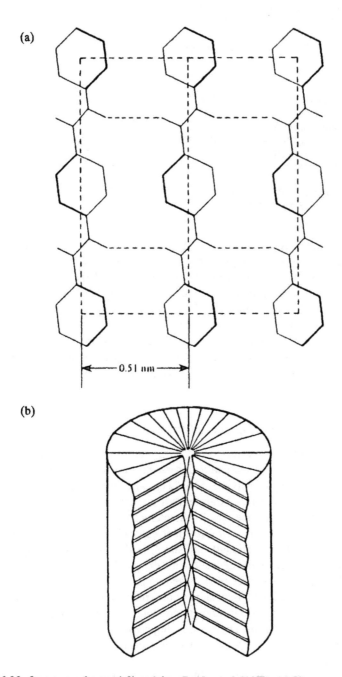

(a)

0.51 nm

(b)

Figure 10.32 Structure of aramid fibre (after Dobb et al. [117]). (a) Planar array of the chain molecules; (b) radial stacking of the planes in the fibre.

Table 10.7 Properties of aramid fibres

Property	Kevlar 49[a]	Kevlar 29[a]	HM-50[b]
Diameter (μm)	11.9	12	12.4
Density (kg/m^3)	1450	1440	1390
Modulus of elasticity (GPa)	125	69	77
Tensile strength (MPa)	2800–3600	2900	3100
Elongation at break (%)	2.2–2.8	4.4	4.2

Notes
a Produced by DuPont Company, USA.
b Produced by Teijin Ltd., Japan.

be of different qualities (Table 10.7). Aramid, because of its stable chain structure, is reasonably resistant to temperature compared with many other synthetic fibres. It is essentially unaffected by temperatures up to 160°C. However, at temperatures higher than 300°C, the fibre may lose most of its strength, and will also creep considerably. Such effects require special evaluation of the composite, particularly from the point of view of fire resistance.

Aramid fibres, in contrast to glass and carbon, fracture in a ductile manner, accompanied by considerable necking and fibrillation. The fibre itself consists of rovings, each composed of a bundle of several thousands of filaments, 10–15 μm in diameter.

10.3.2.2   Mechanical properties of the composite

The properties of cements reinforced with aramid fibres have been studied in composites with short and randomly dispersed fibres, produced either by the spray technique [118] or by premixing [119–121], with fibre contents in the range of 1–5% by volume. The stress–strain curves presented in Figure 10.33 show that these composites exhibit the usual characteristics expected on the basis of the ACK model, with the fibre reinforcement enhancing the ultimate strength and strain over the first crack strength and strain, leading to a tough composite. The effects of fibre content and length show the expected increase in flexural strength (Figure 10.34) and toughness with these two parameters [119]. The increase in fibre content is, of course, associated with a reduction in the flow properties of the fresh mix. The increases in strength and toughness with aramid fibres are not always the same, with some studies showing greater effects than others. A much greater strengthening effect was reported, as expected, for continuous and aligned fibre reinforcement [120].

Treatment of the aramid fibres with polymers has been found to be an efficient way of increasing the flexural properties of the composite. Walton and Majumdar [118] had found that the tensile strength of the aramid composite was only 70% of the expected value, and suggested that this was the result of poor bonding. Wang *et al.* [80] determined the average bond strength to be 2.75 MPa, with a critical

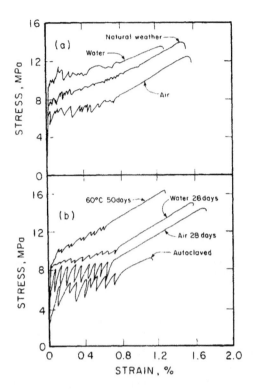

Figure 10.33 Stress–strain curves of cementitious composites reinforced with aramid (Kevlar) fibres produced by the spray method (after Walton and Majumdar [118]). (a) Effect of 2 years of weathering; (b) effect of temperature treatments.

length value of about 3.2 mm. These values are sufficiently high for effective stress transfer of the fibres with lengths of 15–50 mm studied by Walton and Majumdar [118] and Ohgishi *et al.* [119]. However, it is possible that in the composite the real bond is lower, due to ineffective dispersion of the bundled filaments.

Walton and Majumdar [122] studied the tensile creep of Kevlar fibres and the bending creep of cementitious composites made with these fibres (2.4% by volume of short random 2-dimensionally dispersed fibres). The creep coefficients of the aramid fibres themselves were found to be smaller than those of other synthetic fibres (polyethylene, polyvinylchloride and polycarbonate). The creep of the composite was of the same order of magnitude as that expected in a plain mortar matrix.

10.3.2.3 Durability

After two years of ageing (Figure 10.33(a)), and 180 days in 60°C water and 75°C water, the strength and toughness were largely preserved, implying that this

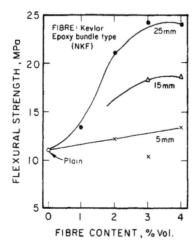

Figure 10.34 Effect of fibre length and content on the flexural strength of mortars reinforced with aramid (Kevlar) fibres (after Ohgishi *et al.* [119]).

composite may not be susceptible to immediate durability problems. The increase in first crack stress and the slight reduction in ultimate tensile strain in water storage may be due to matrix changes leading to increases in fibre–matrix bond. Yet, it should be noted that these fibres may still be susceptible to some alkaline attack, even though the rates may be low enough to provide a sufficiently long life expectancy. The predicted performance of such fibres in an alkaline solution showed 50% residual strength retention after 20 years for unimpregnated fibre bundles, and considerably better performance (by two orders of magnitude) for epoxy impregnation [123]. Ohgishi *et al.* [119] reported a 90% strength retention of the fibres after 10,000 h of immersion in a pH = 12.5 solution. Under similar conditions, the strength retentions of AR glass, carbon and steel fibres were 42.5%, 41.5% and 99.6%, respectively. Wang *et al.* [80] found that aramid fibres removed from a cement matrix showed a drop in fibre strength, especially when stored at a temperature of 50°C.

10.3.3 High strength acrylic fibres

High strength acrylic fibres have been developed with a modulus of elasticity in the range of 14–25 GPa, and tensile strengths up to 1000 MPa [124–126]; that is, the modulus of elasticity is of the same order of magnitude as that of the matrix. The fibres can be produced in the form of discrete short filaments, and can also be woven into various fabrics. The stress–strain curves of these fibres are shown in Figure 10.35, where they are compared with conventional textile acrylic fibres. The increases in tensile strength and modulus of elasticity relative to the textile

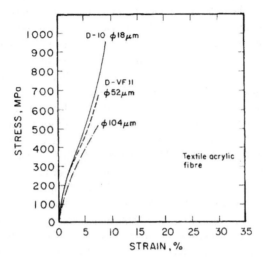

Figure 10.35 Stress–strain curves of acrylic fibres developed for cement reinfored-cement (Dolanit-D-10; D-VF11) and conventional textile acrylic fibres (after Hähne *et al.* [126]).

Table 10.8 Properties of high strength acrylic fibres[a] developed for cement reinforcement (after Hähne [125])

Type	D-10		D-VF11	
Fibre diameter (μm)	13,	18	52, 75,	104
Length (mm)	6,	12	6–24	
Tensile strength (MPa)	900–1000		590–710	410–530
Elongation (%)	8–11		6–9	6–9
Initial elastic modulus (MPa)	17,000–19,500		16,000–18,300	14,200–16,500
Density (kg/m³)			1180	

Note
a Dolanit, produced by Hoechst AG.

fibres are evident. These improved characteristics are accompanied by a reduction in ultimate elongation.

However, the reduced elongation is still in the range of 6–11%, which is considerably greater than that of the matrix. One commercial fibre (Dolanit, produced by Hoechst AG) has been reported to be produced in a range of diameters, from 10 to 100 μm [125]. Preliminary experiments indicated that the larger diameter was needed for mortar and concrete reinforcement, while the smaller diameter was more suitable or asbestos replacement. Properties of one type of commercial fibre are given in Table 10.8.

There is a good bond between the fibre and the cement matrix, which may be due to the shape of the individual filaments, which are characterized as having a 'kidney' or 'dog-bone' shape [125,126]. Hähne *et al.* [126] reported an average 4 MPa bond strength in a pull-out test of 105 μm diameter filaments, with the calculated critical length values ranging from 2 mm to 6 mm of 13 μm to 105 μm filaments. Wang *et al.* [80] reported a similar strong bond of 2.93 MPa, with a 1 mm critical length for 19.2 μm diameter fibres.

Acrylic fibres have been reported to be relatively stable in an alkaline environment. However, some long-term sensitivity to such conditions should not be ruled out. Odler [127] described tests in which the surfaces of the fibres were examined by SEM after exposure in 1M NaOH solution at 20°C. No signs of corrosion could be seen after 3 months. On the other hand, Hähne [125] tested the fibres after exposure to cement extract solution at different temperatures. At 20°C, the tensile strength and elastic modulus were reduced by about 7–15% after several months, but thereafter remained stable, suggesting a reasonable alkali resistance. A marked reduction in properties was observed only at temperatures above 90°C. Similarly, Wang *et al.* [80] reported a small loss in strength over a 60-day period of fibres stored in a cement environment at 22°C, but a significant loss at 50°C. Such results point towards some uncertainty regarding the long-term performance.

Although the properties of acrylic fibres are not as good as those of asbestos, it is still possible to prepare composites by the Hatschek process, and obtain composites which may be considered adequate for replacement of asbestos–cement products (Figure 10.36) [128]. The strength of the acrylic composite is somewhat lower, but its strain capacity is significantly greater [125,128]. For proper replacement of asbestos, the acrylic fibre must be used in combination with processing fibres, sometimes referred to as *filter pulps*. Daniel and Anderson [124] studied the influence of the processing fibre on the reinforcing effect of the acrylic fibre, by evaluating the performance of laboratory-made composites using techniques involving vacuum-dewatering and pressing. The processing fibre type and content

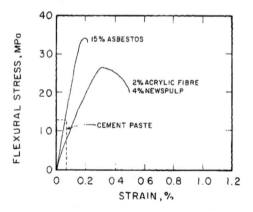

Figure 10.36 Stress–strain curves of asbestos cement and acrylic–cement composite produced to replace asbestos cement (after Gale [128]).

were found to have only a small effect on the flexural strength and toughness index, although the toughness index of composites prepared with polyoxymethylene (POM) processing fibres was somewhat better than that obtained with polyethylene processing fibres. An increase in acrylic fibre content from 1% to 3% was found to lead to a considerable increase in toughness index, but had only a small effect on flexural strength.

The effect of premixing short acrylic fibres using conventional technology has been discussed by Odler [127] and Hähne et al. [126]. The addition of short fibres (2–6 mm in length) resulted in a reduction in workability and this had to be compensated for by an increase in the water content of the mix; the w/c ratio varied accordingly in the range of 0.25–0.70. However, even with this adjustment, the fibre content could not be increased beyond about 3% by volume without being accompanied by clumping and poor dispersion. However, within this range, the fibres led to a considerable increase in flexural strength, which was practically independent of the matrix w/c ratio (Figure 10.37(a)), and were effective in enhancing the strain capacity and toughness (Figure 10.37(b)).

Odler [127] observed that in the composites produced by conventional mixing, the acrylic fibres dispersed into the individual ~ 18 μm diameter filaments, in contrast to glass fibres which remained bundled together in strands after similar mixing. This is probably one of the reasons for the greater efficiency of the acrylic fibres in enhancing the flexural strength of the composite. Odler noted considerable pull-out of fibres that were 2–6 mm long, which may be indicative of relatively poor bond. This is different from the conclusions that can be reached from the pull-out tests of Hähne et al. [126] and Wang et al. [80], who found critical lengths of 1 and 2.5 mm, respectively, for 18 μm diameter filaments.

The presence of 2.5% fibres has also been reported [126] to be effective in reducing the shrinkage cracking induced during setting and hardening, as evaluated by the ring test.

10.3.4   Polyvinyl alcohol (PVA) fibres

High strength PVA fibres have been developed mainly for asbestos replacement [129–132]. The fibres are produced by wet or dry spinning, and boron is added to achieve high strength and stiffness by forming intermolecular bonds. The fibres are surface treated to enhance their compatibility with the matrix and to enable efficient dispersion. The surface treatment, combined with the inherent affinity of this polymer for water, due to the presence of OH^- groups, leads not only to efficient dispersion, but also to a strong bond in the hardened composite. Some properties of different types of high strength PVA fibres are shown in Table 10.9, demonstrating that their modulus of elasticity can be of the same order of magnitude as that of the cementitious matrix.

PVA fibres tend to rupture instead of pull-out of a cementitious matrix, due to the strong chemical bonding and the resulting slip-hardening response during pull-out [133]. Li et al. [134] were able to produce composites with an ultimate tensile strain exceeding 4% and a tensile strength of 4.5 MPa, with a fibre volume of only

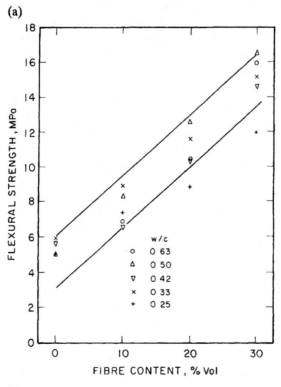

(a)

(b)

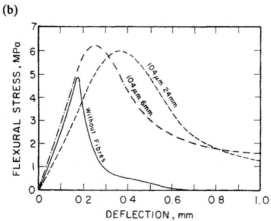

Figure 10.37 Effect of acrylic fibres on the properties of composites produced by conventional mixing. (a) Effect on flexural strength (after Odler [127]); (b) effect on flexural stress vs. deflection curve (after Hähne et al. [126]).

2.0%. The specimens exhibited multiple cracking with the crack widths at ultimate strain limited to below 100 μm. Such composites are classified within the range of high performance FRC, and they are discussed in greater detail in Section 12.2.

PVA FRC can be produced in a number of ways, including premixing and shotcreting. Thin sheets, pipes and other shapes have also been produced using an extrusion process [33,135].

Cement composites reinforced by PVA fibres can be produced with mechanical properties superior to those of asbestos cement, while using smaller contents of PVA fibres than of asbestos fibres (Figure 10.38). This is especially the case for strength and toughness, though the modulus of elasticity is somewhat lower than that of the asbestos–cement composite.

The alkali resistance of PVA fibres can be quite high, as demonstrated in Figure 10.39, showing almost complete retention of strength by PVA fibres (Kuralon, produced by Kurary Co., Japan) after immersion in a cement slurry at 80°C for 14 days. This behaviour is superior to fibres such as polyester and

Table 10.9 Properties of High Tenacity PVA fibres

Property	Type of PVA fibre	
	Kuralon [129]	PVA-I [130]
Density (kg/m^3)	1300	1300
Diameter (μm)	—	10.4
Tensile strength (MPa)	1470	1200–1500
Modulus of elasticity (GPa)	36.3	20–25

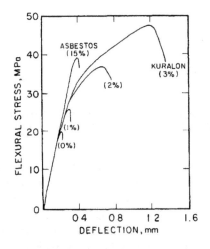

Figure 10.38 Effect of PVA fibre content on the flexural stress vs. deflection curve of a cement composite produced by the Hatschek process in combination with 3% wood pulp (after Hikasa and Genba [129]).

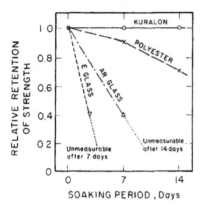

Figure 10.39 Effect of ageing of fibres in cement slurry at 80°C on their strength retention (after Hikasa and Genba [28]).

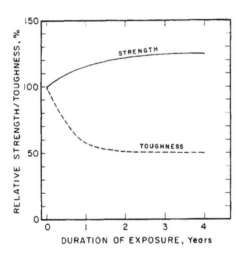

Figure 10.40 Effect of natural ageing on the retention of properties of PVA fibre-reinforced cementitious composite (after Hikasa and Genba [129]).

glass. The fibre is thermally stable, with no strength loss after exposure to temperatures of 150°C, and is insensitive to biological attack [130]. This is reflected in the ageing performance of the composite, which shows no sign of reduction in strength (Figure 10.40). However, at the same time, the toughness of the composite was reduced, particularly within the first year of ageing. This reduction, combined with the small increase in strength, may be due to improved fibre–matrix bond, by mechanisms similar to those suggested for GRC; this must, however, be validated by more detailed studies. It has also been found that the PVA–cement composite has a better freeze–thaw durability than asbestos cement.

References

1. J.J. Zonsveld, 'Properties and testing of concrete containing fibres other than steel', in A. Neville (ed.) *Fibre Reinforced Cement and Concrete*, Proc. RILEM Symp., The Construction Press, Lancaster, England, 1975, pp. 217–226.
2. D.J. Hannant, 'Polymer fibre reinforced cement and concrete', in D.M. Roy, A.J. Majumdar, S.P. Shah and J.A. Manson, *Advances in Cement-Matrix Composites*, Proc. Symp. L, Materials Research Society, Pittsburgh, PA, 1980, pp. 171–180.
3. H. Krenchel and S.P. Shah, 'Synthetic fibres for tough and durable concrete', in R.N. Swamy, R.L. Wagstaffe and D.R. Oakley (eds) *Developments in Fibre Reinforced Cement and Concrete*, Proc. RILEM Symp. Sheffield, RILEM Technical Committee 49-FTR, 1986, Paper 4.7.
4. P.L. Walton and A.J. Majumdar, 'Cement-based composites with mixtures o different types of fibres', *Composites*. 6, 1975, 209–216.
5. N.J. Dave and D.G. Ellis, 'Polypropylene fibre reinforced cement', *Int. J. Cem. Comp.* 1, 1979, 19–28.
6. H. Krenchel and H.W. Jensen, 'Organic reinforcing fibres for cement and concrete', in *Fibrous Concrete*, Proc. Symp. on Fibrous Concrete, The Concrete Society, The Construction Press, Lancaster, England, 1980, pp. 87–98.
7. H. Krenchel and S.P. Shah, 'Applications of polypropylene fibres in Scandinavia', *Concr. Int. Des. & Constr.* 7 (3), 1985, 32–34.
8. S. Mindess and G. Vondran, 'Properties of concrete reinforced with fibrillated polypropylene fibres under impact loading', *Cem. Concr. Res.* 8, 1988, 109–115.
9. A. Bentur, S. Mindess and G. Vondran, 'Bonding in polypropylene fibre reinforced concrete', *Int. J. Cem. Comp. & Ltwt. Concr.* 11, 1989, 153–158.
10. D.J. Hannant and J.J. Zonsveld, 'Polyolefin fibrous networks in cement matrices for low cost sheeting', *Phil. Trans. R. Soc. London A*, 294, 1980, 591–597.
11. D.J. Hannant, J.J. Zonsveld and D.C. Hughes, 'Polypropylene film in cement based materials', *Composites*. 9, 1978, 83–88.
12. A. Vittone, 'Industrial development of the reinforcement of cement based products with fibrillated polypropylene networks, as a replacement of asbestos', in R.N. Swamy, R.L. Wagstaffe and D.R. Oakley (eds) *Developments in Fibre Reinforced Cement and Concrete*, Proc. RILEM Symp., Sheffield, RILEM Technical Committee 49-FTR, 1986, Paper 9.2.
13. ACI Committee 544, *State of the Art Report on Synthetic Fibre-reinforced Concrete*, Draft Document, 2005.
14. D.C. Hughes, 'Stress transfer between fibrillated polyalkene and cement matrices', *Composites*. 15, 1984, 153–158.
15. N. Banthia and N. Nandakumar, 'Crack growth resistance of concrete reinforced with low volume fractions of polymeric fiber', *Journal of Materials Science Letters* 20, 2001, 1651–1653.
16. J.-F. Trottier and M. Mahoney, 'Innovative synthetic fibers', *Concr. Int.* 23(6), 2001, 23–28.
17. S. Altoubat, J.R. Roesler and K.-A. Rieder, 'Flexural capacity of synthetic fiber reinforced concrete slabs on ground based on beam toughness results', in M. di Prisco, R. Felicetti and G.A. Plizzari (eds) *Fiber-Reinforced Concretes, BEFIB 2004*, RILEM Proceedings PRO 39, RILEM Publications, Bagneux, 2004, Vol. 2, pp. 1063–1072.
18. A. Peled and A. Bentur, 'Fabric structure and its reinforcing efficiency in textile reinforced cement composites', *Composites: Part A*. 34, 2003, 107–118.

19. C. Zhang, V.S. Gopalaratnam and H.K. Yasuda, 'Plasma treatment of polymeric fibers for improved performance in cement matrices', *J. Appl. Polym. Sci.* 76, 2000, 1985–1996.

20. L. Tu, D. Kruger and P.A.B. Carstens, 'Effects of the increased surface wettability on the polypropylene-concrete interfacial bonding and the properties of the polypropylene fibre reinforced concrete', in Y. Ohama and M. Puterman (eds) *Adhesion between Polymers and Concrete*, RILEM Proceedings PRO 9, RILEM Publications, Bagneux, 1999, pp. 267–284.

21. A.E. Naaman, S.P. Shah and J.L. Thorne, 'Some developments in polypropylene fibres for concrete', in G.C. Hoff (ed.) *Fiber Reinforced Concrete*, SP-81, American Concrete Institute, Farmington Hills, MI, 1984, pp. 375–396.

22. E.K. Rice, G.L. Vondran and H.O. Kunbargi, 'Bonding of fibrillated polypropylene fibre to cementitious materials', in S. Mindess and S.P. Shah (eds) *Bonding in Cementitious Composites*, Proc. Materials Research Society Symp., Vol. 114, Materials Research Society, Pittsburgh, PA, 1988, pp. 145–152.

23. A. Kelly and C. Zweben, 'Poisson contraction in aligned fiber composites showing pull out', *J. Mater. Sci. Letters.* 11, 1976, 582–587.

24. D.J. Pinchin, 'Poisson contraction effects in aligned fibre composites', *J. Mater. Sci.* 11, 1976, 1578–1581.

25. P. Baggott and D. Gandhi, 'Multiple cracking in aligned polypropylene fibre cement composites', *J. Mat. Sci.* 16, 1981, 65–74.

26. D.J. Hannant, 'Durability of cement sheets reinforced with polypropylene networks', *Mag. Concr. Res.* 35, 1983, 197–204.

27. D.J. Hannant and D.C. Hughes, 'Durability of cement sheets reinforced with layers of continuous networks of fibrillated polypropylene film', in R.N. Swamy, R.L. Wagstaffe and D.R. Oakley (eds) *Developments in Fibre Reinforced Cement and Concrete*, Proc. RILEM Symp., Sheffield, 1986, Paper 7.8.

28. D.C. Hughes and D.J. Hannant, 'Brittle matrices reinforced with polyalkaline films of varying elastic moduli', *J. Mater. Sci.* 17, 1982, 508–516.

29. B.W. Richardson, 'High-volume polypropylene reinforcement for shotcrete', *Concrete Construction.* 35, 1990, 33, 35.

30. V.M. Malhotra, G.G. Carette and A. Bilodeau, 'Mechanical properties and durability of polypropylene fiber reinforced high-volume fly ash concrete for shotcrete applications', *ACI Mater. J.* 91, 1994, 478–486.

31. J.B. Studinka, 'Asbestos substitution in the fiber cement industry', *International Journal of Cement Composites and Lightweight Concrete* 11, 1989, 73–78.

32. A. Bentur and S.A.S. Akers, 'Thin sheet cementitious composites', in N. Banthia, A. Bentur and A. Mufti (eds) *Fiber Reinforced Concrete: Present and Future*, Canadian Society for Civil Engineers, Montreal, 1998, pp. 20–45.

33. Y. Shao, S. Marikunte and S.P. Shah, 'Extruded fibre-reinforced composites', *Concr. Int.* 17 (4), 1995, 48–52.

34. B. Mobasher and C.Y. Li, 'Processing techniques for manufacturing high volume fraction cement based composites', in H. Saadatmanesh and M.R. Ehsani (eds) Proc. 1st International Conference for Composites in Infrastructure, ICCI '96, 1996 pp. 123–136.

35. B. Mobasher and C.Y. Li, 'Mechanical properties of hybrid cement based composites', *ACI Mater. J.* 93, 1996, 284–293.

36. A. Peled and S.P. Shah, 'Parameters related to extruded cement composites', in A.M. Brandt, V.C. Li and I.H. Marshall (eds) *Brittle Matrix Composites 6*, Woodhead

Publications, Cambridge and Zturek Research Scientific Institute, Warsaw, 2000, pp. 93–100.

37. H. Krenchel, 'Fibre concrete – tough and durable', Paper presented at the FIP 9th Congress Colloquium '*What Do We Require from Concrete and What Can We Achieve*', Stockholm, 1982.

38. P. Baggott, 'Polypropylene fibre reinforcement of light weight cementitious matrices', *Int. J. Cem. Comp. & Ltwt. Concr.* 5, 1985, 105–114.

39. G. Spadea and G. Frigione, 'Mechanical and rheological behaviour of polypropylene fibre reinforced concrete', *Il Cemento*. 2, 1987, 173–185.

40. S. Hasaba, M. Kawamura, T. Koizumi and K. Takemoto, 'Resistibility against impact load and deformation characteristics under bending load in polymer and hybrid (polymer and steel) fiber reinforced concrete', in G.C. Hoff (ed.) *Fiber Reinforced Concrete*, ACI SP-81, American Concrete Institute, Farmington Hills, MI, 1984, pp. 187–196.

41. V. Ramakrishnan, G.Y. Wu and G. Hosalli, 'Flexural behavior and toughness of fiber reinforced concretes', *Transportation Research Record 1226*, National Research Council, Washington, DC, 1989, pp. 36–47.

42. P.P. Kraii, 'A proposed test to determine the cracking potenetial due to drying shrinkage of concrete', *Concrete Construction*. 30, 1985, 775–778.

43. Ch. A. Shales and K.C. Hover, 'Influence of mix proportions and construction operations on plastic shrinkage cracking in thin slabs', *ACI Mater. J.* 85, 1998, 495–504.

44. K. Yokoyama, S. Hiraishi, Y. Kasai and K. Kishitani, 'Shrinkage and cracking of high strength concrete and flowing concrete at early ages', in V.M. Malhotra (ed.) Proc. 4th CANMET/ACI International Conference on Superplasticizers and other Chemical Admixtures in Concrete, SP-148, American Concrete Institute, Farmington Hills, MI, 1994, pp. 243–258.

45. P. Soroushian, F. Mirza and A. Alhozaimy, 'Plastic shrinkage cracking of PP fiber reinforced concrete', *ACI Mater. J.* 92, 1995, 553–560.

46. N.S. Berke and M.P. Dallaire, 'The effect of low addition rates of polypropylene fibers on plastic shrinkage cracking and mechanical properties of concrete', in J.I. Daniel and S.P. Shah (eds) *Fiber Reinforced Concrete: Developments and Innovations*, SP-142, American Concrete Institute, Farmington Hills, MI, 1994, pp. 19–42.

47. P.A. Dahl, 'Influence of fibre reinforcement on plastic shrinkage cracking', in A.M. Brandt and I.H. Marshall (eds) *Brittle Matrix Composites – I*, Proc. European Mechanics Colloquium 204, Elsevier Applied Science Publishers, London and New York, 1986, pp. 435–441.

48. I. Padron and R.F. Zollo, 'Effect of synthetic fibres on volume stability and cracking of portland cement concrete and mortar', *ACI Mater. J.* 87, 1990, 327–332.

49. N. Banthia and M. Azzabi, 'Restrained shrinkage cracking in fiber reinforced cementitious composites', *Mater. Struct. (RILEM)*. 26, 1993, 405–413.

50. K. Kovler, J. Sikuler and A. Bentur, 'Restrained shrinkage tests of fibre reinforced concrete ring specimens: effect of core thermal expansion', *Mater. Struct. (RILEM)*. 26, 1992, 231–37.

51. K. Kovler, J. Sikuler and A. Bentur, 'Free and restrained shrinkage of fibre reinforced concrete with low polyprpylene fibre content at early age', in R.N. Swamy (ed.) *Fibre Reinforced Cement and Concrete*, Proc. RILEM,17, E&FN SPON, London and New York, 1992, 91–101.

52. K. Wang, S.P. Shah and P. Phuaksuk, 'Plastic shrinkage cracking in concrete materials – influence of fly ash and fibers', *ACI Mater. J.* 98, 2001, 458–464.

53. R.F. Zollo and J.A. Ilter, 'Plastic and drying shrinkage in concrete containing collated fibrillated polypropylene fibre', in R.N. Swamy, R.L. Wagstaffe and D.R. Oakley (eds) *Developments in Fibre Reinforced Cement and Concrete*, Proc. RILEM Symp., Sheffield, RILEM Technical Committee 49-FTR, 1986, Paper 4.5.

54. Z. Bayasi and M. McIntyre, 'Application of fibrillated polypropylene fibers for restraint of plastic shrinkage cracking in silica fume concrete', *ACI Mater. J.* 99, 2002, 337–344.

55. M.A. Sanjuan, C. Andrarde and A. Bentur, 'Effect of crack control in mortars containing polypropylene fibers on the corrosion of steel in a cementitious matrix', *ACI Mater. J.* 94, 1997, 134–141.

56. H. Krenchel and S.P. Shah, 'Restrained shrinkage tests with polypropylene fiber reinforced concrete', in S.P. Shah and G.B. Batson (eds) *Fibre Reinforced Concrete Properties and Applications*, SP-105, American Concrete Institute, Farmington Hills, MI, 1987, pp. 141–158.

57. R.N. Swamy and H. Stavrides, 'Influence of fibre reinforcement on restrained shrinkage', *J. Amer. Concr. Inst.* 76, 1979, 443–460.

58. R.F. Zollo, 'Collated fibrillated polypropylene fibers in FRC', in G.C. Hoff (ed.) *Fiber Reinforced Concrete*, SP-81, American Concrete Institute, Farmington Hills, MI, 1984, pp. 397–409.

59. V. Ramakrishnan, S. Gollapudi and R. Zellers, 'Performance characteristics and fatigue strength of polypropylene fiber reinforced concrete', in S.P. Shah and G.B. Batson (eds) *Fiber Reinforced Concrete: Properties and Applications*, SP-105, American Concrete Institute, Farmington Hills, MI, 1987, pp.159–177.

60. N. Banthia and A. Dubey, 'Measurement of flexural toughness of fiber-reinforced concrete using a novel technique - Part 1: assessment and calibration', *ACI Mater. J.* 96, 1999, 651–656.

61. N. Banthia and A. Dubey, 'Measurement of flexural toughness of fiber-reinforced concrete using a novel technique - Part 2: performance of various composites', *ACI Mater. J.* 97, 2000, 3–11.

62. A. Badr, K.E. Hassan, I.G. Richardson and J.G. Cabrera, in *Third International Conference on Advanced Composite Materials in Bridges and Structures*, Canadian Society for Civil Engineering, Montreal, 2000, pp. 69–76.

63. N.P. Banthia, S. Mindess and A. Bentur, 'Impact behaviour of concrete beams', *Mater. Struct. (RILEM).* 20, 1987, 293–302.

64. B. Barr and P.D. Newman, 'Toughness of polypropylene fibre reinforced concrete', *Composites.* 16, 1985, 48–53.

65. S. Mindess, N. Banthia and A. Bentur, 'The response of reinforced concrete beams with a fibre concrete matrix to impact loading', *Int. J. Cem. Comp. & Ltwt. Concr.* 8, 1986, 165–170.

66. S. Mindess, A. Bentur, C. Yan and G. Vondran, 'Impact resistance of concrete containing both conventional steel reinforcement and fibrillated polypropylene fibres', *ACI Journal* 86, 1989, 545–549.

67. A.A. Raouf, S.T.S. Al-Hassani and J.W. Simpson, 'Explosive testing of fibre reinforced cement composites', *Concrete.* 1976, 28–30.

68. A.P. Hibbert and D.J. Hannant, 'The design of an instrumented impact machine for fibre concretes', in R.N. Swamy (ed.) *Testing and Test Methods of Fibre Cement Composites*, Proc. RILEM Symp., The Construction Press, Lancaster, England, 1978, pp. 107–120.

69. Z. Bayasi and M.A. Dhareri, 'Effect of exposure to elevated temperature on polypropylene fiber-reinforced concrete', *ACI Mater. J.* 99, 2002, 22–26.

70. F. Dehn and K. Wille, 'Micro analytical investigations on the effect of polypropylene fibres in fire exposed high-performance concrete (HPC)', in M. di Prisco, R. Felicetti and G.A. Plizzari (eds) *Fibre-Reinforced Concretes BEFIB 2004*, RILEM Proceedings PRO 39, *RILEM Publications, Bagneux,* 2004, Vol. 1, pp. 659–668.

71. P. Kalifa, G. Chene and C. Galle, 'High-temperature behaviour of HPC with polypropylene fibres. From spalling to microstructure', *Cem. Concr. Res.* 31, 2001, 1487–1499.

72. F. Dehn and G. König, 'Fire resistance of different fibre reinforced high-performance concrete', in A.E. Naaman and H.W. Reinhardt (eds) *High Performance Fiber Reinforced Cement Composites (HPFRCC4)*, Proc. RILEM PRO 30, *RILEM Publications, Bagneux,* 2003, pp. 189–204.

73. R.V. Velasco, R.D. Toledo Filho, E.M.R. Fairbairn, P.R.L. Lima and R. Neumann, 'Spalling and stress–strain behaviour of polypropylene fibre reinforced HPC after exposure to high temperatures', in M. di Prisco, R. Felicetti and G.A. Plizzari (eds) *Fibre-Reinforced Concretes BEFIB 2004*, Proc. RILEM PRO 39, *RILEM Publications, Bagneux,* 2004, Vol. 1, pp. 699–708.

74. S.Y.N. Chan, X. Luo and W. Sun, 'Mechanical behavior of fibre reinforced high performance concrete after exposure to high temperatures', in P. Rossi and G. Chanvillard (eds) *Fibre Reinforced Concretes (FRC) BEFIB' 2000*, Proc. RILEM PRO 15, *RILEM Publications, Bagneux,* 2000, pp. 521–529.

75. L. Sarvaranta, 'Shrinkage of short PP and PAN fibers under hot-stage microscope', *J. Appl. Polym. Sci.* 56, 1995, 1085–1091.

76. Y.M. Mai, R. Andonian and B. Cotterell, 'Thermal degradation of polypropylene fibres in cement', *Int. J. Cem. Comp.* 4, 1980, 149–156.

77. D.J. Hannant, 'Durability of polypropylene fibers in Portland cement-based composites: eighteen years of data', *Cem. Concr. Res.* 28, 1998, 1809–1817.

78. A.-H.J. Al-Tayyib and M.M. Al-Zahrani, 'Corrosion of steel reinforcement in polypropylene fiber reinforced concrete structures', *ACI Mater. J.* 87, 1990, 108–113.

79. A.-H.J. Al-Tayyib and M.M. Al-Zahrani, 'Use of polypropylene fibers to enhance deterioration resistance of concrete surface skin subjected to cyclic wet/dry sea water exposure, *ACI Mater. J.* 87, 1990, 363–370.

80. Y. Wang, S. Backer and V.C. Li, 'An experimental study of the synthetic fibre reinforced cementitious composites', *J. Mater. Sci.* 22, 1987, 4281–4291.

81. A. Khajuria, K. Bohra and P. Balaguru, 'Long term durability of synthetic fibers in concrete', in V.M. Malhotra (ed.) *Durability of Concrete - 50 Years of Progress*, SP-126, American Concrete Institute, Farmington Hills, MI, 1991, pp. 851–868.

82. P. Balaguru and K. Slattum, 'Test methods for durability of polymeric fibers in concrete and UV light exposure', in D.J. Stevens, N. Banthia, V.S. Gopalratnam and P.C. Tatnall (eds) *Testing of Fiber Reinforced Concrete,* SP-155, American Concrete Institute, Farmington Hills, MI, 1995, pp. 115–136.

83. J. Bijen and E. Geurts, 'Sheet and pipes incorporating polymer film material in cement matrix', in *Fibrous Concrete*, Proc. Symp. on Fibrous Concrete, The Concrete Society, The Construction Press, Lancaster, England, 1980, pp. 194–202.

84. K. Kobayashi and R. Cho, 'Flexural behaviour of polyethylene fibre reinforced concrete', *Int. J. Cem. Comp. & Ltwt. Concr.* 3, 1981, 19–25.

85. H. Nakamura and H. Mihashi, 'Formulation of design criteria for HPFRCC', in H.W. Reinhardt and A.E. Naaman (eds) *High Performance Fiber Reinforced Cement Composites (HPFRCC)*, Proc. RILEM PRO 6, *RILEM Publications, Bagneux*, 1999, pp. 91–100.

86. H.C. Wu and V.C. Li, 'Fibre/cement interface tailoring with plasma treatment', *Cem. Concr. Compos.* 21, 1999, 205–212.

87. L. Yan, R.L. Pendleton and C.H.M. Jenkins, 'Interface morphologies in polyolefin fiber reinforced concrete composites', *Composites Part A.* 29A, 1998, 643–650.

88. V. Ramakrishnan, 'Application of a new high performance polyolefin fiber reinforced concrete in transportation structures', in *TCDC Workshop on Advances in High Performance Concrete Technology and its Applications*, Government of India, Structural Engineering Research Center and United Nations UNDP, Madras, India, 1997.

89. S. Mindess, N. Wang, L.D. Rich and D.R. Morgan, 'Impact resistance of polyolefin fibre reinforced precast units', *Cem. Concr. Compos.* 20, 1998, 387–392.

90. V. Ramakrishnan and C. Sivakumar, 'Performance of polyolefin fiber reinforced concrete under cyclic loading', in V.M. Malhotra, P. Helene, L.R. Prudencio and D.C.C. Dal Molin (eds) *High-Performance Concrete and Performance and Quality of Concrete Structures*. ACI SP-186, American Concrete Institute, Farmington Hills, MI, 1999, pp. 161–181.

91. D. Hull, *'An Introduction to Composite Materials*, Cambridge Solid State Science Series, Cambridge University Press, UK, 1981, 246pp.

92. D.J. Johnson, 'Microstructure of various carbon fibres', in Proc. 1st Int. Conf. on Carbon Fibres, Their Composites and Applications, Plastic Institute, London, 1971, pp. 52–56.

93. K. Nishioka, S. Yamakawa and K. Shirakawa, 'Properties and applications of carbon fibre reinforced cement composites', in R.N. Swamy, R.L. Wagstaffe and D.R. Oakley (eds) *Developments in Fibre Reinforced Cement and Concrete*, Proc. RILEM Symp., Sheffield, RILEM Technical Committee 49-FTR, 1986, Paper 2.2.

94. A. Briggs, 'Review: carbon fibre reinforced cement', *J. Mater. Sci.* 12, 1977, 384–404.

95. S. Akihama, T. Suenaga and T. Nakagawa, 'Properties and application of pitch based carbon fibre reinforced concrete', *Concr. Int. Design and Construction.* 10, 1988, 40–47.

96. J.A. Waller, 'Carbon fibre cement composites', *Civil Engineering and Public Works Review.* 67, 1972, 357–361.

97. A.D. Brown and H.H. Hufford, 'The structural properties of cement pases reinforced with carbon fibre/PVA prepreg', in R.N. Swamy, R.L. Wagstaffe and D.R. Oakley (eds) *Developments in Fibre Reinforced Cement and Concrete*, Proc. RILEM Symp., Sheffield, RILEM Technical Committee 49-FTR, 1986, Paper 2.1.

98. S. Akihama, T. Suenaga, T. Nakagawa and K. Suzuki, 'Influences of fibre strength and polymer impregnation on the mechanical properties of carbon fibre reinforced cement composites', in R.N. Swamy, R.L. Wagstaffe and D.R. Oakley (eds) *Developments in Fibre Reinforced Cement and Concrete*, Proc. RILEM Symp., Sheffield, RILEM Technical Committee 49-FTR, 1986, Paper 2.3.

99. Y. Ohama, M. Amano and M. Endo, 'Properties of carbon fiber reinforced cement with silica fume', *Concr. Int. Design and Construction.* 7 (3), 1985, 58–62.

100. N. Banthia, 'Pitch-based carbon fiber reinforced cement: structure, performance, applications and research needs', *Can. J. Civ. Eng.* 9, 1992, 88–91.

101. J. Sheng, *High Volume Fraction Microfiber Reinforced Cements: Concepts, Strength, Toughness and Durability.* PhD Thesis, Laval University, Quebec, Canada, 1996.

102. T. Ando, H. Sakai, K. Takahashi, T. Hoshiima, M. Awata and S. Oka, 'Fabrication and properties for a new carbon fibre reinforced cement product', in J.I. Daniel and S.P. Shah (eds) *Thin Section Fiber Reinforced Concrete and Ferrocement*, SP-124, American Concrete Institute, Farmington Hills, MI, 1990, pp. 39–60.

103. S. Furukawa, Y. Tsuji and M. Miyamoto, in *Review of the 41st General Meeting/Technical Session (CAJ Review 1987)*, Cement Association of Japan, Tokyo, 1987, pp. 336–339.

104. J. Aveston, R.A. Mercer and J.M. Sillwood, 'Fibre reinforced cements – scientific foundations for speculations', in Proc. Composites Standards Testing and Design, National Physical Laboratory Conference, UK, IPC Science and Technology Press, Guildford, 1974, pp. 93–103.

105. V.C. Li and K. Obla, 'Effect of fiber length variation on tensile properties of carbon-fiber cement composites', *Compos. Eng.* 4, 1994, 947–964.

106. V.C. Li and D.K. Mishra, 'Micromechanics of fiber effect on the uniaxial compressive strength of cementitious composites', in R.N. Swamy (ed.) *Fibre Reinforced Cement and Concrete*, Proc. RILEM, 17, E&FN SPON, London and New York, 1992, pp. 400–414.

107. S. Akihama, T. Suenaga and T. Banno, 'Mechanical properties of carbon fibre reinforced cement composites', *Int. J. Cem. Comp. & Ltwt. Concr.* 8, 1986, 21–34.

108. S. Delvasto, A.E. Naaman and J.L. Thorne, 'Effect of pressure after casting on high strength fibre reinforced mortar', *Int. J. Cem. Comp. & Ltwt. Concr.* 8, 1986, 181–190.

109. A. Briggs, D.H. Bowen and J. Kollek, Proc. 2nd Int. Carbon Fibre Conference, The Plastics Institute, London, 1974, Paper 17.

110. M.A. Ali, A.J. Majumdar and D.L. Rayment, 'Carbon fibre reinforcement of cement', *Cem. Concr. Res.* 2, 1972, 201–212.

111. A. Katz and A. Bentur, 'Mechanisms and processes leading to changes in time in properties of GFRC', *Advances in Cement Based Materials*. 3, 1996, 1–13.

112. P.W. Chen and D.D.L. Chung, 'Carbon-fiber-reinforced concrete as an intrinsically smart concrete for damage assessment during dynamic loading', *J. American Ceramic Society*. 78, 1995, 816–818.

113. Z.Q. Shi and D.D.L. Chung, 'Carbon fiber-reinforced concrete for traffic monitoring and weighing in motion', *Cem. Concr. Res.* 29, 1999, 445–449.

114. S. Wen and D.D.L. Chung, 'Carbon fiber-reinforced cement as a thermistor', *Cem. Concr. Res.* 29, 1999, 961–965.

115. S. Sarkar and M.B. Bailey, 'Structural properties of carbon fibre reinforced cement', in A. Neville (ed.) *Fibre Reinforced Cement and Concrete*, Proc. RILEM Conf., The Construction Press, Lancaster, England, 1975, pp. 361–371.

116. S. Akihama, T. Suenaga and T. Banno, 'The behaviour of carbon fibre reinforced cement composite in direct tension', *Int. J. Cem. Comp. & Ltwt. Concr.* 6, 1984, 159–168.

117. M.G. Dobb, D.J. Johnson and B.P. Saville, 'Structural aspects of high modulus aromatic polyamide fibres', *Philos. Trans. R. Soc. London, Ser. A.* 294, 1980, 483–485.

118. P.L. Walton and A.J. Majumdar, 'Properties of cement composites reinforced with Kevlar fibres', *J. Mater. Sci.* 13, 1978, 1075–1083.

119. S. Ohgishi, H. Ono and I. Tanahashi, 'Mechanical properties of cement mortar pastes reinforced with polyamide fibres', *Trans. Japan Concr. Inst.* 6, 1984, 309–315.

120. P. Konczalski and K. Piekarski, 'Tensile properties of portland cement reinforced with Kevlar fibres', *J. Reinf. Plast. Compos.* 1, 1982, 378–384.

121. A. Nanni, 'Properties of aramid fibre reinforced concrete and SIFCON', *ASCE J. Materials in Civil Engineering.* 2, 1992, 1–15.
122. P.L. Walton and A.J. Majumdar, 'Creep of Kevlar 49 fibre and a Kevlar 49-cement composite', *J Mater. Sci.* 18, 1983, 2939–2946.
123. H.J. Schürhoff and A. Gerritse, 'Aramid reinforced concrete (ARC), aramid fibres of the twaron type, for prestressing concrete', in R.N. Swamy, R.L. Wagstaffe and D.R. Oakley (eds) *Developments in Fibre Reinforced Cement and Concrete*, Proc. RILEM Symp., Sheffield, RILEM Technical Committee 49-FTR, 1986, Paper 2.6.
124. J.L. Daniel and E.D. Anderson, 'Acrylic fibre reinforced cement composites', in R.N. Swamy, R.L. Wagstaffe and D.R. Oakley (eds) *Developments in Fibre Reinforced Cement and Concrete*, Proc. RILEM Symp., Sheffield, RILEM Technical Committee 49-FTR, 1986, Paper 2.8.
125. H. Hähne, 'High tenacity acrylic fibres for composites', Paper presented at the International Man-made Fibres Congress, Austrian Chemical Institute, Dornbirn, Austria, September, 1986.
126. H. Hähne, S. Karl and J.D. Worner, 'Properties of polyacrylonitrile fibre reinforced concrete', in S.P. Shah and G.B. Batson (eds) *Fibre Reinforced Concrete Properties and Applications*, ACI SP-105, American Concrete Institute, Farmington Hills, MI, 1987, pp. 211–223.
127. I. Odler, 'Structure and mechanical properties of Portland cement – polyacrilnitril fibre composites', in S. Mindess and S.P. Shah (eds) *Bonding in Cementitious Composites*, Materials Research Society Symp. Proc., Vol. 114, Materials Research Society, Pittsburgh, PA, 1988, pp. 153–158.
128. D.M. Gale, 'Cement reinforcement with man made fibres', Paper presented at the International Man-Made Fibres Congress, The Austrian Chemical Society, Dornbirn, Austria, September, 1986.
129. J. Hikasa and T. Genba, 'Replacement for asbestos in reinforced cement products – "Kuralon" PVA fibres, properties, structure', Paper presented at the International Man – Made Fibres Congress, Austrian Chemical Institute, Dornbirn, Austria, September, 1986.
130. Z. Zhijiang and C.Y. Tian, 'High tenacity PVA fibres: a suitable alternative for asbestos', Paper presented at the International Man-Made Fibres Congress, Austrian Chemical Institute, Dornbirn, Austria, September, 1986.
131. G. Ji, M. Shi, X. Gao and G. Ye, 'High-strength polyvinyl alcohol fibre applied in cement and other building materials', *Chem. Fibers Int.* 49, 1999, 116–117.
132. T. Saito, T. Horikoshi and H. Hoshiro, 'Progress of PVA fiber reinforced cementitious composites', in A.E. Naaman and H.W. Reinhardt (eds) *High Performance Fiber Reinforced Cement Composites (HPFRCC 4)*, Proc. RILEM PRO 30, RILEM Publications, Bagneux, 2003, pp. 391–398.
133. L.R. Betterman, C. Ouyang and S.P. Shah, 'Fiber matrix interaction in microfiber-reinforced mortar', *Advanced Cement Based Materials.* 2, 1995, 53–61.
134. V.C. Li, S. Wang and C. Wu, Tensile strain-hardening behavior of polyvinyl alcohol engineered cementitious composites (PVA-ECC)', *ACI Mater. J.* 98, 2001, 483–492.
135. Y. Shao and S.P. Shah, 'Mechanical properties of PVA fiber reinforced cement composites by extrusion processing', *ACI Mater. J.* 94, 1997, 555–564.

Chapter 11

Natural fibres

11.1 Fibre classification and properties

The classification and properties of natural fibres for cement reinforcement have been reviewed in several publications [1–5]. Cook [1,2] has suggested four classes of fibres, based on their morphology: stem (or bast), leaf, surface and wood.

Stem or bast fibres are obtained from the stalks of plants, and are freed from the substances surrounding them by a process known as retting, which involves the combined action of bacteria and moisture. This process can be carried out in tanks of heated water, or by a more simple treatment in which the stalks are spread on the ground to allow the effects of bacteria, dew, sun and air to dissolve the material surrounding the fibres. The process is followed by drying, and textile fibres can then be obtained by spinning into a yarn. The best fibres are usually used in the form of full length bundles or strands. Jute and flax fibres are included in this category. Typical properties of jute are given in Table 11.1.

Leaf fibres are obtained from the leaves of plants by a process in which the leaf is crushed and scraped to remove the fibres, followed by drying. Leaf fibres are usually harder, stiffer and coarser in texture than those derived from the stem. In the leaf they provide strength and rigidity, and support water conducting vessels. They are surrounded by cellular tissue and gummy substances, which are separated from the fibres during the processing of the leaf. The most common fibres in this category are sisal, henequen and abaca. Sisal has been used for many years for reinforcing gypsum products. Characteristic properties are provided in Table 11.1.

Surface fibres are found as single cell fibres on the surface of stems, fruits and seeds of plants. Cotton and coir (coconut fibre) are included in this group. Coir (coconut) are the more commonly used fibres of the surface group for cement reinforcement. Coir is obtained from the husk surrounding the nut of the coconut plant. The extraction process includes water soaking the husks to soften them, followed by beating with spikes to separate the long, coarse fibres. The longer fibres are washed, cleaned and dried, and formed into hanks for further manufacture. Typical properties of coconut fibres are provided in Table 11.1.

Wood (cellulose) fibres are relatively short and inflexible, but are usually strong and perform better during long ageing in cement environment. Wood chips are

Table 11.1 Properties of natural fibres

Property	Jute [3]	Sisal [3]	Coconut [3]	Sugar cane [2]	Bagasse [3]
Tensile strength (MPa)	250–350	280–750	120–200	170–290	20
Modulus of elasticity (GPa)	26–32	13–26	19–26	15–29	1.7
Elongation at break (%)	1.5–1.9	3–5	10–25	—	—
Fibre diameter (mm)	0.1–0.2	—	0.1–0.4	0.2–0.4	0.24
Fibre length (mm)	1800–3000	—	50–350	50–300	—
Water absorption (%)	—	60–70	130–180	70–75	78.5

processed in various solutions and subjected to mechanical treatment to extract good quality cellulose fibres in the form of pulp. Cellulose-pulp fibres are then processed into thin sheet composites by various industrial processes, in which the fibre content can be as high as 10%, or even more. In this context the cellulose-pulp fibre may be considered for full or partial asbestos replacement. This topic will be dealt with separately in Section 11.3. Wood fibres derived from bamboo or sugar cane (bagasse) have been used for the production of low cost cement composites. In this case the processing involves crushing of the plant in a roller; with sugar cane some treatment is required to remove the sugar. The properties of sugar cane bagasse are provided in Table 11.1. There is particular interest in bamboo reinforcement, which can be used in the form of fibres after appropriate processing, or as reinforcing rods. This topic has been reviewed by Subrahmanyam [6].

When considering these different fibres, it must be noted that their microstructure is quite complex, and that fibres are composed of many fibre cells, as shown in Figure 11.1(a) for sisal [7,8] . In this particular case, the fibre length is 2–5 mm, its diameter is less than 0.2 mm, and in its cross section there are about 100 cells [9]. The structure of an individual wood cell is presented schematically in Figure 11.1(b), showing the layered fibrils which make up the fibre wall. The orientation of the fibrils in each layer of the wall is different. They are composed mainly of long, oriented cellulose molecules, with a degree of polymerization of ~25,000. The cellulose molecule chains are grouped in long, oriented microfibril units having a thickness of about 0.7 mm and a length of a few μm. The other components of the cell wall, hemi-cellulose and lignin, have much lower degrees of polymerization and are located mainly in the middle lamellae which connects the individual fibre cells together. In the pulp type of processing, in which natural

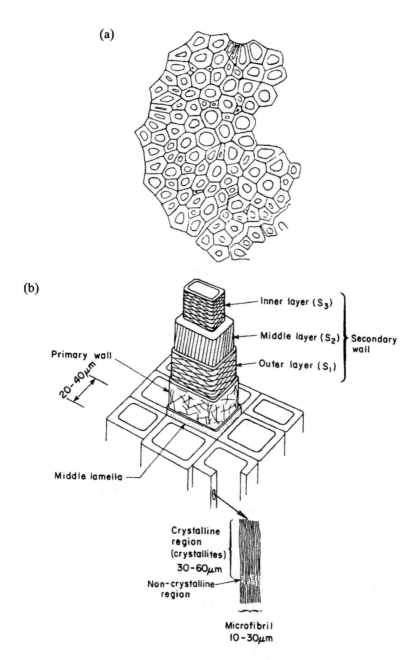

(a)

(b)

Inner layer (S₃)

Middle layer (S₂)

Primary wall

20-40μm

Outer layer (S₁)

Secondary wall

Middle lamella

Crystalline region (crystallites) 30-60μm

Non-crystalline region

Microfibril 10-30μm

Figure 11.1 (a) Cross-section through sisal fibres (after Nutman [7]); (b) structure of the cellulose fibre cells (after Illston et al. [8]).

fibres are used for asbestos replacement, the processing involves breaking many of the fibres into the individual fibre cells. However, in the processing of natural fibres for low cost housing, the fibre retains the cross section shown in Figure 11.1(a). This is a significant difference from both, the mechanical and chemical points of view, since retaining the structure of Figure 11.1(a) implies that most of the hemi-cellulose and lignin is preserved in the reinforcement.

Cook [1,2] has pointed out that in view of these structural characteristics, the properties of the fibre material itself (i.e. the cell walls) can be different by about 70–100% from the properties of the fibre, because the latter consists of an assembly of fibre cells, grouped together in a variety of microstructures. The strength and other properties of the fibre material in the individual cell can be higher than those of the actual fibre, and this often leads to confusion in the data reported in the literature. The differences in the values reported in the two sources which are compared in Table 11.1 for bagasse fibres, probably reflect such influences, as well as perhaps variability in the fibre properties.

The stress–strain curves of the fibres show an ultimate strain in the range of 1–5%, which is much greater than that of the matrix. The curve may be linear up to failure, or it may show non-linear behaviour, depending on the type of fibre. The moisture conditions also have a considerable effect on the mechanical properties. Dry fibres usually exhibit a linear stress–strain curve and a brittle failure, whereas wet fibres may show an increase in ultimate strain and a reduction in modulus of elasticity [10,11]. Stress–strain curves of cellulose fibres conditioned at different relative humidities are shown in Figure 11.2.

11.2   Natural fibres for low cost cementitious composites

Much of the research on the use of natural FRC materials has been motivated by the ready availability of such fibres, which are high in strength. Combined with the simple production processes for making cementitious composites of various shapes, they are potentially suitable for low cost housing applications. This topic has been reviewed in several references [3–5].

In these applications the fibre content is usually less than 5% when applying mixing technologies, but it may be greater when using the technologies of hand lay up of long fibre rovings. The long fibres can be obtained by tying together [12] or by spinning of twines. Hand laying involves the application of a thin mortar layer on a mould, followed by alternate layers of fibres and mortar matrix. The fibres can be rolled into the matrix or worked into it manually, and the process may involve some vibration [2,3]. In the mixing technique, there is a limit to the content and length of fibres that can be incorporated, since as with any other fibres workability is reduced. However, many of the natural FRC composites are intended for the production of thin components such as corrugated sheets and shingles, and for these applications there is a requirement for both plasticity and green strength, that will permit the shaping of the product; for such purposes reduced flow properties are

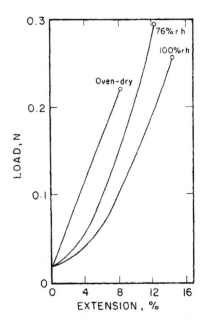

Figure 11.2 The effect of moisture on the stress–strain curves of natural organic fibres (after Mai and Hakeem [10]).

not as detrimental as in the case of conventional concrete. In these components, which have a typical thickness of about 10 mm, the matrix is a cement mortar, and the mix with fibres, or with hand laid fibres, is spread on a mould surface and then shaped. Corrugation can be achieved by pressing between two corrugated sheets [13]. Folded plate roofing panels can also be made [2], and in such processes pressure casting has been found to be useful [2]. Many types of fibres are used, or have been considered for use, in such applications, and they include coconut, sisal, sugar cane bagasse, bamboo, palm, jute, flax, wood chips, akwara, elephant grass, water reed, plantain and musamba. It is beyond the scope of this section to deal with the materials made with each of these fibres. Instead, only the properties and characteristics of a few systems which are representative of this class of composites will be discussed.

In the mixing technology, the increase in fibre content and length is associated with reduced workability. However, more fibres can be incorporated in such systems than in other FRC composites produced by mixing, since the mortar matrix is usually made up of smaller aggregates in order to be suitable for thin sheet components. Also, for components which can be shaped by laying down and tamping, the flow properties need not be as high as those required for mixes which are to be poured into moulds. Thus, most of the work reported for premix natural

FRC composites produced by mixing deals with fibre contents in the range of 1–5% by volume. Many of the reported studies indicate that mixing was carried out manually since this made it easier to overcome the difficulties in mechanical mixing associated with the harsh mix and balling of the fibres [14].

The fibres tend to increase the tensile (flexural) strength and toughness of the hardened composite, the extent of the improvement being dependent on the fibre content and length, as can be seen in Figure 11.3 for jute fibres [14,15]. The strengthening effect of the fibres is evident, as is the existence of both optimum fibre contents and optimum lengths at which maximum improvement is achieved. The decrease in the properties at high fibre volumes and lengths was also observed in other types of reinforcement (e.g. steel, glass) and can be explained on the basis of inefficient compaction and lower density of the mix, which at these fibre volumes is harsher and less workable. Paramasivam et al. [13] reported on a similar study with coconut fibres, in which corrugated sheets were produced by a technique involving the application of 152.1 kN/m² pressure to the green material. Here, too, an optimum fibre content and optimum length was found, yielding a maximum flexural strength of about 22 MPa at a fibre content of 3%, and length of 25 mm.

The significance of casting pressure was evaluated in an attempt to improve the fibre–matrix bond in coir fibre composites [16]. The high initial pressure may offset the reduction in bond which can be caused by the swelling and shrinking of the fibres due to cycles of wetting and drying. The effect of pressure on the flexural strength is shown in Figure 11.4. The optimum casting pressure of 3.1 N/mm² was associated with almost a doubling of the flexural strength. In additional tests at this pressure, the effect of fibre content and length was also similar to that described previously, and the maximum strength was achieved at an optimum length value of ∼37 mm and optimum fibre volume content of ∼4%.

The addition of natural fibres does not always enhance the properties to the same extent reported for jute fibres [15]. With 5% of agava fibres [14], only the toughness was increased, but not the flexural strength (fibre volume of 5%, and fibre length in the range of 50–75 mm). Only when the fibre content was increased above 7% was a flexural strength increase obtained (Figure 11.5). Castro and Naaman [14] found that 75 mm long fibres with a volume content as high as 11% could be added to the mortar mix, to yield the improved performance shown in Figure 11.5, but for that purpose the mortar mix composition had to be adjusted by increasing the w/c ratio or by adding a superplasticizer.

Mwamila [12,17] produced continuous twines of sisal fibres by the spinning together of single fibres. The twine is composed of several strands, each made of a bundle of individual fibres. The bond between the fibres is facilitated by friction and by the spiral form obtained during spinning. In the composites produced by Mwamila the twines were hand laid only at the bottom of a beam, and in that sense his components are not truly a composite material with uniformly distributed fibres. Reinforcement with 2% of twines enhanced the flexural strength

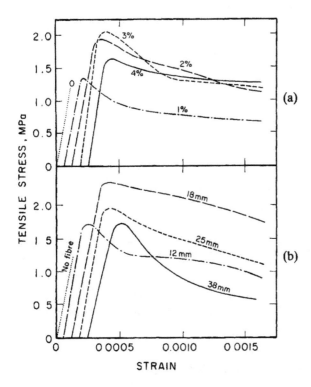

Figure 11.3 The effect of jute fibre length and content on the stress–strain curve in tension of fibre-reinforced cement paste (after Mansur and Aziz [15]). (a) $l = 25$ mm; (b) $V_f = 2\%$.

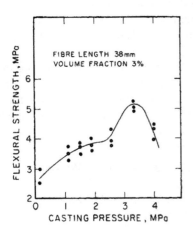

Figure 11.4 Effect of casting pressure on the flexural strength of coir fibre-reinforced cement (after Das Gupta et al. [16]).

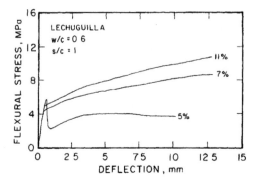

Figure 11.5 The effect of agava (lechuguilla) fibre content on the flexural stress–deflection curve of FRC composites (after Castro and Naaman [14]).

and toughness, but this was accompanied by poor cracking behaviour, in which wide cracks developed. This characteristic could be improved by supplementary reinforcement with short sisal fibres.

11.3 Reinforcement with cellulose-pulp

Single cell cellulose fibres (Figure 11.1(b)) can be derived from wood by the pulping process. The resulting fibres have aspect ratios of about 50, with diameters in the range of 20–60 μm for hardwood, and 30–120 μm for softwood. The lengths are in the range of 0.5–3.0 mm in hardwood and 2.0–4.5 mm in softwood [18]. They are thus different from the natural fibres discussed in Section 11.2, in which each fibre is made up of an assembly of such cells. They will be referred to as *cellulose-pulp*.

The properties of the cellulose-pulp fibres removed from wood depend upon the pulping process, which can be based on chemical treatment, mechanical treatment, or a combination of the two. As shown in Figure 11.1(b), the fibre is cylindrical and hollow, and its properties can be affected by the degree of collapse and its wall thickness. The two extremes of pulping processes are:

1 the fully chemical (kraft) process which has a low yield (~45%) and results in a collapsed, ribbon-like fibre and
2 the high temperature thermomechanical processes (TMP) which give high yield (~90%) and an uncollapsed rod-like fibre.

The TMP process is carried out at temperatures greater than the transition temperature of the lignin binder, and the processed fibre is lignin coated. In the kraft

Table 11.2 Properties of cellulose-pulp and asbestos fibres (after Campbell and Coutts [18])

	Specific gravity	Tensile strength (MPa)	Specific strength (MPa)
Cellulose, kraft (*P. radiata*)[a,b]	1.5	500	333
Cellulose, TMP (*P. radiata*)[b,c]	0.5	125	250
Asbestos	2.6	700	269

Notes
a Collapsed fibre.
b Based on 'average fibre' with external diameter of 38 μm internal diameter of the hollow core (lumen) of 33 μm.
c The uncollapsed fibre-cell wall substance strength is be the same as for the delignified fibre.

process, the lignin, which has an adverse effect on the strength and colour of the cellulose fibre, is removed and the end product is largely delignified. Different grades of kraft pulp can be made, depending on the end use, with the unbleached grades containing more lignin. Fibres of intermediate characteristics can be produced by various refining processes to control the yield as well as the extent of collapse and delignification. An additional source of cellulose fibres is recycled materials from kraft paper and newspaper [19]. Kraft paper is made of softwood while newspaper fibres are a mixture of softwood and hardwood, with a higher content of the latter. The newspaper fibres are subjected to more extensive processing of mechanical beating which may create fine microfibrils.

The strengths and densities of cellulose-pulp fibres produced by the two extreme pulping methods are presented in Table 11.2 where they are compared with asbestos. Although the strength and other properties of the cellulose-pulp fibre are inferior to those of many other fibres, such as asbestos, they are highly cost effective [18,20]. This, combined with their compatibility with processes for producing asbestos cement, such as the Hatschek process, made the cellulose-pulp fibres an attractive alternative to asbestos reinforcement. As a result of intensive research and development, cellulose-pulp fibres are now used in many places as full or partial replacement for asbestos in cement composites [18,20–22]. In the partial replacements, the cellulose functions as a processing fibre, with the other fibre, which is usually stiffer and stronger, providing the reinforcing effect in the hardened composite. In this section, composites with cellulose-pulp reinforcement only will be considered. Here, the fibres serve both functions, processing and reinforcing. The evaluation of the reinforcing effects is usually determined by testing composites made in a full-scale Hatschek process or pilot plant, or specimens prepared in the laboratory by a simulation process involving preparation of a slurry followed by vacuum-dewatering in combination with pressing.

11.3.1 Mechanical properties

The mechanical properties of cellulose fibres are all inferior to those of asbestos, and this will result in a composite of lower strength and stiffness, as can be seen from Figure 11.6. However, the strength values obtained with cellulose fibres are sufficiently high that the fibres have been adopted for various applications, and they have the advantage of increasing the strain capacity. The approach to be taken with fibre replacements such as cellulose is that the properties of asbestos–cement composites will not be duplicated. However, if the properties of these composites are properly exploited, and their advantages such as higher strain capacity are taken into account in the design, they may do equally well, or even better, in performing the functions of the more traditional asbestos composites [22,23].

In considering the applicability of cellulose fibres, as well as other natural fibres, to cement reinforcement, attention must be given to the potentially deleterious effects of certain fibre species. The hydration and setting of the cementitious matrix may be retarded by various organic compounds, sugar in particular, which are present in natural fibres. From this point of view, fibres produced by alkaline chemical pulping are superior; with fibres made by mechanical pulping, the hydration and setting of the matrix adjacent to the fibres may be adversely affected.

In using cellulose-pulp fibres, attempts have been made to optimize the properties of the composites by changing the fibre content and their quality, through modifications in the pulping process as well as by surface treatments to enhance fibre–matrix interactions. The effects of fibre content on strength show a maximum in the range of 8–12% by mass (Figure 11.7(a)). This has been observed by various investigators [18,20,24,25] and can be explained as with asbestos fibres: the increase in strength due to an increase in fibre content is offset by the increase

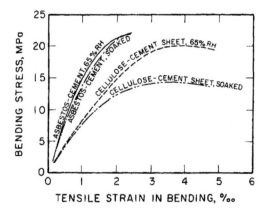

Figure 11.6 Typical stress–strain curves in flexure of composites produced by the Magnani process, comparing the behaviour of asbestos and cellulose FRC composites (after Fordos and Tram [20]).

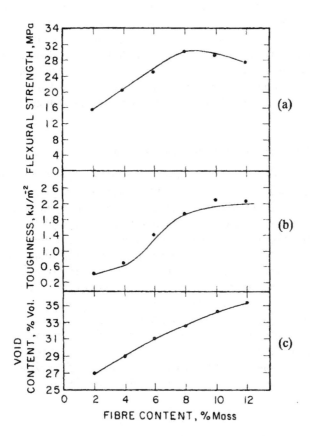

Figure 11.7 Effect of cellulose-pulp fibre content on the flexural strength, toughness and void content of air cured composites (after Coutts and Warden [24]).

in void content (Figure 11.7(c)); beyond the optimum fibre content, the detrimental effect due to the porosity increase is greater than the reinforcing influence of the added fibre. However, when the toughness is considered, the reduction at higher fibre contents is less evident (Figure 11.7(b)), similar to the observation for asbestos cement. It should be noted that the maximum observed in the flexural strength vs. fibre content curve for cellulose-pulp composites is not as sharp as that reported for asbestos cement (Figure 9.7). This may be the result of the effect of the fibre itself or differences in the method of composite preparation in the various studies.

Cellulose-pulp FRC composites with flexural strengths in excess of 20 MPa can be readily produced [18,20,24–27], with about 10% fibres by mass, using processes applied in the asbestos–cement industry. From the point of view of flexural strength, this is a sufficiently high value for the use of these composites as substitutes for asbestos cement.

The strength and other mechanical properties are dependent on the nature of the fibre processing and the moisture content during testing. The effect of the latter parameter is clearly demonstrated in Figure 11.6 and is associated with the hygroscopic nature of the fibres. The influence of moisture is exhibited in strength as well as volume stability; these will be considered separately in Section 11.4.1.

11.3.2   Effect of fibre type and fibre processing

Campbell and Coutts [18] have shown that considerable differences in the strength of the cementitious composite are obtained with fibres processed by the chemical (kraft) and thermomechanical (TMP) pulping processes (Figure 11.8). In the kraft process, the more efficient removal of the hemicellulose and lignin results in a cell wall which is less stiff, and the fibre collapses into a ribbon-like structure in the composite [28]. In semi-chemical and mechanical treatments, where more lignin remains in the fibre it retains its cylindrical structure and the lumen (core cavity) remains open [28]. The more compact nature of the collapsed kraft fibre may be the cause for its better reinforcing effect. Davies *et al.* [28] demonstrated that with chemically treated cellulose-pulp the mode of failure was governed to a greater extent by fibre fracture, rather than fibre pull-out with fibres which were not treated chemically. They suggested that the greater amount of pull-out in the latter case was associated with poor bond, due to the detrimental influence of the extractive which affected the hydration of the matrix in the vicinity of the fibre.

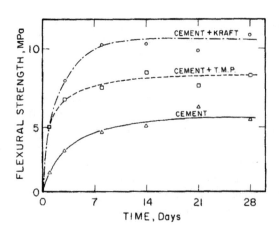

Figure 11.8 Effect of the type of pulping process (kraft vs. TMP) on the development of flexural strength in cellulose-pulp FRC composites (after Campbell and Coutts [18]).

Table 11.3 Effect of fibre type on the properties of autoclaved cellulose-pulp FRC composite, prepared with 8% fibres by volume and tested at 22°C/50% RH (after Coutts [26,29])

Reinforcing fibre	Aspect ratio	Flexural strength (MPa)	Impact toughness (kJ/m^2)	Density (kg/m^3)
Pinus radiata (soft wood)	~75	22.6	1.90	1320
Eucalyptus regnans (hard wood)	~50–60	21.7	1.05	1270
Phromium tenax (flax) plant	~190	23.2	0.84	1320

The comparisons between different types of fibres that were treated similarly indicate that the composites produced may have a similar flexural strength, but may be quite different in toughness (Table 11.3) [26,29]. These differences cannot be explained on the basis of density (which is similar in all the specimens), or the geometry (aspect ratio) of the fibre, which is highest for the fibre which led to the composite of lowest toughness (Table 11.3). Similar trends were reported by Coutts [30] for softwood and hardwood fibres prepared with room temperature cured matrix. There, the better performance of the softwood fibres was attributed to the higher fibre aspect ratio.

The effect of refinement of the cellulose-pulp by additional mechanical treatment (beating) was studied by Coutts *et al.* [23,31–33]. The need for refinement stemmed from an attempt to improve the processing properties of the fibre in the Hatschek process, to increase its ability to retain particles. The mechanical effects during the refining treatment led to shortening of the fibres, as well as to external and internal fibrillation [32]. The fibrillation of the refined fibres enabled them to form a web capable both of retaining the particulate matrix to wrap around cement grains, and maintaining a sufficient drainage rate for processing in the Hatschek machine. The refinement was also found to affect the mechanical properties of the hardened composite. There was an optimum refinement level for flexural strength. Extensive refinement results in peeling of the external fibrils to form fines, as well as shortening of the fibres. However, impact toughness decreased continuously as the degree of refinement* increased (Table 11.4). The reduction in toughness with a higher degree of refinement may result from the reduction in fibre length, which reduces the pull-out energy involved in fracture. It should be noted that the effects of the refinement of the fibres (as shown in Table 11.4 and discussed here), are not general, and depend on the type of cellulose-pulp fibre. Coutts [26] found that with New Zealand flax (*Pharmium tenax*) fibres, the properties of the composites were approximately the same, regardless of the degree of refinement

* The degree of refinement was estimated by the Canadian Standard Freeness Tester (CSF), which measures the ease with which water drains away from a mat formed from the fibres.

Table 11.4 Effect of refinement on the mechanical properties of cellulose-pulp FRC composite (after Coutts and Ridikas [31])

Freeness CSF (ml)	Flexural strength (MPa)	Impact toughness (kJ/m²)	Density (kg/m³)
730	15.0	1.90	1320
680	19.9	1.25	1320
565	18.4	0.95	1320
230	17.9	0.75	1340

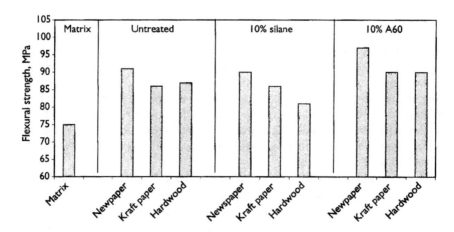

Figure 11.9 Effect of fibre type and treatment on the flexural strength of cellulose fibre–cement composites (after Lin et al. [33]).

(beating). The trend observed with the *Pinus Radiata* fibres in Table 11.4 was not reproduced with the flax.

Comparison between the flexural strength of composites with different cellulose fibres was presented by Lin *et al.* [34], showing that newspaper fibres provided the highest flexural strength (Figure 11.9). The enhanced strength was attributed to the presence of fibrils in the newspaper fibres, induced by their more intensive processing. These microfibrils are more entangled and enable to induce mechanical enhanced mechanical bonding and interlocking with growing hydration products [19,34].

11.3.3 Surface treatments

The effect of surface treatment of the cellulose-pulp fibres with various coupling agents to enhance the fibre–matrix interaction was reported in several studies [34–40]. Coutts and Campbell [35] and Blankenhorn *et al.* [36–38] evaluated a range of silanes, alkoxides of titanium, acrylic emulsions as well as sodium and potassium silicates. The treatments could provide an increase in flexural strength and toughness over untreated fibres, but the changes were not very large and sometimes could lead to a reduction in properties (e.g. Figure 11.9).

The effect of bleaching of kraft cellulose fibres was also studied by Mai *et al.* [39,40]. The need for bleaching results from the concern that the small amount of lignin retained after the kraft pulping can inhibit the curing during autoclaving of composites with a cement/silica matrix. Bleaching led to an increase in flexural strength and modulus of elasticity, and reduced the toughness (Table 11.5). This was associated with a different failure mode, in which the bleached fibre composite exhibited a more unstable and brittle behaviour with a sharper load drop after the maximum load was reached during a flexural test (Figure 11.10). Differences of this kind are probably the result of the greater stiffness and reduced toughness of the bleached fibres. An alternative explanation may be based on bond considerations. Bleaching tends to reduce the content of lignin and other chemicals which may interfere with the bond, in particularly in autoclaved products [33]. These chemicals may have a retarding effect on the matrix in the vicinity of the fibres.

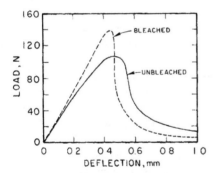

Figure 11.10 Typical load–deflection curve in bleached and unbleached cellulose-pulp autoclaved FRC composites (after Mai *et al.* [39]).

Table 11.5 Effect of bleaching on the mechanical properties of cellulose-pulp autoclaved FRC cement (after Mai *et al.* [39])

Property	Unbleached	Bleached
Modulus of elasticity (GPa)[a]	7.97	10.38
Flexural strength (MPa)[a]	19.23	22.4
Fracture toughness (kJ/m^2)[a]	0.48	0.34

Note
a Tested oven dry; the same trends, but with different values observed when testing immersed in water.

The improved bonding due to bleaching is expected to result in higher strength and stiffness, but lower toughness.

11.3.4   Effect of matrix composition and processing

The matrix composition, in particular room temperature vs. autoclave curing, may lead to differences in the properties of the composite. Contradictory trends have been reported. Coutts [24,25] reported that room temperature curing resulted in higher flexural strength but did not greatly affect the toughness (Table 11.6). Akers and Studinka [41], on the other hand, found that the autoclaved composite had a much higher flexural strength, but was significantly lower in toughness. The trends in Coutts' results may be related to the lower density of the autoclaved composite. Microstructural analysis of Akers and Studinka's composites [42,43] suggested that the transition zone around the fibres in the room temperature cured composite was much more open than that of the autoclaved specimen, and this was accompanied by a greater amount of fibre pull-out in the room temperature cured composite, but a greater proportion of brittle fibre failure in the autoclaved composite (Figure 11.11). A denser matrix in the fibre–matrix transition zone can lead to higher bond, resulting in higher strength, lower toughness and a greater extent of fibre failure by fracture rather than by pull-out.

Thus, within each set of data (Coutts and Akers), there is an internal consistency with microstructural characteristics, which may account for the effect of autoclaving. However, the differences between the two sets of data may suggest that the influences of additional parameters, in addition to curing, should be resolved. This may be related to the mode of specimen preparation; in Coutts' case, they were prepared by a special laboratory technique, while the composites produced by Akers were made in a mini-Hatschek. Such effects may also account for the differences in the microstructural characteristics at the interface observed by Coutts [44] and Bentur and Akers [42,43]: Coutts reported that in both the autoclaved and room temperature cured composites, the matrix in the transition zone was dense, similar

Table 11.6 Effect of curing on the properties of cellulose-pulp FRC composites

	After Coutts [24,25]		After Akers and Studinka [41]	
	Room temperature	Autoclave	Room temperature	Autoclave
Flexural strength (MPa)	29.2	21.3	18.3	23.0
Toughness	2.28[a]	1.92[a]	2.8[b]	0.07[b]
Density (kg/m^3)	1470	1300	1770	1610

Notes
a Energy (kJ/m^2).
b Ultimate strain (%).

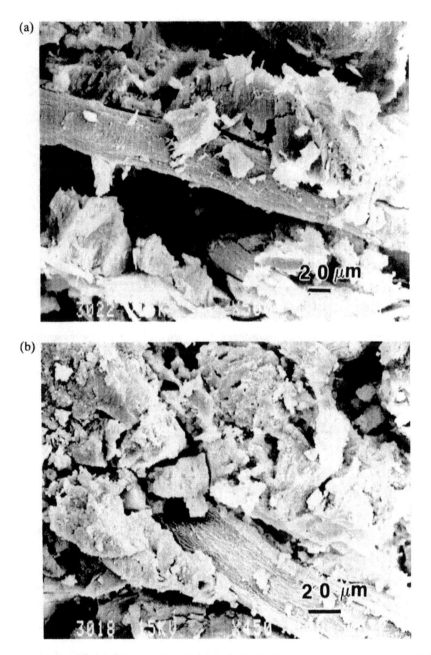

Figure 11.11 (a) Open matrix along a fibre, (b) typical pulled-out fibre, in an unaged composite cured in a normal environment [42].

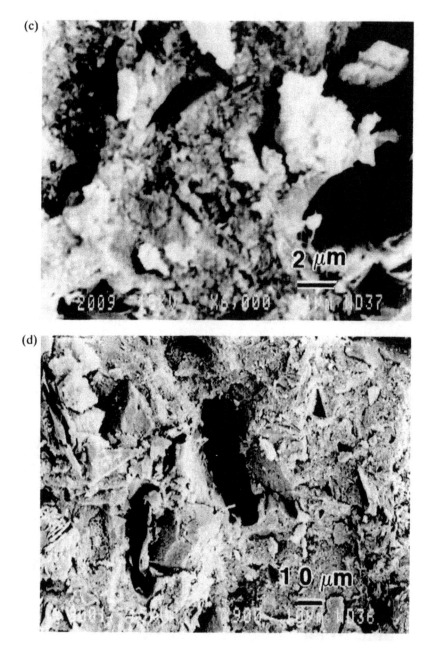

Figure 11.11 Continued. (c) dense matrix and the intimate contact forms with a fibre [43] and (d) brittle fibre failure, in an unaged autoclaved composite.

to that described in Figure 11.11(b) for an autoclaved composite; Bentur and Akers [42,43], however, found a considerable difference in the transition zone obtained in the two types of curing, which was discussed earlier and shown in Figure 11.11. It should be noted, that in both studies, no large deposits of CH were seen in the vicinity of the cellulose. This is similar to the observations of the interfaces in asbestos–cement produced by similar methods.

Milestone and Suckling [45] developed a novel method to determine the effect of autoclaving on the cellulose fibres and concluded that degradation of these fibres occurs in the precuring stage when the cement sets and hardens, before autoclaving. This shows up in the degree of polymerization of the cellulose which was 5850 before the incorporation in the matrix, and reduced to 4340 during the precuring stage. In the autoclave, an additional small decline to 3670 was reported. During autoclaving there is movement of species between the fibres and the matrix, with

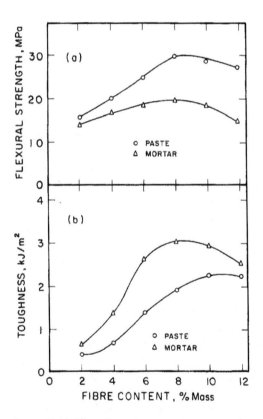

Figure 11.12 The effect of matrix composition (paste vs. mortar) in room temperature cured cellulose-pulp composites on the (a) flexural strength and (b) toughness (after Coutts [46]).

Ca^{++} ions moving into the fibre layer and organic material into the surrounding matrix.

Coutts [46] also studied the effects of the matrix composition on room temperature cured composites, comparing a Portland cement matrix with a mortar matrix. The paste matrix gave a stronger composite (Figure 11.12(a)) but with a lower toughness (Figure 11.12(b)). These differences can be attributed to a lower bond in the mortar composite, due to less effective interfacial contact between the fibre and matrix. This may lead to more fibre failure by pull-out in the mortar, thus increasing the toughness but reducing the strength.

Silica fume was shown to enhance the flexural strength of cellulose fibre reinforced cement, with an increase from 15.2 MPa to 20.9 MPa by the addition of 6% silica fume [19]. Moulding pressure and high shear mixing were also reported to have a marked positive influence on flexural strength, enhancing it by 50% and more [19]. The beneficial effect of pozzolans and silica fume were also reported by Soroushian and Marikunte [47], but the influences were smaller compared with those of Li *et al.* [19].

11.4   Mechanics of natural fibre reinforced cementitious composites

11.4.1   *Effect of moisture content*

Natural fibres are very sensitive to changes in moisture content, which affects both the mechanical properties of the fibre and its dimensions. On wetting, the natural fibre loses stiffness and gains ductility (Figure 11.2). The swelling and shrinking of the fibres during wetting and drying, with strains greater than those of the matrix, may lead to changes in the contact pressure across the interface, thus causing variations in the actual bond. As a result, the overall properties of the composite are very sensitive to moisture content, to a greater extent than other fibre reinforced cementitious systems, in which the fibres are not hygroscopic. Much of the work on the effect of moisture content was carried out with cellulose-pulp fibres but similar effects would be expected with other natural fibres. Generally [25,26,39], as can be seen from the results of the bending tests in Figure 11.6, an increase in moisture content is associated with reductions in modulus of elasticity and strength, but an increase in toughness. These changes are much greater than those occurring with asbestos fibres (Figure 11.6), and therefore most of the moisture sensitivity of cellulose-pulp composites can be attributed to the hygroscopic nature of the fibres. This effect is much greater than that induced by the hygroscopic nature of the cement matrix. Such trends occur over the entire fibre content range, but their magnitude depends upon the moisture content (Figure 11.13). Coutts *et al.* [25,31] pointed out the need to define and standardize the moisture conditions during testing. As can be seen from Figures 11.6 and 11.13, the influence of moisture on the test results is much greater than with other FRC systems.

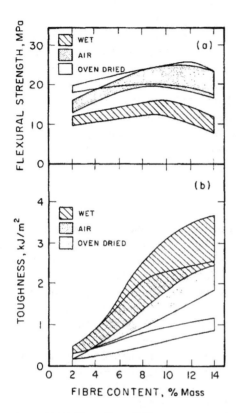

Figure 11.13 The effect of fibre content and moisture preconditioning on (a) the flexural strength and (b) toughness of cellulose-pulp FRC composites (after Coutts [25]).

The changes in mechanical properties due to moisture content are accompanied by changes in the mode of failure, with fibre pull-out seen more frequently in wet composites. The failure in the dry composites is dominated by fibre fracture [39,44] with only a small extent of pull-out. The change in mechanical properties and fracture mode with moisture can be explained in terms of changes in fibre properties and the fibre–matrix bond; at lower moisture contents the fibre is stiffer and more brittle, and the bond is stronger [39,44]. According to composite materials theory, these two effects should lead to higher strength and lower toughness, as observed in the dried composite. Fibre failure in the wet composite is accompanied by a reduction in cross-sectional area, and twisting and unravelling of the fibres [39] all of which are associated with the ductile behaviour of the wet fibres. In the dry composite there was no significant reduction in the cross-sectional area of the broken fibre [39], and the outer layer of the fibres was stripped away rather than debonded at the interface [32], indicating a strong bond. The strong bond in the dry state may be the result of hydrogen bridges [48].

Coutts [25] noted that under ambient conditions the composite maintains most of the higher strength, typical of an oven dried material but is not as brittle as in the oven-dry state (Figure 11.13). He therefore suggested that a number of micromechanisms may be effective; although wetting results in a loss in hydrogen bonding, the fibre swells, and this should lead to frictional resistance. Mai *et al.* [39] suggested that the outcome of these two opposing effects with respect to bond is unclear. The influence of moisture content on the mechanical properties may be mainly the result of its effect on the properties of the fibre, and not necessarily on the fibre–matrix bond.

11.4.2   Interfaces and bond

The discussion in Section 11.4.1 indicated the significance of the fibre failure mode (pull-out vs. fracture) in determining the performance of the composite, and its dependency on the fibre–matrix interaction. There is, however, some controversy regarding the dominant mode of the cellulose-pulp fibre failure [10,27,39,41,48]. Part of this problem can be resolved by considering the effect of moisture content. However, one should be cautious in trying to establish general trends which suggest that one or the other mode is the dominant one under certain conditions. Another parameter which plays an important role is the microstructure of the matrix in the vicinity of the fibres, which depends on the production technology and curing, and which may vary in its porosity. As may be seen from Figure 11.11, a more open microstructure may lead to a failure by pull-out, while a denser one results in fibre fracture. Such differences are also exhibited in the mechanical properties of the composites. Thus, the transformation of failure mode from fibre pull-out to fibre fracture occurs with age [28] due to the densification of the matrix around the fibres in a 90-day old composite.

When a dense microstructure is developed at the interface, considerable bond strength can be generated, and it has been estimated by pull-out tests that the shear stress at the onset of debonding can be of the order of 10 MPa [48]. However, indirect evaluation of the bond strength by regression analysis of strength vs. fibre content curves (using composite materials equations) has led to bond values in the range of 0.35–0.45 MPa [27]. There is not necessarily a contradiction here; as indicated in the discussion in Chapter 3, the lower value represents an average frictional bond, while the shear stress at the onset of debonding is a maximum value at the tip of a stress concentration profile. Although such values are smaller than those reported for steel (Table 3.2) they are sufficiently high to develop useful reinforcing effects, as evidenced by the flexural strengths of the cellulose-pulp composites, which can exceed 20 MPa. Morrissey *et al.* [49] have estimated, on the basis of pull-out tests of sisal slivers, that the critical length is 30 mm, which corresponds to a critical aspect ratio of 110 ± 50. Similar critical length values, in the range of 20–55 mm, were reported by Cook [2] for natural fibres. This is in the range of fibre lengths used for cement reinforcement, thus indicating that the geometry and bond strength are in the range which can provide useful reinforcement.

11.4.3   Modelling of mechanical behaviour

Andonian *et al.* [27] used the composite materials approach, in the form of the rule of mixtures, to account for the strength properties of the composite. They used the same concepts as those described for asbestos–cement composites (Eqs 9.8 and 9.9, Chapter 9), namely, that strength can be calculated as the sum of the effects of the matrix and the fibres; the fibre contribution is governed by pull-out and is a function of $\tau \ell/d$, while the matrix contribution is a function of the strength of a void-free matrix, σ_{mo}, multiplied by its solid content, $(1 - V_0)$. Therefore,
   For tensile strength:

$$\sigma_{cu} = \sigma_{mu}(1 - V_0)V_m + 2\eta\,\tau\,V_f\ell/d \tag{11.1}$$

For flexural strength:

$$\sigma_b = \frac{\alpha}{\beta}(\sigma_b)_{mo}(1 - V_0)V_m + 2\eta\alpha\,\tau\,V_f\ell/d \tag{11.2}$$

Similarly, the modulus of elasticity in tension, E_t, and bending, E_b, is:

$$E_b, E_t = E_{mo}(1 - V_0)V_m + \eta E_f V_f \tag{11.3}$$

The efficiency factor η was taken as 0.41 after Romualdi and Mandel [50]. The values α and β, which are the ratios of bending strength to tensile strength of the composite and matrix, respectively, were found to be 2.96 and 2.81, respectively. The properties of the void-free matrix were determined in separate tests of the matrix only, with a void content of 23%, from which tensile strength, bending strength and modulus of elasticity for $V_0 = 0$ were calculated as 9.71, 27.27 and 30 ± 7 MPa, respectively. The average aspect ratio ℓ/d, evaluated from microscopical analysis was about 135. Substituting these parameters in Eqs 11.1–11.3 permitted the prediction of the mechanical properties as a function of fibre content. Best fit for the strength curves was obtained for bond strength values in the range of 0.35–0.45 MPa.
   Das Gupta *et al.* [16] applied a somewhat different composite materials model, similar to the one described in Section 4.3, which is also based on the law of mixtures. They included in their calculation the effects of orientation and length and developed different relations for fibres with lengths greater than, and shorter than the critical length. Their model was used to analyse pastes reinforced with short, randomly distributed coir fibres. The average fibre diameter was 0.119 mm and the fibre–matrix bond determined by pull-out tests was found to be 1.5 MPa. Comparison of their experimental and analytical results is provided in Figure 11.14. The agreement is reasonably good, up to a fibre content of 5%, but the model does

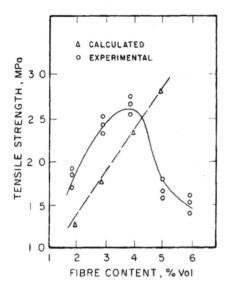

Figure 11.14 Relations between experimental and calculated tensile strength in coir fibre reinforced cement (after Das Gupta *et al.* [16]).

not predict the decline beyond this volume, which is probably associated with a reduction in bond and matrix strength due to poor compaction. Part of this effect could be predicted by Andonian *et al.* [27], since they considered the effect of an increase in void content on the contribution of the matrix. Their data do not show a marked reduction in properties at high fibre contents, probably due to a different method of specimen preparation. It should be noted that, while Andonian *et al.* [27] assumed fibre pull-out, analysis of the fractured surfaces in later work suggested that a large proportion of the fibres may have failed by fibre fracture [39].

Fracture mechanics concepts have also been applied to predict the behaviour of natural fibre FRC. Fracture parameters were found to be dependent on the moisture content, with wet composites having higher toughness values [40,51]. Wet composites were also found to be notch-insensitive, suggesting that LEFM cannot adequately model the behaviour of such composite [51]. This was probably the result of the cracking mode, in which the crack path was tortuous, with some fibres failing by pull-out.

Hughes and Hannant [52] examined the effects of moisture on the first crack stress, by considering the reduction in the modulus of elasticity of the wet fibre, and using the concepts of crack arrest when the fibre spacing is less than the critical flaw size [52]. Their results, assuming a fibre modulus of elasticity of 4 and 40 GPa for wet and dry conditions, respectively, are presented in Table 11.7. If the bond is assumed to remain constant, at about 0.5 MPa (as suggested by Andonian *et al.* [27]), the matrix failure strain should increase in the wet state. If the bond increases

Table 11.7 Effect of moisture content and the resulting assumed change in the fibre modulus of elasticity on the calculated matrix failure strain and the fibre stress across a stable flaw (after Hughes and Hannant [52])

Moisture state	Fibre modulus[a] (GPa)	Bond strength[a] (MPa)	Matrix failure strain[b] (%)	Maximum fibre stress across a stable flaw[b] (MPa)
Wet	4	0.5	0.074	27
Dry	40	0.5	0.106	96
Dry	40	2.0	0.156	172

Notes
The unreinforced matrix failure strain is assumed to be 0.05%.
a Assumed.
b Calculated.

from 0.5 MPa to an assumed value of 2.0 MPa, the matrix failure strain would be expected to double. The increase in bond on drying may be a result of hydrogen bonding, which is more readily generated in the dry state [44]. The transition of the fracture mode from fibre fracture in the dry state to fibre pull-out in the wet state may also be associated with the effects calculated in Table 11.7, in which a large stress can be developed in the fibre prior to matrix cracking, leading to fibre failure before the matrix fails [53].

A micromechanical model to predict the effect of moisture and ageing on the flexural properties of cellulose–cement composites was presented by Kim *et al.* [54]. The model was based on the fracture mechanics concept of fibre bridging across a crack which develops in flexure, similar to the models discussed in Section 4.7. Fibre and matrix characteristic values were identified, based on data in the literature, and were used to model the effect of fibre type (refined and unrefined), the moisture conditions and ageing of a thin sheet composite with 10% fibre content (Table 11.8). The model could account for the influence of moisture conditions, the increase in first crack strength and the reduction in post-cracking load-bearing capacity upon drying. The effect of drying was demonstrated to be the result of an increase in fibre modulus, interfacial adhesional bond and snubbing coefficient, while interfacial friction and fibre strength remained approximately the same in the dry and wet conditions. The main difference between the refined and unrefined fibres is the length, which is reduced by refinement. The relation between the characteristic bridging stress and crack opening could be deduced from this model, to show the influence of moisture conditions and ageing (Figure 11.15). The high bridging stresses at low crack opening of the dry composite reflects its high cracking stress.

11.5 Long-term performance

The long-term performance with natural fibres can be affected by two processes: environmentally induced changes which may be greater than those in cementitious

Table 11.8 Parameters for modelling the behaviour of cellulose–cement composite by a fracture mechanics based micromechanics model (after Kim *et al.* [54])

Parameter	Refined			Unrefined		
	Dry	Wet	Aged	Dry	Wet	Aged
Fibre length (mm)	2.5	2.5	2.5	4	4	4
Fibre diameter (μm)	30	30	30	30	30	30
Fibre modulus (GPa)	30	10[a]	40[a]	35[a]	12[a]	47[a]
Fibre strength (MPa)	550	550	550	650	650	650
Frictional bond (MPa)	0.8[a]	0.8[a]	3.0[a]	0.6[a]	0.6[a]	2.3[a]
Interfacial adhesional (chemical) energy (J/m^2)	3.0[a]	1.0[a]	3.5[a]	3.0[a]	1.0[a]	3.5[a]
Fibre snubbing coefficient	0.8[a]	0.5[a]	0.8[a]	0.8[a]	0.4[a]	0.7[a]
Matrix modulus (GPa)	15.4	13.5	18.0	14.9	12.2	20.3

Note
a Estimated values.

systems because of the hygroscopic nature of the fibres, and variations in mechanical properties over time which may be associated with reduced strength and toughness. These two effects are independent of each other, but they both may lead to undesirable results such as increased sensitivity to cracking. However, in properly designed components, and adequately formulated and treated composites, these effects may be minimized or even eliminated.

11.5.1 Dimensional changes

In thin sheet cementitious components, in which the cement content is usually high, the matrix is very prone to volume changes due to variations in the moisture conditions. This may be compounded by the presence of the hygroscopic natural fibres, which can aggravate the extent of volume changes due to the swelling and shrinking of the fibres themselves. These dimensional changes are greater than those occurring in conventional concrete, and the shrinkage strains can exceed 0.15% [55] compared with less than 0.05% in concrete. Frequent length changes of this kind, induced by environmental influences, can lead to distortion of the shape of the components as well as to cracking in restrained panels, in particular near connections which are not sufficiently flexible. Such changes may not be accompanied by a reduction in strength and toughness, yet they can be detrimental to the overall performance if they are not accounted for.

Unfortunately, not much attention has been paid to these types of effects, and long-term performance has usually been treated only from the point of view of durability related to mechanical properties. Dimensional changes due to moisture

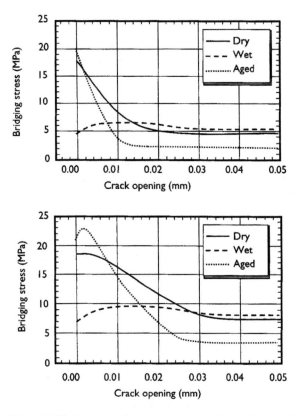

Figure 11.15 Stress-crack opening curves ($\sigma_b - \delta$) for refined (a) and unrefined (b) cellulose–cement composites, as a function of moisture state and ageing (after Kim *et al.* [54]).

movement in relation to long-term performance have been studied mainly in wood–cement particle boards, which were once produced with a magnesite matrix and are currently made also with a Portland cement matrix, and about 20% by weight of wood flakes [56]. Curing is usually achieved by heat treatment at atmospheric pressure, or autoclaving. Residual strains in particle boards and cellulose-pulp FRC composites have been studied after exposure to moisture cycling either by using the standard French accelerated V313 test [21,56] (in which each cycle consists of 72 h of water soaking at 20°C, 24 freezing at −12°C and 72 h of drying at 70°C), or by using wetting–drying cycles in which each cycle consisted of water immersion at 20°C and drying at 40°C or 105°C [55]. Tests of this kind indicate that length changes of about 0.20% (20°C–40° C wetting–drying [55]) and 0.30% (V313 test [21]) can occur, without being accompanied by a reduction in strength

[21,55–57]. Only when the composites are subjected to much harsher cycling (20 − 50°C) was a reduction in strength observed [55]. Sharman and Vautier [21] found that carbonation of autoclaved cellulose-pulp fibre composites increased the residual expansion measured after moisture cycling using the V313 test. The increase was not large and it was not accompanied by a reduction in strength.

11.5.2   Changes in mechanical properties

Exposure of natural fibre components to natural weathering may result in changes in mechanical properties which may be deleterious, that is, reductions in strength and toughness. The data in the open literature are contradictory, since different trends have been reported. Sometimes, weathering was associated with a reduction in properties while in other tests, it was not associated with any decrease in performance, with strengthening effects even taking place. This apparent discrepancy is the result of the fact that the ageing of these composites involves different mechanisms, and the outcome of exposure depends on the type of fibres and the processing of the composite. Some of the mechanisms to be considered with regard to the ageing of natural fibre FRC are:

- degradation of fibres due to alkaline attack;
- degradation of fibres due to biological attack;
- changes in the matrix and petrification (mineralization), which may lead to improved fibre–matrix bond, and to fibre strengthening and stiffening.

Overview of degradation mechanism of this kind was presented in Chapter 5. In the following sections additional input which is characteristic of the nature of natural fibres will be discussed.

11.5.2.1   Degradation of fibres in alkaline environment

Natural fibres can suffer various degrees of degradation when exposed to an alkaline environment. However, the extent of the attack, as determined by strength loss of the fibres can be quite different, ranging from rapid strength decline to hardly any strength reduction, depending on the type of fibres (Figure 11.16) [58]. Wide differences in the strength retention of natural fibres after immersion in alkaline solutions, similar to those shown in Figure 11.16, were also reported by Singh [59], who found a small drop in strength in coir fibres and a sharp drop in the strength of sisal, hemp and jute. Castro and Naaman [14] also reported strength reduction in natural fibres immersed in various solutions.

The mechanisms of the alkaline degradation of natural fibres were discussed extensively by Gram [9,60,61] with particular reference to sisal fibres. The two mechanisms considered were:

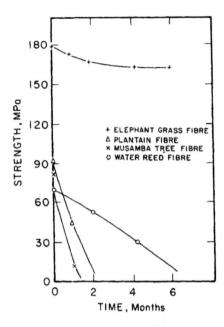

Figure 11.16 The effect of immersion in alkaline solution on the tensile strength of different natural fibres (after Lewis and Mirihagalia [58]).

1   A 'peeling off' effect, in which the end of the molecular chain is unhooked and end groups are continuously liberated. This is the result of the reaction between the reducing end group and OH^- ions. The degradation effect due to this mechanism is probably small, since its rate at temperatures below 75°C is low, and its effect on the long cellulose molecule (with a degree of polymerization of 25,000) is thus small.

2   Alkaline hydrolysis, which causes the molecular chain to divide, and the degree of polymerization to be reduced significantly. Hemicellulose and lignin are particularly sensitive to this degradation effect.

Gram suggested that in sisal fibres the main mechanism of alkaline attack is associated with the dissolution and decomposition of the hemicellulose and lignin in the middle lamella, thus breaking the link between individual fibre cells, as shown schematically in Figure 11.17. The breakdown of the long fibre into small unit cells would lead to the loss of its reinforcing efficiency. Velpari *et al.* [62] found similar effects in an SEM study of jute fibres immersed in a cement extract solution with a pH of about 13. Thirty days of immersion resulted in a reduction in tensile strength of the fibres from 50 to 12 MPa. It was observed that lignin was leaching out, and the fibre gradually broke down and fibrillized as predicted by Gram.

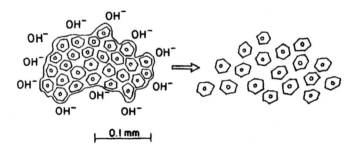

Figure 11.17 Schematic description of the degradation of sisal fibres in concrete, by the decomposition and dissolution of the middle lamellae in the alkaline porewater (after Gram [9]).

It should be noted that Gram [9] and Singh [59] found that the extent of alkaline attack was greater in a calcium hydroxide solution than in a sodium hydroxide solution, although the latter has a higher pH. It may be that the presence of Ca^{++} ions leads to some additional degradation. Singh speculated that this may be associated with crystallization of lime in the pores and spaces in the fibre.

In cellulose-pulp fibres, in which the chemical treatment resulted in removal of the middle lamella, including the lignin and hemicellulose, the durability in an alkaline environment should be improved [59]. Campbell and Coutts [18] and Harper [57] indicated their favourable experiences with such fibres from the point of view of durability. Sharman and Vautier [21] carried out a systematic study of the durability of cellulose-pulp FRC autoclaved composite. They did not observe any decline in flexural and tensile strengths after accelerated testing in water at 50°C, and concluded that corrosion of the fibre by alkaline attack is unlikely. They too interpreted this effect as being due to the removal of lignin and hemicellulose during the processing of the fibres.

Natural weathering was found to lead to a loss in strength and toughness in sisal FRC [9,63], as a result of the chemical degradation of the fibres in the alkaline environment [62]. Treatment of the matrix with silica fume, to reduce its alkalinity, was shown to be an effective means of slowing down or even preventing the strength loss and embitterment (Figure 11.18). With 45% replacement, the loss in toughness was eliminated, and this was correlated with the reduced alkalinity of the matrix. With fly ash replacement, however, no significant improvement was obtained, due to its limited effect on the alkalinity. On the other hand, a marked improvement was observed when using an alumina cement matrix [9,60]. Treatments of sisal fibres with water repellant agents (formine and stearic acid) led to some improvement in durability, but it was not as effective as that obtained with silica fume replacement [9,60].

Although the data of Sharman and Vautier [21] seem to be conclusive, one should note the results of Gram [9] and Bergstrom and Gram [63] for sisal fibres.

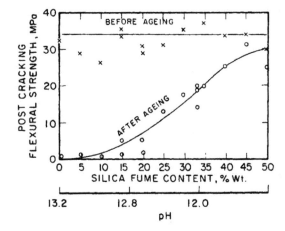

Figure 11.18 The effect of replacing Portland cement with silica fume on the retention of post-cracking strength of a composite with sisal fibres after accelerated ageing by 120 cycles of wetting and drying (after Bergström and Gram [63]).

In their work, the 50°C immersion test did not lead to reduced strength and toughness of the composite, though the fibres themselves lost strength when immersed in a lime solution. On the other hand, in outdoor exposure, and in accelerated tests based on wetting and drying, strength and toughness loss was as were observed and attributed to alkaline attack. Thus, one may question the validity of the hot water immersion test as an accelerated test for natural fibre FRC composites [9,63]. It may be that the transport of OH^- ions to the fibres under such conditions proceeds at a slower rate than when exposed to wetting and drying. It should be noted that Akers and Studinka [41] found a reduction in the degree of polymerization of cellulose-pulp fibres removed from aged composites, but this reduction was not necessarily accompanied by a decrease in the strength of the composite. Thus, while alkali attack may be present, its influence may be small, or other processes might compensate for it. This will show up only when testing the actual composite, but not necessarily during testing of the fibre itself by immersion in various solutions.

11.5.2.2   Biological attack on the fibres

There is a possibility of bacterial and fungal attack in semi-moist and in alternate wetting and drying conditions. Lewis and Mirihagalia [58] studied the susceptibility to rot of natural fibres by exposing them to wetting and drying cycles. They found considerable differences between the performance of the various fibres (Figure 11.19), with the biological attack occurring at the water–air interface.

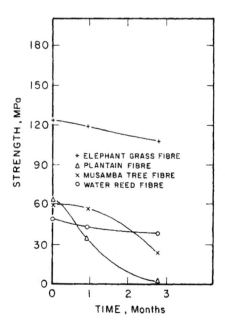

Figure 11.19 The effect of exposure to wetting–drying cycles on the strength of different natural fibres (after Lewis and Mirihagalia [58]).

On the other hand, tests of actual composites under conditions which would have been expected to promote biological attack indicated the immunity of the composite to this problem [11,57,64]. This can be attributed to the alkaline nature of the matrix, which can provide resistance to this kind of degradation. However, some concern has been raised regarding the effect of carbonation on the immunity to biological attack. Sharman and Vautier [21] studied this effect in a carbonated autoclaved cellulose-pulp FRC composite, exposed to a fungal cellar. Some reduction in strength of the carbonated composite was observed, but it was not accompanied by any biological attack.

11.5.2.3 Microstructural changes

The ageing performance of the actual composite may be different from that of the fibres, due to matrix effects, and to influences associated with the fibre–matrix interactions leading to microstructural changes.

Cellulose–pulp composites are generally less sensitive to degradation effects [21,57,65]. This is in part due to the better alkaline resistivity of the treated pulp. Natural and accelerated weathering can lead to improvement or decline in the properties of the cellulose–cement composites, depending on the type of ageing. Water immersion has little influence on the properties [9,66,67], Figure 11.20,

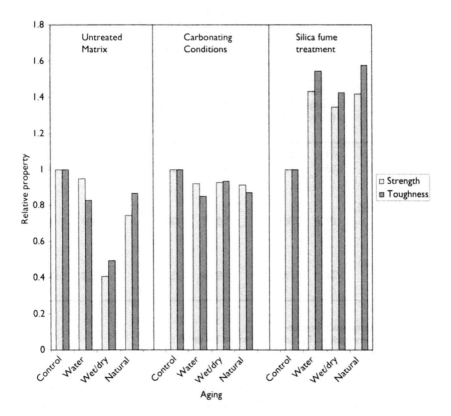

Figure 11.20 Effect of matrix composition and ageing on the strength and toughness relative to the control of sisal–cement composite: normal-untreated matrix, carbonation curing of composite and silica fume slurry treatment of fibres (adopted from the data of Filho *et al.* [67]). Control-28 days of normal curing; water-322 days in water; wet/dry-46 cycles; natural-outdoor exposure.

suggesting that alkali attack is practically not an issue. On the other hand, wetting and drying were reported to increase flexural strength in the presence of carbonating conditions and reduce toughness [41] (Table 11.9), as well as increase in the stiffness during weathering [41,66]. This is different than ageing in wetting/drying conditions without carbonation, where the flexural strength remained the same or decreased (Table 11.9), and the increase in stiffness was smaller. Both ageing conditions however led to reduction in toughness, which tended to be greater under carbonating conditions. There are strong indications that the accelerated ageing in carbonating conditions better simulates the natural ageing [41–43,68], as can be seen in Table 11.9.

Several studies of the mechanisms involved in such changes were reported [42,43,65,67,69], with an attempt to correlate them with the changes in the

Table 11.9 Effect of natural and accelerated weathering on the changes in the mechanical properties of cellulose-pulp FRC composite cured at room temperature (after Akers and Studinka [41])

	Room temperature curing			Autoclave curing		
	Flexural strength (MPa)	Modulus of elasticity (GPa)	Strain at failure (%)	Flexural strength (MPa)	Modulus of elasticity (GPa)	Strain at failure (%)
Unaged	16.4	10.9	2.8	23.0	13.2	0.07
Accelerated ageing[a] (normal environment)	12.3	14.8	0.04	20.2	13.5	0.05
Accelerated ageing[b] (CO_2 rich environment)	23.9	18.9	0.05	25.6	13.1	0.05
Natural ageing[c]	25.1	18.0	0.05	27.2	18.6	0.04

Notes
a Wetting–drying cycles, 3 months.
b Wetting–drying cycles in CO_2, rich environment, 3 months.
c 5 years.

mechanical properties during ageing. Bentur and Akers [42,43] reported that the naturally weathered composites and the specimens exposed to accelerated carbonation (Table 11.9) were both carbonated to such an extent that no CH was detected. This was accompanied by a marked change in the microstructural characteristics of the matrix at the interface, and in the fibre itself [42,43,69]. The latter became petrified, with reaction products filling the fibre cavity, and possibly impregnating the fibre cell wall. These characteristics are shown in Figure 11.21(a). Slight etching with HCl showed the cellulose fibre circumference engulfing a cavity filled with reaction products (Figure 11.21(b)). The reaction products have been slightly etched from within the fibre core, to reveal the petrified morphology. The hydration products filling the fibre lumen became carbonated, as well as the fibre–matrix interface which was previously occupied by gel [69]. These changes were accompanied by a reduction in porosity, suggesting densening of the microstructure [68,69]. By contrast, in an unaged composite, or a composite exposed to wetting and drying cycles not accompanied by carbonation, the fibre was not petrified and its hollow morphology could clearly be observed at the fracture surface (Figure 11.22).

Petrification of the fibre apparently increased its strength and rigidity, and led to the increase in strength and modulus of elasticity shown in Table 11.9, for natural ageing and accelerated ageing in CO_2. The effect of increased rigidity, combined

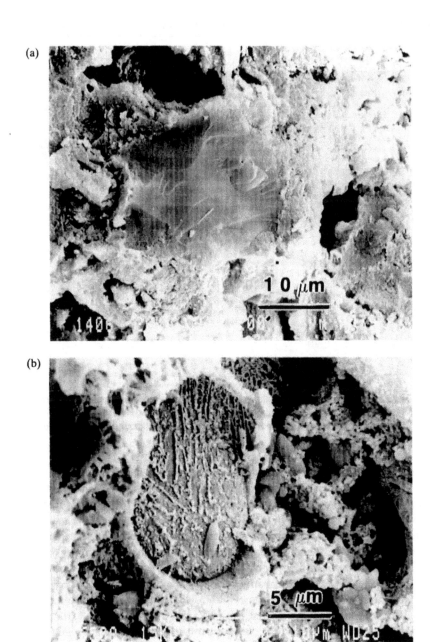

Figure 11.21 Microstructure of cellulose-pulp FRC composites cured at room temperature after accelerated ageing in CO_2 environment (after Bentur and Akers
[42]). (a) Petrified fibre and the dense matrix around it; (b) petrified fibre
after slight etching with HCl, revealing the fibre cell wall and the products
deposited in its core.

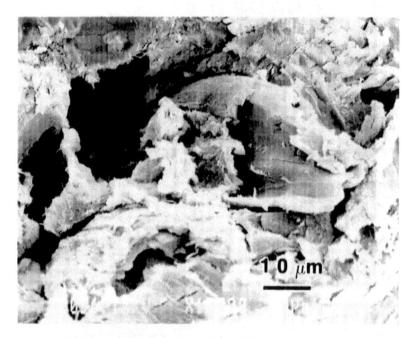

Figure 11.22 Microstructure of cellulose-pulp FRC composite cured at room temperature, after accelerated ageing (wetting–drying) in normal environment, showing the hollow nature of the fibre and the dense matrix around it (after Bentur and Akers [42]).

with the increased density of the matrix at the interface, can account for the reduction in toughness which accompanied the increase in strength. This was associated with a change in the mode of fracture from fibre pull-out in the unaged composite to fibre fracture in the aged composite. It should be noted that in the system in which ageing did not result in carbonation and petrification, the strength did not increase, but the toughness decreased, probably due to an increase in density of the matrix at the interface, which took place even under non-carbonating conditions. The micromechanical model of Kim *et al.* [54] took such influences into account in the parameters of the aged material (Table 11.8), by having higher fibre–matrix bond in the aged composite (increase from 0.8 to 3.0 MPa and from 0.6 to 2.3 MPa in the refined and unrefined fibre, respectively). These changes are exhibited in the stress-crack opening relations of the aged composite (Figure 11.15), and could simulate the changes in mechanical properties due to ageing.

Mohr *et al.* [70] suggested the coexistence of all of these microstructural effects, with each dominating at a different ageing stage. Based on accelerated wetting/drying ageing tests of Kraft pulp cement composite, they concluded that fibre–matrix debonding is the first to occur (up to 2 cycles), followed by precipitation of hydration products at the former interface (up to 10 cycles) and finally embrittlement due to mineralization of the fibre (beyond 10 cycles).

It is interesting that cellulose-pulp tested as single fibres do not fracture simply [28], but rather fail by separation between the layers of the secondary fibre wall. In contrast, in the cement composite the fibre fracture seems to be clean, without layered separation within the fibre. It has been suggested that the cement matrix may have entered the fibre and led to additional internal bonding in the cell wall. This may be the first step in the petrification process described by Bentur and Akers [42].

In view of these mechanisms, one may expect that matrix modification to control microstructure may change the ageing trends. Filho *et al.* [67] demonstrated the feasibility of such an approach by carbonation curing and silica fume treatment, in which the fibres were immersed in silica fume slurry prior to their incorporation in the composite (Figure 11.20). Strength and toughness loss were observed in wet/dry accelerated ageing and in natural ageing. Carbonation curing almost eliminated these losses, while treatments in silica fume slurry resulted in improvement in properties during ageing. A 30% silica fume replacement was also reported to improve the durability performance of recycled fibre cement composite [66].

This petrification mechanism may also account for the reported increase in strength after natural ageing and accelerated ageing in CO_2, of a cellulose-pulp FRC composite [21,64]. Strengthening effects of this kind are expected to be dependent on the nature of the cementitious matrix and its tendency to undergo carbonation, which involves apparent deposition of reaction products in the fibre. The Akers and Studinka [41] and Bentur and Akers [43] data were for room temperature cured composites, while Sharman and Vautier's [21] results were for autoclaved composites. This suggests that the strengthening and carbonation effects are independent of matrix composition. However, further tests on autoclaved composites [41,42] showed no change in strength on ageing, in contrast to the behaviour of non-autoclaved material. This may be due to a more limited carbonation. Sharman and Vautier [21] on the other hand, reported that for room temperature cured asbestos composites, ageing did not lead to carbonation to the extent observed in autoclaved materials. On the face of it, there is an apparent contradiction and inconsistency between these results. However, carbonation is also dependent on environmental conditions. In an autoclaved matrix carbonation takes place more readily at high humidities [71] whereas in room temperature cured cementitious materials, the highest carbonation rates will be at intermediate humidities. Sharman and Vautier [65] pointed out that their weathering tests took place in New Zealand, where the relative humidity is high, thus promoting carbonation in the autoclaved composite, and slowing it down in the room temperature cured material. The specimens studied by Akers and Studinka [41] and Bentur and Akers [42] were exposed to lower relative humidities, thus promoting more carbonation in the room temperature cured composite.

This discussion indicates the complex nature of the ageing mechanisms involved in natural fibre reinforced cements. It shows, however, that in cellulose-pulp composites it is possible to achieve materials whose strength not only stays stable with ageing but might even increase under carbonating conditions. However, the stability or increase in strength may be accompanied by reduced toughness. In addition, in the carbonated composite there is a tendency for somewhat higher moisture movement [21].

References

1. D.J. Cook, 'Natural fibre reinforced concrete and cement – recent developments', in D.M. Roy, A.J. Majumdar, S.P. Shah and J.A. Manson (eds) *Advances in Cement-Matrix Composites*, Proc. Symposium L, Materials Research Society Annual Meeting, Pittsburgh, PA, 1980, pp. 251–258.
2. D.J. Cook, 'Concrete and cement composites reinforced with natural fibres', in *Fibrous Concrete*, Proc. Symp. The Concrete Society, The Construction Press, London, 1980, pp. 99–114.
3. M.A. Aziz, P. Paramasivam and S.L. Lee, 'Concrete reinforced with natural fibres', in R.N. Swamy (ed.) *Concrete Technology and Design, Vol. 2: New Reinforced Concretes*, Surrey University Press, 1984, pp. 107–140.
4. M.A. Aziz, P. Paramasivam and S.L. Lee, 'Prospects for natural fibre reinforced concrete in construction', *Int. J. Cem. Comp. & Ltwt. Concr.* 1, 1981, 123–132.
5. P.N. Balaguru and S.P. Shah, 'Alternative reinforcing materials for less developed countries', *Int. J. Res. Tech.* 3, 1985, 87–105.
6. B.V. Subrahmanyam, 'Bamboo reinforcement for cement matrices', *Int. J. Res. Tech.* 3, 1985, 141–195.
7. K.F.J. Nutman, 'Agave fibres part 1: morphology, histology, length and fineness; grading problems', *Empire J. of Exp. Agriculture.* 1, 1937, 75–92.
8. J.M. Illston, J.M. Dinwoode and A.A. Smith, *Concrete, Timber and Metals*, Van Nostrand Reinhold, New York, 1979.
9. H.E. Gram, *Durability of Natural Fibres in Concrete*, Research report, Swedish *Cem. Concr. Res.* Institute, Sweden, Stockholm, 1983.
10. Y.W. Mai and M.I. Hakeem, 'Slow crack growth in cellulose fibre cements', *J. Mater. Sci.* 19, 1984, 501–508.
11. O.J. Uzomoka, 'Characteristics of akwara as a reinforcing fibre', *Mag. Concr. Res.* 28, 1976, 162–167.
12. B.L.M. Mwamila, *Low Modulus Reinforcement of Concrete – Special Reference to Sisal Twines*, Swedish Council for Building Research, Report DIO, 1984.
13. P. Paramasivam, O.K. Nathan and N.C. Das Gupta, 'Coconut fibre reinforced corrugated slabs', *Int. J. Cem Comp. & Ltwt. Concr.* 6, 1984, 19–28.
14. J. Castro and A.E. Naaman, 'Cement mortar reinforced with natural fibres', *J. of Ferrocement.* 11, 1981, 285–301.
15. M.A. Mansur and M.A. Aziz, 'A study of jute fibre reinforced cement composites', *Int. J. Cem. Comp. & Ltwt. Concr.* 4, 1982, 75–82.
16. N.C. Das Gupta, P. Paramasivam and S.L. Lee, 'Mechanical properties of coir reinforced cement paste composites', *International journal for Housing Science and its Applications.* 2, 1978, 391–406.

17. B.L.M. Mwamila, 'Natural twines as main reinforcement in concrete beams', *Int. J. Cem. Comp. Ltwt. Concr.* 7, 1985, 11–20.
18. M.D. Campbell and R.S.P. Coutts, 'Wood fibre reinforced cement composites', *J. Mater. Sci.* 15, 1980, 1962–1970.
19. X. Li, M.R. Silsbee, D.M. Roy, K. Kessler and P.R. Blankenhorn, 'Approaches to improve the properties of wood fiber reinforced cementitious composites', *Cem. Concr. Res.* 24, 1994, 1558–1566.
20. Z. Fordos and B. Tram, 'Natural fibres as reinforcement in cement based composites', Paper 2.9 in R.N. Swamy, R.L. Wagstaffe and D.R. Oakley (eds) *Developments in Fibre Reinforced Cement and Concretes*, Proc. RILEM Symp., Sheffield, 1986.
21. W.R. Sharman and B.P. Vautier, 'Accelerated durability testing of autoclaved wood fibre reinforced cement sheet composites', *Durab. Bldg. Mat.* 3, 1986, 255–275.
22. H. Krenchel and H.W. Jensen, 'Organic reinforcement fibres for cement and concrete', in *Fibrous Concrete*, Proc. Symp. The Concrete Society, The Construction Press, London, 1980, pp. 87–98.
23. A. Bentur and S.A.S. Akers, 'Thin sheet cementitious composites', in N. Banthia, A. Bentur and A. Mufti (eds) *Fiber Reinforced Concrete: Present and Future*, Canadian Society for Civil Engineering, Montreal, Canada, 1998, pp. 20–45.
24. R.S.P. Coutts and P.G. Warden, 'Air cured wood pulp, fibre cement composites', *J. Mater. Sci.* 4, 1985, 117–119.
25. R.S.P. Coutts, 'Autoclaved beaten wooden fibre-reinforced cement composites', *Composites.* 15, 1984, 139–143.
26. R.S.P. Coutts, 'Flax fibres as a reinforcement in cement mortars', *Int. J. Cem. Comp. Ltwt. Concr.* 5, 1983, 257–262.
27. R. Andonian, Y.W. Mai and B. Cotterell, 'Strength and fracture properties of cellulose fibre reinforced cement composites', *Int. J. Cem. Comp.* 1, 1979, 151–158.
28. G.W. Davies, M.D. Campbell and R.S.P. Coutts, 'A SEM study of wood fibre reinforced cement composite', *Holzforschung.* 15, 1981, 201–204.
29. R.S.P. Coutts and A.J. Michell, 'Wood pulp fibre-cement composites', *J. Appl. Polym. Sci.* 37, 1983, 829–844.
30. R.S.P. Coutts, 'Eucalyptus wood fibre-reinforced cement', *J. Mater. Sci. Letters* 6, 1987, 955–957.
31. R.S.P. Coutts and V. Ridikas, 'Refined wood fibre-cement products', *Appatita.* 35, 1985, 395–400.
32. R.S.P. Coutts and K. Kightly, 'Microstructure of autoclaved refined wood-fibre cement mortars', *J. Mater. Sci.* 17, 1982, 1801–1806.
33. R.S.P. Coutts, 'A review of Australian research into natural fibre-cement composites', *Cem. Concr. Compos.* 27, 2005, 518–526.
34. X. Lin, D.M. Silsbee and D.M. Roy, 'The microstructure of wood fiber reinforced cementitious composites', in S. Diamond, S. Mindess, F.P. Glasser, L.W. Roberts, J.P. Skalny and L.D. Wakeley (eds) *Microstructure of Cement-Based Systems/Bonding and interfaces in Cementitious Materials*, Materials Research Society Symp. Proc. Vol. 370, Materials Research Society, Pittsburgh, PA, pp. 487–495.
35. R.S.P. Coutts and M.D. Campbell, 'Coupling agents in wood fibre reinforced cement composites', *Composites.* 10, 1979, 228–232.
36. P.R. Blankenhor, M.R. Silsbee, B.D. Blankenhorn, M. DiCola and K. Kessler, 'Temperature and moisture effects on selected properties of wood fiber-cement composites', *Cem. Concr. Res.* 29, 1999, 737–741.

37. P.R. Blankenhor, B.D. Blankenhorn, M.R. Silsbee, M. DiCola and K. Kessler, 'Effect of surface treatments on mechanical properties of wood fiber-cement composites', *Cem. Concr. Res.* 31, 2001, 1049–1055.

38. J.L. Pehanich, P.R. Blankenhor and M.R. Silsbee, 'Wood fiber surface treatment level effects on selected mechanical properties of wood fiber-cement composites', *Cem. Concr. Res.* 34, 2004, 59–65.

39. Y.W. Mai, M.I. Hakeem and B. Cotterell, 'Effects of water and bleaching on the mechanical properties of cellulose fibre cements', *J. Mater. Sci.* 18, 1983, 2156–2162.

40. Y.W. Mai and M.I. Hakeem, 'Slow crack growth in bleached cellulose fibre cements', *J. Mater. Sci. Letters*. 3, 1984, 127–130.

41. S.A.S. Akers and J.B. Studinka, 'Ageing behavior of cellulose fibre cement composites in natural weathering and accelerated tests', *Int. J. Cement Composites and Lightweight Concrete.* 11, 1989, 93–97.

42. A. Bentur and S.A.S. Akers, 'The microstructure and ageing of cellulose fibre reinforced cement composite cured in normal environment', *Int. J. Cem. Comp. Ltwt. Concr.* ll, 1989, 99–109.

43. A. Bentur and S.A.S. Akers, 'The microstructure and ageing of cellulose fibre reinforced autoclaved cement composites', *Int. J. Cem. Comp. Ltwt. Concr.* 11, 1989, 111–115.

44. R.S.P. Coutts, 'Fibre-matrix interface in air cured wood-pulp fibre-cement composites', *J. Mater. Sci. Letters* 6, 1987, 140–142.

45. N.B. Milestone and I. Suckling, 'Interactions of cellulose fibers in an autoclaved cement matrix', in K. Kovler, J. Marchand, S. Mindess and J. cWeiss (eds) *Concrete Science and Engineering*, Proc. RILEM Symp., PRO 36, RILEM Publications, Bagneux, France, 2004, pp. 153–164.

46. R.S.P. Coutts, 'Air cured wood pulp, fibre/cement mortars', *Composites.* 18, 1987, 325–328.

47. P. Soroushian and S. Marikunte, 'High performance cellulose fiber reinforced cement composites', in H.W. Reinhardt and A.E. Naaman (eds) *High Performance Fiber Reinforced Cement Composites*, Proc. RILEM Int. Symp., E&FN SPON, London and New York, 1992, pp. 84–99.

48. R.S.P. Coutts and P. Kightly, 'Bonding in wood fibre-cement composites', *J. Mater. Sci.* 19, 1984, 3355–3359.

49. F.E. Morrissey, R.S.P. Coutts and P.V.A. Grossman, 'Bond between cellulose fibres and cement', *Int. J. Cem. Comp. Ltwt. Concr.* 7, 1985, 73–80.

50. J.P. Romualdi and J.A. Mandel, 'Tensile strength of concrete affected by uniformly distributed and closely spaced short lengths of wire reinforcement', *J. Amer. Concr. Inst.* 6, 1964, 657–670.

51. S. Mindess and A. Bentur, 'The fracture of wood fibre reinforced cement', *Int. J. Cem. Comp. Ltwt. Concr.* 4, 1982, 245–250.

52. D.C. Hughes and D.J. Hannant, 'Reinforcement of Griffith flaws in cellulose reinforced cement composites', *J. Mater. Sci. Letters* 4, 1985, 101–102.

53. D.J. Hannant, D.C. Hughes and A. Kelly, 'Toughening of cement and other brittle solids with fibres', *Phil. Trans. Roy. Soc. London Ser. A*, 310, 1983, 175–190.

54. P.J. Kim, H.C. Wu, Z. Lin, V.C. Li, B. deLhoneux and S.A.S. Akers, 'Micromechanics-based durability study of cellulose cement in flexure', *Cem. Concr. Res.* 29, 1999, 201–208.

55. A. Bentur and S. Mindess, 'Effect of drying and wetting cycles on length and strength changes of wood fibre reinforced cement', *Durability of Building Materials*. 2, 1983, 37–43.
56. J.M. Dinwoodie and B.H. Paxton, *Wood Cement Particleboard – a Technical Assessment*, Information Paper IP 4/83, Building Research Establishment, England, 1983.
57. S. Harper, 'Developing asbestos-free calcium silicate building boards', *Composites*. 13, 1982, 123–138.
58. G. Lewis and P. Mirihagalia, 'Natural vegetable fibres as reinforcement in cement sheets', *Mag. Concr. Res.* 31, 1979, 104–108.
59. S.M. Singh, 'Alkali resistance of some vegetable fibres and their adhesion with portland cement', *Research and Industry.* 15, 1985, 121–126.
60. H.E. Gram, 'Durability studies of natural organic fibres in concrete, mortar or cement', Paper 7.1 in R.N. Swamy, R.L. Wagstaffe and D.R. Oakley (eds) *Developments in Fibre Reinforced Cement and Concrete*, Proc. RILEM Symp., Sheffield, 1986.
61. H.E. Gram, *Methods for Reducing the Tendency towards Embrittlement in Sisal Fibre Concrete*, Nordic Concrete Research, Publ. No. 5, Oslo, 1983, 62–71.
62. V. Velpari, B.E. Ramachandran, T.A. Bashkaron, B.C. Pai and N. Balasubramanian, 'Alkali resistance of fibres in cement', *J. Mater. Sci. Letters.* 15, 1980, 1579–1584.
63. S.G. Bergstrom and H.E. Gram, 'Durability and alkali-sensitive fibres in concrete', *Int. J. Cem. Comp. Ltwt. Concr.* 6, 1984, 75–80.
64. U.N. Sinha, S.N. Dutta, B.P. Chaliha and M.S. Iyenger, 'Possibilities of replacing asbestos in asbestos cement sheets by cellulose pulp', *Indian Concr. J.* 49, 1975, 228–237.
65. W.R. Sharman and B.P. Vautier, 'Durability studies on wood fibre reinforced cement sheet', Paper 7.2 in R.N. Swamy, R.L. Wagstaffe, and D.R. Oakley (eds) *Developments in Fibre Reinforced Cement and Concrete*, Proc. RILEM Symp., Sheffield, 1986.
66. P. Soroushian, S. Zahir, J.P. Won and J.W. Hsu, 'Durability and moisture sensitivity of recycled wastepaper-fiber-cement composites', *Cem. Concr. Res.* 16, 1994, 115–128.
67. R.D.J. Filho, K. Ghavami, G.L. England and K. Scrivener, 'Durability of sisal fibre reinforced mortar composite', in P. Rossi and G. Chanvillard (eds) *Fibre Reinforced Concrete (FRC) – BEFIB'2000*, Proc. RILEM 5th Int. Symp., RILEM Publications, Bagneux, France, 2000, pp. 655–664.
68. R. MacVicar, L.M. Matuana and J.J. Balatinecz, 'Aging mechanisms in cellulose fiber reinforced cement composites', *Cem. Concr. Compos.* 21, 1999, 189–196.
69. B.J. Pirie, F.P. Glasser, C. Schmitt-Henco and S.A.S. Akers, 'Durability studies and characterization of the matrix and fibre-cement interface of asbestos-free fibre-cement products', *Cem. Concr. Compos.* 12, 1990, 233–244.
70. B.J. Mohr, H. Nanko and K.E. Kurtis, 'Durability of kraft pulp fibre-cement composites to wet/dry cycling', *Cem. Concr. Compos.* 27, 2005, 435–448.
71. ACI Committee 515, High pressure steam curing, modern practice and properties of autoclaved products, *J. Amer. Concr. Inst.* 62, 1965, 869–908.

Chapter 12

High performance systems

12.1 Introduction

There is continued interest and much innovative effort to develop new FRC systems with enhanced performance in terms of the technical properties or the cost/effectiveness of the FRC application. In order to achieve such goals greater degrees of freedom in the design of the composites have been exercised, including modifications of the matrix, the fibres, the production process, and combinations of all three. Such composites cannot be simply classified in terms of the composition of the reinforcing fibres, and should be treated from the point of view of the strategy applied to achieve the enhanced performance.

High performance–high ductility FRC composites are usually defined as materials with a strain hardening behaviour in tension or deflection hardening in bending. This implies that upon reaching the first crack during loading (either in tension or bending) additional straining of the composite (increase in strain in tension and increase in deflection in bending) will require an increase in load, as seen schematically in the classification in Figure 12.1 [1]. A strain hardening composite in tension will be by default a deflection hardening composite in bending. However, deflection hardening in bending will not necessarily exhibit strain hardening in tension; it was shown in Section 4.7 that deflection hardening in bending (i.e. strengthening beyond the matrix cracking strength) can be obtained when the post-cracking tensile strength is greater than about 1/3 of the tensile strength of the matrix. This implies that a composite can be strain softening in tension but still deflection hardening in bending if its residual post-cracking tensile strength is greater than about 1/3 of the tensile strength of the matrix.

Several general terms have been used to describe this family of materials, most notably high performance fibre reinforced cement composites (HPFRCC) [1] and ductile fibre reinforced cement composites (DRFRCC) [2]. The Japanese committee [2] suggested nomenclature that defines DFRCC as the whole family of materials in this class, both strain hardening in tension and deflection hardening in bending, whereas HPFRCC, according to the Japanese classification refers only to strain hardening in tension.

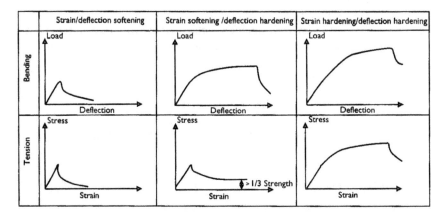

Figure 12.1 Schematic description of strain softening, strain hardening (in tension) and deflection hardening (in bending) in FRC composites [1].

A variety of experimental and commercial systems have been developed to achieve strain/deflection hardening composites, using different strategies, all of which are based on the mechanical and micromechanical principles which govern the behaviour of FRC. These principles were outlined in Chapter 4, and their extension to high performance FRC will be discussed in Section 12.2.

The different types of high performance systems will than be treated separately, based on classification in terms of the performance and the strategy to achieve it, to include the following topics: monofibre systems with normal strength matrix, monofibre systems with high strength matrix, hybrid FRC, slip hardening fibres and FRC with polymer–cement matrices.

12.2 Mechanics of high performance FRC

The basic principle to obtain a strain hardening FRC is to design a composite where the fibre content exceeds the critical volume, as outlined in Section 4.3, Eqs 4.29, 4.30, 4.36–4.38 and summarized in Figure 4.11. Although the principles may seem simple, in practice it is difficult to meet the requirements embedded in it (e.g. difficulties of incorporating large quantities of fibres in the matrix, limitations on enhancing fibre–matrix bond). Therefore, most of the developments of such composites are based on optimization of the system from the mechanical and production points of view. These attempts are to a large extent based on some analysis of the conditions required to achieve this high performance behaviour, and those are outlined in this section, based on a composite mechanics approach and fracture mechanics. Three categories of principles might be identified: composite mechanics of fibre reinforcement, matrix control and hybrid fibre reinforcement.

An extension of the approach outlined in Section 4.3, based on mechanics of composites, was presented by Naaman [1], to provide an overview of the conditions which need to be met to achieve strain and deflection hardening. The following outline is based on this paper. In the case of strain hardening (tensile loading), revised relations for the tensile post-cracking strength, σ_{pc}, and the first cracking strength, σ_{cc}, of the composite were developed:

$$\sigma_{cc} = \sigma_{mu}(1 - V_f) + \alpha_1 \alpha_2 \tau_{fu} V_f \ell/d \tag{12.1}$$

where V_f is the fibre volume content; ℓ, the fibre length; d, the fibre diameter; ℓ/d, the aspect ratio; σ_{mu}, the tensile strength of the matrix; τ_{fu}, the average fibre–matrix bond strength; α_1, the coefficient representing the faction of bond mobilized at the first matrix cracking and α_2, the efficiency factor of fibre orientation in the uncracked state.

$$\sigma_{pc} = \lambda_1 \lambda_2 \lambda_3 \tau V_f \ell/d \tag{12.2}$$

where λ_1 is the expected pull-out length ratio (1/4 from probability considerations); λ_2, the efficiency factor of orientation in the cracked state and λ_3, the group reduction factor associated with the number of fibres pulling out per unit area (the density of fibre crossing).

For strain hardening to occur, the post-cracking strength should be bigger than the cracking strength:

$$\sigma_{pc} > \sigma_{cc} \tag{12.3}$$

Substituting Eqs 12.1 and 12.2 into 12.3, and solving it for critical fibre volume for tensile strain hardening, yields the following relation:

$$V_f \geq (V_f)_{critical-tension} = \frac{1}{1 + (\tau_{fu}/\sigma_{mu})(\ell/d)(\lambda_1 \lambda_2 \lambda_3 - \alpha_1 \alpha_2)} \tag{12.4}$$

A graphical description of Eq. 12.4 for some characteristic values is provided in Figure 12.2, showing the relations between critical fibre volume and fibre characteristics (bond and aspect ratio). It can clearly be seen that for typical aspect ratios of steel fibres, in the range of 50–100, the critical fibre volume ranges from about 1–15%, depending on the fibre–matrix bond, with a higher bond leading to a smaller fibre content.

Naaman [1] extended the treatment to flexural loading, using the principles outlined in Section 4.7, considering different assumptions for the strain and stress distribution in the cross section in bending. The criterion for deflection hardening is that the bending moment resistance in the post-cracking zone, M_{pc}, should be bigger than the moment of resistance at the first crack in bending, M_{cc}. The first crack moment can simply be calculated assuming a linear–elastic behaviour in the cross section, with maximum tensile stress equal to the tensile first crack strength of the composite, σ_{cc}. In the post-cracking zone, the moment of resistance can be

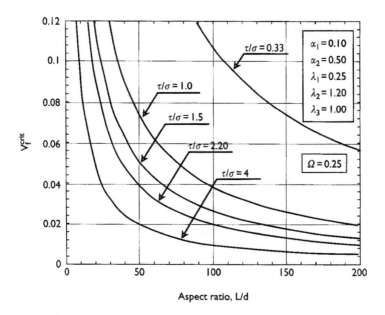

Figure 12.2 Relations between critical fibre volume to achieve tensile strain hardening and fibre characteristics (bond and aspect ratio), based on Eq. 12.4 (after Naaman [1]).

calculated assuming either (i) a perfectly plastic behaviour in tension, that is, a rectangular block of stress distribution in tension, as shown in Figure 4.43, with a tensile stress of $\bar{\sigma}_{pc}$ or (ii) a triangular stress distribution in tension having a tensile stress at the neutral axis which is equal to the tensile post-cracking strength of the composite, σ_{pc} and decreasing linearly to zero at the bottom of the cross section. The resulting conditions for deflection hardening are as follows:

For perfectly plastic behaviour:

$$\bar{\sigma}_{pc} \;\geq\; \frac{1}{3}\sigma_{cc} \tag{12.5}$$

For triangular stress distribution:

$$\sigma_{pc} \geq \sigma_{cc} \tag{12.6}$$

The value of $\bar{\sigma}_{pc}$ is the average of the stress estimated over the whole crack mouth opening in the beam under flexure, and its value is between the cracking stress in tension, σ_{cc}, and the post-cracking strength in tension, σ_{pc}. The actual value will depend to a large extent on the length of the fibre relative to the crack opening in bending. In any case, considering these boundaries, Eqs 12.5 and 12.6 can be

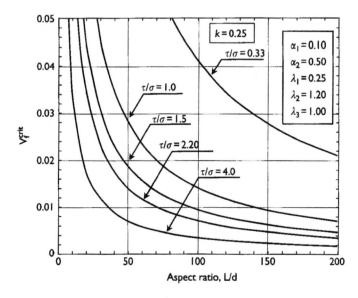

Figure 12.3 Relations between critical fibre volume to achieve deflection hardening and fibre characteristics (bond and aspect ratio), based on Eq. 12.8 (after Naaman [1]).

combined as:

$$\sigma_{pc} \geq k\sigma_{cc} \tag{12.7}$$

where k is a coefficient ranging between 1/3 and 1. This is consistent with the treatment of bending in Section 4.7 in which it was shown that the moment of resistance in bending in the post-cracking zone could exceed by a factor of about 3 to the moment of resistance at the first crack. These conditions can now be combined together to calculate the critical fibre volume for deflection hardening:

$$(V_f)_{critical-bending} = \frac{k}{k + (\tau/\sigma_{mu})(\ell/d)(\lambda_1\lambda_2\lambda_3 - k\alpha_1\alpha_2)} \tag{12.8}$$

Figure 12.3 provides a graphical presentation of this equation for characteristic composite values. It shows, as expected, that the critical fibre volume required for deflection hardening is smaller than for strain hardening (compare curves in Figures 12.3 and 12.2).

It should be noted that Eq. 12.8 can be viewed as a general one, outlining the conditions for strain hardening as well as deflection hardening: for $k = 1$ it is applicable for strain hardening and for $1/3 < k < 1$ it is relevant for deflection hardening.

12.2.1   Monofibre systems with normal strength matrix

The conditions set forth in Eqs 12.4 and 12.8 to obtain high performance FRC imply the need for the design of the fibres to achieve this purpose (fibre content, bond and aspect ratio). Li *et al.* [3] developed a complementary approach based on fracture mechanics in which the design of the matrix is also taken into consideration as one of the means to obtain cost-effective (e.g. low fibre content) strain hardening composites (for additional details see Section 4.4.3). The critical fibre volume, $(V_f)_{crit}$ can be calculated in terms of the fibre properties and the composite fracture mechanics toughness and characteristic crack opening:

$$(V_f)_{crit} = \frac{12 J_{tip}}{f \tau_{fu} \, (\ell/d) \, \delta_0}$$
(12.9)

where J_{tip} is the composite crack tip toughness; ℓ, the fibre length; d, the fibre diameter; f, the snubbing factor; τ_{fu}, the frictional bond strength and δ_0, the crack opening at maximum bridging stress, expressed in Eq. 12.10.

$$\delta_0 = \frac{\tau_{fu} l^2}{E_f d \, (1 + \eta)}$$
(12.10)

where E_f is the fibre modulus of elasticity

$$\eta = (V_f E_f) / (V_m E_m)$$

where V_f, V_m are fibre and matrix volume fraction, respectively and E_f, E_m are fibre and matrix moduli of elasticity, respectively.

The composite crack tip toughness in Eq. 12.9 is related to the matrix fracture toughness, K_m:

$$J_{tip} = \frac{K_m^2 \, (1 - V_f) \left(1 - v_m^2\right)}{E_m}$$
(12.11)

where v_m is the matrix Poisson ratio and $(1 - V_f)$, the a factor which takes into account the reduced crack front dimension due to the presence of fibres.

For an approximation in which $(1 - v_m^2)$ equals 1:

$$J_{tip} = \frac{K_m^2}{E_m}$$
(12.12)

The dependency of critical fibre volume on fibre–matrix bond and the composite crack-tip toughness is demonstrated in Figure 12.4, for ultra-high density polyethylene (Spectra) fibre, with modulus of 117 GPa and 38 μm diameter. It clearly shows that to maintain a fibre content below 2% there is a need for a bond strength of about 1 MPa. In order to accommodate fibres with bond of about 0.5 MPA, which is typical to many polymeric fibres, a fracture toughness below

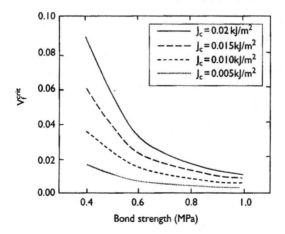

Figure 12.4 Relations between critical fibre volume and bond strength for differ-
ent levels crack tip toughness, based on Eq. 12.9 ($E_f = 117$ GPa, $l_f = 12.7$ mm, $d_f = 38$ μm, $g = 2.0$, $E_m = 25$ GPa) (after Li et al. [3]).

0.01 kJ/m² is required. Li *et al.* [3] thus explored the optimization of the system
by controlling the bond as well as the matrix to keep its crack-tip toughness sufficiently low; this obviously implies control of the matrix properties to be below
some critical strength value. This approach is consistent with lowering the composite first cracking strength to comply with the requirements set out in Eqs 12.3
and 12.7.

For optimization of the matrix Li *et al.* [3] considered at the same time various
parameters, in particular w/c ratio and sand content. The role of the sand content
turns out to be quite important for any given w/c ratio: higher sand content increased
the matrix fracture toughness, and had little influence on the bond strength with
the Spectra fibre. The similar influence of reducing the w/c ratio from 0.45 to
0.35 was also identified, leading to effects which are of the same nature as those
obtained by increases in the sand content. Thus control of the fracture toughness
of the matrix, by limiting the sand content, enables one to obtain a strain hardening
composite with fibre contents as low as 2%.

The theoretical requirement for strain hardening behaviour, based on these considerations, can be shown on a diagram of matrix toughness vs. bond strength
(Figure 4.23), demonstrating that systems in which the toughness remains sufficiently low can exhibit strain hardening with fibre contents lower than 2%. Within
this context it should be noted that the matrices with higher sand content, which
were less favourable for a strain hardening composite, exhibited higher compressive strength and modulus, which were induced by the presence of sand.
A more comprehensive account of the fracture mechanics approach involved in
this concept is provided in Section 4.4, Figures 4.19 and 4.20.

12.2.2 Monofibre systems with high strength matrix

A totally different concept, of using very dense and low w/c ratio matrices, for obtaining high performance FRC has been advanced in several systems with a variety of trade names, such as fibre-reinforced densified small particle systems (known as DSP or Densit – Densified cement ultra-fine-particle-based materials) [4,5,6] and RPC (Reactive Powder Concrete) [7]. There is an apparent contradiction between the two approaches outlined, with respect to control of the matrix properties – one limiting its properties (Li et al. [3]), with the other calling for a high performance–low w/c ratio matrix [5,7]. One should note however that in both cases the type of fibres used are quite different. Li et al. [3] refer to microfibres of small diameter of about 40 μm, in which the bond is not too high, about 0.5 MPa. In this system there is sufficient anchoring effect due to the small diameter of the fibres to provide strain hardening if the matrix is not too strong, and also a smooth transition from the linear elastic to the quasi-plastic stages of loading. The systems with the dense matrix use a meso-size steel of about 0.1–0.2 mm. The dense and controlled microstructure ensures sufficient bond enhancement in order to get the required anchoring effect, to provide strain or deflection hardening in a dense matrix; for example, the load-bearing capacity of the fibre, which is controlled by bond, exceeds that of the dense matrix. Some aspects of the mechanics of the dense matrix system are highlighted later.

The two approaches for obtaining high performance–high ductility FRC reflect different strategies for achieving enhanced toughness, by balancing between the strength properties of the matrix and the fibre bond strength. The nature of the fibres and their bond should be adjusted and simultaneously tailored to that of the matrix, with different 'tailoring' for normal strength and high strength matrices. Mihashi et al. [8] used the terms 'low initial bond design' and 'high initial bond design'.

The principles stated for RPC [7] include four elements to account for the high performance: (i) increased matrix homogeneity which is achieved by lack of coarse aggregates; (ii) high density which is achieved by carefully controlled grading of the sand, fillers, cement and silica fume, and whenever possible pressure casting; (iii) improvement of microstructure by heat treatment and (iv) steel fibres to improve ductility. The role of matrix homogeneity as outlined in (i) earlier has been questioned, as it was demonstrated in several experimental and analytical studies (see Section 12.3) that high performance, similar to that obtained with small aggregates (0.6 mm) could be achieved also with larger particles, in the size range of 5–8 mm [9,10]. In view of such reports, it is suggested here that the use of the high density matrix with well-graded fine particles is probably intended to a large extent (i) to enable the incorporation of 2–3% by volume of meso-steel fibres using normal mixing process (i.e. rheological control) and (ii) to provide enhanced steel–matrix bond to permit mobilization of the reinforcing effect of the short steel fibres, to obtain a strain hardening composite, that is, provide a system where the strength in the post-cracking zone is higher than that of the high strength matrix. This implies that the system is tailored so that the enhancement of the efficiency of the steel fibres, in terms of their content and bond, is greater than that of the increase in the matrix strength, thus eventually providing the conditions for strain or deflection hardening.

Support for the first effect (enhanced bonding) can be found in a study by Hoy and Bartos [11] who reported that increasing the fines content in the mix can enable incorporation of higher fibre content to achieve optimal packing and workability.

Support for the second effect is provided in a model developed by Tjiptoboro and Hansen [12], who compared the behaviours of conventional FRC and high performance FRC with a highly dense matrix. Their model is based on the balance between the energy required to *open* up the first crack formed during loading, E_{1-2}, and the energy required to *generate* new microcracks after the first crack formation, E_2 (see Section 4.5.4 for more details). If E_{1-2} is bigger than E_2, multiple cracking will occur, accompanied by strain hardening. This strain hardening will take place up to the stage where the cumulative energy involved in the generation of new microcracks ($\sum E_2, E_3, E_4, \ldots$) is greater than E_{1-2}. In this model it was assumed that the energy required to generate new microcracks is always equal ($E_2 = E_3 = \ldots E_n$). The critical fibre volume is the volume of fibres satisfying the condition:

$$E_{1-2} = E_2$$

The critical fibre volume according to this model is:

$$(V_f)_{ef-crit} = \frac{\gamma_m + (\gamma_m^2 + A)^{1/2}}{A} \tag{12.13}$$

where

$$A = 2\gamma_m + \frac{l}{r}\left(\frac{11}{48}\tau_{fu}^2\frac{\ell}{E_f r}\right) \tag{12.14}$$

$$(V_f)_{crit} = 2V_{ef-cr} \tag{12.15}$$

where $(V_f)_{ef-crit}$ is the effective critical fibre volume; $(V_f)_{crit}$, the critical fibre volume; γ_m, the matrix fracture energy, $G_m/2$ and τ_{fu}, the frictional bond strength.

The input parameters for the normal and high density matrix composites are provided in Table 12.1. The calculated critical fibre volume for the dense matrix (DSP) is 3.8%, whereas for the conventional matrix composite it is 15%. This is the range of the reinforcement in dense FRC systems such as DSP and RPC, which provide strain and deflection hardening (see Section 12.3).

Tjiptoboro and Hansen [12] attributed the marked reduction in the critical fibre content to the significant increase in the interfacial bond (1 vs. 5 MPa) and the debonding energy of the fibre, G_{II} (2.5 N/m vs. 120 N/m). They suggested that these differences are the result of changes in the microstructure at the fibre–matrix interface, due to the reduced w/c ratio and the dense packing of the matrix particles, consisting of cement and small fillers only, which practically eliminate

Table 12.1 Characteristic materials parameters used for the modelling of normal strength matrix and highs strength–low w/cm matrix composites (DSP matrix) (after Tjiptoboro and Hansen [12])

Materials	Fibre-reinforced DSP	Normal strength FRC
Matrix		
Compressive strength, MPa	175	20[a]
Modulus of elasticity, GPa	49.1	21[a]
Frictional bond stress, τ_f, MPa	5[a]	1[a]
Fracture energy, G_m, N/m	120	20[a]
Interfacial fracture energy, G_{II}, N/m	120[a]	2.5[a]
w/cm	0.18	0.4–0.6
Steel fibre		
Modulus of elasticity, GPa	200	200
Length, mm	6	25
Diameter, mm	0.15	0.5

Note
a Estimates, based on literature and other sources.

the transition zone (see Chapter 4). This is consistent with several studies of the nature of bonding in such matrices. Guerro and Naaman [13] showed that the use of microsand of 14 μm diameter is very effective in increasing bond of macrofibres, by a factor of two or more. This is in agreement with the reports of Shannang *et al.* [14] showing a similar effect in a DSP matrix, and of Chan and Chu [15] for RPC, where the frictional bond of smooth steel fibres was shown to be as high as 6 MPa, twice to three times higher than in normal mortar. Sun *et al.* [16] reported even higher bond values for steel fibres in RPC, 6–7.5 MPa for normal curing, 12.1–13.8 MPa for steam curing (90°C) and 12.9–14.2 for autoclave curing (200°C, 1.7 MPa pressure). Moreover, Chan and Chu [15] demonstrated that in the presence of silica fume the increase in the pull-out energy was greater than the increase in pull-out load, reflecting the observation that when the matrix was sufficiently dense and contained fine particles, the pull-out curve changed from slip softening to slip hardening. The issue of bond in RPC was also addressed by Orange *et al.* [17], showing bond values of 4.8 and 10 MPa for 0.1 and 0.2 mm diameter fibres. They attributed the high bond to the compressive hydraulic pressure developed around the fibres due to the shrinkage of the matrix (addressed as 'clamping stresses' in Section 3.2.4). Such clamping would be expected to be higher for larger diameter fibres.

12.2.3   Concepts of hybrid reinforcement

Rossi *et al.* [18] considered the crack evolution in an FRC composite in terms of three stages of crack formation: microcrack formation prior to peak load, coalescence of microcracks into one macrocrack at the peak load, and thereafter propagation of a macrocrack. They suggested that in the first stage the steel fibres limit microcrack propagation (e.g. similar to the concept of crack arrest in Section 4.4.1) and when the macro-cracks form and propagate, the fibres bridge over them, in an action which has similarity to that of conventional reinforcement. They addressed this mode of fibre action in terms of effects at two levels: the material (microcracking) and structure (macrocracking). Based on this analysis the concept of hybrid reinforcement was proposed, in terms of a large volume of short steel fibres to control microcracking and long fibres to control macrocracking, to induce ductility at the structural level. This was developed into a concept called multimodal fibre reinforced cement composite [19]. For this discussion the shape of the fibres will be classified into three categories, depending on the cross-sectional dimensions of the fibres:

- Microfibres, where the cross section is less than 100 μm, which is the size range of the cement particles.
- Mesofibres, with cross-sectional dimensions in the range of 0.1 to 0.3 mm.
- Macrofibres, larger than 0.3.

In most of these systems, the aspect ratio of the fibre is kept similar, about 50–200, and thus the length of the microfibres is usually in the range of 3–10 mm, while the macrofibres are usually longer than 20 mm. The small size of the microfibres enables denser packing of the cement particles around them, providing a potentially denser transition zone and higher interfacial bond.

The various concepts outlined earlier have led to different strategies for achieving high performance FRC systems, which were tailored with different types of fibres and matrices, to be compatible with various production methods. The range of composites which were developed, or are currently being investigated, are outlined in the following sections, with each describing composites of different concepts: high content fibre reinforcement (SIFCON, SIMCON – Section 12.3), dense matrix composites (fibre-reinforced DSP, RPC, DUCTAL® – Section 12.4), matrix with upper limit of fracture toughness (ECC – Section 12.5), systems with hybrid fibre reinforcement – 12.6, systems with specially engineered fibres – Section 12.7 and systems with cement–polymer matrix – Section 12.8.

12.3   High volume–high aspect ratio fibre reinforcement

The special nature of the cementitious matrix, which consists initially of discrete particles, makes it extremely difficult to incorporate a large volume of fibres, or fibres of high aspect ratio (exceeding 100). In production technologies based on mixing, the maximum content of fibres that can be incorporated is less than about 2% by volume, and when using special techniques, such as the Hatschek process, or spraying, the volume content can be increased to 5–12% when using short fibres.

These are considerable limitations when attempting to maximize the reinforcing potential of the fibres.

A method for producing high volume–high aspect ratio fibre–cement composites was developed by Lankard *et al.* [20–22], based on preparation of a preplaced steel fibre bed which is infiltrated with a fluid cement slurry. In this system, called SIFCON (Slurry Infiltrated Fibre CONcrete), the placement of the fibres can be accomplished by hand or by fibre dispensing units. The fibre content in the preplaced bed is controlled by the fibre length and aspect ratio; higher fibre loading can be obtained with the aid of vibration of the bed during filling. Fibre volumes as high as 20% can be achieved [22]. With such high fibre contents, it is possible to increase the flexural strength and toughness by more than an order of magnitude, compared with the unreinforced matrix, or to a matrix reinforced with a low fibre volume (Figure 12.5). Impressive gains were also achieved for compressive strength.

In the preparation of these composites, special attention is required to avoid non-uniform fibre distribution. Yet, because of the nature of the preparation of the preplaced bed, there is a tendency for preferred orientation, which depends on the size of the mould relative to the fibre length, and the method of place-ment. The effect of orientation can have a considerable influence on the properties (Figure 12.6) [23]. The type and size of the steel fibres may also affect the properties of the composite [23].

Naaman and Homrich [24] studied the strengthening mechanisms in SIFCON and found that the mode of failure was by fibre pull-out without fibre fracture. The tensile strength achieved with hooked and deformed fibres was roughly similar,

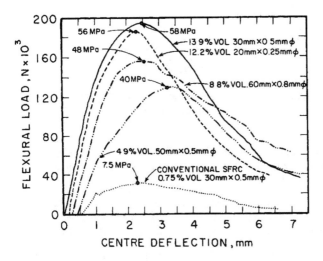

Figure 12.5 Effect of fibre content in SIFCON product on the load–deflection curve of the composite (after Lankard [22]).

about 15 MPa (Table 12.2). The modulus of elasticity was much lower in the hooked fibre composite (Table 12.2). It was hypothesized that this may have to do with the rougher surface of the deformed fibre which provides a better restraint to shrinkage, thus reducing pre-loading shrinkage cracking which can lead to a reduction of the modulus of elasticity.

The mix composition of the slurry is particularly important, to ensure efficient infiltration during production, as well as to control the properties of the hardened concrete. The slurry can be made either of paste or mortar with fine sand particles. Usually, superplasticizers are needed to improve the flow properties, and in many instances the mix contains fly ash and silica fume [22,23,25]. Changes in the composition of the matrix were found to have a considerable effect on the properties of the hardened composite [23].

It is interesting to note that in spite of the high cement content of the matrix, the shrinkage of SIFCON is an order of magnitude less than that of the matrix, not exceeding the shrinkage of normal concrete (Figure 12.7). This is due to the high fibre content, as well as to the nature of the reinforcing bed, in which the fibres

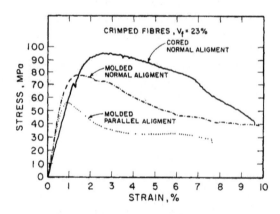

Figure 12.6 Effect of fibre orientation on the behaviour of SIFCON product under compression (after Homrich and Naaman [23]).

Table 12.2 Tensile properties of SIFCON composites with hooked and deformed fibres in a 0.26 w/c ratio matrix (after Naaman and Homrich [24])

Type of fibre	Fibre content % vol.	Modulus of elasticity GPa	Maximum stress, MPa	Strain at maximum stress, %	Stress at 2% equivalent strain, MPa
Hooked	12.1	4.6	15.7	1.21	13.7
Deformed	13.8	13.9	16.1	0.68	11.6

form contact with each other, generating an effect which was termed by Homrich and Naaman [23] as 'fibre interlock'. It should be noted however that shrinkage could lead to non-linearity in the stress–strain curve, which is associated with the presence of some internal cracks prior to loading [22].

In the SIFCON technology the fibres are placed into the mould to prepare the pre-bed reinforcement. This is a tedious operation which may require labour-intensive placing on site. Alternative methods have been developed, producing a mat of the fibres, which can then easily be transported as rolls, and readily be placed and impregnated with the slurry. With such mats it is possible to use fibres of higher aspect ratio and to have a better control of orientation, thus enabling the reduction in the critical fibre content. This advantage comes on top of the ease of handling and the better quality control in production of such composites and their site application. Two types of products of this kind have been reported in the literature, one consisting of steel wool [26] and the other of conventional steel fibres such as those used in SIFCON. The composite produced for the latter type of mat is called SIMCON, Slurry Infiltrated Mat CONcrete [27].

The steel wool consists of fibres with a length exceeding 50 mm and diameters of 100–130 μm, that is, an aspect ratio of 400–500 (Figure 12.8) [26]. This is much greater than the aspect ratio of conventional steel fibres. Infiltration of such wools with a 0.3 w/c slurry yielded composites containing 2–3% vol. of fibres, in which the flexural strength was as high as 25 MPa, and the toughness index was about an order of magnitude greater than that of the matrix (Figure 12.9) [28]. These properties are comparable with those of asbestos cement (a somewhat lower flexural strength but a higher toughness index) and they were achieved with only a modest fibre content. In such composites, an increase in the fibre volume over an optimum value results in reduction in properties, due to difficulties in

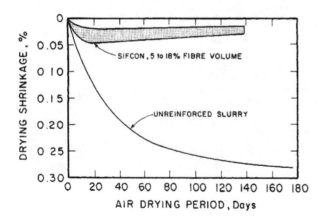

Figure 12.7 Drying shrinkage of the SIFCON composite and of the matrix slurry, at 23°C/50% RH (after Lankard [21]).

(a)

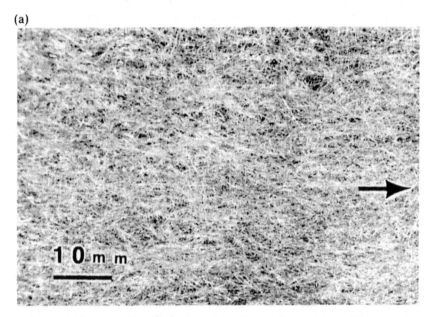

(b)

Figure 12.8 Micrographs of steel wool at different magnifications. The longitudinal direction of the mat is marked by an arrow in (a) (after Bentur and Cree [26]).

(c)

(d)

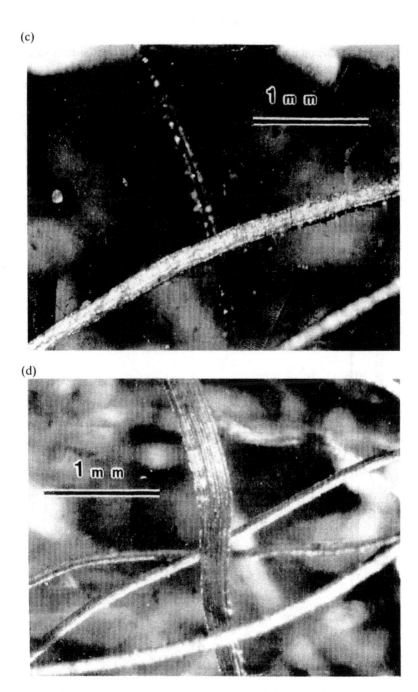

Figure 12.8 Continued.

compaction. This optimum value is dependent on the production process, and it might be increased by using more fluid slurries and more intensive compaction techniques (i.e. higher pressure in combination with dewatering). The high flexural strength which is achieved with only a modest fibre loading reflects the high efficiency associated with the high aspect ratio of the fibre incorporated in this composite. This could be seen in the failure mode of the composite which was characterized by fibre fracture rather than by fibre pull-out, thus utilizing their strength (Figure 12.10(a)). However, this did not lead to a brittle response, since the steel fibres failed in a ductile mode, as can clearly be seen by their necking (Figure 12.10(b)). Thus, in such composites, the strength as well as the ductility of the fibres is efficiently utilized, achieving the 'plastic-release' mechanisms proposed by Morton and Groves [29] and Bowling and Groves [30] (Section 3.2.4).

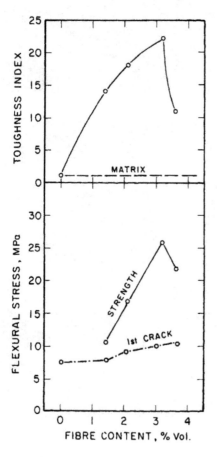

Figure 12.9 Effect of the steel wool content on the flexural properties of the composite (after Bentur [28]).

(a)

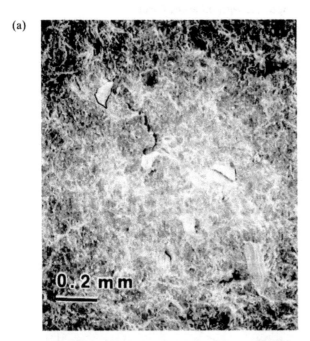

(b)

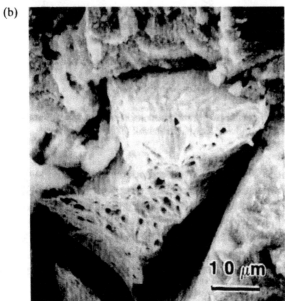

Figure 12.10 Fracture of steel wool-reinforced cementitious composite (after Bentur [28]): (a) Fibre fracture; (b) high magnification, showing fracture of steel fibre in the wool by plastic yielding and necking.

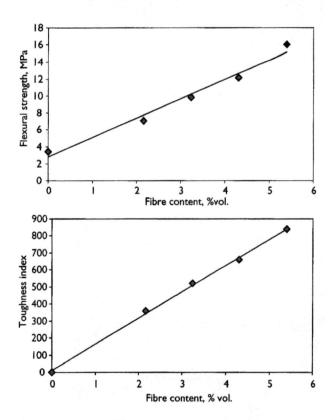

Figure 12.11   Relations between fibre content and mechanical properties in SIMCON
system (after Krstulovic-Opara and Malak [31]).

The tensile properties of SIMCON were reported by Krstulovic-Opara and
Malak [31,32], and it was shown, as expected, that the efficiency of SIMCON
was much greater than that of SIFCON. With fibre contents in the range of 3–5%,
the tensile strength obtained was in the range of 10–16 MPa (Figure 12.11). For
a similar strength of about 16 MPa, 14% of fibres were required in a SIFCON
system. This behaviour reflects a very high efficiency of the fibres, as can be
concluded by analysing the results in [31] in terms of the rule of mixtures, which
leads to the following relation:

$$\sigma_{tu} = \sigma_m V_m + 0.44 \sigma_f V_f \tag{12.16}$$

The equation was based on the experimental results in [31] (Figure 12.11), taking
the tensile strength of the fibres as 551 MPa, which is the value recommended by
the manufacturer.

The high efficiency of the SIMCON system is the result of a mat consisting of high aspect ratio fibres (241 mm long, 0.33 mm diameter, 713 aspect ratio) with 60–70% alignment induced in the production of the mat. The linear relations observed in Figure 12.11 suggest that the mechanics of composites can be applied to account for the strengthening of the fibres. Krstulovic-Opara and Malak [31] suggested that a portion of the fibres will fracture, while other will pull out, in contrast to a SIFCON system where all the fibres pull out. In a subsequent paper, Krstulovic-Opara and Malak [32] carried out a micro-mechanical analysis based on energy conservation principles and assumed that failure occurs by fibre debonding and pull-out. They demonstrated that an increase in fibre content to 5.39% resulted in more than doubling of the strain at first crack.

Oluokum and Malak [33] systematically evaluated the properties of SIMCON composites, and studied numerous effects such as orientation, aspect ratio and volume content. They concluded that the high concentration of fibres and high bond results in a small spacing between the fibres and enhanced interaction and load transfer between them. This type of behaviour is characteristic of a composite with continuous reinforcement, and therefore the behaviour is independent of the aspect ratio in the range studied (713 and 500).

The flexural properties of SIMCON were reported by Bayasi and Zeng [34], using the JCI (Japan Concrete Institute) standard for toughness index evaluation, for deflections of $\ell/25$, $\ell/50$ and $\ell/150$. The toughness at $\ell/150$ of the span (2 mm deflection) is considered as the standard value for conventional FRC. However, since the peak deflection of SIMCON in this test is about 2 mm, it was suggested that higher deflections may better account for the properties of this composite. Comparison of the toughness values for deflection of 2 mm (1/150 of the span), 6 mm (1/50 of the span) and 12 mm (1/25 of the span) are presented in Figure 12.12, showing that the toughness index is highest for the $\ell/50$ deflection.

A variety of applications of SIFCON and SIFCOM have been considered, and studies were carried out to support such uses. Of greatest interest have been rehabilitation of structures, seismic resistant components, blast-resistant structures, and thin precast products [35–41]. For some of these applications the dynamic properties of the SIFCON or SIMCON, both as materials and as part of structural components, in combination with conventional reinforcement, were studied and modelled [35–38]. As expected, they were found to be superior to those of other cement composites, because of the high toughness induced by the large fibre volume. In view of the expected uses of SIMCON in combination with reinforcing steel, the interaction and bonding between this composite and reinforcing bars was studied [39,40]. Much better interaction than in conventional reinforced concrete was reported, which shows up as an increase in the bond strength and the energy absorbed during pull-out as well as reduced cracking in a reinforced member.

A modified SIMCON system, called DUCON [42] was evaluated for overlay and floor applications. In this system the reinforcement consists of mat layers, rather than discrete fibres, allowing a better control of the reinforcement, as well as production of cages with cavities.

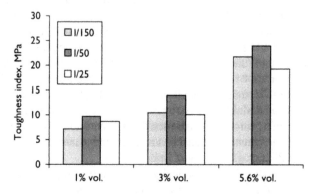

Figure 12.12 Effect of fibre content in SIMCON composite on the JCI fracture toughness factor for different deflections, 1/25,1/50,1/150 (adapted from Bayasi and Zeng [34]).

12.4   Mono-fibre systems with high density matrix

Several types of high performance fibre reinforced cement systems having extremely dense matrix and strain hardening behaviour were developed, and some of them have already been applied, at least in demonstration projects. A variety of names have been given to the different composites in this class, such as CRC (compact reinforced composite), which is essentially a matrix of fibre reinforced DSP combined with dense steel bar reinforcement, RPC (reactive powder concrete), DUCTAL® (which is a newer commercial generation of RPC) and CARDIFRC®. A common feature of all of these composites is the establishment of an extremely dense matrix with w/c smaller than 0.20 and incorporation of 2–6% of meso-steel fibres with diameter in the range of 0.1–0.2 mm and length of 5–15 mm. The matrix design is based on the principles of optimized dense packing of a mix with particles smaller than few millimetres, in combination with HRWR (high range water reducer, or superplasticizer). The combination of optimal packing and the effective water dispersant leads to a workable mix with an extremely low w/c ratio, which is self levelling in nature. This mix can incorporate several percentages by volume of the meso-steel fibres, and maintain its workability. Concepts of optimum packing and high workability are described in several references, for example [11,43].

The theoretical background and the mechanics concept of such composites was discussed in Section 12.2.2, and in this section the properties reported and the characteristics of different systems will be reviewed.

A common feature of all of these composites is the dense matrix made of well-graded particles, ranging from about 0.1 μm silica fume to sand particles with maximum size in the range of 0.6–4 mm. In several reports [9,10] the effect of the maximum size of the graded aggregate, up to 8 mm, was evaluated. The main differences in the concept of the different composites in this class is the maximum size of the graded particles, about 0.6 mm in RPC/DUCTAL® and 4 mm in DSP/CRC. Properties of different materials that may be classified in this category are shown in Table 12.3 [9,12,44] for mixes with different gradations, characterized with maximum grain sizes of 0.6, 4 and 8 mm. The effect of curing conditions (room temperature, steam curing and autoclaved curing) are shown in Table 12.3. Common features in these systems, in addition to the range of w/c ratio and fibre content outlined earlier, is a relatively large cement content in the range of 750–1000 kg/m^3 and compressive strength in the range of 100–160 MPa. The strength can be increased up to about 250 MPa by thermal curing. The flexural strength is in the range of 20–60 MPa, with the higher values obtained by thermal curing.

Recommendations for design, testing and evaluation of such composites were developed in France [45]. Using these recommendations, Chanvillard and Rigaud [46] evaluated the tensile and flexural properties of DUCTAL® using notched specimens. Elastic–plastic behaviour in tension was observed, up to crack widths

Table 12.3 Properties of composites with low w/cm high strength matrices

Ingredient		DSP, [12]	RPC, [44]		RPC with range of aggregates, [9]	
			20° and 90°C curing	Autoclave curing	Fillers only	Graded aggregate
Cement		750 kg/m^3	I	I	934 kg/m^3	937 kg/m^3
Silica fume		179	0.23–0.25	0.23	234	235
Filler, < 1.0 mm		572[a]	1.1–1.49[b]	0.89[c]	1030[d]	—
Sand fraction (1–4 mm)		460–690	—	—	—	—
Graded aggregate (up to 8 mm)		—	—	—	—	1031
Water		169	0.17–0.19	0.19	215	200
Fibre content		3–12% vol.	1–3% vol.	1–3% vol.	2.4%	2.4%
w/cm		0.18	0.14–0.15	0.15	0.18	0.17
Compressive strength, MPa	20°C curing	—	170–230[e]	—	~160	~160
	90°C curing	—		—	~160	~160
	Autoclave curing	—	—	490–680	~190–210	~190–210
Flexural strength, MPa	20°C curing	~10–40	30–60[e]	—	45.7	35.5
	90°C curing	—		—	48.3	40.6
	Autoclave curing	—	—	45–141	60.1	40.7

Notes
a 33% of the filler particles were smaller than 250 μm; the rest are between 250 μm and 1mm.
b The filler for the 90°C curing contained 26% of particles of about 10 μm average size and the rest 150–600 μm size; the filler for the 20°C curing consisted only of the 150–600 μm size particles.
c The filler for the autoclave curing contained 44% of particles of about 10 μm average size and the rest 150–600 μm size.
d 150–400 μm size range.
e Lower range for the 20°C curing.

of about 0.35 mm. Calculations based on the rule of mixtures for the first crack in tension suggest that the contribution of the fibres to the elastic limit is about 9% (~1 MPa). It was suggested that this is an indication that in DUCTAL® the fibres contribute at the materials scale (enhanced first crack) and at the structural level (enhanced post-crack strength).

The elastic limit in bending and tension reported by Chanvillard and Rigaud [46] were 18.8 and 11.5 MPa, respectively. The difference between the two was explained by scale effect which does not exist in an elastic–brittle material. The scale effect is associated with the damage in bending ahead of the crack tip; if this zone is large relative to the cross section, the flexural strength is expected to be higher [12,47]. The scale effect is of significance in the design of thin-walled elements [12,47] and it also shows up in the results obtained in the testing of beams in bending, as seen for the flexural strength values and the ratio between flexural and tensile strength (Table 12.4). This is consistent with the data reported by Collepardi et al. [9] showing flexural strength values of $40 \times 40 \times 160$ mm beams to be much higher than those of $150 \times 150 \times 600$ mm beams. Since the use of such materials is expected to be in thin cross-section composites, the values obtained with smaller cross sections of about 40 mm seem to be relevant.

The first composite in this class was developed on the basis of the DSP concept in which the matrix is made of a mix of cement, 20% silica fume and graded fine aggregate with maximum size of 4 mm [5,6,48]. With an appropriate dispersing agent workable mixes of low w/c ratio of about 0.20 or somewhat lower can be prepared. These mixes can maintain a good workability and incorporate up to 6% of straight, smooth steel fibres of 6–12 mm length and 0.2–0.4 mm diameter. These composites were used as a matrix for a heavily reinforced steel bar component known as CRC (compact reinforced concrete) for a variety of structural applications [49]. The matrix itself possesses superior mechanical properties, having compressive strengths in the range of 100–130 MPa. The composites had flexural strength values of 15–25 MPa and tensile strength of 10–20 MPa, depending on the fibre content.

Lange-Kornbak and Karihaloo [10,50] developed a generalized model to determine the optimum composition and properties for strain hardening FRC,

Table 12.4 Scale effect based on experimental result of flexural and tensile tests of high strength–low w/c ratio matrix FRC composite, DUCTAL® (after Chanvillard and Rigaud [46])

Specimen size, mm	Flexural strength, MPa	Flexural/tensile strength ratio
$40 \times 40 \times 160$	19.0	1.76
$70 \times 70 \times 280$	16.3	1.51
$100 \times 100 \times 400$	15.3	1.42
Direct tension	10.8	—

based on fracture mechanics concepts. The model was a general one, and was applied in particular to dense matrix composite such as DSP. The following conditions were identified for optimization of the properties of the composite [10]:

1   For optimum strength the fibre–matrix bond strength, τ, should be maximized and the brittleness, B, should be minimized,

$$B = \tau_v^2 d^2 / E_f G_{db}^i \tag{12.17}$$

where E_f is the fibre modulus of elasticity; G_{db}^i, the fracture energy of fibre–matrix interface and τ_v, the shear stress at this interface at the onset of strain softening.

2   For optimization of pre-peak deformation (strain hardening) capacity, the ductility, D, should be maximized,

$$D = V_f' \tau_n' / K_{Ic}^m \tag{12.18}$$

where τ_v' is the shear stress at onset of pullout, which is a function of τ_v and B; $\tau_v' = f(\tau_v, B)$ and K_{Ic}^m, the effective fracture toughness of the FRC matrix.

3   To optimize post-peak deformation (tension softening) capacity, the value of $V_f \tau_f / K_{Ic}^i$ should be maximized where K_{Ic}^i is the fracture toughness of the fibre–matrix interface.

A design for optimization of the composition based on these principles, to achieve high resistance to cracking and maximize tensile strength for a given compressive strength was carried out. The composition meeting these requirements included a matrix with 70% aggregate of maximum size of about 4 mm, and fracture toughness of the paste, $K_{Ic,p}$ of about 0.44 MPa·m$^{1/2}$. This is consistent with the properties determined experimentally with fibre reinforced RPC with a maximum particle size in the range of 0.23–5 mm (w/c ratio of 0.20) [51]:

•   The compressive strength with particle size of 5 mm was similar to that obtained with the fine aggregate of 0.23 mm.
•   The tensile strength was only slightly lower.
•   The values for tensile strength for the mixes with 0.23 and 1.18 mm particles were similar, and only slightly higher than those of mixes with 5 mm aggregate.

These concepts were extended in a series of studies by Karihaloo and co-workers [52,53] to provide models based on micromechanical considerations which can be used to optimize the performance of the composite in terms of the composition and properties of the matrix and fibres. This methodology provided the basis for a high performance FRC with dense matrix, which was patented and called CADIFRC® [54]. This composite, similar to others in this class, is based on maximizing the density of the matrix by a continuous grading of its particles in the range of 0.5–2 mm, and reinforcing it by 0.15 mm diameter steel fibres with length of 6 and 13 mm. The constitutive equations used to describe and optimize the composite address the different stages of the loading of the composite:

- The strain hardening zone, after the first crack and up to the maximum stress, is treated in terms of the arresting influence of the short fibres on the growth of microcracks which nucleate at this stage.
- The progressive reduction in stiffness in this zone is due to nucleation of microcracks [55].
- When the strain hardening capacity is exhausted, some of the fibres debond and microcracks coalesce, cracks open up, and at a certain stage a reduction in load will occur, leading to strain softening.

Tjiptobroto and Hansen [12] investigated these composites with steel fibre reinforcement of 0.15 mm diameter and 6 mm length, at contents in the range of 3–6% by volume. The flexural behaviour demonstrated high ductility, with a considerable strain hardening effect as shown in Figure 12.13. The strain capacity of the matrix was 150 microstrain and it increased to 2000 microstrain in the 12% fibre composite.

In the RPC composite a great emphasis is given to the properties of the matrix, to maintain its homogeneity by limiting the size of the quartz to 0.6 mm and improving the microstructure by thermal curing. The combination of these two characteristics was proposed to be the key element for the superior behaviour, as seen by the high compressive and flexural strengths [44,56]. A detailed study of RPC microstructure and properties was presented by Richard and Cheyrezy [44,56]. Their mix design was optimized for density at low w/c ratios, in the range of 0.08–0.20, with a binder containing 25% silica fume by weight of cement (based on consideration of the optimum packing of the binder itself, in addition to the optimization of the whole mix, that is, the binder as well as the aggregates whose maximum size was 0.6 mm). Heat curing and its combination with pressing of the fresh paste through the whole setting period, resulted in reduction of porosity and enhanced homogeneity of the microstructure.

The significance of homogeneity and the associated need for a matrix with a maximum quartz size of 0.6 mm was questioned [9]. Collepardi et al. [9] studied an RPC system with graded aggregates of a size of up to 8 mm, keeping the w/c ratio in the same range (0.20–0.25) which was slightly higher than in [44,56]. The results indicated that the mechanical properties were largely retained, implying that homogeneity is not necessarily a key factor [9]. It was suggested in the discussion in Section 12.2.2 that it is the bond obtained in such well-graded systems which is of key importance.

The long-term performance of such composites has been evaluated, especially in view of the potential benefits of the dense microstructure. Although the system is dense, shrinkage may be significant due to autogenous shrinkage which is an inherent characteristic of the low w/c ratio matrix. Values of 300–400 microstrain

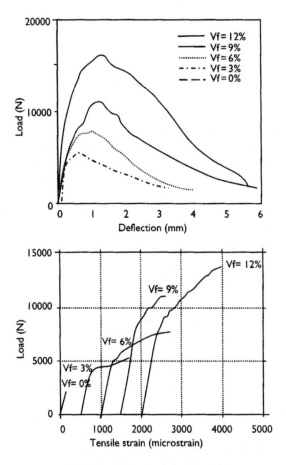

Figure 12.13 Effect of fibre content on load–strain curves in tension and load–deflection curves in bending of steel fibre composites with DSP matrix (after Tjiptobroto and Hansen [12]).

for autogenous shrinkage, were reported [57,58] and it was suggested that this provides a major contribution to the overall shrinkage of RPC cured at normal temperature (600 microstrain in [9]). This shrinkage is consistent with internal desiccation at 7 days to 88%RH for a 0.14 w/c composite. This high shrinkage (which shows up in the normal cured composite but not in the heat treated one) is not necessarily an issue, and Habel [57] reported that the stresses that may be potentially generated under restrained conditions can be relaxed by the early age viscoelastic response of the material; the creep at this stage can amount to 60% of the free autogenous shrinkage.

The early age shrinkage can be drastically reduced, by an order of magnitude, in autoclave curing, leading to a very stable material [9,58]. This is consistent with the drastic reduction in the creep of autoclaved cured composite, with specific creep values declining by an order of magnitude from about 35×10^{-6} MPa^{-1} specific creep of the room temperature cured composite [9]. These characteristics demonstrate the benefits of the heat curing which is applied in RPC/DUCTAL®.

The superior resistance to a variety of degrading effects was reported by Bonneau et al. [59] for RPC: freeze–thaw – durability factor equal or greater than 1; scaling – cumulative mass of scaling residue less than 0.03 kg/m^2 (compared with a maximum of 0.80 kg/m^2 allowed in the standards); chloride permeability (ASTM C1202) – less than 10 Coulomb (from a practical point of view this is impermeable concrete; a range of 500–1000 Coulomb is considered as high performance concrete). The permeability under load conditions, in which cracks are induced, was evaluated by Charron et al. [60] showing that at strains of up to 0.13% the permeability was as low as that of the matrix. This is an indication of the durability potential of such composites under load.

Several modifications of the matrix and fibres in RPC were reported, in attempt to optimize its cost-effectiveness [16]. Replacement of about 50% of the cement with fly ash and slag, keeping a w/c ratio of 0.15, resulted in composites with properties similar to 'regular' RPC: flexural strength in the range of about 30–50 MPa, for fibre contents in the range of 2–4%; the higher range values are for the higher fibre contents and heat curing. The interfacial bond strength in these systems was also high, reaching 14–16 MPa for the various curing regimes. Freeze–thaw durability and chloride penetration resistance were similar or better than 'conventional' RPC.

A new generation of RPC in which the differences between compressive and flexural characteristics are reduced was developed by improving the fracture toughness of the matrix, using non-isotropic fillers or natural inorganic microfibres [61]. Mica flakes and wollastonite microfibres at 2% and 10% content, respectively, enabled a marked increase in the fracture toughness of the matrix from about 10 J/m^2 to about 18 and 30 J/m^2. The additives used substituted for the sand fraction in the matrix. These modifications enable an increase in the flexural strength of the new generation composites (called DUCTAL – A) from about 25 to 35 MPa in conventional curing. The use of wollastonite microfibres as a means for reinforcement of the cement matrix was reported also by Low and Beaudoin [62–64]. They demonstrated that these fibres, which are about 25–30 μm in cross section and 0.4–0.6 mm in length can increase the flexural strength of a 0.35 w/cm ratio mix from about 8.5–9.3 MPa to 20.9–24.9 MPa when added at a content of 11.5% by volume. Higher contents result in a slight decrease in strength. The increased flexural strength was also accompanied by an increase in toughness as could be estimated from the enhanced post-peak load-bearing capacity. However, over time (360 days in water), the toughness tended to reduce while flexural strength remained constant. This change occurred in spite of the fact that the fibres were stable chemically, and it may be accounted for by microstructural effects which take place more readily in stiff microfibres (see Section 5.2). Such reduction in toughness may have some effect on DUCTAL-A, although their influence may be minimized due to the presence of macrofibres.

An alternative method for improvement was based on the incorporation of sub-micron additives intended to improve fibre–matrix bond [61]. A variety of additives such as precipitated silica, calcium phosphates and polymers were used in DUCTAL-B formulations. A combination of matrix toughening and bond enhancement means were applied in DUCTAL-C, leading to flexural strength values of about 50 MPa [61].

A range of applications of these materials have been considered and demonstrated such as shell structures [65] and bridge deck slabs [66,67]. In applications of this kind the mechanical properties as well as durability are considered.

A different approach for achieving strain hardening composite using small microfibres in a dense matrix was developed, based on special processing of the matrix by means such as extrusion [68–71]. These special processing techniques generate adequate matrix microstructure and fibre–matrix bonding, and they can thus be viewed as alternatives to the methods reviewed earlier which are based on rheological control (through particle size distribution and superplasticizers), without the need for special production techniques. Extensive studies of the mechanical properties and microstructure of such composites, in particular PVA fibre extruded systems investigated by Shah and his co-workers [68–71], were used to identify the mechanism and processes controlling these composites. These systems consisted of a 0.26–0.28 w/c matrix with different contents of sand, and were reinforced with PVA microfibres (3% vol. content, 14 μm diameter and 41 GPa modulus). With such systems strain hardening could be readily achieved, having tensile and flexural strength values as high as 9.1 and 33.8 MPa, respectively (Table 12.5). These values are considerably higher than those obtained by casting techniques (Table 12.5).

The main microstructural characteristics of the extruded system which account for the high performance are uniform dispersion of the fibres and lower porosity in general, in particular at the interface. These characteristics lead to enhanced fibre efficiency, especially the much higher bonding which is induced by 'peeling' of the fibre during pull-out (Figure 12.14(b)) and resulting in mechanical anchoring. The pulled out fibre did not exhibit this unique failure mechanism, and the SEM observations suggested that its bonding was probably controlled by interfacial shear (Figure 12.14(a)). These characteristics could account for the influence of the fibre length and sand content on the performance of the composite. An increase in fibre length from 2–6 mm resulted in reduction in toughness (Table 12.5), which was attributed to the high bond in the extruded system; it was suggested that 6 mm is longer than the critical length, leading to a greater amount of fibre fracture at

Table 12.5 Properties of composites with 3% PVA fibres, produced by extrusion and casting (after Shah et al. [68])

Processing	Fibre length, mm	Tensile properties		Flexural strength, MPa
		Strength, MPa	Toughness[a]	
Extruded	2	9.1	96	33.8
	6	7.3	49	25.0
Cast	2	2.8	29	20.1
	6	5.3	59	29.2

Note
a Area under the stress–strain curve, up to 1.2% strain $\times 10^{-2}$.

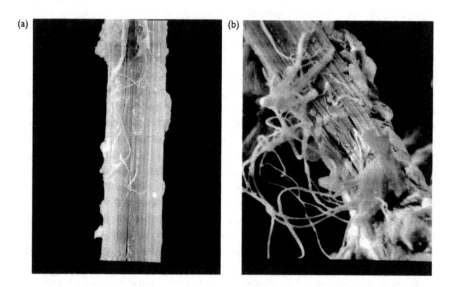

(a)   (b)

Figure 12.14 PVA fibre surface after testing a composite produced by (a) casting and (b) extrusion (after Shah et al. [68]).

the expense of fibre pull-out [68]. Addition of fly ash could reduce the detrimental effect of the increase in length by reducing the fibre–matrix bond [71]. The trend for lower performance with longer fibres in the extruded composite is opposite to the one observed in the cast composite (Table 12.5). In the latter case the level of bonding is lower, and increase in length is therefore beneficial. The addition of sand, even at small contents of 6% and 12% by weight of cement, was found to lead to a reduction in the properties of the composite [70], and this was accounted for by the detrimental effect of sand on the dispersion of the fibres. It should be noted that the extrusion resulted in some alignment of the fibres [70,72], which may also be one of the causes for the enhanced mechanical performance.

The extrusion process was found to be effective in providing strain hardening behaviour with microfibres other than PVA, such as cellulose, polyacrylonitrile, glass and polyethylene terephthalate (PET) [73,74]. In these fibres, there was a need to adjust the fly ash content, to levels of up to 80% replacement, to optimize the flexural strength and toughness, for the same reason outlined earlier for the longer PVA fibres [72]: the bond without fly ash was too high, leading to failure by fibre fracture rather than pull-out. With the fly ash adjustment flexural strength as high as 30 MPa was reported.

12.5   Mono-fibre systems with normal strength matrix

Concepts for obtaining strain hardening composites with normal strength matrix and moderate fibre content of about 2% by volume were developed by Li and co-workers [75–77]. Fracture mechanics and micromechanics concepts were used to optimize the performance of the composite and provide tools for tailoring the composition and properties of the fibres and matrix. Composites developed based on this concept were called by Li 'Engineered Cementitious Composites' (ECC) [75]. A central element in this analysis is the characteristic bridging stress-crack opening curve (Figure 4.16) which can be modelled and predicted as a function of the fibre content, modulus, angular distribution and interfacial properties [78] (for more details see Section 4.4).

Two guiding concepts were identified [75–77]:

1   'Strength criterion' to enable strain hardening with a modest amount of fibres. Eqs 12.9–12.12 outline the quantification of this criterion, which requires that the first crack strength will not exceed the maximum bridging stress of the crack in the characteristic stress–crack opening function.
2   'Energy criterion' to assure that steady-state multiple cracking will occur, that is, the crack which develops is going to be of the type labelled 'steady-state flat crack' and not 'Griffith crack'. The quantification of this criterion and the parameters to control it are discussed in detail in Section 4.4.3.

An essential element required to meet these criteria is to control the bond strength: weak bond will result in low bridging strength and will compromise the first criterion, while bond which is too high will lead to a maximum bridging stress at small crack opening and will compromise the energy criterion.

The two criteria show up clearly in characterization of the behaviour of this composite by fracture mechanics testing, in which distinction between the bridging effect and the microcracked zone ahead of the crack (fracture process zone) are characterized (Figure 12.15). In such tests the total fracture energy, J_t, and the contributions to it by the fibre bridging effect, J_b, and the microcracked damage, J_m, can be determined. In Figure 12.16 the results of such characterization for PE fibre composites are presented, as a function of fibre content. It can be seen that there is a need for a critical volume of more than about 0.4–0.8% of fibres to obtain a large effect of microcracking ahead of the crack tip, both with respect to the magnitude of the energy (J_m) and the area which the microcracked zone covers in front of the crack (the values in parenthesis in Figure 12.16). This is consistent with tensile tests of similar composites showing that this was approximately the critical volume required to obtain a ductile behaviour of strain hardening.

Another factor which was found to be of significance for maximizing the ductility for a given fibre content is the distribution of the fibres [77]. Uneven distribution was found to reduce the ultimate strain for the composite, and a linear relation could be established between the distribution factor (a value between 0 and 1, with 1 being ideally uniform distribution): for 2% PVA fibre composite in a 0.42 w/c ratio matrix, the ultimate tensile strain varied in the range of about 2–7% with a change in the distribution factor from about 0.64 to 0.79 (ultimate strain increased linearly with the distribution coefficient).

Earlier studies with this composite were based on the use of high performance polyethylene fibre (ultra-high molecular weight polyethylene) [78–82]. However, in view of the high cost of this fibre, alternative options were considered, and PVA

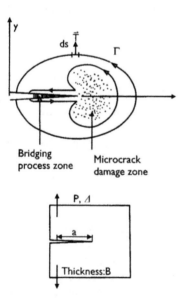

Figure 12.15 Schematic description of the fibre bridging effect (bridging process zone) and microcracking (microcrack damage zone) in fracture mechanics evaluation (after Maalaj et al. [76]).

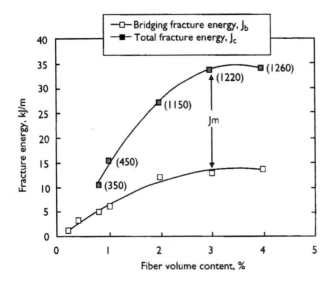

Figure 12.16 Effect of polyethylene fibre volume content on the total energy
expanded in fracture and the contribution of the bridging process
zone and the microcrack damage zone (see Figure 12.15); the values
in paranthesis represent the area of the microcracked zone (after
M. Maalej et al. [76]).

Table 12.6 Properties of PVA and ultra high density polyethylene
for making ECC type composites (after Matsumoto
et al. [82])

Property	PVA	Ultra high density PE
Length, mm	12	8
Diameter, μm	37.7	12
Modulus of elasticity, GPa	36.7	75
Tensile strength, MPa	1610	2790
Interfacial bond strength, MPa	2.01	0.66

was found to be suitable in terms of effectiveness and cost [82]. The properties of
the two fibres are provided in Table 12.6.

Both types of fibres, PE and PVA, could provide strain hardening at modest
fibre contents of 2–3% by volume, but the PE had some advantage in its ability to
provide a greater extent of multiple cracking. This difference could be correlated
with the mode of failure, in which the PVA fibre composite exhibited fibre fracture
while the PE composite was characterized by a greater extent of fibre pull-out.
The fraction of fibre fracture was 70% for the PVA and 35% for the PE [82].

This difference was mainly the result of the higher bond developed in the PVA, with part of it being of the chemical (adhesional) type [83]. With this effect in mind studies for tailoring the PVA fibre–matrix interfacial bond were carried out, to optimize the bond in PVA to achieve 'saturated' multiple cracking with crack spacings of about 2 mm [83,84] using a modest fibre content of 2% by volume. The presence of chemical (adhesional) bond changes the shape of the bridging stress-crack opening curve leading to a reduction in the complementary energy (the shaded area in Figure 4.20). Such a reduction may lead to conditions where the energy criterion is not satisfied (the fracture energy of the matrix should be smaller than the complementary area at the peak bridging stress in Figure 4.20 in order to induce multiple cracking and strain hardening; for more details see Section 4.4.3). In a study of the bond of PVA in a 0.45 w/c ratio matrix, the chemical (adhesional) bond and frictional bond were determined, as a function of oiling treatment, to reduce the bond and adjust it to a target range to provide the strain hardening composite (Figure 12.17). The treatment changes the pull-out failure of the PVA from one which is characterized by rupture and severe delamination to one where delamination is prevented and the pull-out is by slip. The influences of such oiling treatment on the overall behaviour of the composite are shown in Figure 12.18.

The unique properties of the composite, the enhanced mechanical load-bearing capacity and the control of cracking (i.e. extremely thin cracks under extensive deformations) are potentially attractive for a range of applications, in which the composite can provide mechanical reinforcement as well as protection from aggressive environment. Several applications have been studied and evaluated: incorporation of the material as a layer around the reinforcing bars in reinforced concrete components [85,86], a sprayable formulation to be applied in new structures for protection [87] and as a means for repair and retrofit of existing structures [88,89]. An optimum mix for spray application was developed, having a w/c ratio of 0.46 and sand/cement ratio of 0.8 [88]. Analytical modelling tools for structural design were investigated [89–91]. Special structures, such as dampers for structural control [92] and precast panels [93] were evaluated. To facilitate application and increase the cost-effectiveness, formulations in which part of the cement was replaced with fly ash and slag were investigated [94]. In these formulations, when the content of the supplementary material is high, the strength may be reduced but the toughness increased. The feasibility of producing lightweight composite by incorporation of voids (air entrainment, polymeric micro-hollow-bubbles, perlite and glass micro-bubbles) was also investigated [95].

12.6   Hybrid fibre systems

The combination of two or more types of fibres (hybrid reinforcement) has been studied and applied with the objective of optimizing the overall system to achieve

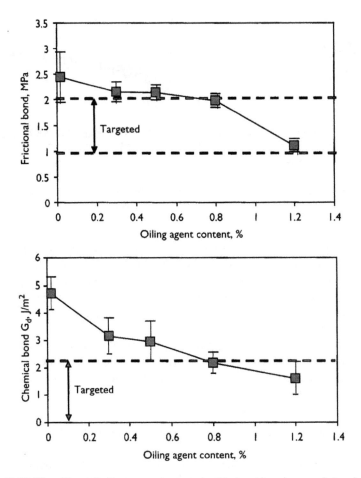

Figure 12.17 The effect of oiling treatment on the frictional bond, τ_0, and the chemical bond, G_d, in relation to the target values to achieve multiple cracking and strain hardening (after Li et al. [83]).

synergy, whereby the overall performance of the composite would exceed that induced by each of the fibres alone. Banthia and Gupta [96] classified the synergies into three groups, depending on the mechanisms involved:

• *Hybrids based on fibre constitutive response*, where one fibre is stronger and stiffer and provides strength, while the other is more ductile or can readily undergo considerable slippage to provide toughness at high strains and crack openings.

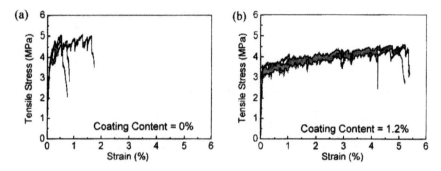

Figure 12.18 Effect of oiling treatment on the stress–strain curves of composites with 2% PVA fibres in 045 w/c ratio matrix (after Li *et al.* [83]).

- *Hybrids based on fibre dimensions* where one fibre is small (micro or mesofibre) and provides microcrack control at earlier stages of loading to arrest microcracks and delay their coalescence, thus promoting enhanced first crack and strength; the other fibre, which is bigger (macrofibre), provides the bridging mechanism across macrocracks, and induces toughness at high strains and crack openings.
- *Hybrids based on fibre function* where one type of fibre (the primary fibre) induces strength or toughness in the hardened composite, while the second type of fibre provides the fresh mix properties suitable for processing (the processing fibre).

The hybrid can thus be made of fibres of different properties, or having different shapes.

Earlier studies and publications addressed hybrid systems for applications in thin sheet composites, in particular for asbestos–cement replacements. In recent years the concept has been extended for high performance–high ductility FRC, which can be produced by simple mixing.

As outlined in Section 12.2.3 these new insights are based on adjusting the fibres to the cracking mechanisms which control the composite at different stages of loading [18,19]: (i) arrest and bridging of microcracks at early stages of loading to increase the cracking strength and obtain strain or deflection hardening and (ii) bridging of macrocracks which develop later on to induce ductility into the composite. The fibres for these purposes could be either made from one material, but with different geometries [18,19] (e.g. small steel microfibres for microcrack control and long macrosteel fibres for macrocrack bridging) or be composed of different materials, such as PVA and PE microfibre for microcrack control and deformed steel fibres for macrocrack bridging [96]. These newer concepts were applied for the development of high performance–high ductility FRC with strain hardening behaviour (2–10% fibre volume) and for crack control in conventional concretes (less than about 1% fibre by volume). This section will deal separately with each of the systems, depending on their end application: thin sheet composites, high performance–high ductility composites and reinforcement of conventional concretes.

12.6.1 Hybrid fibre systems in thin sheet composites

The concept of hybrid reinforcement consisting of primary and processing fibres was developed in conjunction with asbestos replacement [97,98]. In this case it was difficult to develop synthetic fibres that would simultaneously provide the reinforcing effect and the filtering and solid retention characteristics which are needed in the Hatscheck process. These types of combination were discussed in Section 9.6, dealing with asbestos replacement.

The combination of different types of fibres to optimize the performance of thin sheet composite in the hardened state, with respect to strength and toughness, has been studied by various investigators [97–102], using asbestos, carbon, steel or glass to achieve strength, and polypropylene and polyethylene to improve toughness.

In composites with one type of fibre, the high modulus fibres tend to increase strength with only modest improvements in toughness (Figure 12.19), while the low modulus fibres lead to a considerably higher toughness with hardly any improvement in strength. In fact, in the latter case, there is usually a tendency for a decline in the load-bearing capacity immediately following the first crack, and the fibres become effective in supporting additional load only after considerable extension or deformation of the composite. In a hybrid composite, however, the additive influence can be seen (Figure 12.20). Kobayashi and Cho [100] concluded that in steel and polyethylene combinations the shortcomings of both fibres were offset and only the advantages of each were exhibited. Similar conclusions might be drawn from the data of Hasaba et al. [102] for polypropylene and steel hybrid composites. Walton and Majumdar [99] reported similar effects with carbon and asbestos fibres to which polypropylene was added. They demonstrated that the improved toughness (impact resistance) obtained by the low modulus polymer fibres in the hybrid composite was retained even after accelerated ageing for one year at 60°C. This retention was found to be of significance in a glass fibre–polypropylene hybrid thin sheet composite [99]. Glass fibres increase strength and toughness but a great part of the latter is lost after ageing. The presence of the polypropylene considerably increased the toughness retention after ageing and this approach is potentially suitable for dealing with the embrittlement of GRC.

Mai et al. [103] analysed the hybrid effects in polypropylene–cellulose composites by comparing the experimental tensile strength and modulus of elasticity with the calculated values derived from the rule of mixtures. They used equations similar to those developed for asbestos and cellulose composites (Eqs 9.1–9.9, Section 9.42, Eqs 11.1–11.3, Section 11.4.3), except that the term describing the

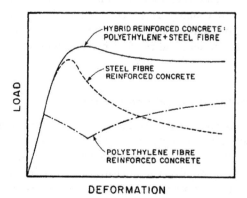

Figure 12.19 Schematic description of the load–deformation curve of a hybrid composite with high modulus (steel) and low modulus fibre (polyethylene)
(after Kobayashi and Cho [100]).

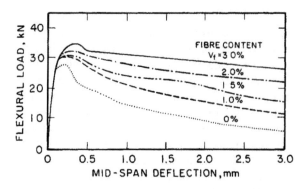

Figure 12.20 Load–deflection curves of hybrid composites with high modulus fibre
(steel) and increasing content of low modulus polyethylene (PE) fibre
(after Kobayashi and Cho [100]).

contribution of the fibres was the sum of the contributions of the two types of
fibres, with their respective properties and volume contents. They indicated also
that the rule of mixtures must be applied with caution to hybrid effects, since
the efficiency coefficients may be different from those in a monofibre composite. In particular, they pointed out that the rule of mixtures should not be applied
to toughness prediction, unless the fracture mechanisms and interaction of such
mechanisms in the hybrid composite are accounted for.

Synergy in hybrid reinforcement of thin sheet composites made from continuous
fibrillated polypropylene net and glass fibre reinforcement in the form of fabric and

chopped strand was reported by Xu and Hannant [104]. A schematic description of the properties of composites from the individual reinforcement and the combined one is shown in Figure 12.21. The mono-reinforcements exhibited a three stage stress–strain curve in accordance with the ACK model, while the hybrid showed a four stage curve. The synergy between the two is exhibited in stage 2 and 3 of the hybrid curve, and this is quantified in Figure 12.22, showing that at intermediate tensile strains of about 1.5% the load-bearing capacity of the hybrid is greater than the sum of the two components. The synergy at this stage was explained by stabilization of the debonding cracks which develop in glass fibre reinforced cement, and this is attributed to the presence of the polypropylene. As a result the two reinforcements carry at this range higher stresses than each of them would in a monofibre composite alone. This effect comes on top of the enhancement of the first crack stress which is greater than would be expected by simple addition of the effect of each fibre. First crack stress increased from about 2.6–4.0 MPa to about 8 MPa in the hybrid reinforcement with 5% fibre content. Similar influences, showing even greater synergistic effects were obtained in flexure [105]. Qualitative and quantitative modelling of these effects in tension, based on the concepts of the ACK model, and considering the nature of the multiple cracking and their modification were developed by Hannant and co-workers [104,106].

The combined effect of high modulus polyethylene fibre which can be considered as the reinforcing fibre and fibrillated polypropylene pulp, which is effective for processing, was reported by Souroshian et al. [107] showing positive interactions between the two with respect to hardened composite properties.

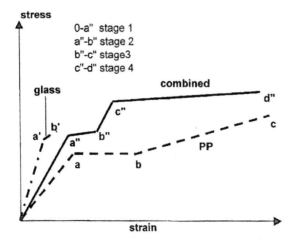

Figure 12.21 Schematic description of the effect of hybrid reinforcement (glass, polypropylene, hybrid) on the stress–strain diagram (from Xu and Hannant [104]).

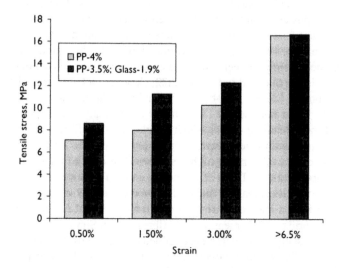

Figure 12.22 The effect of mono- and hybrid reinforcement on the post-cracking stress-bearing capacity at different strains (after Xu and Hannant [104]).

Peled *et al.* [108] evaluated the use of PP and PVA microfibres as supplements to glass in reinforced composites produced by an extrusion process, as a means to improve the overall mechanical behaviour and the retention of toughness after ageing. Hybrid combinations (total of 5% vol. fibre reinforcement) of 40:20:40 and 40:0:60 glass/PP/PVA provided a composite with strength similar to 100% glass reinforcement but with a significant improvement in toughness. Toughness retention after accelerated ageing was also improved with the hybrid reinforcement, eliminating toughness loss after 6 weeks of ageing in water at 60°C.

12.6.2 High performance–high ductility concretes

As indicated earlier, new initiatives and concepts were developed in the 1990s and beyond, to use the concept of hybrid fibre for producing high performance–high ductility FRC, which could be manufactured by simple mixing. Various terminologies have been used for this concept in addition to hybrid reinforcement, such as multi-scale fibre reinforcement [18] and multimodal reinforcement [19]. This was suggested as a complementary concept to other methodologies of producing such composites, which were based on adjustment and tailoring the matrix to a single type of fibre as outlined in Sections 12.1.1 and 12.2.2. The hybrid fibre reinforcement approach relies much more on engineering and tailoring of the fibres, to adjust them to the cracking and fracture processes in the matrix.

Rossi [19] reported the development of a composite with about 40 MPa flexural strength reinforced with 5% vol. of meso-steel fibres (0.25 mm diameter and 5 m length) and 2 % vol. macrofibres (hooked, 0.3 mm diameter and 25 mm length) with a low w/c matrix of 0.156. Similar ranges of properties were reported by Katz *et al.* [109] in a study to optimize the reinforcing system for a deflection hardening composite. The fibre system which gave the optimum performance in terms of minimizing the fibre content and providing maximum mechanical performance (flexural strength and toughness) consisted of 1% vol. of deformed steel macrofibres (30 mm length, 0.375 mm diameter) and 4.9% vol. of meso-steel fibres (6 mm long 0.16 mm diameter), Figure 12.23; increasing the macrofibre content above 1% led to workability problems, while the use of mesofibres only, which enabled a workable mix with fibre contents as high as 9% was less efficient in providing reinforcement compared with the optimal hybrid system. In order to increase the composite strength from a level of about 30–40 MPa to 60 MPa (with room temperature curing) a hybrid of three fibres was developed in LCPC France, called CEMTEC®$_{multiscale}$ [110,111]. The content of the fibres in this composite is higher, 11% vol. and the strength levels achieved are about 60 MPa flexural strength [111] and 20 MPa tensile strength [110].

Markovic *et al.* [112,113] evaluated a composite with a similar matrix, of 0.20 w/c ratio, using hybrid reinforcement of meso-steel fibres (0.16–0.2 mm diameter and 6–13 mm length), and hooked macrofibres of (0.5–0.71 mm diameter and 40–60 mm length). Optimum properties of 45 MPa flexural strength and 48,550 J/m^2 fracture energy were obtained with a hybrid of 2% vol. of mesofibres and 1% vol. of macrofibres. The small diameter and length of the mesofibre provided a system with sufficiently small spacing to control microcracking at early

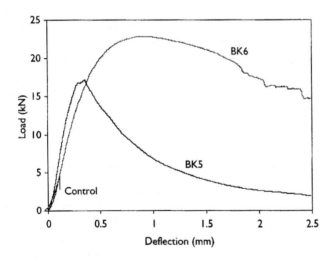

Figure 12.23 Effect of mono(BK5)- and hybrid (BK6) steel fibre reinforcement on the load–deflection curves of composites with low w/c ratio matrix (after Katz *et al.* [109]).

stages of loading, while at advanced stages, when bigger cracks develop, they were less efficient and the macrofibres took the role of crack bridging [114]. The effectiveness of the hooked macrofibres in such composites was estimated by evaluating the relative number of fibres in which the failure of the fibre was accompanied by deformation of the hook; deformation of the hook maximizes the pull-out resistance and therefore such failure mode is indicative of an 'active' fibre. Fibres which only slip were considered as only partially active [113]. When the reinforcement was with hooked fibres only, 15% of the fibres were fully active; in the hybrid composite the effectiveness increased and 32% of the fibres were fully active [112]. One may consider this effect as an additional benefit of hybridization in such systems, which can provide synergistic influence. This is consistent with the observations of Shannag *et al.* [115] showing an increase in fibre–matrix bond in systems reinforced with meso-steel fibres of 0.19 mm diameter and length of 6, 12 and 18 mm; the increase in bond was observed in both conventional and DSP matrices, but was much more significant in the DSP matrix and for the longer fibres.

The systems described earlier, used a hybrid steel fibre system, consisting of meso and macrofibres. Systems in which the smaller fibre in the hybrid reinforcement was a microfibre were reported in several studies [116–118]. The microfibres were in most cases of a different composition than the macrofibre (usually steel). The microfibres evaluated were alumina [116], carbon [116,117], PVA [117,118], ultra high density polyethylene [117,118] and steel [117]. The characteristics of such fibres are presented in Table 12.7.

The matrix used in [116] was 0.3 w/b ratio, with the macrofibre being polypropylene and the microfibres alumina and carbon. Reinforcement with 8% polypropylene gave higher strength and toughness, but was not efficient in stabilizing microcracks, as seen by developments of relatively large crack opening displacements at early stages of loading. Replacement of 4% of the polypropylene with microfibres provided much greater stability at the microcracking stage, although the composite was less effective in its load-bearing capacity at greater crack opening. The combination of 1% of carbon, PVA, PE (Spectra fibre) and steel microfibres, on top of 2% of twisted steel fibre (Torex) [117], resulted in a marked

Table 12.7 Properties of microfibres evaluated for reinforcement of high strength low w/c matrix FRC (from [116–118])

Property	Alumina [116]	Carbon [116]	Carbon [117]	PP [116]	PVA [117,118]	Ultra high density PE [117,118]	Micro-steel [109]
Length, mm	0.762	1.0	6.35	12	4–18	6–15	1–5
Diameter, μm	2.5	25	10	35 × 250	15–40	12–38	50
Tensile strength, MPa	1725	2600	2825	340–500	700–1600	2585–2770	—
Modulus of elasticity, GPa	105	230	241	8.5–12.5	40–139	88–117	200

improvement of the load-bearing capacity of the composite at small deflections as well as in large ones (pre- and post-peak loads). The enhanced performance in these zones was higher with the PVA (40 μm), PE (38 μm) and carbon (9 μm) than with the steel (50 μm) and the smaller PVA fibre (15 μm). With these combinations tensile and flexural strength of about 8 and about 30 MPa, respectively, were reported [117] with a hybrid of 1% microfibre and 2% macrofibre. It was suggested that the enhanced performance due to the presence of the microfibres is the result of their influence on the mechanical properties of the matrix (microcrack control) and bonding of the macrofibre. This is consistent with the analysis earlier for the steel–fibre hybrid systems.

The nature of cracking of such systems with steel macrofibres (1% vol.) and polyethylene (PE) and PVA microfibres (1% vol.) was reported in [118,119] for a 0.40–0.50 w/c ratio matrix. The performance obtained with PE was somewhat better, exhibiting higher tensile strength (about 4 vs. 3 MPa), and a better post-cracking behaviour, showing some strain hardening and better pseudo-plastic behaviour. This was accompanied by more multiple cracking in the PE composite with cracks which are much better distributed and less localized than in the PVA [118]. In a further study of the cracking in the PE hybrid composite [119] it was shown that microcracks of about 1–15 μm are present before loading; the process of multiple cracking as a function of increased stress was characterized.

The significance of crack control in such composites was highlighted in several studies [120–122]. It was demonstrated that in systems of this kind the development of a macrocrack can be considerably delayed, and even at relatively high loading (i.e. high stress/strength ratio and high strains) the system exhibits only very small microcracks, usually less than 100 μm wide [118,119]. Charron et al. [120] studied the permeability through such cracks which were obtained after loading of a notched composite (of the CEMTEC®$_{multiscale}$ type, with w/c ratio of 0.125 and 6% vol. steel fibre) to predefined deformation levels. The relation between the strain after loading and the equivalent water permeability is shown in Figure 12.24, demonstrating that up to a residual strain of 0.13% water permeability is maintained at low levels ($K < 10^{-9}$ m/s); since the specimens studied were 100 mm long, and the strain was localized in the notched section, this strain is equivalent to a crack opening of 130 μm. This value is much larger than the 50 μm limiting one reported for normal concrete [121]. Such characteristics highlight the advantage of such composites for structures where a high level of durability performance is required.

12.6.3 Performance enhancement and crack control in normal strength concretes

The concept of hybrid reinforcement can be also applied for normal strength concrete to control cracking and strain softening. The concepts are similar to the ones outlined in Section 12.2.3, but the content of fibres required is less than about

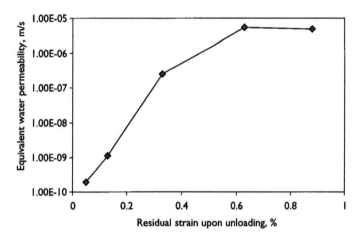

Figure 12.24 The effect of the residual strain (after unloading a high performance FRC with dense matrix) on the equivalent permeability to water (after Charron *et al.* [120]).

1% by volume, since strain and deflection hardening are not required. The effects of hybrid reinforcement using steel macrofibre and steel microfibre (22 μm diameter, 6 mm long) and PVA (14 μm, 12 mm long) were studied by Lawler *et al.* [122] and their effect on the mechanical properties of mortar (0.45 w/c ratio) are shown in Figure 12.25. Similar trends to those of Figure 12.25 were reported by Granju *et al.* [123] for a hybrid with 0.49% vol. polypropylene structural fibre (modulus of 3.3 GPa) and 0.135% vol. amorphous corrosion resistant steel fibre (having a geometry which is in between micro and mesofibre cross section of 0.03×1.6 mm and 30 mm length). In both studies the presence of the microfibres in the hybrid system clearly shows up at small displacements, where the transition across the first crack becomes smooth (indicative of microcrack control to eliminate unstable coalescence into bigger cracks), as well as at large displacements where the load-bearing capacity in the post-peak zone is bigger than the simple addition of the capacity of each of the fibres alone. The marked influence on the mode of crack formation was confirmed in a later study by Lawler *et al.* [124] in which the cracking at different stages of loading was characterized: in the unreinforced concrete and the one reinforced by macrofibres only the crack opening was dominated by a single crack that became quite large, for example, once this crack opens up no new cracks develop. In the hybrid systems with microfibres, as deformation of the composite increased, the growth of the main crack was smaller than that of the systems without microfibres, and it was accompanied by the formation of additional small cracks. In the macrofibre system, at a nominal strain of 0.08%, crack width as large as 90 μm was recorded, while at a similar strain in the hybrid system with PVA, the largest crack width recorded was less than 40 μm.

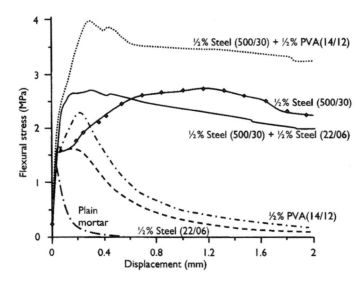

Figure 12.25 Effect of mono- and hybrid reinforcement on the flexural properties of concrete with low level reinforcement (less than 1% fibre by volume), (a) Flexural stress–displacement curves; (b) flexural stress-crack mouth opening displacement, after Lawler *et al.* [122].

Similar trends of the effect of microfibre at 0.5% vol. reinforcement were reported by Yao *et al.* [125] for carbon microfibre, showing a pseudo-plastic behaviour for system with 0.2% carbon microfibre (7 μm diameter, 5 mm long) and 0.3% vol. steel (0.5 mm diameter, 30 mm long), with a relatively smooth transition in the load–deflection curve around the matrix cracking stress. Qian and Stroeven [126] demonstrated the enhanced efficiency of hybrid reinforcement of 0.4 w/c ratio concretes, showing that a system of steel macrofibre and polypropylene microfibre could be used to obtain a flexural reinforcing effect at a fibre content of 0.75% vol., similar to that obtained with a higher content (0.9% vol.) of mono-steel macrofibres.

The change in the nature of the concrete with hybrid fibre has implications not only on the mechanical performance which is relevant to applications such as pavements and slabs on grade [127,128], but also on the permeability of loaded (and cracked) concrete, which is relevant to durability performance and serviceability. The change in the cracking characteristic could account for the marked improvement in the permeability performance of hybrid-reinforced concrete under loading conditions, as seen in Figure 12.26 [124]: in the hybrid composite with PVA, the permeability was much smaller for an equal strain (Figure 12.26(a)). Sun *et al.* [129] showed reduction by more than an order of magnitude in the permeation coefficient of concrete reinforced with hybrid fibre systems, consisting of combinations of up to 2% by volume of steel macrofibre with microfibres of polypropylene

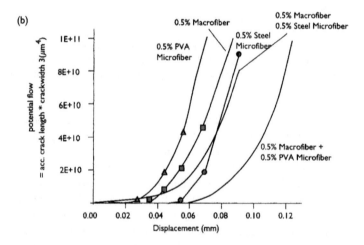

Figure 12.26 Effect of displacement upon loading in tension of mono- and hybrid fibre reinforced concretes on their permeability (a) and the potential flow quantified in terms of crack length and width (b) (after Lawler *et al.* [122,124]).

(10 μm diameter) and PVA (15 μm diameter). However, in this study the permeation was evaluated on well-cured specimens, and the improvement could not be related to improved crack control; it was attributed to a favourable change in the pore structure.

The significance of crack control in reduction of permeability and the consequences to durability were observed for strain hardening composites (Sections 12.3–12.6; Figure 12.24) as well as for strain softening composites discussed in this section, when means are taken to control the cracks at small crack openings

(Figure 12.26(a)). Several studies addressed the mechanisms involved and quantified them. Lawler *et al.* [124] explained the superior performance of the reinforced concrete in terms of a function related to the crack width, which is the cumulative crack length multiplied by the cube of crack width ($\sum$ length · width^3), called the 'potential flow' (Figure 12.26(b)). Curves of the potential flow vs. displacement in loading showed reasonable correlation with the curves of the measured flow rate vs. displacement (Figure 12.26). These trends reflect the observation that for a similar displacement, the cracks formed in the hybrid composite are of smaller width and are therefore less permeable to water. The implications for the enhanced permeability of the hybrid composite are quite evident. Mihashi *et al.* [130] analysed systems which exhibited behaviour similar to that reported in Figure 12.26 in terms of the relations developed for flow through cracks in rock mechanics, which are based on simulation of flow through parallel plates:

$$q_0 = \xi \frac{gI\ell w^3}{12v} \tag{12.19}$$

where q_0 is the rate of water flow through an idealized smooth crack, m^2/s; g, the gravity acceleration; I, the pressure gradient, h/d, where h is the height of the fluid column on the inlet side, and d is the crack length in the flow direction (i.e. the thickness of the specimen in a permeability test); ℓ, the crack length at right angle to flow direction w, the crack width, v, the kinematic viscosity, m^2/s and ξ, the a coefficient between 0 and 1 which accounts for the tortuosity of the crack; that is, accounts for the fact that the crack in the concrete cannot be described as an ideal smooth parallel plate shaped slit.

Mihashi *et al.* [130] obtained flow curves similar to those in Figure 12.26, and calculated the coefficient ξ from Eq. 12.19 and the experimental data for q_0. A linear relation between ξ and crack width was obtained, but with reducing slope (i.e. lower ξ values) for the fibrous concretes (Figure 12.27). The increase in ξ with crack width and the decrease in the presence of fibres suggest changes in the microstructure of the crack, which is a function of crack width, fibre content and type. These changes are associated probably with characteristics such as tortuosity, discontinuities and branching. The linear relation of ξ with crack width suggests that the effect of crack width on permeability in concretes is a function of the fourth power of the width.

12.7 Slip hardening fibres

Engineering of the fibre and the interface to obtain a slip hardening behaviour of the fibre during pull-out is an additional means that has been studied and used to obtain composites of enhanced performance. Changing the shape of the fibre is almost a 'routine' means taken to improve the bond in general, and to obtain a strain hardening pull-out in particular. This is achieved by inducing mechanical bonding (anchoring) into the system, as is often done by changing the geometry

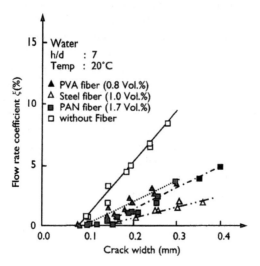

Figure 12.27 Effect of crack width and fibre reinforcement on the flow rate coefficient, ξ, in Eq. 12.19 (after Mihashi et al. [130]).

of the fibre (e.g. hooks, crimps). When the geometry is properly adjusted, slip hardening behaviour can be obtained, as was already demonstrated in Section 3.4 and shown graphically for hooks and crimps in Figures 3.31 and 3.33.

More advanced means have been recently suggested to enhance this response in the fibres, by formation of more complex shapes, which are not linear in nature, but rather 2D and 3D dimensional. Naaman [131–133] developed the concept of twisting polygonal fibres, and optimizing this effect by using geometries which before twisting, have a shape which is intrinsically of a higher surface area. He defined a parameter of 'fibre intrinsic efficiency ratio' (FIER) which is the ratio of the bonded lateral surface area of the fibre, to its cross-section area. The ratio can be calculated per unit length or total length:

$$\text{FIER} = \psi \ell / A \tag{12.20}$$

where ψ is the perimeter; ℓ, the length of fibre and A, the cross-section area of fibre.

In order to normalize the cross-section size, the relative FIER value was used, defined as the FIER of the fibre relative to the FIER of a circular cross-section fibre having the same area. Examples are provided in Figure 12.28 for a simple shape of a triangle or a more complex shape of substantially triangular fibre. Twisting of a polygonal fibre provides a fibre with a profile similar to a screw (Figure 12.29), with a drastically enhanced pull-out resistance. Increase in the pull-out resistance and slip hardening bond could be correlated with the density of the ribs, or the pitch of the screw type fibre (Figure 12.30). Equivalent average bond strengths as high as 15 MPa were reported for a fibre obtained by twisting a triangular cross-section.

A different type of mechanical bonding was obtained by using clip and circular type geometries [134,135]. The fibre pull-out resistance could be maximized by optimizing the radius of curvature at the edges of the clip to the length [134]. The latter provides the mechanical anchoring characteristics. The performance of such a fibre was shown to be better than a corrugated one, and it provides a quasi-plastic pull-out behaviour as well as increased maximum load resistance.

An additional means for inducing slip hardening bond behaviour is based on densification in the matrix around the fibre, in combination with the presence of fine fillers, as has already been demonstrated in Section 12.2. A demonstration of the efficiency of this means was provided in Section 12.2.2 showing the marked influence of matrix modification in high strength cementitious systems, to obtain bond values with straight steel fibres exceeding 10 MPa.

A different mechanism for obtaining slip hardening behaviour was reported for polymer fibres, in conditions which promote abrasion of the fibre surface and its peeling, leading to special mechanical anchoring effects between the fibre and the matrix. Geng and Leung [136] reported such an effect for nylon fibres which were abraded during pull-out, allowing water to penetrate into the fibre and induce clamping due to the swelling of the hydrophilic fibre. Shah *et al.* [68] obtained a similar effect with PVA fibre in a composite produced by extrusion (Figure 12.14(b)), but not with a cast composite. In the extruded composite the

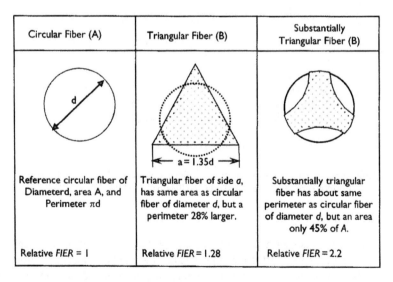

Figure 12.28 Shapes of fibres and the relative FIER values (after Naaman [133]).

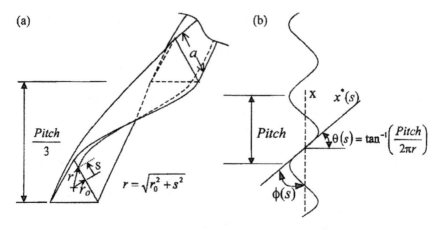

Figure 12.29 The shape and characteristic geometrical parameters of twisted fibre (a) 3D view; (b) angle at contact surface (after Sujivorakul and Naaman [132]).

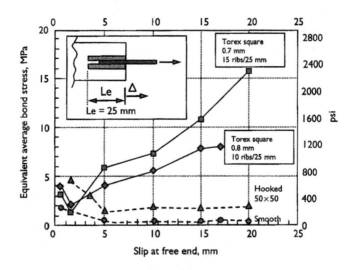

Figure 12.30 Effect of the shape of the twisted fibre and its rib density of the equivalent bond stress at increasing slip (after Naaman [131]).

interface was much denser and tighter, leading probably to abrasion and peeling of the fibre, resulting in an effective improved mechanical bonding (for additional details see Section 12.4). The nature of the debonding in Figure 12.14(b) resembles the sleeve and core mechanisms described in Section 13.2.

12.8   FRC with polymer–cement matrices

The mechanics of FRC composites is very much dependent on the strength of the matrix, its modulus of elasticity and the fibre–matrix bond. Modification of these characteristics can be particularly useful for optimizing the efficiency of the fibres. A drastic change in the matrix properties can be achieved by combining the matrix with polymers, in various forms. Some attempts have been made to explore this approach, and there are some promising results, but the full potential of this combination is far from being realized.

Combinations with polymers can be obtained by impregnation of the hardened composite (polymer impregnated concrete – PIC) or with a polymer latex which is added during mixing (latex modified cement – LMC) [137]. Both types of polymer incorporation would be expected to increase the matrix strength and the fibre–matrix bond. However, their influence on the other matrix properties is quite different. In PIC, the matrix has a higher modulus of elasticity and becomes more brittle, while in LMC the matrix will be more ductile with a lower modulus of elasticity.

Hughes and Guest [138] concluded that the impregnation of steel fibre-reinforced mortar led to a considerable increase in the fibre efficiency when considering strength and toughness, and this could be attributed to improved bond. Thus, in combination with 2% by volume of steel fibres, the brittle PIC showed a considerable toughening, by a factor of about 70. Jamrozy and Sliwinski [139] estimated an increase in fibre–concrete matrix bond by a factor of 5 due to polymer impregnation. Similar conclusions were reached by Akihama et al. [140] when evaluating the effect of impregnation on autoclaved carbon fibre reinforced cement. Impregnation resulted in a reduction of the critical fibre length from about 1.5 mm to 0.5 mm, and the shape of the stress–strain curve changed accordingly, showing less pseudo-plastic behaviour and higher strength (Figure 12.31).

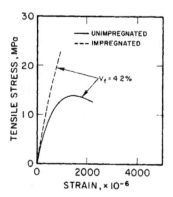

Figure 12.31 Effect of polymer impregnation on the stress–strain curve of carbon fibre reinforced cement (after Akihama et al. [140]).

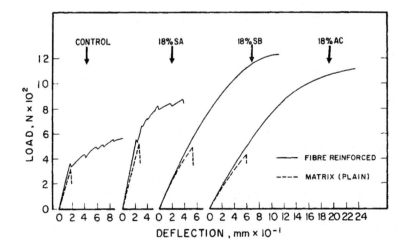

Figure 12.32 Load–deflection curves in bending of polymer modified matrices (broken lines) and fibre-reinforced polymer modified cements (full lines) made of Portland cement (control) and cements modified with 18% saran (SA), styrene–butadiene (SB) and acrylic (AC) latex (after Bentur [146]).

The addition of polymer latex to improve the mechanical properties of FRC composites has been studied by several investigators [141–146]. The strength and toughness were shown to be a linear function of the fibre content and the polymer content (in the range of 0–20% polymer content by weight of cement), with the composite properties increasing with these two parameters [143]. However, an increase in the polymer content above an optimal value could lead to a reduction in the properties of the composite [146], and this is similar to the reduction in the strength of the matrix alone as the polymer content increases beyond the optimum. The presence of the polymer latex affected the shape of the load–deflection curve in bending, causing a smoother transition across the first crack stress, and reducing the extent of discontinuities in the curve in the post-cracking zone [146] (Figure 12.32). These effects were also dependent on the type of polymer and the defoaming agent. The improvement in the properties by the combination of the fibres and the polymer latex often exceeded the cumulative improvement induced by each of these alone, thus indicating a synergistic effect. This may be due to a marked increase in the efficiency of the fibres induced by the effect of the polymer latex on increasing the extensibility of the matrix as well as its bond with the fibres. These two effects may cause a considerable change in the mechanics of the system in the post-cracking zone, as is suggested from the curves in Figure 12.32. They show many fewer discontinuities (i.e. micro cracks) in the curves of the system with polymer latex, in particular the more extensible ones (styrene butadiene and acrylic).

The incorporation of polymers is often advantageous for improving properties other than strength and toughness. Polymer impregnation and latex modification are known to have a sealing effect on the matrices [137], thus improving their durability and watertightness. These characteristics can be potentially useful for improving the durability of FRC composites. Polymer modification has been extensively applied in glass fibre reinforced cements (see Chapter 8). Polymer latex modification can also reduce the shrinkage of the composite [144]. It should be noted that this type of polymer also improves the workability of the fresh mix, and may thus offer the potential advantage of more uniform dispersion of the fibres in composites prepared by mixing, as well as facilitate the incorporation of greater fibre contents.

References

1. A.E. Naaman, 'Strain hardening and deflection hardening fiber reinforced cement composites', in A.E. Naaman and H.W. Reinhardt (eds) *Fourth International Workshop on High Performance Fiber Reinforced Cement Composites (HPFRCC 4)*, RILEM Publications, Bagneux, France, 2003, pp. 95–113.
2. JCI-DFRCC Committee, 'DRFRCC terminology and application concepts', *J. Advanced Concrete Technology*. 1, 2003, 335–340.
3. V.C. Li, D.K. Mishra and H.C. Wu, 'Matrix design for pseudo-hardening fibre reinforced cementitious composites', *Mater. Struct.* 28, 1995, 586–595.
4. L. Hjorth, 'Development and application of high-density cement based materials', *Philos. Trans. R. Soc. London Ser.* A 310, 1983, 167–173.
5. H.H. Bache, *Densified Cement/Ultrafine-Fine-Particle-Based Materials*, CBL Report No. 40, Aalborg Portland, Denmark, 1981.
6. H.H. Bache, *Compact Reinforced Composite, Basic Principles*, CBL Report No. 41, Aalborg, Denmark, 1987.
7. Dugat, N. Roux and G. Bernier, 'Mechanical properties of reactive powder concretes', *Mater. Struct.* 29, 1996, 233–240.
8. H. Mihashi, J.P.B. Leite and A. Kawamura, 'Multi-mechanism design for developing highly ductile cementitious composites', in M. Di Prisco, R. Felicetti, and G.A. Plizzari (eds) *Fibre Reinforced Concrete – BEFIB 2004*, Proc. RILEM Symp., PRO 39, RILEM Publications, Bagneux, France, 2004, pp. 515–525.
9. S. Collepardi, L. Coppol, R. Troli and M. Collepardi, 'Mechanical properties of modified reactive powder concrete', in V.M. Malhotra (ed.) *Superplasticizers and Other Chemical Admixtures*, Proc. 5th CANMET/ACI International Conference, ACI SP – 173, American Concrete Institute, 1997, pp. 2–21.
10. D. Lange-Kornbak and B.L. Karihaloo, 'Optimum design of high performance fibre-reinforced cement composites', in H.W. Reinhardt and A.E. Naaman (eds) *Third International Workshop on High Performance Fiber Reinforced Cement Composites (HPFRCC 3)*, RILEM Publications, Bagneux, France, 1999, pp. 65–74.
11. C.W. Hoy and P.J.M. Bartos, 'Interaction and packing of fibers: effects of the mixing process', in H.W. Reinhardt and A.E. Naaman (eds) *Third International Workshop on High Performance Fiber Reinforced Cement Composites (HPFRCC 3)*, RILEM Publications, Bagneux, France, 1999, pp. 181–191.

12. P. Tjiptobroto and W. Hansen, 'Tensile strain hardening and multiple cracking in high-performance cement-based composites containing discontinuous fibers', *ACI Mater. J.* 90, 1993, 16–25.

13. P. Guerrero and A.E. Naaman, 'Effect of mortar fineness and adhesive agents on pullout response of steel fibers', *ACI Mater. J.* 97, 2000, 12–20.

14. M.J. Shannang, R. Brincker and W. Hansen, 'Interfacial (fiber-matrix) properties of high-strength mortar (150MPA) from fiber pullout', *ACI Mater. J.* 83, 1996, 480–486.

15. Y.W. Chan and S.H. Chu, 'Effect of silica fume on steel fiber bond characteristics in reactive powder concrete', *Cem. Concr. Res.* 34, 2004, 1167–1172.

16. W. Sun, S. Liu and J. Lai, 'Study on the properties and mechanisms of ultra-high performance ecological reactive powder concrete (ECO-RPC)', in A.E. Naaman and H.W. Reinhardt (eds) *Fourth International Workshop on High Performance Fiber Reinforced Cement Composites (HPFRCC 4)*, RILEM Publications, Bagneux, France, 2003, pp. 409–417.

17. G. Orange, J. Dugat and P. Acker, 'DUCTAL: New ultrahigh performance concrete, Damage resistance and micromechanical analysis', in P. Rossi and G. Chanvillard (eds) *Fibre Reinforced Concrete – BEFIB 2000*, Proc. 5th Int. RILEM Symp., RILEM PRO 15, RILEM Publications, Bagneux, France, 2000, pp. 181–790.

18. P. Rossi, P. Acker and Y. Malier, 'Effect of steel fibres at two different stages: the material and the structure', *Mater. Struct.* 20, 1987, 436–439.

19. P. Rossi, 'High performance multimodal fiber reinforced cement composites (HPM-FRCC): the LCPC experience', *ACI Mater. J.* 94, 1997, 478–483.

20. D.R. Lankard, 'Slurry infiltrated fiber concrete (SIFCON): properties and applications', in J.F. Young (ed.) *Very High Strength Cement-Based Materials*, Proc. Materials Research Society Symp., Vol. 42, Materials Research Society, Pittsburgh, PA, 1985, pp. 277–286.

21. D.R. Lankard and J.K. Newell, 'Preparation of highly reinforced steel fiber reinforced concrete composites', in G.C. Hoff (ed.) *Fiber Reinforced Concrete*, ACI SP-81, American Concrete Institute, Detroit, MI, 1984, pp. 287–306.

22. D.R. Lankard, 'Preparation, properties and applications of cement-based composites containing 5 to 20 percent steel fiber', in S.P. Shah and A. Skarendahl (eds) *Steel Fiber Concrete*, Proc. US–Sweden Joint Seminar, Elsevier Applied Science Publishers, Barking, Essex, 1985, pp. 199–217.

23. J. Homrich and A. Naaman, 'Stress-strain properties of SIFCON in compression', in S.P. Shah and G.B. Batson (eds) *Fiber Reinforced Concrete*, in *Properties and Applications*, ACI SP-105, American Concrete Institute, Detroit, MI, 1987, pp. 283–394.

24. A.E. Naaman and J.R. Homrich, 'Tensile stress-strain properties of SIFCON', *ACI Mater. J.* 86, 1989, 244–251.

25. P. Balaguru and J. Kendzulak, 'Flexural behavior of slurry infiltrated fiber concrete (SIFCON) made using condensed silica fume', in V.M. Malhotra (ed.) *Fly Ash, Silica Fume, Slag and Natural Pozzolans in Concrete, ACI SP-91*, American Concrete Institute, Detroit, MI, 1986, pp. 1216–1229.

26. A. Bentur and R. Cree, 'Cement reinforced with steel wool', *Int. J. Cem. Comp. & Ltwt. Concr.* 9, 1987, 217–223.

27. L.E. Hackman, M.B. Farrell and O.O. Dunham, 'Slurry infiltrated mat concrete (SIMCON)', *Concr. Int.* 14, 1992, 53–56.

28. A. Bentur, Unpublished results.

29. J. Morton and G.W. Groves, 'Large work of fracture values in wire reinforced brittle matrix composites', *J. Mater. Sci.* 10, 1975, 170–172.
30. J. Bowling and G.W. Groves, 'The debonding and pull-out of ductile wires from a brittle matrix', *J. Mater. Sci.* 14, 1979, 431–442.
31. N. Krstulovic-Opara and S. Malak, 'Tensile behavior of slurry infiltrated mat concrete (SIMCON)', *ACI Mater. J.* 94, 1997, 39–46.
32. N. Krstulovic-Opara and S. Malak, 'Micromechanical tensile behavior of slurry infiltrated mat concrete (SIMCON)', *ACI Mater. J.* 94, 1997, 373–384.
33. A.F. Oluokun and S.A.J. Malak, 'Some parametric investigations of the tensile behavior of slurry infiltrated mat concrete (SIMCON)', in H.W. Reinhardt and A.E. Naaman (eds) *Third International Workshop on High Performance Fiber Reinforced Cement Composites (HPFRCC 3)*, RILEM Publications, Bagneux, France, 1999, pp. 271–297.
34. Z. Bayasi and J. Zeng, 'Flexural behavior of slurry infiltrated mat concrete (SIMCON)', *ASCE Journal of Materials in Civil Engineering.* 9, 1997, 194–199.
35. A.E. Naaman, J.K. Wight and H. Abdou, 'SIFCON connections for seismic resistant frames', *Concr. Int.: Design and Construction.* 9, 1987, 34–39.
36. P. Balaguru and J. Kendzulak, 'Mechanical properties of slurry infiltrated fiber concrete (SIFCON)', in H.W. Reinhardt and A.E. Naaman (eds) *Third International Workshop on High Performance Fiber Reinforced Cement Composites (HPFRCC 3)*, RILEM Publications, Bagneux, France, 1999, pp. 247–268.
37. N. Krstulovic-Opara, 'Use of SIMCON in seismic retrofit and new construction', in H.W. Reinhardt and A.E. Naaman (eds) *Third International Workshop on High Performance Fiber Reinforced Cement Composites (HPFRCC 3)*, RILEM Publications, Bagneux, France, 1999, pp. 629–647.
38. N. Krstulovic-Opara, V. Kilar and L. Krstulovic-Opara, 'Modeling and use of SIMCON and other high performance fiber-composites for increasing structural resistance to extreme loadings', in A.E. Naaman and H.W. Reinhardt (eds) *Fourth International Workshop on High Performance Fiber Reinforced Cement Composites (HPFRCC 4)*, RILEM Publications, Bagneux, France, 2003, pp. 167–177.
39. A.E. Naaman, H.W. Reinhardt and C. Fritz, 'Reinforced concrete beams with a SIFCON matrix', *ACI Struct. J.* 89, 1002, 79–88.
40. A.M. Hamza and A.E. Naaman, 'Bond characteristics of deformed reinforcing steel bars embedded in SIFCON', *ACI Mater. J.* 93, 1996, 578–588.
41. R. Breitenbucher, 'High performance fibre concrete SIFCON for repairing environmental structures', in H.W. Reinhardt and A.E. Naaman (eds) *Third International Workshop on High Performance Fiber Reinforced Cement Composites (HPFRCC 3)*, RILEM Publications, Bagneux, France, 1999, pp. 585–594.
42. S. Hauser and J.D. Worner, 'DUCON, a durable overlay', in H.W. Reinhardt and A.E. Naaman (eds) *Third International Workshop on High Performance Fiber Reinforced Cement Composites (HPFRCC 3)*, RILEM Publications, Bagneux, France, 1999, pp. 603–615.
43. F. de Larrard, *Concrete Mixture Proportioning: A Scientific Approach, Modern Concrete Technology Series*, E&FN SPON, London and New York, 1999.
44. P. Richard and M. Cheyrezy, 'Composition of reactive powder concretes', *Cem. Concr. Res.* 25, 1995, 1501–1511.
45. AFGC Recommendations, Ultra High Performance Fibre-Reinforced Concretes, Interim Recommendations, AFGC – Association Francaise de Genie Civil, Bagneux, France, 2002.

46. ·G. Chanvillard and S. Rigaud, 'Complete characterization of tensile properties of DUCTAL UHPFRC according to the French recommendations', in H.W. Reinhardt and A.E. Naaman (eds) *Fourth International Workshop on High Performance Fiber Reinforced Cement Composites (HPFRCC 4)*, RILEM Publications, Bagneux, France, 2003, pp. 21–34.

47. H. Stang, 'Scale effects in FRC and HPFRCC structural elements', in A.E. Naaman and H.W. Reinhardt (eds) *Fourth International Workshop on High Performance Fiber Reinforced Cement Composites (HPFRCC 4)*, RILEM Publications, Bagneux, France, 2003, pp. 245–257.

48. H.H. Bache, 'Principles of similitude in design of reinforced brittle matrix composites', in H.W. Reinhardt and A.E. Naaman (eds) *2nd High Performance Fiber Reinforced Cement Composites (HPFRCC2)*, RILEM Conference, E&FN SPON, London and New York, 1992, pp. 39–56.

49. B. Aarup, 'A special fibre reinforced high performance concrete', Paper F-13, in *Advances in Concrete through Science and Engineering*, Hybrid-Fiber Session, Int. Conference, Evanston, RILEM Publications, CD-ROM, Bagneux, France, 2004.

50. D. Lange-Kornbak and B.L. Karihaloo, 'Design of fiber reinforced DSP mixes for minimum brittleness', *Advanced Cement Based Materials*. 7, 1998, 89–101.

51. B.L. Karihaloo and K.M.B. De Vriese, 'Short-fibre reinforced reactive powder concrete', in H.W. Reinhardt and A.E. Naaman (eds) *Third International Workshop on High Performance Fiber Reinforced Cement Composites (HPFRCC 3)*, RILEM Publications, Bagneux, France, 1999, pp. 53–63.

52. B.L. Karihaloo and J. Wang, 'Mechanics of fibre-reinforced cementitious composites', *J. Advanced Engineering Materials*. 2, 2000, 19–34.

53. D. Lange-Kornbak and B.L. Karihaloo, 'Tension softening of short-fibre-reinforced cementitious composites', *Cem. Concr. Compos.* 19, 1997, 315–319.

54. D.S.P. Benson and B.L. Karihaloo, 'CARDIFRC – Manufacture and constitutive behavior', in A.E. Naaman and H.W. Reinhardt (eds) *Fourth International Workshop on High Performance Fiber Reinforced Cement Composites (HPFRCC 4)*, RILEM Publications, Bagneux, France, 2003, pp. 65–79.

55. B.L. Karihaloo, J. Wang and M. Grzybowski, 'Doubly periodic arrays of bridged cracks and short-fibre-reinforced cementitious composites', *J. Mech. Phys. Solids.* 44, 1996, 1586–1656.

56. M. Cheyrezy, V. Maret and L. Frouin, 'Microstructural analysis of RPC (Reactive Powder Concrete)', *Cem. Concr. Res.* 25, 1995, 1491–1500.

57. K. Habel, '*Structural Behavior of Elements Combining Ultra-High Performance Fibre Reinforced Concretes (UHPFRC) and Reinforced Concrete*', PhD Thesis, EPFL, Lausanne, 2004.

58. M. Behloul, A. Durukal, J.F. Batoz and G. Chanvillard, 'Ultrahigh –performance concrete technology with ductility', in M. Di Prisco, R. Felicetti, and G.A. Plizzari (eds) *Fibre Reinforced Concrete – BEFIB 2004*, Proc. RILEM Symposium, PRO 39, RILEM publications, Bagneux, France, 2004, pp. 1281–1290.

59. O. Bonneau, M. Lachemi, E. Dallaire, J. Dougat and P.C. Aitcin, 'Mechanical properties and durability of two industrial reactive powder concretes', *ACI Mater. J.* 94, 1997, 286–290.

60. P.J. Charron, E. Denarie and E. Bruhwiler, 'Permeability of UHPFRC under high stresses', Paper F-12, in *Advances in Concrete through Science and Engineering*,

Hybrid-Fiber Session, Int. Conference, Evanston, RILEM Publications, CD-ROM, Bagneux, France, 2004.

61. G. Orange, P. Acker and C. Vernet, 'A new generation of UHP concrete: DUCTAL, damage resistance and micromechanical analysis', in H.W. Reinhardt and A.E. Naaman (eds) *Third International Workshop on High Performance Fiber Reinforced Cement Composites (HPFRCC 3)*, RILEM Publications, Bagneux, France, 1999, pp. 101–111.

62. M.P. Low and J.J. Beaudoin, 'Mechanical properties of high performance cement binders reinforced with wollastonite micro-fibres', *Cem. Concr. Res.* 22, 1992, 981–989.

63. M.P. Low and J.J. Beaudion, 'Flexural strength and microstructure of cement binders reinforced with wollastonite micro-fibres', *Cem. Concr. Res.* 23, 1993, 905–916.

64. M.P. Low and J.J. Beaudoin, 'Stability of Portland cement-based binders reinforced with natural wollastonite micro-fibres', *Cem. Concr. Res.* 24, 1994, 874–884.

65. S.M. Adeeb, H. Nowodworski, K. Rosiak, V.H. Perry, J. Kroman, G. Tadros, T.G. Brown and N.G. Shrive, 'Modeling of the world's first DUCTAL architectural shell structure', Paper F-23, in *Advances in Concrete through Science and Engineering*, Hybrid-Fiber Session, Int. Conference, Evanston, RILEM Publications, CD-ROM, Bagneux, France, 2004.

66. G. Rosati, 'High performance concrete applications in precast and prestressed concrete bridge slabs', in H.W. Reinhardt and A.E. Naaman (eds) *Third International Workshop on High Performance Fiber Reinforced Cement Composites (HPFRCC 3)*, RILEM Publications, Bagneux, France, 1999, pp. 651–660.

67. A.E. Naaman and K. Chandrangsu, 'Innovative bridge deck system using high-performance fiber-reinforced cement composites', *ACI Mater. J.* 101, 2004, 57–64.

68. S.P. Shah, A. Peled, C.M. Aldea and Y. Akkaya, 'Scope of high performance fiber reinforced cement composites', in H.W. Reinhardt and A.E. Naaman (eds) *Third International Workshop on High Performance Fiber Reinforced Cement Composites (HPFRCC 3)*, RILEM Publications, Bagneux, France, 1999, pp. 113–129.

69. Y. Shao and S.P. Shah, 'Mechanical properties of PVA fiber reinforced cement composites fabricated by extrusion process', *ACI Mater. J.* 94, 1997, 555–564.

70. Y. Akkaya, A. Peled, D. Picka and S.P. Shah, 'Effect of sand addition on properties of fiber-reinforced cement composites', *ACI Mater. J.* 97, 2000, 393–400.

71. A. Peled and S.P. Shah, 'Parameters related to extruded cement composites', in A.M. Brandt, V.C. Li and I.H. Marshall (eds) *Brittle Matrix Composites 6*, Woodhead Publications, Warsaw, 2000, pp. 93–100.

72. X. Qian, X. Zhou, B. Mu and Z. Li, 'Fiber alignment and property direction dependency of FRC extrudate', *Cem. Concr. Res.* 33, 2003, 1575–1581.

73. A. Peled, M.F. Cyr and S.P. Shah, 'High content of fly ash (class F) in extruded cementitious composites', *ACI Mater. J.* 97, 2000, 509–517.

74. S. Shao, J. Qui and S.P. Shah, 'Microstructure of extruded cement-bonded fiberboard', *Cem. Concr. Res.* 31, 2001, 1153–1161.

75. V.C. Li, 'On engineered cementitious composites (ECC) – A review of the material and its applications', *J. of Advanced Concrete Technology.* 1, 2003, 215–229.

76. M. Maalej, T. Hashida and V.C. Li, 'Effect of fiber volume fraction on the off-crack-plane fracture energy in strain-hardening engineered cementitious composites', *J. American Ceramic Society.* 78, 1995, 3369–3375.

77. S.I. Torigoe, T. Horikoshi, A. Ogawa, T. Saito and T. Hamad, 'Study on evaluation method for PVA fiber distribution in engineered cementitious composite', *J. Advanced Concrete Technology.* 1, 2003, 265–268.

78. Z. Lin and V.C. Li, 'On interface property characterization and performance of fiber reinforced cementitious composites', *Concr. Sci. Eng.* 1, 1999, 173–184.

79. V.C. Li, 'Engineered cementitious composites – tailored composites through microme-chanical modeling', in N. Banthia, A. Bentur and A. Mufti (eds) *Fibre Reinforced Concrete: Present and the Future*, Canadian Society for Civil Engineering, Montreal, 1998, pp. 64–97.

80. V.C. Li and T. Hashida, 'Engineering ductile fracture in brittle matrix composites', *J. Mater. Sci. Letters.* 12, 1993, 898–901.

81. M. Maalej, V.C. Li and T. Hashida, 'Effect of fiber rupture on tensile properties of short composites', *ASCE J. Engineering Mechanics.* 121, 1995, 903–913.

82. T. Matsumoto, P. Suthiwarapirak and T. Kanda, Mechanisms of multiple cracking and fracture of DFRCC under fatigue flexure, *Journal of Advanced Concrete Technology.* 1, 2003, 299–306.

83. V.C. Li, C. Wu, S. Wang, A. Ogawa and T. Saito, 'Interface tailoring for strain-hardening polyvinyl alcohol- engineered cementitious composite (PVA-ECC)', *ACI Mater. J.* 99, 2002, 463–472.

84. V.C. Li, S. Wang and C. Wu, 'Tensile strain-hardening behavior of polyvinyl alcohol engineered cementitious composite (PVA-ECC)', *ACI Mater. J.* 98, 2001, 483–492.

85. M. Maalej and V.C. Li, 'Introduction of strain-hardening engineered cementi-tious composites in design of reinforced concrete flexural members for improved durability', *ACI Struct. J.* 92, 1995, 167–176.

86. V.C. Li and G. Fisher, 'Interaction between steel reinforcement and engineered cemen-titious composites', in H.W. Reinhardt and A.E. Naaman (eds) *Third International Workshop on High Performance Fiber Reinforced Cement Composites (HPFRCC 3)*, RILEM Publications, Bagneux, France, 1999, pp. 361–369.

87. T. Kanda, T. Saito, N. Sakata and M. Hiraishi, 'Tensile and anti-spalling prop-erties of direct sprayed ECC', *J. Advanced Concrete Technology.* 1, 2003, 269–282.

88. Y.Y. Kim, H.J. Kong and V.C. Li, 'Development of sprayable engineered cementitious composites', in H.W. Reinhardt and A.E. Naaman (eds) *Third International Workshop on High Performance Fiber Reinforced Cement Composites (HPFRCC 3)*, RILEM Publications, Bagneux, France, 1999, pp. 233–243.

89. Y.Y. Kim, H.J. Kong and V.C. Li, 'Design of engineered cementitious composite suitable for wet-mixture shotcreting', *ACI Mater. J.* 100, 2003, 511–518.

90. P. Kabele, 'New developments in analytical modeling of mechanical behavior', *J. Advanced Concrete Technology.* 1, 2003, 253–264.

91. P. Kabele, S. Takeuchi, K. Inaba and H. Horii, 'Performance of engineered cemen-titious composites in repair and retrofit analytical estimates', in H.W. Reinhardt and A.E. Naaman (eds) *Third International Workshop on High Performance Fiber Rein-forced Cement Composites (HPFRCC 3)*, RILEM Publications, Bagneux, France, 1999, pp. 617–627.

92. H. Fukuyama and H. Suwada, 'Experimental response of HPFRCC dampers for structural control', *J. Advanced Concrete Technology.* 1, 2003, 317–326.

93. K.E. Kesner and S.L. Billington, 'Experimental response of precast infill panel connections and panels made with DFRCC', *J. Advanced Concrete Technology.* 1, 2003, 327–333.

94. G. Song and G. Van Zijl, 'Tailoring ECC for commercial application', in M. Di Prisco, R. Felicetti and G.A. Plizzari (eds) *Fibre Reinforced Concrete - BEFIB 2004*, Proc. RILEM Symp., PRO 39, RILEM Publications, Bagneux, France, 2004, pp. 1391–1400.

95. S. Wang and V.C. Li, 'Lightweight engineered cementitious composites (ECC)', in *Fibre Reinforced Concrete - BEFIB 2004*, M. Di Prisco, R. Felicetti and G.A. Plizzari (eds), Proc. RILEM Symp., PRO 39, RILEM Publications, Bagneux, France, 2004, pp. 379–390.

96. N. Banthia and R. Gupta, 'Hybrid fiber reinforced concrete (HyFRC): fiber synergy in high strength matrices', *Mater. Struct.* 37, 2004, 707–716.

97. J. Studinka, 'Replacement of asbestos in the fiber cement industry – state of substitution, experienced up to now', paper presented at the *International Man Made Fibres Congress*, Austrian Chemical Institute, Dornbirn, Austria, 1986.

98. C. Bleiman, M. Bulens and P. Robin, 'Alternatives for substituting asbestos in fibre cement products', in *High Performance Roofing Systems*, Proc. Conf. Plastics & Rubber Institute, Plastics and Rubber Institute, London, 1984, pp. 8.1–8.12.

99. P.L. Walton and A.J. Majumdar, 'Cement-based composites with mixtures of different types of fibres', *Composites.* 6, 1975, 209–216.

100. K. Kobayashi and R. Cho, 'Flexural characteristics of steel fibre and polyethylene fibre hybrid - reinforced concrete', *Composites.* 13, 1982, 164–168.

101. Y. Ohama and M. Endo, 'Properties of hybrid fibre reinforced polymer modified concrete', Paper 2.14 in R.N. Swamy, R.L. Wagstaffe and D.R. Oakley (eds) *Developments in Fibre Reinforced Cement and Concrete*, Proc. RILEM Symp., Sheffield, RILEM Technical Committee 49-FTR, 1986.

102. S. Hasaba, M. Kawamura, T. Koizumi and K. Takemoto, 'Resistibility against impact load and deformation characteristics under bending load in polymer and hybrid (polymer and steel) fiber reinforced concrete', in G.C. Hoff (ed.) *Fiber Reinforced Concrete, ACI SP-81*, The American Concrete Institute, Detroit, MI, 1984, pp. 187–196.

103. Y.W. Mai, R. Andonian and B. Cotterell, 'On polypropylene – cellulose fibre-cement hybrid composite', in A.R. Bunsell (ed.) Advances in Composite Materials, Paris, Pergamon Press, Oxford and New York, 1980, pp. 1687–1699.

104. G. Xu and D.J. Hannant, 'Synergistic interaction between fibrillated polypropylene networks and glass fibres in a cement-based composite', *Cem. Concr. Compos.* 13, 1991, 95–106.

105. G. Xu and D.J. Hannant, 'Flexural behavior of combined polypropylene network and glass fibre reinforced cement', *Cem. Concr. Compos.* 14, 1992, 51–61.

106. M. Kakemi and D.J. Hannant, 'Mathematical model for tensile behavior of hybrid continuous fibre cement composites', *Composites.* 26, 1995, 637–643.

107. S. Soroushian, A. Tili, A. Alhozaimy and A. Khan, 'Development and characterization of hybrid polyethylene fiber reinforced cement composites', *ACI Mater. J.* 90, 1993, 182–190.

108. A. Peled, M. Cyr and S.P. Shah, 'Hybrid fibers in high performances extruded cement composites', in M. Di Prisco, R. Felicetti, and G.A. Plizzari (eds) *Fibre Reinforced*

Concrete – BEFIB 2004, Proc. RILEM Symposium, PRO 39, RILEM Publications, Bagneux, France, 2004, pp. 139–148.

109. A. Katz, A. Bentur, D. Dancygier, D. Yankelevsky and D. Sherman, 'Ductility of high performance cementitious composites', in K. Kovler, J. Marchand, S. Mindess and J. Weiss (eds) *Concrete Science and Engineering, A Tribute to Arnon Bentur*, RILEM Proceeding PRO 36, RILEM Publications, Bagneux, France, 2004.

110. C. Boulay, P. Rossi and J.L. Tailhan, 'Uniaxial tensile test on a new cement composite having a hardening behavior', in M. Di Prisco, R. Felicetti, and G.A. Plizzari (eds) *Fibre Reinforced Concrete – BEFIB 2004*, Proc. RILEM Symp., PRO 39, RILEM Publications, Bagneux, France, 2004, pp. 61–68.

111. P. Rossi, A. Arca, E. Parant and P. Fakhiri, 'Bending and compressive strength of new cement composite', *Cem. Concr. Res.* 35, 2005, 27–33.

112. I. Markovic, J.C. Walraven and J.M.G. van Mier, 'Development of high performance hybrid fibre concrete', in A.E. Naaman and H.W. Reinhardt (eds) *Fourth International Workshop on High Performance Fiber Reinforced Cement Composites (HPFRCC 4)*, RILEM Publications, Bagneux, France, 2003, pp. 277–300.

113. I. Markovic, J.C. Walraven and J.G.M. van Mier, 'Tensile response of hybrid-fibre concrete', in M. Di Prisco, R. Felicetti, and G.A. Plizzari (eds) *Fibre Reinforced Concrete – BEFIB 2004*, Proc. RILEM Symp., PRO 39, RILEM Publications, Bagneux, France, 2004, pp. 1341–1352.

114. J.G.M. Van Mier, 'Cementitious composites with high tensile strength and ductility through hybrid fibres', in M. Di Prisco, R. Felicetti, and G.A. Plizzari (eds) *Fibre Reinforced Concrete – BEFIB 2004*, Proc. RILEM Symp., PRO 39, RILEM Publications, Bagneux, France, 2004, pp. 219–236.

115. M.J. Shannag, R. Brincker and W. Hansen, 'Pullout behavior of steel fibers from cement-based composites', *Cem. Concr. Res.* 27, 1997, 925–936.

116. B. Mobasher and C.Y. Li, 'Mechanical properties of hybrid cement-based composites', *ACI Mater. J.* 93, 1996, 284–292.

117. C. Sujivorakul and A.E. Naaman, 'Ultra high-performance fiber-reinforced cement composites using hybridization of twisted steel and micro fibers', in M. Di Prisco, R. Felicetti, and G.A. Plizzari (eds) *Fibre Reinforced Concrete – BEFIB 2004*, Proc. RILEM Symp., PRO 39, RILEM Publications, Bagneux, France, 2004, pp. 1401–1410.

118. A. Kawamata, H. Mihashi and H. Fukuyama, 'Properties of hybrid fiber reinforced cement-based composites', *Journal of Advanced Concrete Technology.* 1, 2003, 283–290.

119. K. Otsuka, H. Mihashi, M. Kiyota, S. Mori and A. Kawamata, 'Observation of multiple cracking in hybrid FRCC at micro and meso levels', *Journal of Advanced Concrete Technology.* 1, 2003, 291–298.

120. J.P. Charron, E. Denarie and E. Bruhwiler, 'Permeability of UHPFRC under high stresses', Paper F-12, in *Advances in Concrete through Science and Engineering*, Hybrid-Fiber Session, Int. Conference, Evanston, RILEM Publications, CD-ROM, Bagneux, France, 2004.

121. K. Wang, D.C. Jansen, S.P. Shah and A.F. Karr, 'Permeability study of cracked concrete', *Cem. Concr. Res.* 27, 1997, 433–439.

122. J.S. Lawler, D. Zampini and S.P. Shah, 'Permeability of cracked hybrid fiber-reinforced mortar under load', *ACI Mater. J.* 99, 2002, 379–385.

123. J.L. Granju, V. Sabathier, M. Alcantra, G. Pons and M. Mouret, 'Hybrid fibre reinforcement of ordinary or self-compacting concrete', in M. Di Prisco, R. Felicetti,

and G.A. Plizzari (eds) *Fibre Reinforced Concrete – BEFIB 2004*, Proc. RILEM Symp., PRO 39, RILEM Publications, Bagneux, France, 2004, pp. 311–1320.

124. J.S. Lawler, T. Wilhelm, D. Zampini and S.P. Shah, 'Fracture processes of hybrid fiber-reinforced mortar', *Mater. Struct.* 36, 2003, 197–208.

125. W. Yao, V. Li and K. Wu, 'Mechanical properties of hybrid fiber-reinforced concrete at low fiber volume fraction', *Cem. Concr. Res.* 33, 2003, 27–30.

126. C.X. Qian and P. Stroeven, 'Development of hybrid polypropylene-steel fibre–reinforced concrete', *Cem. Concr. Res.* 30, 2000, 63–69.

127. M.A. Mitlenberger, E.A. Attiogbe and B. Bissonnette, 'Behavior of conventional reinforcement and a steel-polypropylene fiber blend in slabs-on-grade', Paper F-16, in *Advances in Concrete through Science and Engineering*, Hybrid-Fiber Session, Int. Conference, Evanston, RILEM Publications, CD-ROM, Bagneux, France, 2004.

128. A. Meda, G.A. Plizzari and L. Sorelli, 'Fracture properties of concrete reinforced with hybrid fibers', Paper F-1 in *Advances in Concrete through Science and Engineering*, Hybrid-Fiber Session, Int. Conference, Evanston, RILEM Publications, CD-ROM, Bagneux, France, 2004.

129. W. Sun, H. Chen, X. Luo and H. Qian, 'The effect of hybrid fibers and expensive agent on the shrinkage and permeability of high-perfromance concrete', *Cem. Concr. Res.* 31, 2001, 595–601.

130. H. Mihashi, T. Nishiwaki and J.P. de B. Leite, 'Effectiveness of crack control on durability of HPFRCC', in A.E. Naaman and H.W. Reinhardt (eds) *Fourth International Workshop on High Performance Fiber Reinforced Cement Composites (HPFRCC 4)*, RILEM Publications, Bagneux, France, 2003, pp. 437–450.

131. A.E. Naaman, 'Fiber with slip hardening behavior', in H.W. Reinhardt and A.E. Naaman (eds) *Third International Workshop on High Performance Fiber Reinforced Cement Composites (HPFRCC 3)*, RILEM Publications, Bagneux, France, 1999, pp. 371–385.

132. C. Sujivorakul and A.E. Naaman, 'Modeling bond components of deformed fibers in FRC composites', in A.E. Naaman and H.W. Reinhardt (eds) *Fourth International Workshop on High Performance Fiber Reinforced Cement Composites (HPFRCC 4)*, RILEM Publications, Bagneux, France, 2003, pp. 35–48.

133. A.E. Naaman, 'Engineered steel fibres with optimal properties for reinforcement of cement composites', *Journal of Advanced Concrete Technology.* 1, 2003, 241–252.

134. P. Rossi and G. Chanvillard, 'A new geometry of steel fibre for fibre reinforced concretes', in H.W. Reinhardt and A.E. Naaman (eds) *2nd High Performance Fiber Reinforced Cement Composites (HPFRCC2)*, RILEM Conference, E&FN SPON, London and New York, 1992, pp. 129–139.

135. O.C. Choi and C. Lee, 'Flexural performance of ring-type steel fiber-reinforced concrete', *Cem. Concr. Res.* 33, 2003, 841–849.

136. Y. Geng and C.K.Y. Leung, 'Damage evolution of fiber/mortar interface during fiber pullout', in S. Diamond, S. Mindess, F.P. Glasser, L.W. Roberts, J.P. Skalny and L.D. Wakeley (eds) *Microstructure of Cement-Based Systems/Bonding and interfaces in Cementitious Materials*, Materials Research Society Symp. Proc. Vol. 370, Materials Research Society, Pittsburgh, PA, 1995, pp. 519–528.

137. ACI Committee 548, *State of the Art Report – Polymers in Concrete*, American Concrete Institute, Detroit, MI.

138. B.P. Hughes and J.E. Guest, 'Polymer modified fibre reinforced cement composites', Polymers in Concrete. Proceedings of the First International Congress on Polymer Concretes, 1975. The Construction Press, Lancaster, 1976, pp. 85–92.

139. Z. Jamrozy and J. Sliwinski, 'Properties of steel fibre reinforced concrete impregnated with methyl methacrylate', *Int. J. Cem. Comp.* 1, 1979, 117–124.

140. S. Akihama, T. Suenaga, H. Nakagawa and K. Suzuki, 'Influences of fibre strength and polymer impregnation on the mechanical properties of carbon fibre reinforced cement composites', Paper 2.3 in R.N. Swamy, R.L. Wagstaffe and D.R. Oakley (eds) *Developments in Fibre Reinforced Cement and Concrete*, Proc. RILEM Symp., Sheffield, RILEM Technical Committee 49-FTR, 1986.

141. Y. Ohama and M. Endo, 'Properties of hybrid fibre reinforced polymer-modified concrete', Paper 2.14 in R.N. Swamy, R.L. Wagstaffe and D.R. Oakley (eds) *Developments in Fibre Reinforced Cement and Concrete*, Proc. RILEM Symp., Sheffield, RILEM Technical Committee 49-FTR, 1986.

142. K.T. Iyengar, T.S. Nagaraj and B.K. Rao, 'Superplasticized natural rubber latex modified steel fibre reinforced concretes', Paper 3.3 in R.N. Swamy, R.L. Wagstaffe and D.R. Oakley (eds) *Developments in Fibre Reinforced Cement and Concrete*, Proc. RILEM Symp., Sheffield, RILEM Technical Committee 49-FTR, 1986.

143. Y. Ohama, S. Kan and M. Miyara, 'Flexural behaviour of steel fiber reinforced polymer modified concrete', *Trans. Japan Concr. Inst.* 4, 1982, 147–152.

144. Y. Ohama, M. Miyara and S. Kan, 'Drying shrinkage of steel fiber reinforced polymer modified concrete', *Trans. Japan Concr. Inst.* 4, 1982, 153–158.

145. P.S. Mangat and R.N. Swamy, 'Properties of polymer modified plain and fibre reinforced concrete', Polymers in Concrete. Proceedings of the First International Congress on Polymer Concretes, 1975. The Construction Press, Lancaster, 1976, pp. 296–299.

146. A. Bentur, 'Properties of polymer latex-cement-steel fibre composites', *Int. J. Cem. Comp. & Ltwt. Concr.* 3, 1981, 283–289.

Chapter 13

Continuous reinforcement

13.1   Structure and properties

Continuous reinforcement of cementitious matrices is particularly attractive for fabrication of thin elements, where cement paste or mortar is impregnated into a fabric. Earlier interest in this kind of reinforcement was driven by the need to develop new thin sheet components, that could serve as replacements for asbestos cement, or provide thin sheets with improved performance, especially with regard to toughness [1–20]. New types of reinforcements were studied and developed for these purposes, and the mechanical properties of the composites as well as production technologies were explored.

The lay up of several layers of mats was developed by Hannant and Zonsveld [9] and Hannant [10]. With a sufficiently high fibre content, flexural strengths in the range of 20–40 MPa can be achieved (Figure 13.1). This method has been used to produce flat and corrugated sheets (Keer and Thorne [11]; Baroonian et al. [12]). A comparison between asbestos fibre corrugated sheets and fibrillated polypropylene corrugated sheets is shown in Figure 13.2 (Keer and Thorne [13]). In such products care should be taken to place some of the oriented mats in the perpendicular direction, in order to obtain a component in which the properties at different orientations do not differ much. Keer [3] has also described a production method in which thin layers of the cementitious matrix are deposited on a moving belt, followed by laying the continuous reinforcement onto the layers of slurry. This technique requires the use of special dewatering and compaction devices. Alternatively, the matrix may be sprayed onto the continuous reinforcement.

A comparison between the properties of polypropylene reinforced cement and asbestos cement flat sheets is given in Figure 13.3. In tension, the strength of the asbestos cement is higher, but the polypropylene composite is much tougher and possesses a much higher strain capacity. As a result, in flexural loading the strength of the polypropylene composite is slightly higher than that of the asbestos cement, both for flat sheets and corrugated sheets (Figure 13.2). However, the polypropylene composite had a lower first crack strength, or limit of proportionality (LOP), that is, the fibres were not very effective in increasing the first crack stress. Ponding of water in the corrugations at the crack zone while the sheets were

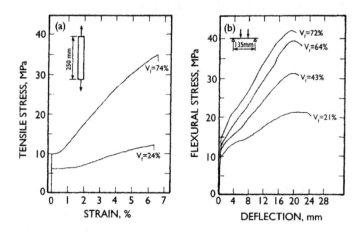

Figure 13.1 The effect of fibre content on the tensile stress–strain curve and calculated bending stress–deflection curve of Netcem composite consisting of layers of fibrillated continuous polypropylene network in a mortar matrix (after Vittone [14]).

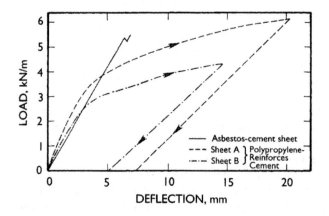

Figure 13.2 Load–deflection curves in bending of corrugated sheets prepared with asbestos fibres and fibrillated polypropylene network, Netcem (after Keer and Thorne [13]).

under load showed some dampness at the underside but no water droplets [11,13]. This was attributed to the ability of the polypropylene to control the crack width, as well as to some additional effect of crack healing.

An analysis of the behaviour of cracked and uncracked corrugated sheets made with both asbestos and polypropylene reinforcement was presented by Baroonian et al. [12]. They demonstrated that the effects of cracks on any corrugation under

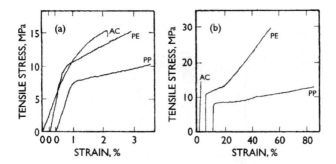

Figure 13.3 Stress–strain curves in tension of cement composites prepared with asbestos fibres (12% mass), fibrillated polypropylene network (Netcem) and polyethylene (both 18% by volume) (a) low strain range; (b) high strain range (after Bijen and Geurtz [15]).

concentrated load is local, that is, the effects of the deformation of a cracked corrugation on adjacent corrugations is small. Thus, from a load performance point of view, the polypropylene product is not inferior to asbestos cement. Hibbert and Hannant [16] concluded that the toughness (i.e. the energy absorbed by the composite up to fracture) was much greater in the polypropylene composite than in glass and asbestos composites, when compared at equal fibre contents in the range of 5–10% by volume. Although this high energy absorption is achieved at the expense of large deformations, this may be acceptable in practice when the toughness is required to provide protection from the consequences of collapse under transient loads; large deflections may be acceptable in these circumstances, particularly in applications such as roofing sheet.

Additional developments in fibrillated polypropylene composites were reported by Xu, Hannant and co-workers [8,21], who applied a fibrillated polypropylene network in the form of layered opened nets (12 net layers in [21]), with the majority in the longitudinal direction and the rest in transverse orientation, and impregnated them by hand, or by a special mechanized system developed for that purpose [8]. These systems could be fabricated from polypropylene only, or from a hybrid of polypropylene nets combined with glass fabrics and chopped glass fibres. Flat thin sheets, as well as corrugated ones could be produced [8].

Swamy and Hussin [6,7] used polypropylene woven fabric to make flat and corrugated sheets and demonstrated that deflection hardening could be obtained even with the low modulus reinforcement at relatively low yarn contents of a few per cent. Components reinforced with continuous yarns produced by filament winding processes were reported by Mobasher and co-workers [17,18]. Systems which could be produced manually or mechanically (Figure 13.4), using computer-controlled operation were developed. They could be applied for making components such as laminates and pipes. The concepts of filament winding were

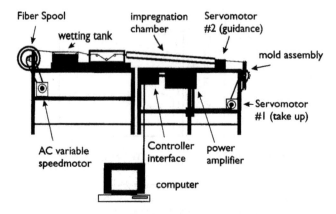

Figure 13.4 Schematic side view of a mechanized, computer controlled filament winding process (after Mobasher and Pivacek [18]).

further extended by Peled and Mobasher [19,20] to a pultrusion process which could be used for impregnation of fabrics.

More recent developments were driven by the potential of using the high performance fibres in technical fabrics for making a variety of thin sheet components having structural merit (e.g. profiles) [22] as well as on-site application for repair and retrofit, taking advantage of the mechanical properties of the composite as well as its potential for providing durability protection [23,24]. Depending on the nature of the reinforcement, composites which are pseudo-plastic in nature and exhibit strain/deflection hardening behaviour have been developed. The first type, which tend to be more elastic–plastic in nature, are reinforced by lower modulus–high extension materials such as polypropylene [3], while the strain hardening types use higher modulus materials such as glass and aramid [25].

This section further deals with the mechanics of continuous reinforcement based on the use of modern textile yarns, made of an assembly of yarns consisting of numerous filaments with a diameter of 10–50 μm. In these systems several hundreds or thousands of filaments are assembled into a strand, and several strands are assembled into a roving. The filaments are usually made of glass or a polymeric material, and are characterized by high strength or ductility, or both. The treatment in this section will refer only to reinforcements of this kind, and will not include discussion of ferrocement, which one may include in the category of continuous reinforcement. The reinforcement in ferrocement is of a different nature, steel wire mesh, which is incorporated in a mortar matrix. This material is used for a variety of applications and its properties and application are documented quite well in several recent publications [26,27].

The parameters controlling the performance of continuously reinforced composites are more numerous and complex than those of fibre reinforced cement, because the reinforcement can assume a more complex geometry and properties than conventional fibres. The geometry of the fabric can be 2D or 3D. Some geometries of 2D fabrics are presented in Figure 13.5 [28]. The terminology of textile fabrics refers to the warp and weft directions, one representing the 'main' yarn, and the other representing those which bind the main yarns together in the fabric. In a woven fabric the yarns are undulating, one over the other, and essentially the yarns in each direction can be described as crimped. In the weft insertion knitted fabrics the warp yarns are straight, and the weft yarns, which bind them together, can assume different geometries, depending on the nature of the production process. An example of a fabric where the warp and weft yarns are linked by stitches is presented in Figure 13.6 [29]. The geometry in 3D can assume many different complex shapes, depending on the objective of the reinforcement. An

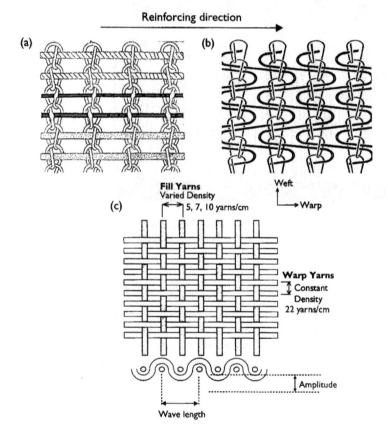

Figure 13.5 Structure of fabrics: (a) weft insertion knit, (b) short weft knit and (c) woven (plain weave (after Peled and Bentur [28])).

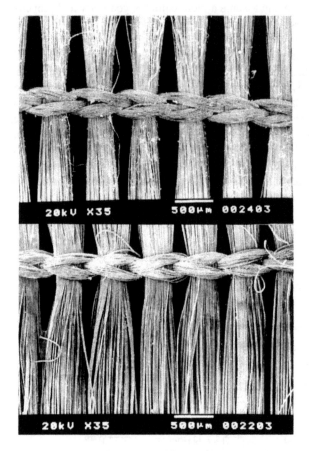

Figure 13.6 Structure of knitted fabrics from yarns of different numbers of filaments (a) Kevlar fabric with 325 filaments; (b) polyethylene with 900 filaments (after Peled and Benur [29]).

example of 3D spacer fabric, in which the yarn in the third dimension serves as a spacer to link and position 2D fabrics is shown in Figure 13.7.

In considering the properties of the composite there is a need to take into account the sensitivity to the mode of production of the composite, especially whether it is manual or mechanical. Mechanical processes will usually provide better impregnation, and they are applied mostly in an industrial operation. Therefore they are more applicable for production of components such as thin sheets and various profiles. Manual production is particularly useful in applications of repair and retrofit, where a fabric is placed on the surface to be repaired or strengthened, and thereafter impregnated with paste, mortar or fine (small) aggregate concrete

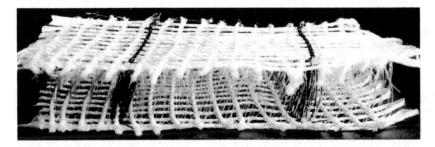

Figure 13.7   3D spacer fabric (after institut für & Textiltechnit at RWTH Aacher University, Prof Th. Gries, A. Roye, D. Mecit).

which is essentially a better graded mix with a maximum particle size characteristic of sand.

The reinforcement itself should be considered at three levels: the mechanical properties of the individual filaments, the geometry of the yarn and the structure of the fabric.

13.2   Yarns: structure and properties

The filaments in the yarn can be made of glass and a variety of polymeric materials such as polyethylene (high and low modulus), polypropylene, PVA, aramid or carbon. The range of properties of such materials is outlined in Table 1.1. Most of these high performance filaments are produced by filament drawing processes, such as those described for glass in Section 8.2 which includes also their incorporation into strands and yarns. The nature of the yarn, which is an assembly of such filaments, and the sizing applied on the filaments will have a significant influence on the stress transfer between the filaments in the yarn. The bonding and strength of the yarn is not a simple and straightforward function of the characteristics of the individual filaments. Therefore, in modelling and accounting for the reinforcing efficiency of the yarn there is a need to look into the micromechanics within this unit, as well as its interaction with the surrounding matrix. Some of these concepts have been already discussed in Section 8.4, in conjunction with the reinforcement by glass fibres which are made of multifilament strands.

The strength of a strand can be quite different from that of the filaments from which it is composed. Jesse and Churbach [31] studied the strength of AR glass strands and the filaments which make them up, by evaluating 10 commercial yarns obtained from different manufactures. The strength of the individual filaments was about the same, 2000 MPa, regardless of their source. However, the strength of the strands was considerably lower, and varied over a wide range of 444–1476 MPa. The straightforward explanation for this observation is the inherent variability in the strength of the filaments: when testing a strand composed of several hundreds to several thousands of filaments, the strength of the strand is controlled by the weakest filament. Another contributing factor is the lack of ideal alignment of all the filaments in the strand, and as a result some of them contribute only partially to the load-bearing capacity of the strand. Within this context, one has to consider also the stress transfer between the filaments in the strand, and to what extent a fractured filament can effectively transfer its load to a large number of surrounding filaments, and not only to the ones immediately adjacent. The latter effect is dependent on the frictional interactions between neighbouring filaments. It was shown that when analysing the strength of a composite prepared from glass multifilament yarns, there is no correlation between the composite strength and the glass filament strength, while a reasonable correlation was obtained with the strength of the yarn [31]. This issue should be kept in mind when modelling such composites, and one should not rely on the strength of the filament which can be considerably different from that of the strand.

The bonding of the basic unit, the individual filament, was studied by special pull-out tests [32–36], which were developed to accommodate a small filament which is embedded in a cement matrix at a length of few millimetres. The range of bond strengths for carbon and glass filaments is 1–15 MPa. Banholzer and Brameshuber [34] evaluated the influence of the surface treatment on filaments which were extracted from commercial AR glass strands: uncoated filaments, filaments with insoluble sizing and filaments treated by plasma. The pull-out curves were analysed in terms of adhesional and frictional bond. An adhesional bond component was observed in untreated filaments and ones treated by plasma, with values of about 5 and 12.5 MPa for the two, respectively. The frictional bond was about 5 MPa. However, application of a non-soluble silane type coupling agent prevented the formation of an adhesional bond. The failure in these filaments was at the interface or sizing layer, and there were no remnants of cementitious particles on the pulled-out fibre.

It should be recognized that the bonding of the multifilament unit, strand and roving, cannot be simply described in terms of the bond of a single filament. As already outlined in Chapter 8, dealing with glass fibre reinforcement, there is a need to address the difference between the external filaments in the strand, which are more tightly bonded to the matrix, and the internal ones, which are usually loosely held to each other by frictional forces. The interaction between the multifilament strand and the cementitious matrix is quite unique in the sense that the particulate matrix (~ 10 μm cement grains) cannot readily penetrate into the spaces between the filaments. This is different from a polymer matrix, which is a viscous fluid during the processing stage, and can more easily impregnate the spaces in between the filaments. More than that, because of the special nature of the cement matrix, the nature of matrix built up in the yarn can change over time due to precipitation from the pore fluid and further hydration. These changes can be a cause for variation in the properties over time, leading to durability problems which are not necessarily the result of chemical instability of the reinforcement. Such issues are encountered with alkali-resistant (AR) glass (see Chapters 5 and 8 for more details). Thus, the need to deal with the microstructure and the resulting micromechanics of the reinforcing unit itself is of a special need in cement composites.

The mechanics of the pull-out of a strand was studied by special techniques to evaluate individual filaments within the strand. Zhu and Bartos [37] used for that purpose a micro-push off technique, which was used to load individual filaments in the strand and determine their resistance. Banholzer and Brameshuber [32–34] developed a novel test in which pull-out loading and imaging of the individual filaments could be carried out simultaneously, to observe the filaments which were fractured during the test. Both studies confirmed earlier reports (for more details see Sections 8.4 and 8.5) of a sleeve-core failure mechanism, where the external filaments (sleeve), which are better bonded to the matrix, tend to fracture, while the internal ones (core), undergo pull-out (Figure 13.8), in a mode which was described by Bartos [38] as a telescopic mode of pull-out. Banholzer and Brameshuber [32–34] demonstrated the parallel trend between the pull-out resistance and the number of filaments which remain active (i.e. the non-fractured filaments) as a

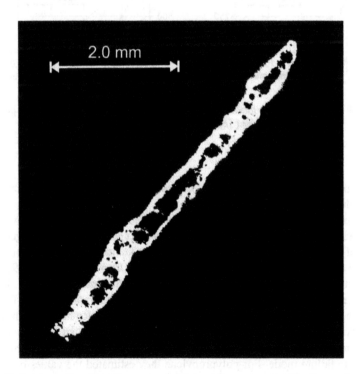

Figure 13.8 Image of a glass fibre strand during different stages of pull-out, showing a telescopic failure, with white pixels representing broken filaments (after Banholzer [32]).

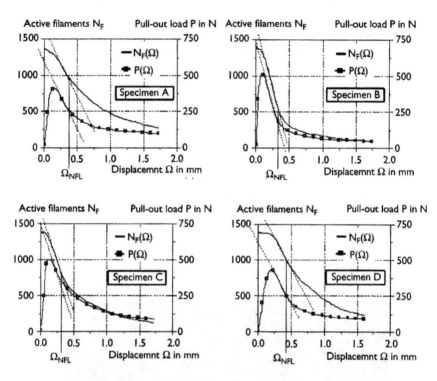

Figure 13.9 Active filaments (unbroken) and load as a function of displacement in pull-out test (after Banholzer [32]).

function of slip (Figure 13.9), and on that basis provided a schematic description of the failure mechanism (Figure 13.10).

Based on these concepts, a variety of models have been advanced to deal with the yarn's internal structure and bonding, in particular in Germany, where a large-scale project of 'Textile Reinforced Concrete' has been launched [22]. The modelling to a large extent is based on concepts developed by Ohno and Hannant [39] for a continuous fibrillated polypropylene network for reinforcement of cement. It was shown there, that although the reinforcing unit is a mono-yarn of about 100 μm in size, shear failure can occur within the yarn, which was then modelled in terms of a sleeve and core (Figure 13.11). Ohno and Hannant applied this concept of bonding to the ACK model and modified it to accommodate a fibre exhibiting sleeve-core failure mode. For polypropylene they estimated the values of τ_f, τ_{is} and τ_i (see Figure 13.11) to be 2, 1 and 5 MPa, respectively, and applied them to model the behaviour of the polypropylene composite according to the ACK model. They modified the model to accommodate the sleeve-core fibre, identifying and

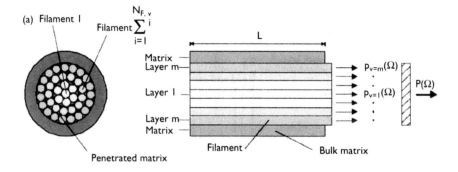

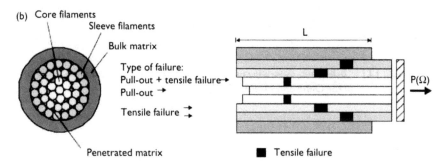

Figure 13.10 Schematic description of the mode of a strand failure: (a) loading and (b) failure (after Banholzer [32]).

defining two multiple crack regions, first and second, as shown schematically in Figure 13.12.

The concept was extended to a multifilament glass strand, using idealized models describing the strand in terms of several layers. Hegger *et al.* [40] developed an analytical treatment based on a ring model, assuming that the bond between the rings can be simulated by an increasing function, showing negligible bond at the centre, with an increasing value as one approaches the perimeter. A cube distribution function was found to best represent pull-out results, and in a 50 layer simulated strand, the bond in layers 15–35 was about 55% of the full bond achieved in the perimeter (Figure 13.13). From this model and the simulation of the actual load-displacement curve obtained in a pull-out test, they concluded that in the strand they studied, about 65% of the filaments were active to different degrees at the peak load, while the others were activated only in the post-peak zone. In an additional level of simulation of the actual behaviour in a composite, they followed the Ohno and Hannant [39] model, of sleeve and core (i.e. two layers rather than 50) and modelled the behaviour by a finite element analysis, in which the

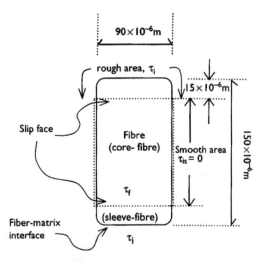

Figure 13.11 Schematic description of a fibre exhibiting sleeve and core behaviour (after Ohno and Hannant [39]).

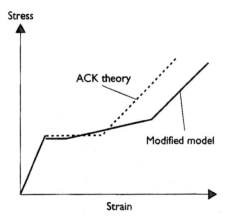

Figure 13.12 Schematic description of the stress–strain curve for a conventional fibre and one exhibiting sleeve-core failure (after Ohno and Hannant [39]).

strand was treated as a two layer parallel chains, where the core chain was bonded to the matrix by non-linear bond elements. A smeared bond law was developed for this simulation.

A multiple layer approach for modelling the failure of the strand was also developed by Banholzer and Brameshuber [32–34], based on a study of the penetration

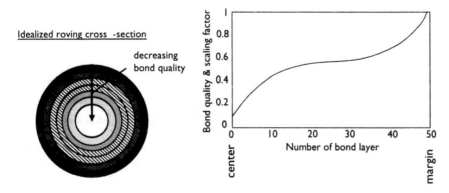

Figure 13.13 Idealized description of a strand in terms of layers and the variation of bond between the layers (after Hegger *et al.* [40]).

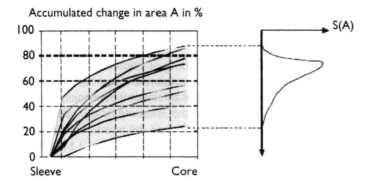

Figure 13.14 Reduction in the area occupied by the matrix as a function of the position in the strand, from the outer filaments (sleeve) inward (core) (after Banholzer [32]).

of the matrix into the spaces between the filaments. Based on special 3D imaging techniques, they could estimate the reduction of the space which is occupied by the matrix, as one moves from the sleeve to the core. Obviously, in view of the nature of the system, there is considerable variability, and the reduction in the occupied space was described in terms of a band (Figure 13.14). Based on such a function, the microstructure of the strand was described in an idealized model (Figure 13.15) and the pull-out was than modelled based on the concept shown in Figure 13.13.

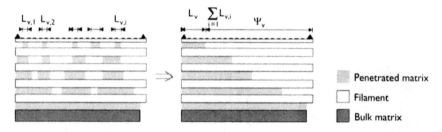

Figure 13.15 Idealized representation of the microstructure of a strand, in terms of penetration of the matrix in between the filaments (after Banholzer [32]).

13.3 Bonding of fabrics

The special geometry of the fabric may lead to different types of mechanical anchoring processes, associated with effects such as filling of the spaces between perpendicular yarns, which form the fabric, as well as the crimping of the yarns in woven fabric. Thus, the bond of the fabric cannot be simply described and accounted for by the bonding of the individual strands, and there is a need to consider another level of complexity, on top of the multifilament nature of the yarn.

Kruger *et al.* [41] analysed the bonding of fabrics using models developed for deformed reinforcing bars, where mechanical anchoring effects are quite significant. They considered the effect of mechanical interaction, τ_m and the frictional component, τ_f and the effect of the 3D stress field on these values. This effect could be represented by a parameter Ω, by which the characteristic values of τ_m and τ_f should be multiplied. The parameter Ω is a function of three contributions:

$$\Omega = \Omega_s \cdot \Omega_c \cdot \Omega_{cyc} \tag{13.1}$$

Ω_s is the influence of the yielding of the reinforcement; in the case of fabric $\Omega_s = 1$; Ω_c, the influence of lateral stresses induced by stresses in the concrete and in the reinforcement; Ω_{cyc}, the influence of cyclic loading on bond.

They argued that the main parameter to consider is Ω_c, and this in turn, is a function of the radial stresses in the matrix and the strains in the reinforcement:

$$\Omega_c = 1.0 + \tanh\left[\alpha_r \cdot \frac{\overline{\sigma_R}}{0.1 \cdot f_f} - \alpha_f \mu_s \left(\varepsilon_f - \varepsilon_{\rho,0}\right) \frac{1}{1 - r_f^2/(r_f + h_f)^2}\right] \tag{13.2}$$

where α_r is the factor controlling the influence of the radial concrete stress (set for 1 in the calculation in ref. [41]; $\overline{\sigma_R}$, the average radial stress in the concrete in the vicinity of the reinforcement; f_c, the uniaxial compressive strength of the concrete; μ_f, the Poisson's ratio of the reinforcement; ε_f, the reinforcement strain; $\varepsilon_{p,0}$, the strain due to prestressing of the reinforcement; r_f, the radius of reinforcement; h_f, the constant representing the surface roughness of the reinforcement; α_f, the a parameter controlling the influence of the roughness of the reinforcement, h_f; Ω_c can theoretically vary between 0 and 2.

In an analysis of test results of bonding of fabrics from different materials (glass, carbon, aramid) they concluded that in a reinforcement with a rough surface, the bonding is mainly influenced by the radial stresses in the surrounding cementitious matrix, while in a reinforcement which is smoother, the influence of the reinforcement becomes greater. They also quoted reports that impregnation of the fabric with epoxy results in higher bond and suggested that this may be due to two effects: (i) the ribbed surface formed by the binding yarns in the warp knitted fabric, which are fixed by the epoxy and (ii) the change in the roving diameter over its length.

Analysis such as the one carried by Kruger *et al.* [41] highlights the need to consider and evaluate the nature of the mechanical bonding (anchoring) mechanism that may be invoked by the different geometries and structure of available yarns. A series of studies by Peled and Bentur [42–47] identified some of the parameters, and demonstrated their practical significance. In these studies the influence of the structure of the fabric was investigated by the evaluation of the pull-out curves of single yarns and yarns which are part of a fabric (Table 13.1). When one considers the bonding of the straight yarns in Table 13.1, the superior behaviour of the high quality yarns is clearly evident, and can be accounted for by their higher modulus of elasticity (see Section 3.2.4). However, when the bonding in the fabric is considered, the bond changes quite drastically, decreasing in the weft insertion knitted fabric and increasing in the woven fabric. This change, which is induced in addition to the inherent bonding characteristic of the straight yarn, is the result of influences due to the geometrical shape of the fabric, as outlined below.

The increase in the bonding of the yarn in the woven fabric is to a large extent the result of the crimped geometry which is induced when the yarn is positioned in a woven fabric, providing enhanced bonding by mechanical effects. These effects were outlined in Section 3.4 and quantitative relations were developed (Eq. 3.51 in Section 3.4), showing that the main parameter controlling the bond in the crimped structure is the 'wave amplitude' of the crimp, analogous to the lug height (or surface roughness h_f in Eq. 13.1) in deformed reinforcing bars. Yet, there is still another contribution which can be quantified, when one compares the bonding of a straight yarn, a crimped yarn and a yarn in a fabric (Figure 13.16). The additional enhancement of the bonding of the woven fabric, over the one induced by the crimped shape, may be due to another mechanical anchoring effect, which is provided by the fill yarns in the weft direction, when the fabric is embedded in the cement matrix (Figure 13.17(a)). These mechanisms can account for the increase in bond with the increase in the fill density in the woven fabric (Figure 13.17(b)): higher density implies a more intensely crimped geometry

Table 13.1 Effect of fabric geometry on the pull-out resistance of a yarn from the fabric embedded in a cement matrix, in comparison with the pull-out of a straight yarn (after Peled and Bentur [45,46])

Yarn type		Modulus of elasticity (MPa)	Number of filaments in a bundle	Bond per unit external bundle surface (MPa)		Bond of yarn in a fabric relative to the bond of a straight yarn[b] (%)
				Single straight yarn	Fabric	
Woven, polyethylene	7 yarns/cm[a]	1760	1	0.17	1.2	700%
	5 yarns/cm[a]				0.73	430%
Knit weft insertion	PP	6900	100	3.5	2.8	80%
	HDPE	55,000	900	11.5	1.8	15%

Notes
a Density of yarns perpendicular to reinforcing direction.
b Calculated per single filament.

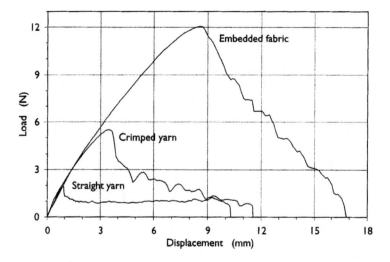

Figure 13.16 Load displacement curves in pull-out of a straight yarn, a crimped yarn untied from a woven fabric and the fabric itself, all embedded in a cement matrix (after Peled *et al.* [42]).

(i.e. greater number of 'waves' per unit length), and greater density of the anchoring fill yarns.

The reduction in the bonding of the yarn in the case of the knit weft insertion fabric (Table 13.1) can be accounted by two influences which are related to the

(a)

(b)

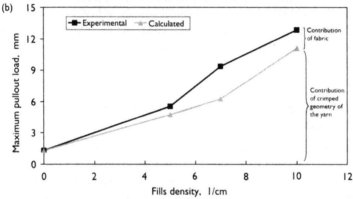

Figure 13.17 Woven fabric structure and bonding: (a) crimped and anchored woven fabric structure and (b) the contribution of the crimping and the fabric to the bond a woven fabric in a cement matrix (after Peled *et al.* [42,43]).

presence of the stitches and the multifilament nature of the yarn: (i) the presence of the stitches which bind the filaments together, limiting the penetration of the matrix especially in the stitched zone and (ii) the stitch itself, which is not well impregnated with the matrix, causing the intersection to be a zone of weakness from the point of view of bond. The fact that the reduction in the bond in the fabric was greater in the high density polyethylene (HDPE) relative to the polypropylene (PP) (Table 13.1) might be attributed to the much higher number of filaments in the HDPE bundle (900 vs. 100 filaments per bundle in the HDPE and PP, respectively).

In the case of a short weft knit fabric the structure of the yarns is quite complex (Figure 13.6), making it difficult or impossible to set up a test for characterization of the pull-out of a single yarn from a fabric which is embedded in a cement matrix. However, indirect indication of the influence of the fibre structure on the nature of bonding may be obtained by considering the relative strength efficiency

Table 13.2 Relative flexural efficiency factor (efficiency of yarn in a fabric relative to its efficiency when laid as single yarns for reinforcement of the same matrix), after Peled and Bentur [44–46]

Woven, low modulus PE		Knitted short weft	Knitted weft insertion	
5 yarns/cm	7 yarns/cm	Low modulus PE	Low modulus PP	Low density PE
1.36	1.64	4.15	0.70	0.74

of the reinforcement when evaluated in flexural tests (efficiency of the fabric reinforcement relative to the efficiency of the reinforcement by same straight yarn). This relative efficiency was much higher for the knitted short weft reinforcement (Table 13.2), probably reflecting mechanical bonding effects associated with the complex geometry of the yarn in the knitted short weft fabric; this is to be compared with the straight geometry in the knitted weft insertion fabric.

13.4   Reinforcing effects and mechanisms in the composites

The reinforcing mechanisms of cement matrices by fabrics are essentially similar to the ones outlined in Section 12.2 for high performance FRC. Most of the fabric reinforcements provide systems where the content and the spacing of the reinforcement are such that strain and deflection hardening are obtained.

Mobasher *et al.* [48] and Stang *et al.* [49] demonstrated the significance of microcrack control and crack arrest in systems with uniaxial aligned polypropylene yarns produced by a pultrusion technique. Investigation of the composites in tension included characterization of the stress–strain curve and the damage developed, by means such as measurements of crack density and acoustic emission. Crack arrest could be clearly seen by the observed increase in the first crack stress, for example from about 9.6–10.5 MPa to 13–16.5 MPa in 8% and 11% volume reinforcement, respectively. The arresting influence of the yarns was clearly observed after the first crack, by a calculation of the load carried by the fibre and the matrix at different levels of strains; even at strains as high as 2%, the matrix had still a considerable contribution to the load-bearing capacity of the composite, amounting to 30–40% of the load carried by the composite. This was consistent with characterization of the cracking, showing that dispersed microcracks developed as load increased, rather than local macrocracks.

When analysing the reinforcing mechanisms in the composite, there is also a need to take into account the special geometry of the fabric, which may induce strengthening or weakening influences which cannot be predicted by simple consideration of yarn–matrix interactions of unidirectional longitudinal reinforcement.

Influences of this kind may be induced by modifications in the bonding mechanisms which are the result of the special shape of the yarns, and some of those were outlined in Sections 3.4 and 13.3 based on the studies of Peled *et al.* [42–47]. The flexural behaviour of the composite with the woven fabric is better than the 'reference' composite reinforced with the same straight yarn (Figure 13.18(a)), while the opposite trend is observed for the weft insertion knit fabric (Figure 13.18(b)). This difference reflects the influence of the fabric structure on bonding, as outlined in Section 13.3. The flexural behaviour was quantified in terms of an efficiency factor, which was defined as the ratio between flexural strength and the tensile reinforcing potential of the yarn reinforcement, (strength of yarn) × (volume content). These factors are presented in Figure 13.19, showing the positive influence of the fabric in geometries in which the warp (reinforcing) yarns have a complex geometrical shape which is not a straight one (woven and short weft insertion knit), and the negative influence in systems where the warp yarns remain straight (weft insertion knit). The fact that some of the efficiency factors are greater than 1 is a reflection of the definition of the efficiency factor in this case, which implies that the reinforcing efficiency takes into account the enhancement in tensile strength as well as the ductility, since flexural efficiency was compared with the tensile reinforcing potential of the yarns.

The curves in Figure 13.18(a) indicate that if the bonding is enhanced (e.g. woven fabric in the case of polyethylene (PE) reinforcement), deflection hardening behaviour can be obtained, even with the low modulus PE yarn (1.76 GPa). Treatments to improve bond in low modulus fabric (PP of 3.5 GPa modulus) were also reported as a means taken to enhance the toughness in a cement-reinforced mesh [50].

Trends of the kind outlined above suggest the need for special considerations when making the choice for the optimal fabric reinforcement. For example, when one considers the effect of density (Figure 13.20), the trends could be quite different in the knitted short weft fabrics and in the woven one. In the knitted weft insertion fabric the increase in density leads to reduction in flexural strength, due to the negative effect of the stitches. In the case of woven fabrics, increased density is beneficial, because of the increase in the mechanical bonding which it provides; however, from a certain point on it has a negative effect (the maximum in the woven curve in Figure 13.20) due probably to inefficient impregnation of a fabric which is too dense.

Mu and Meyer [51] studied the effect of orientation in cements reinforced with AR glass and PVA fabrics. They developed a model to calculate the post-crack bending capacity at different orientations, as a function of the orientation angle and the post-crack bending capacity in the weft and warp directions. The model could predict the orientation effect of the different fabrics, and it was independent of the nature of the fabric. They suggested that the influence of the fabric itself is already expressed in the bending capacity in the warf and weft directions.

An additional factor which needs to be considered is the mode of fabrication of the composite, in relation to the structure of the fabric and yarn. Pultrusion (relative

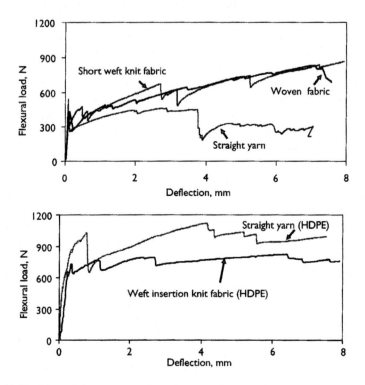

Figure 13.18 Effect of the structure of the fabric on the load–deflection curve, in comparison with straight yarn reinforcement, (a) woven and short weft knit; (b) weft insertion knit (after Peled and Bentur [44–46]).

to hand lay up-casting) was found to enhance the tensile performance of a multifilament yarn composite but did not have a marked influence on a monofilament fabric-reinforced composite (polyethylene) (Figure 13.21) [52]. This difference in behaviour could be attributed to the ability of the pultrusion process to better impregnate the multifilament yarn and especially the zone around the stitches which is more difficult to penetrate. In the case of monofilament reinforcement, the casting as well as the pultrusion resulted in a similar interface, as there was no need to apply special means to consolidate the matrix close to the reinforcement surface. Bond testing of the different fabrics is consistent with this explanation, showing that pultrusion doubled the bond of the multifilament PP fabric (from 1.55 MPa to 2.91 MPa for embedded length of 7.6 mm, in [52]).

Within this context of the production process, there is a need to consider also parameters related to the impregnation of the fabric. In one report the pultrusion

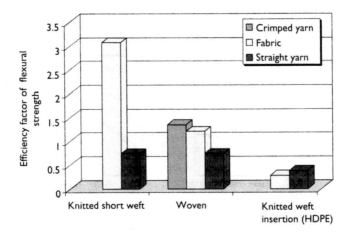

Figure 13.19 Flexural strength efficiency factors for different fabrics, untied crimped yarns and straight yarns (Peled and Bentur [44–46]).

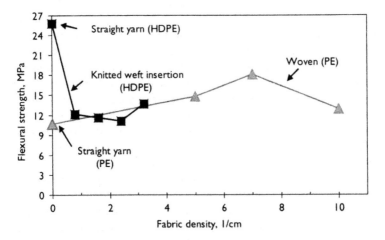

Figure 13.20 Effect of density of yarns perpendicular to the load direction on the flexural strength of woven and knitted weft insertion fabrics in a cement composite (after Peled and Bentur [44–46]).

was not found to have an effect on an AR glass fabric [52] in spite of the multifilament nature of the reinforcement. This could be correlated with the fact that the strands were impregnated with epoxy, thus eliminating the possibility or need for forcing the matrix in between the filaments. However, when AR glass reinforcement was not impregnated with epoxy, the penetration of the matrix into the bundle

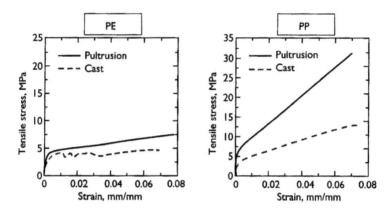

Figure 13.21 Effect of casting and pultrusion production processes on the stress–strain curves of polypropylene (multifilament) and polyethylene (monofilament) fabric cement composites (after Peled *et al.* [52]).

by the pultrusion process was found to be effective in controlling the mechanical behaviour [53]. It was resolved in this study that the composition of the pultruded matrix is of importance, and a matrix in which about 60% of the cement was replaced by fly ash gave the best performance in tension. The improved performance could be correlated with smaller crack spacing during the tensile test, which is indicative of improved bonding. It was suggested that this enhanced performance was the result of improved rheological properties of the mix with fly ash.

In considering the properties of the composite, one has to take into account influences on the structural, macro-scale level. One important characteristic is the layered structure of the composite, which is essentially a laminate where the reinforcement in each lamina may assume different orientations. This can be the case in fabric reinforcement as well as in filament winding. Models to predict and account for such characteristics in cement composites were developed [54–56]. Xu *et al.* [25] evaluated the influence of combining layers of continuous reinforcements from different types of yarns, polypropylene, glass and PVA. PVA–polypropylene hybrid was less efficient at strains up to 1%, but thereafter the composite demonstrated superior behaviour because of a synergy between the two yarns, where both participate in carrying the loads. The influence of lamination was reported also by Pivacek *et al.* [17] who developed laminated composites using polypropylene and glass rovings by a process of filament winding. They obtained composites with tensile strength which could be as high as 50 MPa using 5% AR glass fibre. They optimized the overall behaviour of the composite by changing the stacking sequence of the laminates, to obtain ultimate strain of 2% and more for glass reinforcement and 8% for polypropylene reinforcement. Synergy of a different type was reported by Naaman and Chandrangsu [57] for combinations of meshes and

Table 13.3 Effect of prestressing of epoxy impregnated carbon
textile fabric on the flexural properties of cement
composite (after Reinhardt et al. [58])

	No prestress	Prestress (MPa)		
		1.5	2.5	3.0
First crack stress, MPa	12.3	12.5	13.1	17.1
Flexural strength, MPa	19.6	24.8	36.1	44.7

short fibres in composites consisting of Kevlar, Spectra and carbon meshes. A strain hardening composite could be obtained with only 1.15% if reinforcement, providing a flexural strength of 27 MPa. Addition of 1% of PVA microfibre led to an effective additional increase of the flexural strength to 39 MPa and reduction by 55% of the spacing between the multiple cracks, as well as an improvement of the first crack stress from 5.3 to 8 MPa. This improvement by the hybrid mesh–fibre reinforcement was seen in the Kevlar and Spectra meshes, but not in the carbon.

An additional consideration on the macro-scale is the potential for improvement in the overall behaviour by prestressing of the fabric. Several reports analysed the potential for the increase in the load-bearing capacity of the composite, using this technique for AR glass, carbon and aramid fabrics [58–60]. The effects of prestressing were considered in terms of their influence on the fabric itself, mainly the elimination by prestressing of the initial strain in the woven fabric (which is essentially stretching of the weaved geometry), so that the stiff behaviour of the yarns themselves could be materialized [58]. In such high performance systems, the advantage of impregnating the yarns with epoxy, to assure efficient bonding of the internal filaments was evaluated and demonstrated. The combined influence of prestressing and impregnation was demonstrated by Reinhardt et al. [58,59] for carbon fabric (Table 13.3), showing a marked improvement in the first crack strength (limit of proportionality) and the flexural strength.

References

1. ACI Committee 549, Report on Thin Reinforced Cementitious Products ACI 549.2R-04, American Concrete Institute, USA, 2004.
2. A. Bentur and S.A.S. Akers, 'Thin sheet cementitious composites', in N.Banthia, A.Bentur and A.Mufti (eds) Fiber Reinforced Concrete: Present and Future, Canadian Society for Civil Engineers, Montreal, Canada, 1998, pp. 20–45.
3. J.G. Keer, 'Performance of non-asbestos fiber cement sheeting', in J.I. Daniel, S.P. Shah (eds) Thin Section Fiber Reinforced Concrete and Ferrocement, ACI SP-124, American Concrete Institute, Farmington Hills, NJ, 1990, pp. 19–38.

4. M. Schupack, 'Thin sheet glass and synthetic fabric reinforced concrete 60–120 pound pcf density', in J.I. Daniel and S.P. Shah (eds) *Thin Section Fiber Reinforced Concrete and Ferrocement*, ACI SP-124, American Concrete Institute, Farmington Hills, NJ, 1990, pp. 421–436

5. A. Dubey and J. Reicherts, 'lighweight cement boards in exterior wall construction and finish systems', in M. Di Prisco, R. Felicetti and G.A. Plizzari (eds) *Fibre Reinforced Concrete – BEFIB 2004*, Proc. RILEM Symp., PRO 39, RILEM, Bagneux, France 2004, pp. 477–492.

6. R.N. Swamy and M.W. Hussin, 'Continuous woven polypropylene mat reinforced cement composites for applications in building construction', in P. Hamlin and G.Verchery (eds) *Textile Composites in Building Construction, Part 1*, Editions Pluralis, Paris, 1990, pp. 57–67.

7. R.N. Swamy and M.W. Hussin, 'Woven polypropylene fabrics – an alternative to asbestos for thin sheet applications', in R.N. Swamy and B. Barr (eds) *Fibre Reinforced Cement and Concrete, Recent Developments*, Elsevier Applied Science, UK, 1989, pp. 90–100.

8. G. Xu, S. Magnani and G. Mesturini, 'Hybrid polypropylene-glass/cement corrugated sheets', *Composites Part A*. 27A, 1996, 459–466.

9. D.J. Hannant, 'Durability of cement sheets reinforced with polypropylene networks', *Mag. Concr. Res.* 35, 2004, 197–204.

10. D.J. Hannant, J.J. Zonsveld and D.C. Hughes, 'Polypropylene film in cement based materials', *Composites*. 9, 1978, 83–88.

11. J.G. Keer and A.M. Thorne, 'Performance of polypropylene reinforced corrugated cement sheeting', *Composites*. 16, 1985, 28–32.

12. A. Baroonian, J.G. Keer, D.J. Hannant and P. Mullord, 'Theoretical and experimental performance of asbestos cement and polypropylene network reinforced cement corrugated sheeting', in R.N. Swamy, R.L. Wagstaffe and D.R. Oakley (eds) *Developments in Fibre Reinforced Cement and Concrete*, Proc. RILEM Symp., Sheffield, RILEM Technical Committee 49-FTR, 1986, Paper 4.8.

13. J.G. Keer and A.M. Thorne, 'Performance of polypropylene-reinforced cement sheeting elements', in G.C. Hoff (ed.) *Fiber Reinforced Concrete*, ACI SP-81, American Concrete Institute, Farmington Hills, MI, 1984, pp. 213–231.

14. A. Vittone, 'Industrial development of the reinforcement of cement based products with fibrillated polypropylene networks, as a replacement of asbestos', in R.N. Swamy, R.L. Wagstaffe and D.R. Oakley (eds) *Developments in Fibre Reinforced Cement and Concrete*, Proc. RILEM Symp., Sheffield, RILEM Technical Committee 49-FTR, 1986, Paper 9.2.

15. J. Bijen and E. Geurtz, 'Sheet and pipes incorporating polymer film material in cement matrix', in *Fibrous Concrete*, Proc. Symp. on Fibrous Concrete, The Concrete Society, The Construction Press, UK, 1980, pp. 194–202.

16. A.P. Hibbert and D.J. Hannant, 'Toughness of cement composites containing polypropylene films compared with other fibre cements', *Composites*. 13, 1982, 393–399.

17. A. Pivacek, G.I. Haupt and B. Mobasher, 'Cement based cross-ply laminates', *Advanced Cement Based Materials*. 6, 1997, 144–152.

18. B. Mobasher and A. Pivacek, 'A filament winding technique for manufacturing cement based cross ply laminates', *Cem. Concr. Compos.* 20, 1998, 405–415.

19. A. Peled, B. Mobasher and S. Sueki, 'Technology methods in textile cement-based composites', in K. Kovler, J. Marchand, S. Mindess and J. Weiss (eds) *Concrete Science and*

Engineering, A Tribute to Arnon Bentur, Proc. RILEM PRO 36, RILEM Publications, Bagneux, France 2004, pp. 187–202.

20. A. Peled and B. Mobasher, 'Cement based pultruded composites with fabrics', in A.M. Brandt, V.C. Li and I.H. Marshall (eds) *Brittle Matrix Composites (BMC7)*, Proc. 7th Int. Symp., Warsaw, Woodhead Publishing Limited, Cambridge, 2003, pp. 505–514.

21. G. Xu and D.J. Hannant, 'Synergistic interaction between fibrillated polypropylene networks and glass fibres in a cement-based composite', *Cem. Concr. Compos.* 13, 1991, 95–106.

22. M. Churbach, S. Ortlepp, A. Bruckner, M. Kratz, P. Offermann and T. Engler, 'Development of large-sized thin-structured TRC façade element' (in German) *Beton und Stahkbetonbau.* 98, 2003, 345–350.

23. R. Ortlepp, S. Ortlepp and M. Curbach, 'Stress transfer in the bond joint of subsequently applied textile reinforced concrete strengthening', in M. Di Prisco, R. Felicetti and G.A. Plizzari (eds) *Fibre Reinforced Concrete – BEFIB 2004*, Proc. RILEM Symp., PRO 39, RILEM Publications, Bagneux, France, 2004, pp. 1483–1494.

24. A. Peled and A. Bentur, 'Cement impregnated fabrics for repair and retrofit of structural concrete', in M. Di Prisco, R. Felicetti and G.A. Plizzari (eds) *Fibre Reinforced Concrete–BEFIB 2004*, Proc. RILEM Symp., PRO 39, RILEM Publications, Bagneux, France 2004, pp. 313–323.

25. G. Xu, S. Magnani and D.J. Hannant, 'Tenile behavior of fiber-cement hybrid compsoites containing polyvinyl alcohol fiber yarns', *ACI Mater. J.* 95, 1998, 667–674.

26. A.E. Naaman, *Ferrocement and Laminated Composites*, Techno Press 3000, USA, 2000.

27. ACI Committee 549, *State of the Art Report on Ferrocement*, ACI Document 549R-97, American Concrete Institute, USA, 1997.

28. A. Peled and A. Bentur, 'Fabric structure and its reinforcing efficiency in textile reinforced cement composites', *Composites: Part A.* 34, 2003, 107–118.

29. A. Peled and A. Bentur, 'Reinforcement of cementitious matrices by warp knitted fabrics', *Mater. Struct.* 31, 1998, 543–550.

30. A. Roy, T. Gries and A. Peled, 'Spacer fabric for thin walled oncrete elements', in *Fibre Reinforced Concrete – BEFIB 2004*, M. Di Prisco, R. Felicetti and G.A. Plizzari eds, Proc. RILEM Symposium, PRO 39, RILEM, Bagneux, France, 2004, pp. 1505–1514.

31. F. Jess and M. Curbach, 'Strength of continuous AR-glass fiber reinforcement of cementitious composites', in A.E. Naaman and H.W. Reinhardt (eds) *Fourth International Workshop on High Performance Fiber Reinforced Cement Composites (HPFRCC 4)*, RILEM Publications, Bagneux, France, 2003, pp. 337–348.

32. B. Banholzer, 'Bond of strand in a cementitious matrix', *Mater. Struct.* (in press).

33. B. Banholzer, 'Bond Behaviour of a Multi-Filament Yarn Embedded in a Cementitous Matrix', in *Schriftenreihe Aachener Beiträge zur Bauforschung*, Institut für Bauforschung der RWTH Aachen, Diss., 2004.

34. B. Banholzer and W. Brameshuber, 'Tailring of AR-glass filament/cement based matrix bond – analytical and experimental techniques', in v M. Di Prisco, R. Felicetti and G.A. Plizzari (eds) *Fibre Reinforced Concrete – BEFIB 2004*, Proc. RILEM Symp., PRO 39, RILEM Publications, Bagneux, France 2004, pp. 1443–1452.

35. A. Katz, V.C. Li and A. Kazmer, 'Bond properties of carbon fibers in cementitious matrix', *ASCE J. Materials in Civil Engineering.* 7, 1995, 125–128.

36. A. Katz and V.C. Li, 'Bond properties of micro-fibers in cementitious matrix', in S. Diamond, S. Mindess, F.P. Glasser, L.W. Roberts, J.P. Skalny and L.D. Wakeley (eds) *Microstructure of Cement-Based Systems/Bonding and interfaces in Cementitious*

Materials, Materials Research Society Symp. Proc. Vol. 370, Pittsburgh, PA, USA, 1995, pp. 529–537.

37. W. Zhu and P. Bartos, 'Assessment of interfacial microstructure and bond properties in aged GRC using novel microindentation method', *Cem. Concr. Res.* 27, 1997, 1701–1711.

38. P. Bartos, 'Brittle matrix composites reinforced with bundles of fibres', in J.C. Maso (ed.) *From Material Science to Construction Materials*, Proc. RILEM Symp., Chapman and Hall, 1987, pp. 539–546.

39. S. Ohno and D.J. Hannant, 'Modeling the stress-strain response of continuous fiber reinforced composites', *ACI Mater. J.* 91, 1994, 306–312.

40. J. Hegger, O. Bruckermann and R. Chudoba, 'A smeared bond-slip relations for multi-filament yarns embeddd in fine concrete', in M. Di Prisco, R. Felicetti and G.A. Plizzari (eds) *Fibre Reinforced Concrete – BEFIB 2004*, Proc. RILEM Symp., PRO 39, RILEM Publications, Bagneux, France, 2004, pp. 1453–1462.

41. M. Kruger, J. Ozbolt and H.W. Reinhardt, 'A new 3D discrete bond model to study the influence of bond on the structural performance of thin reinforced and prestressed concrete plates', in A.E. Naaman and H.W. Reinhardt (eds) *Fourth International Workshop on High Performance Fiber Reinforced Cement Composites (HPFRCC 4)*, RILEM Publications, Bagneux, France, 2003, pp. 49–63.

42. A. Peled, A. Bentur and D. Yankelevsky, 'Effect of Woven Fabrics Geometry on The Bonding Performance of Cementitious Composites: Mechanical Performance', *Advanced Cement Based Materials Journal*. Vol. 7, No. 1, 1998, pp. 20–27.

43. A. Peled, A. Bentur and D. Yankelevsky, 'Flexural performance of cementitious composites reinforced with woven fabrics', *Journal of Materials in Civil Engineering*. 11, 1999, 325–330.

44. A. Peled and A. Bentur, 'Geometrical characteristics and efficiency of textile fabrics for reinforcing composites', *Cem. Concr. Res.* 30, 2000, 781–790.

45. A. Peled and A. Bentur, 'Fabric structure and its reinforcing efficiency in textile reinforced cement composites', *Composites, Part A.* 34, 2003, 107–118.

46. A. Bentur and A. Peled, 'Cementitious composites reinforced with textile fabrics', in H.W. Reinhardt and A.E. Naaman (eds) *Workshop on High Performance Fiber Reinforced Cement Composites (HPFRCC 3)*, RILEM Publications, Bagneux, France, 1999, 16–19 May, pp.31–40.

47. A. Bentur and A. Peled, 'Optimization of the structure and properties of low modulus polymer fabrics for cement reinforcement', in A.M. Brandt, V.C. Li and I.H. Marshal (eds) *Brittle Matrix Composites 6 (BMC6)*, Proc. Int. Conf., Woodhead Publishing Limited, Cambridge, 2000, pp. 430–438.

48. B. Mobasher, H. Stang and S.P. Shah, 'Microcracking in fiber reinforced concrete', *Cem. Concr. Res.* 20, 1990, 665–676.

49. H. Stang, B. Mobasher and S.P. Shah, 'Quantitative damage characterization in polypropylene fiber reinforced concrete', *Cem. Concr. Res.* 20, 1990, 540–558.

50. B. Mu, C. Meyer and S. Shimanovich, 'Improving the interface bond between fiber mesh and cementitious matrix', *Cem. Concr. Res.* 32, 2002, 783–787.

51. B. Mu and C. Meyer, 'Flexural behavior of fiber mesh-reinforced concrete with glass aggregate', *ACI Mater. J.* 99, 2002, 425–434.

52. A. Peled, A. Bentur and B. Mobasher, 'Pultrusion versus casting processes for the production of fabric-cement composites', in M. Di Prisco, R. Felicetti and G.A. Plizzari (eds) *Fibre Reinforced Concrete – BEFIB 2004*, Proc. RILEM Symposium, PRO 39, RILEM Publications, Bagneux, France, 2004, pp. 1495–1504.

53. B. Mobasher, A. Peled and J. Pahilajan, 'Pultrusion of fabric reinforced high fly ash blended cement composites', in M. Di Prisco, R. Felicetti and G.A. Plizzari (eds) *Fibre Reinforced Concrete – BEFIB 2004*, Proc. RILEM Symp., PRO 39, RILEM Publications, Bagneux, France, 2004, pp. 1473–1482.

54. B. Mobasher, 'Modeling of cemnt based composite laminates', in A.E. Naaman and H.W. Reinhardt (eds) *Fourth International Workshop on High Performance Fiber Reinforced Cement Composites (HPFRCC 4)*, RILEM Publications, Bagneux, France, 2003.

55. B. Mobasher, A. Pivacek and G.J. Haupt, 'Cement based cross-ply laminates', *Journal of Advanced Cement Based Materials.* 6, 1997, 144–152.

56. J. Hegger and V. Stefan, 'Textile reinforced concrete under biaxial loading', in M. Di Prisco, R. Felicetti, and G.A. Plizzari (eds) *Fibre Reinforced Concrete – BEFIB 2004*, Proc. RILEM Symp., PRO 39, RILEM Publications, Bagneux, France, 2004, pp. 1463–1472.

57. A.E. Naaman and K. Chandrangsu, 'Bending behavior of laminated cementitious composites reinforced with fiber-reinforced polymeric meshes (FRP) and fibers', in A. Peled, S.P. Shah and N. Banthia (eds) *High-Performance Fiber Reinforced Concrete Thin Sheet Products*, ACI – SP 190, American Concrete Institute, 2000, pp. 97–116.

58. H.W. Reinhardt, M. Kruger and U. Grosse, 'Concrete prestressed with textile fabric', *Journal of Advanced Concrete Technology.* 1, 2003, 231–239.

59. H.W. Reinhardt and M. Krueger, 'Prestressed concrete plates with high strength fabric', in M. Di Prisco, R. Felicetti and G.A. Plizzari (eds) *Fibre Reinforced Concrete – BEFIB 2004*, Proc. RILEM Symp., PRO 39, RILEM Publications, Bagneux, France, 2004, pp. 187–196.

60. C. Meyer and G. Vilkner, 'Glass concrete thin sheets prestressed with aramid fiber mesh', in A.E. Naaman and H.W. Reinhardt (eds) *Fourth International Workshop on High Performance Fiber Reinforced Cement Composites (HPFRCC 4)*, RILEM Publications, Bagneux, France, 2003, pp. 325–336.

Chapter 14

Applications of fibre reinforced concrete

14.1 Introduction

Since its introduction into the marketplace in the late 1960s, the use of fibre reinforced concrete has increased steadily. As of 2001, approximately 80 million m^3 of FRC were produced annually, with the principal applications being slabs on grade (60%), fibre shotcrete (25%), precast members (5%), with the remainder of the production distributed amongst a number of other specialty products and structural forms. Unfortunately, at the time of writing (2005), fibres are still very little used in truly structural applications, despite their obvious effectiveness under seismic and other forms of dynamic loading, their ability to control crack widths, and so on. This is largely because most structural design codes for concrete, such as ACI 318, *Building Code Requirements for Structural Concrete*, are based primarily on concrete *strength* as the principal design criterion; they do not consider *toughness*. That is, they are concerned primarily with the peak loads that a structure might withstand, rather than the 'post-peak' behaviour. However, it is precisely in this post-peak region that the fibres become most effective. Thus, there would need to be significant changes in our design philosophy, and of course in the codes themselves, for fibres to be used properly either in conjunction with conventional steel reinforcement or alone.

According to Fischer [1], while 'The advantages of using FRC in structural members have been demonstrated in academic and industrial research activities in the past few decades, however, at present the unique features of FRC are not considered in the Building Code for Structural Concrete (ACI 318, 2002).' The situation is similar in the UK. According to Barr and Lee [2], 'The benefits of FRC in construction are well known. However, the potential of the material is not reflected in the number of applications in actual industry practice in UK. This stems from the lack of widely accepted standards for test and design'. Nonetheless, as will be seen later, there have been some significant attempts (primarily in Europe) to write design codes which do incorporate the use of fibres.

14.2   Structural design considerations

14.2.1   General considerations

It is unlikely that fibres will ever be used alone in large structural members; such members would also have to contain conventional reinforcing bars. Thus, in this section, the use of FRC will be discussed for structural members containing both fibres and conventional reinforcement. In such members, the fibres act in two ways:

1   They permit at least a portion of the tensile strength of the FRC to be used in design, since the matrix no longer loses its load-carrying capacity at first crack.
2   They improve the bond between the matrix and the reinforcing bars [3,4] by inhibiting the growth of cracks emanating from the deformations (lugs) on the bars.

It is relatively easy to deal with the increased tensile strength in structural analysis. The effect of the improved bond strength is much more difficult to quantify. However, it has been shown, for instance, that under reversed cyclic loading [5] specimens containing steel fibres exhibited much better anchorage bond characteristics, and a decreased rate of crack development, compared with specimens without fibres.

The use of an appropriate FRC matrix can increase the ultimate moment and ultimate deflection of conventionally reinforced beams [6,7]; the higher the tensile stress carried by the FRC, the higher the ultimate moment. However, it has also been shown [6] that if compression steel is also used, the beneficial effects of the fibres are reduced. Since the role of the fibres is primarily to provide tensile capacity in the bottom portion (tension side) of a beam, it has been suggested that, it may not be necessary to provide fibres throughout the full depth of a reinforced concrete beam. For reasons of economy, it may be sufficient to add fibres only in the bottom half of a beam. However, Bentur and Mindess [8] showed that, with a steel fibre volume of 1.5%, partial fibre reinforcement (to 1/2 of the beam depth) increased the ultimate load by 32%, while full depth fibre reinforcement increased the ultimate load by about 55%. Thus, there are benefits to having fibres even in the compression half of a beam. Typical load vs. deflection curves for these tests are shown in Figure 14.1.

The addition of fibres to reinforced concrete tension members can also considerably reduce the tensile stresses in the reinforcing bars, and reduce the crack widths, as shown in Figure 14.2 for steel fibres [9].

Perhaps of greater structural significance, there is considerable evidence (e.g. [10–15]), that fibres can be particularly effective in providing reinforcement

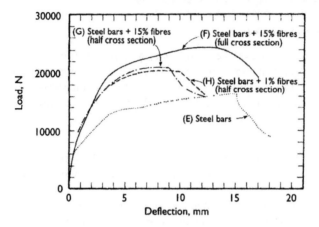

Figure 14.1 Typical load vs. deflection curves of conventionally reinforced concrete and concrete reinforced with both steel bars and steel fibres [8].

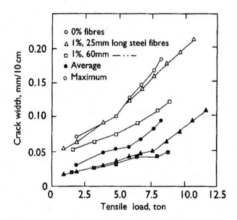

Figure 14.2 Relationship between tensile load and both average and maximum crack widths in reinforced concrete made with SFRC [9].

against shear stresses in conventionally reinforced concrete. There are a number of reasons for this:

1   the randomly distributed fibres are spaced much more closely than the conventional bars and stirrups;
2   both the first crack and ultimate tensile and flexural strengths are increased by fibre additions;
3   the shear-friction strength is increased.

Turning to compression members, Nagasaka [16] found that steel fibres were also very effective in increasing both the shear capacity and the energy absorption in reinforced concrete columns. The fibres were apparently more effective than the conventional hoop reinforcement. Campione *et al.* [17,18] and Campione and Mindess [19] also demonstrated the effectiveness of fibres in compression members.

The literature on the use of fibres in different types of structural members is too extensive to be summarized here. It is clear that the properties of virtually all reinforced concrete elements, or structural systems, can be enhanced by the presence of fibres. However, each structural system must be evaluated separately to determine how effective (and economical) fibre reinforcement will be.

It must be noted that all these results must still be treated with caution. The improvements in properties have, for many cases, not yet been quantified sufficiently for practical design applications. In what follows in this section, some of the approaches towards developing rational design procedures, suitable for inclusion in building codes, are described.

14.2.2 Shear strength

In terms of material behaviour, it has been found [20] that under pure shear, FRC fails in a ductile manner, while plain concrete fails in a brittle manner. The shear strength of FRC containing 1% by volume of steel fibres was found to be about 20% higher than that of the plain concrete. These observations help to explain the enhancement of the shear behaviour of structural elements containing fibres.

Swamy and Bahia [10] found that in the range of steel fibre volumes up to about 1%, there was a linear relationship between the flexural strength of the SFRC and the shear strength of reinforced beams. Indeed, Lim *et al.* [7,11] have suggested that fibres can replace conventional stirrups in whole or in part, as long as parity in the shear reinforcement is maintained. Later, Adebar *et al.* [14] and Mindess *et al.* [15] showed that increasing the fiber content of reinforced beams both increased the shear strength and reduced the crack widths. Longer fibres led to more ductility than shorter ones, but no change in shear strength. Similarly, Narayanan and Darwish [12] showed that steel fibres greatly increase the ultimate shear capacity in prestressed concrete as well.

Based on his own tests, and those of others, Sharma [13] proposed the following expression for predicting the average shear stress, τ_c, in conventionally reinforced SFRC beams. (The typographical error in the original paper has been corrected here):

$$\tau_c = (2/3)f_t'(d/a)^{0.25} \tag{14.1}$$

where (d/a) is the effective depth to shear span ratio and f_t', is the splitting tensile strength of the SFRC, which depends upon the fibre type.

More recently, RILEM Committee TC 162-TDF, *Test and Design Methods for Steel Fibre Reinforced Concrete*, has prepared recommendations for the shear design of members containing both conventional reinforcement and SFRC [21,22]. This is based on the 'standard method' for shear in Eurocode 2 [23], extended to include the contribution to shear resistance of the steel fibres. Using this method, the design shear resistance, V_{Rd3}, of a section is :

$$V_{Rd3} = V_{cd} + V_{fd} + V_{wd} \tag{14.2}$$

where V_{cd} is the shear resistance of a member without shear reinforcement [24]; V_{wd}, the contribution of the shear reinforcement [23] and V_{fd}, the contribution of the steel fibres; given by:

$$V_{fd} = 0.7 k_f k_1 \tau_{fd} b_w d \text{ (Newtons)} \tag{14.3}$$

where
  k_f is the factor taking into account the contribution of flanges in a T-section
  $= 1 + n(h_f/b_w)(b_f/d)$ and $k_f \le 1.5$ where h_f is the flange height (mm); b_f, the flange width (mm); b_w, the web width (mm); d, the beam depth (mm) and n, the $(b_f - b_w)/h_f \le 3; \le \ 3b_w/h_f$.
  $k_1 = 1 + (200/d)^{1/2} \ \le 2$
  τ_{fd} is the design value of the increase in shear strength due to the steel fibres
  $= 0.12 f_{Rk,4}$ (MPa) where $f_{Rk,4}$ is a measure of the residual strength in bending. The fibre type and dosage must be such that $f_{Rk,4}$ is at least 1 MPa.
  A similar expression has been recommended in the Italian draft standard UNI U73041440, *Design, Production and Control of Steel Fibre Reinforced Structural Elements*[25].
  Many other shear strength equations have also been proposed over the years, (e.g. [26]) but have not found their way into any existing structural design codes. A number of these have been reviewed by Batson and Kim [27].

14.2.3 Flexural strength

Henager and Doherty [28] developed a method for analysing reinforced concrete beams with a SFRC matrix. In order to compute the moment capacity, they added the tensile strength contribution derived from the steel fibres to that from the reinforcing bars, as shown in Figure 14.3.
  This leads to an equation for the nominal moment strength, M_n, of the form

$$M_n = A_s f_y (d - a/2) + \sigma_t b(h - e)(h/2 + e/2 - a/2) \tag{14.4}$$

$$e = [\varepsilon(\text{fibres}) + 0.003](c/0.003) \tag{14.5}$$

$$\sigma_t(\text{MPa}) = 0.00772 \ell/d v_f F_{be}) \tag{14.6}$$

where ℓ is the fibre length; d, the fibre diameter; v_f, the volume per cent of fibres; F_{be}, the fibre bond efficiency, which ranges from 1.0 to 1.2; a, the depth of

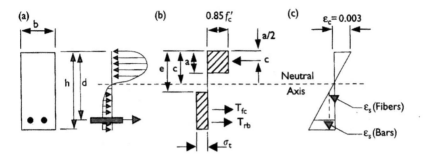

Figure 14.3 Stress and strain distributions across a reinforced SFRC beam. (a) assumed stress distribution; (b) equivalent stress blocks; (c) strain distribution [28].

the equivalent rectangular block; b, the beam width; c, the depth to neutral axis; h, the depth of beam to centre of tensile reinforcement; e, the distance from top of beam to top of the tensile stress block of the SFRC; ε_s, the f_y/E_s of the steel reinforcing bars; ε_f, the σ_f/E_c of the fibres developed at pull-out; σ_t, the tensile stress in the SFRC; T_{fc}, the tensile yield of the SFRC $= \sigma_t b(h - e)$ and T_{rb}, the tensile yield force of the reinforcing bars $= A_s f_y$.

Essentially the same equation is suggested in the report of ACI Committee 544 [29].

For this analysis, the maximum strain that the SFRC was able to withstand was taken as 0.003, but this is probably on the conservative side; later work by Swamy and Al-Ta'an [30] suggests that a value of 0.0035 is more realistic. It should be noted that, in the Henager and Doherty [28] formulation, the SFRC accounts for only about 5–15% of the tensile stress; the remainder is still carried by the reinforcing bars.

More recently, based on extensive theoretical and experimental studies, RILEM Committee TC162-TDF [21,22] developed a more detailed procedure for design for flexural loading, based on the European pre-standard [23] for design of concrete structures, but now including the post-peak behaviour of the SFRC. The SFRC parameters were determined using a load vs. CMOD (crack mouth opening displacement) diagram obtained in 3-point bending, as shown in Figure 14.4 (RILEM TC162-TDF [31].

From this curve, the limit of proportionality, $f_{fct,L}$, is calculated as

$$f_{fct,L} = (3F_L L)/(2bh_{sp}^2) \text{ (MPa)} \tag{14.7}$$

where L is the specimen span (mm); b is the specimen width (mm) and h_{sp} is the distance between tip of notch and top of specimen (mm).

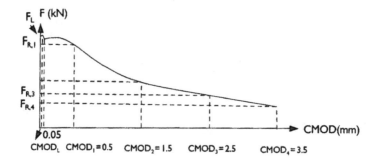

Figure 14.4 Load vs. CMOD diagram (after RILEM TC162-TDF [31]).

As well, four residual flexural strengths, at CMOD values of 0.5, 1.5, 2.5 and 3.5 are calculated:

$$f_{R,i} = (3F_{R,i}L)/(2bh_{sp}^2) \text{ (MPa)} \tag{14.8}$$

The proposed $\sigma-\varepsilon$ diagram, and the associated size factor κ_h, are shown in Figure 14.5 [22]. It is assumed that the $\sigma-\varepsilon$ relationship in compression for SFRC is the same as that for plain concrete.

The principal assumptions made in this method are as follows:

1 Plane sections remain plane.
2 The stresses in the SFRC in both tension and compression are derived from Figure 14.5.
3 The stresses in the reinforcing bars are derived from an idealized bilinear stress–strain diagram.
4 For members in pure uniaxial compression, the compressive strain in the SFRC is limited to 0.002; in bending, it is limited to 0.0035.
5 For SFRC containing conventional reinforcement, the strain is limited to 0.25 at the position of the reinforcement, as shown in Figure 14.6.

A very similar approach to the above has also been followed in Germany [32].

A somewhat different approach is being considered in Italy [25,33]. In this approach, three different types of SFRC behaviour are defined, as shown in Figure 14.7.

In this figure, P_1 is the first crack load of the SFRC specimen; Δ_1, the deflection corresponding to P_1; P_{max}, the maximum load reached after first crack; δ_u, the deflection corresponding to P_{max}; P_{eq}, the equivalent post-peak load; δ_d, the ultimate design deflection for the SFRC and P_{res}, the residual load corresponding to δ_d.

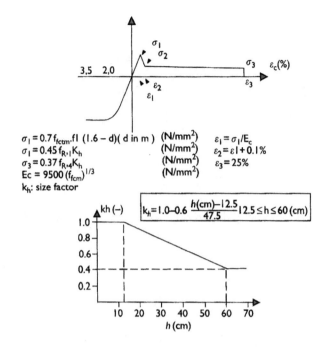

$\sigma_1 = 0.7 f_{fctm.fl} (1.6 - d)(\text{ d in m })$ (N/mm^2) $\varepsilon_1 = \sigma_1/E_c$
$\sigma_1 = 0.45 f_{R,1} K_h$ (N/mm^2) $\varepsilon_2 = \varepsilon_1 + 0.1\%$
$\sigma_3 = 0.37 f_{R,4} K_h$ (N/mm^2) $\varepsilon_3 = 25\%$
$E_c = 9500 (f_{fcm})^{1/3}$ (N/mm^2)
k_h: size factor

$$k_h = 1.0 - 0.6 \frac{h(\text{cm}) - 12.5}{47.5} \quad 12.5 \leq h \leq 60 \,(\text{cm})$$

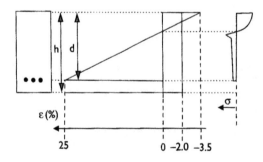

Figure 14.5 $\sigma{-}\varepsilon$ diagram and size factor (after RILEM TC162-TDF [22]).

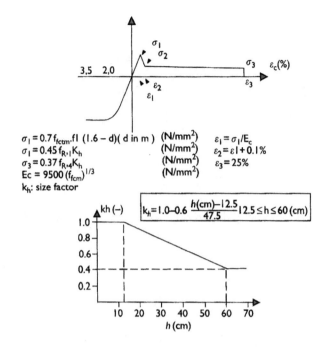

Figure 14.6 Stress and strain distributions in a conventionally reinforced SFRC beam (after RILEM TC162-TDF [22]).

For SFRC in tension, the $\sigma{-}\varepsilon$ models shown in Figure 14.8(a) for Type A behaviour and in Figure 14.8(c) for Types B and C behaviour are then applied, with

$$\varepsilon_1 = f_{1td}/E_{Fm} \tag{14.9}$$

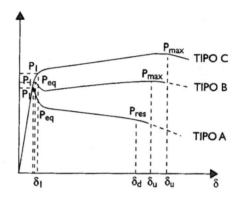

Figure 14.7 Flexural (or tensile) load vs. deflection (P–δ) curves for three types of SFRC [25].

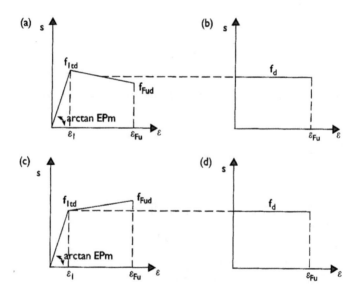

Figure 14.8 Idealized σ–ε models for SFRC in tension [25].

The cracking (ultimate) strains are given by

$$\varepsilon_{Fu} = \delta_d/l_{cs} \text{ for Type A} \tag{14.10}$$

$$= \delta_u/l_{cs} \text{ for Types B and C} \tag{14.11}$$

Alternately, the simplified $\sigma-\varepsilon$ models of Figure 14.8(b) and Figure 14.8(d) may be used, where

$$f_d = 0.33 f_{Ftf} \leq f_{ltd} \tag{14.12}$$

and f_{Ftf} is the flexural strength of the SFRC.
The characteristic length, l_{cs}, is selected on the basis of

$$l_{cs} = \min(s, y) \tag{14.13}$$

$$s = (50 + 0.25 k_1 k_2 \varphi / \rho) 50 / \lambda \, (\text{mm}) \tag{14.14}$$

$$\lambda = \ell / d_F \geq 50 \, (\text{aspect ratio}) \tag{14.15}$$

where s is the crack spacing ($= y$ for unreinforced elements); y, the distance from the neutral axis to the bottom of the beam, evaluated neglecting the tensile strength of the SFRC; ℓ, the fibre length; d_F, the equivalent fibre diameter; φ, the reinforcing bar diameter; k_1, 0.8 for deformed bars, and 1.6 for plain bars; k_2, 0.5 for bending when $y \leq h$, and 1.0 for tension or when $y < h$ and ρ, the reinforcement ratio within the tension area of the cross section.

With the assumptions described above, the beam may then be analysed.

It might be added that RILEM TC162-TDF [34] has also developed a design procedure based on fracture mechanics principles (see also Section 4.4.3 for some more details). Other design methods for FRC containing reinforcing bars are currently (as of 2005) under development, but these have not yet found their way into the building codes.

In closing this section, it must be pointed out that fibre reinforced concrete *is* being used, at least to a limited degree, in a variety of structural applications, despite the absence of explicit design provisions in building codes. This will undoubtedly continue, and should hasten the development of the appropriate codified design provisions.

14.2.4 Seismic applications

As stated by Parra-Montesinos [35], 'The excellent ductility, energy dissipation capacity, and damage tolerance exhibited by fibre reinforced cement composites makes them attractive materials for use in critical regions of earthquake-resistant structures.' This is particularly so for the high performance (strain-hardening) FRCs now being developed. For instance, work by Katzensteiner *et al.* [36] showed that modified SFRC frames maintained better joint integrity, with minimum concrete spalling, than conventionally detailed frames, even during simulated high-magnitude earthquakes. The SFRC frames also absorbed more energy. Kosa and Goda [37] also found that the use of SFRC or Ductal® could greatly increase

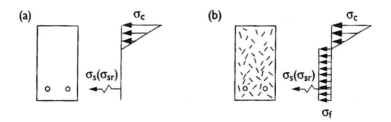

Figure 14.9 Assumed stress distribution for (a) plain concrete; (b) SFRC [41].

the deformation capacity of bridge piers subjected to seismic loading. However, as Parra-Montesinos [35] went on to say, there were some research needs that had to be addressed before FRC materials could be used in practice in seismic–resistant structures:

- Development of constitutive relations for FRC's under reversed cyclic loading;
- Development of guidelines for the design of critical regions of seismic-resistant structures incorporating FRC;
- Close interaction between researchers, structural engineers and contractors.

Fortunately, this type of activity is already well underway (e.g. [38–40]).

14.2.5 Crack control

As has already been mentioned many times in this book, one of the principal functions of the fibres in FRC is to control both the extent of cracking, and the crack widths. This may be accomplished with a high enough volume percentage of fibres in unreinforced FRC, or by the appropriate combination of fibres and conventional reinforcement. The calculation of the design crack width has been described by Vandewalle [41,42]. It is similar to the way in which crack widths for ordinary reinforced concrete are calculated (European pre-standard, 1991), except that the tensile strength in the SFRC matrix after cracking is not equal to zero, but to σ_f as shown in Figure 14.9 [41].

Here, $\sigma_f = 0.45\,f_{Rm,1}$
where $f_{Rm,1}$ is the average residual flexural strength of the SFRC at the moment when a crack is expected to occur.

The design crack width, w_k, is then given by:

$$w_k = \beta s_{rm}\ \varepsilon_{sm} \tag{14.16}$$

where s_{rm} is the average final crack spacing; ε_{sm}, the the allowable mean steel strain in the reinforcement and β, the coefficient relating the average final crack width to the design value.

The crack spacing may be calculated as:

$$s_{rm} = (50 + 0.25k_1k_2 \quad \phi_b/\rho_r \ (50/\ell d) \tag{14.17}$$

where $50/(\ell/d) \quad \leq 1$; φ_b is the bar size (mm); k_1, the coefficient related to the bond properties of the bars; k_2, the coefficient related to the form of the strain distribution; ρ_r, the the effective reinforcement ratio $A_s/A_{c,eff}$ where A_s is the area of the reinforcement contained within the effective tension area $A_{c,eff}$; ℓ, the length of steel fibres and d, the diameter of steel fibres.

A similar formulation of a cracking model has been described by Balazs and Kovacs [43].

A detailed description of the calculation of the minimum reinforcement required to control crack formation in SFRC is given in Vandewalle [42].

14.3   Pavements and slabs on grade

14.3.1   General considerations

As stated earlier, slabs on grade constitute about 60% of the FRC production. While these are not strictly 'structural' applications, a number of rational design methods have been developed for FRC pavements and slabs on grade, for both steel and synthetic fibres. It should, however, be noted that the design methods referred to here have been developed by the fibre manufacturers themselves. While the principles of these design methods may be generalized, the numerical constants used in these methods are only applicable to the particular fibre type for which the method was originally developed.

All the methods are based on two considerations:

1   An analysis of the stresses in the concrete pavement and in the soil subgrade, using the elastic methods developed by Westergaard [44,45] and/or the yield line theories developed by Losberg [46], Meyerhof [47] and Baumann and Weisgerber [48]. The theory of beams on elastic foundations [49] may also be used for some cases of line loading. (More recently, Meda *et al.* [50] developed a non-linear fracture mechanics approach to the analysis and design of SFRC pavements, but this has not yet been applied in practice).

2   The observation that fibre reinforcement of slabs avoids their brittle collapse; once the first crack appears, considerable stress redistribution begins due to the ability of the fibres to 'tie' the cracks together. Thus the slabs can carry further loads until a collapse mechanism occurs when the fibres are no longer able to bridge the cracks effectively.

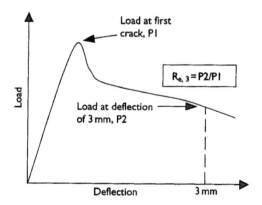

Figure 14.10 Calculation of the equivalent flexural strength ratio, $R_{e,3}$ (after [51]).

In designing an FRC slab, the appropriate equations mentioned in [1] above are used, but modified for the toughness, or the residual (post-cracking) tensile or flexural strength of the FRC. It should be noted that this modification depends on the specific test procedure used to characterize the FRC (see Chapter 6), and again may not be easy to generalize.

14.3.2 Design based on concrete society TR 34

In the UK, the toughness of the FRC is included in the design calculations for concrete floor slabs [51]. The ductility of the FRC is characterized using the Japanese JCI [52] standard JSCE-SF4 (which was described in detail in Chapter 6). The measure of ductility used is the $R_{e,3}$ (equivalent strength ratio) value, which is a ratio of the load at a deflection of 3 mm (i.e. span/150 of a beam with a span of 450 mm) to the load at first crack, as shown in Figure 14.10. Using $R_{e,3}$, the residual positive bending moment capacity, M_p, is calculated as [2]:

$$M_p = \frac{f_{ctk,fl}}{\gamma_c}(R_{e,3})(h^2/6)$$

where $f_{ctk,fl}$ is the characteristic flexural strength of plain concrete; h, the slab depth and γ_c, the partial safety factor for concrete.
It is assumed that the onset of cracking at the top of the slab is the limiting design criterion. Since fibres do not particularly affect the cracking stress, the negative bending moment capacity, M_n, is simply calculated as

$$M_n = \frac{f_{ctk,fl}}{\gamma_c}(h^2/6) \tag{14.18}$$

Table 14.1 Beam toughness parameters [53]

Specimen ID	100 × 100 × 350 mm Flexural beam			150 × 150 × 550 mm Flexural beam		
	Flexural strength (MPa)	$f_{e,2}$ (MPa)	$R_{e,2}$(%)	Flexural strength (MPa)	$f_{e,3}$ (MPa)	$R_{e,3}$(%)
Plain concrete	5.17	0.1		4.73	0.13	
			2.8			3.0
0.32% synthetic fibre	4.31	1.5	35.0	4.69	1.02	24.0
0.48% synthetic fibre	4.63	2.2	47.3	4.82	1.9	39.0
0.35% steel fibre	4.67	3.2	69.6	4.68	2.0	43.0

Notes
The $R_{e,2}$ and $R_{e,3}$ values are the equivalent flexural strengths; the $f_{e,2}$ and $f_{e,3}$ values are the residual strengths at deflections of 2 mm and 3 mm, respectively.

A similar approach was adopted by Altoubat *et al.* [53]. However, to illustrate the importance of the test procedures in determining the modification factor for the effect of the fibres, an examination of Table 14.1 from their study is instructive. The $R_{e,2}$ values listed in Table 14.1 were obtained in an analogous manner to the $R_{e,3}$ values, but on smaller 100 x 100 x 350 mm beams. For the FRCs, the $R_{e,2}$ values are consistently greater than the $R_{e,3}$ values, though both test sizes are permissible. Thus, the precise way in which the equivalent flexural strength is calculated may have a significant effect upon the results.

The Grace Construction Products [54] recommendations for design of slabs incorporating their synthetic fibres, and the Bekaert [55] design procedures for their steel fibres are both also based on the Concrete Society TR34 [51] recommendations. However, it is likely that other companies or jurisdictions will, or already have, developed design procedures for FRC slabs on grade based on other considerations.

14.3.3 Fibres vs. welded wire mesh

There has been a long-standing controversy regarding the relative effectiveness of fibres and welded wire fabric (WWF) in slabs, with the producers and proponents of each type of reinforcement arguing that theirs is better than the other [56,57]. The prime function of the WWF is to hold the concrete together once it has started to crack, and a properly designed FRC mix will also serve this function. In fact, there appears to be little difference between SFRC slabs and WWF reinforced slabs, in terms of strength, failure mode, crack initiation and propagation, and load vs. deflection behaviour [58]. The caveat is that the WWF must be correctly placed, at or slightly above the mid-height of the slab, and this is often difficult to achieve in practice. On the other hand, Sorelli *et al.* [59] found that the crack patterns did differ between the two types of reinforcement. The SFRC slab showed the same crack pattern as the plain concrete, while the WWF slab led to narrower and more diffuse crackings that developed in the radial direction. They also found that SFRC slabs were less effective for loads applied at a corner or an edge. Trottier *et al.* [57] reported that properly placed WWF outperforms synthetic fibres in these respects because the latter have a lower modulus of elasticity than the steel or the concrete.

Fibres do impart other properties to the concrete, in particular a reduction in plastic shrinkage cracking. However, compared with both steel and synthetic fibres, WWF has been found to reduce the maximum crack width [60], even though the age at which cracking first appeared was not prolonged. Thus, when choosing between fibres and WWF, it is necessary to define carefully the performance that is most important in a particular application. Indeed, for some applications, it may be that a combination of fibres and WWF will provide the best solution.

14.4   Fibre shotcrete

14.4.1   Introduction

Fibre reinforced shotcrete is mortar or concrete containing discontinuous discrete fibres that is pneumatically projected at high velocity onto a surface, as shown in Figure 14.11. Indeed, using high speed photography, it has been shown that particles in a shotcrete stream may move at speeds in excess of 100 km/h [61]. Steel fibre shotcrete was first placed experimentally in 1971 by D.R. Lankard of the Battelle Memorial Institute. Since then fibre shotcretes using both steel and synthetic fibres have been used extensively worldwide, at fibre volumes up to about 2%. As stated earlier, fibre shotcrete now constitutes about 25% of the total production of FRC, the major applications being for repair and rehabilitation of concrete structures, and for tunnel linings. Its uses are described in some detail in an American Concrete Institute report [62].

Conventional shotcrete techniques (either wet or dry) may be used, with the equipment modified so that the fibres can be mixed with the other concrete constituents. Alternatively, the fibres may be supplied to the spray nozzle in a separate hose before being mixed with the concrete. Such a system has been developed by AB Besab (Sweden); their system is shown in Figure 14.12 [63].

Fibre shotcrete is particularly effective because the very nature of the application ensures that the fibres are preferentially aligned in two dimensions. It has been shown by Morgan and Mowat [64] that steel fibre shotcrete, for instance, can provide better load-bearing capacity than plain wire mesh-reinforced shotcrete at small deflections, and at least equal capacity at large deflections (Figure 14.13).

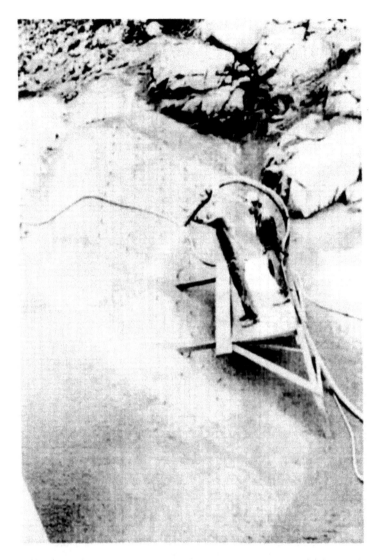

Figure 14.11 Steel fibre shotcreting. Photograph courtesy of B.H. Levelton and Assocoiates, Vancouver, Canada.

14.4.2 Properties of fibre shotcrete

In a number of respects, the properties of fibre shotcrete may be quite different from those of 'ordinary' FRC [65]. In particular:

1 Aggregates in fibre shotcrete tend to be smaller than those in ordinary FRC.

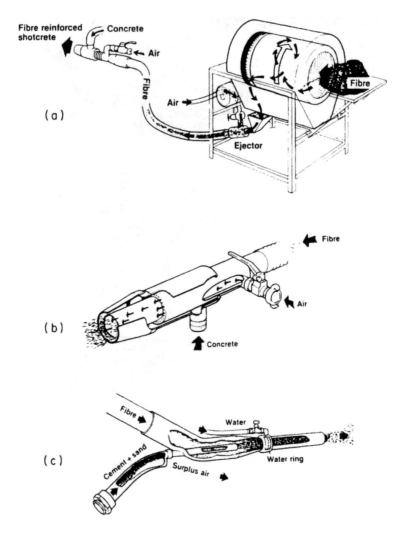

Figure 14.12 (a) The BESAB system for the production of steel fibre shotcrete; (b) wet spray nozzle and (c) dry spray nozzle [63].

2   The aggregate content of fibre shotcrete is much lower than that of ordinary FRC (~30% by mass, compared with ~50–60%), because of both placement difficulties and high rebound (see later).

3   Fibre shotcrete has much higher *in situ* cement contents, approaching 600–700 kg/m^3 in some dry process applications.

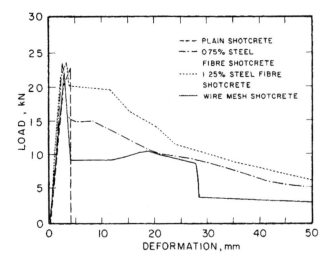

Figure 14.13 Load vs. deflection curves for steel fibre shotcrete panels, compared with plain shotcrete and wire mesh shotcrete [28].

4   Because of the pneumatic compaction of the fibre shotcrete, the internal voids tend to be different from those in cast FRC, in terms of both distribution and range of sizes.

Because of the fact that fibre shotcrete is often not fully consolidated, and because of rebound effects, the water/cement ratio 'law' that governs the strength of cast concrete does not work very well, and so there is generally a poor correlation between w/c and compressive strength, as shown in Figure 14.14 [65].

14.4.3   Specifications and tests for fibre shotcrete

In general, fibre shotcrete may be specified and tested in the same manner as ordinary FRC and plain shotcrete. However, the European Federation of Producers and Contractors of Specialist Products for Structures [66] have suggested some particular specifications and tests for what they term 'sprayed concrete'. In particular, they recommend that the *toughness* of the fibre shotcrete be specified either by a *residual strength class* (from a beam test) or an *energy absorption class* (from a plate test), though they go on to state that the two tests will not give values that are comparable.

Five residual strength classes are defined in terms of the shape of the load vs. deflection curve of a beam of dimensions 75 mm high, 125 mm wide, and

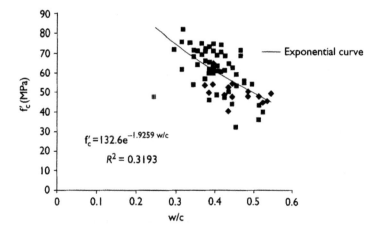

Figure 14.14 Dry-mix shotcrete compressive strength vs. w/c [65].

Table 14.2 Energy absorption requirements [66]

Toughness classification	Energy absorption in joule for deflection up to 25 mm
a	500
b	700
c	1000

600 mm long, tested in third-point loading on a span of 450 mm. These classes are shown in Figure 14.15; the table defines the four points that define the boundaries between the classes. This is a variant of the 'Template' approach described earlier in Section 6.3.3.

The energy absorption classes are given in Table 14.2 [66]. The values are obtained from the area under the load vs. deflection curve of the EFNARC [66] plate test described in Section 6.33, out to a deflection of 25 mm. According to Papworth [67], there is a good correlation between the EFNARC plate test and the ASTM C1550 centrally loaded round panel test described in detail in Section 6.3.3.

14.4.4   Fibre rebound

The process of shotcreting results in a considerable loss of material due to rebound. In dry-mix shotcrete, the total loss of material is generally in the range of 30–40% [68], and even in wet-mix shotcrete the losses are generally of the order of

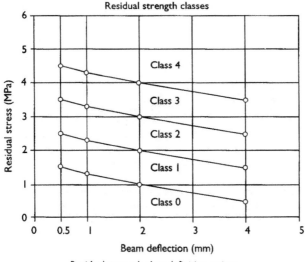

Figure 14.15 Residual strength classes and definition points [66].

Deformation class	Beam deflection (mm)	Residual stress (MPa) for strength Class			
		1	2	3	4
	0.5	1.5	2.5	3.5	4.5
Low	1	1.3	2.3	3.3	4.3
Normal	2	1.0	2.0	3.0	4.0
High	4	0.5	1.5	2.5	3.5

10–15% [69]. However, in percentage terms, the rebound of steel fibres is always greater than the total material rebound, as may be seen in Figure 14.16 [65].

In dry-mix shotcrete, the amount of rebound appears to be related to the fibre geometry [70]. For a given diameter, shorter fibres lead to lower rebound; for a given length, larger diameters lead to less rebound. However, in wet-mix shotcrete, the fibre rebound does not appear to be related to the geometry [69]. As shown in Figure 14.16, the use of mineral admixtures can modify the amount of rebound, with larger particles leading to larger rebound values.

14.4.5 Applications of fibre shotcrete

As stated earlier, both steel fibres and synthetic fibres have been used successfully in fibre shotcrete, though steel fibres tend to lead to greater improvements in toughness than do synthetic fibres [71]. The most rapidly growing area of application

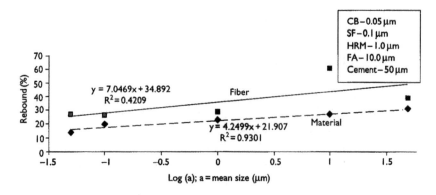

Figure 14.16 Material and fibre rebound as a function of mineral admixture particle size; CB = carbon black; SF = silica fume; HRM = high reactivity metakaolin; FA = fly ash [65].

for fibre shotcrete is in ground support, for tunnel linings and in mining operations. There have been many applications of this technique in mining operations in Canada, Australia, South Africa and elsewhere. Design guidelines for use of fibre shotcrete in ground support have been provided by Papworth [67]. He found that in cases where deflections must be limited, steel fibres would prove to be more economical than synthetic fibres. However, where high deflections are permissible, synthetic fibres would generally be more economical. Indeed, in these cases they are the only acceptable fibres (except for temporary works) because of the potential of rusting of steel fibres in wide cracks. As well, fibre shotcrete is now often used in tunnel linings (e.g. [72,73]).

Fibre shotcrete is also particularly well suited for repairs to deteriorated concrete structures. The key to successful fibre shotcrete repairs is proper preparation of the substrate, including removal of any damaged material. If this is not done, it is all too common to find perfectly good fibre shotcrete repair material lying beside the structure after debonding from an improperly prepared surface. As well, the surface of any structural or reinforcing steel should be blast cleaned to achieve a good bond.

Finally, it is absolutely essential that the nozzlemen be properly trained, since it is they who control the final quality of the fibre shotcrete application. In North America, the American Concrete Institute has developed detailed policies for the certification of shotcrete nozzlemen [74]. For major projects, it is wise to require all the nozzlemen proposed for the shotcrete applications to shoot preconstruction test panels, constructed with the same reinforcing and other details as specified for the structural repair itself. This enables them to become familiar with the materials and procedures to be used on the jobsite, and permits an evaluation of their skills.

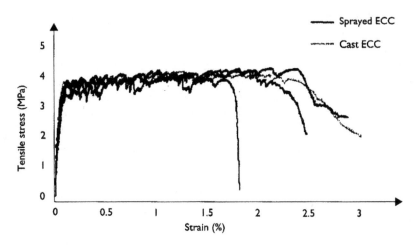

Figure 14.17 Uniaxial stress vs. strain curves for ECC [77].

Two good examples (out of a great many reported in the literature) of the use of fibre shotcrete in repairs in Canada are the rehabilitation of berth faces at the Port of St. John [75], and repairs at the Port of Montreal [76].

It has also been possible to develop very high performance fibre shotcretes [77], comparable in performance to cast engineered cementitious composites (ECC). Though such materials require close control of the particle size distribution, and of the necessary chemical and mineral admixtures, the resulting materials exhibit proper pumpability and shootability in the fresh state, and strain-hardening behaviour in the hardened state, as shown in Figure 14.17.

14.5   Self-compacting FRC

It has generally been assumed that the addition of fibres reduces the workability of concrete, since the fibres tend to inhibit flow. This is indeed true for ordinary FRC mixes [78]. The addition of fibres changes the structure of the aggregate skeleton. In particular, the packing density decreases, requiring a higher fines content in order to compensate for this effect [79]. If proper attention is paid to the mix design, FRC can not only have the same workability as plain concrete with the same w/c ratio, but can also be made to be self-compacting. However, self-compacting FRC is a rather unforgiving material, with little margin for change in the mixture proportions in order to avoid segregation, and to maintain the property of self-compaction. In addition, these mix designs will tend to decrease the randomness of the fibre orientation.

In order to produce self-compacting FRC, careful attention must be paid to the total particle size distribution of the mix, and to the choice of appropriate admixtures, both chemical and mineral. For example, to produce their self-levelling and self-compacting FRC mixtures, Ambroise *et al.* [80] used a polycarboxylate superplasticizer, a mixture of a suspension of starch and a suspension of precipitated silica as a viscosity agent, and powdered limestone to increase the amount of very fine material. They found that shorter fibres are better than longer ones in order to maintain the appropriate rheological properties. As well, PVA fibres required about 10% more water than steel fibres to maintain the same flow characteristics. Similarly, Pereira *et al.* [81] used a blend of three different aggregates, fly ash, a limestone filler, and a polycarboxylate superplasticizer to ensure self-compacting properties in steel FRC.

On the other hand, Markovic *et al.* [82] were able to produce self-levelling FRC, containing a mixture of short straight and long hooked-end fibres, at a w/c ratio of 0.2, using only a superplasticizer and silica fume as admixtures, though with a maximum aggregate size of only 1 mm. They could use fibre contents of up to about 4% by volume without a large impact on the workability. However, they too found that their materials were very sensitive to the proportioning; a slight increase in w/c could lead to segregation of the fibres, implying the necessity of a high degree of quality control.

Granju *et al.* [83] also made self-compacting FRC using a blend of amorphous metal fibres and polypropylene fibres (i.e. hybrid fibre reinforcement). They could achieve self-compaction using a maximum aggregate size of 10 mm, a superplasticizer, and a fine mineral filler. Their results suggest an improved fibre–matrix bond in the self-compacting mixes.

Taking the process a step further, Borsa *et al.* [84] were able to produce a lightweight aggregate self-compacting FRC mix, using terracotta lightweight aggregate, a maximum aggregate size of 8 mm, microsilica, limestone filler, a superplasticizer and an air entraining agent.

From the studies cited above, the mechanical properties of self-compacting FRC show the strength and toughness that one would normally associate with various FRC mixes. In addition, Walter *et al.* [85] found a good bond between a thin self-compacting steel fibre concrete overlay cast on a steel plate, even though the only measure taken to prepare the steel surface was sandblasting.

Clearly, the use of this technology, which is still in its infancy, will continue to grow, despite the initial high cost of the material, and the stringent quality control requirements. These can be more than offset by the quality of the resulting product, and by the ability to use FRC in very difficult placements.

14.6   Precast concrete

In many precast concrete components the conventional reinforcement is non-, or semi-structural, and is intended mainly for crack control. The cracking in question

Figure 14.18 Septic rank made of polymer fibre reinforced concrete, courtesy of Grace Construction Products.

could be the result of environmental loading (temperature changes, moisture movement and wind) or accidental loads, such as those occurring during transportation or assembly of the precast unit on site. Components of this kind are usually of smaller dimensions and are characterized by a complex shape, such as pipes, septic tanks (Figure 14.18), cable ducts, thin roofing elements and non-load-bearing wall elements and tunnel linings.

The advantage of the fibre reinforcement is that they can provide performance which is equivalent to that of conventional mesh, as well as simplify the production process of the component, especially those of complex shape, by eliminating the need to place a mesh or a cage and cast the concrete around it. This saving is particularly effective in cutting down labour costs. Additional advantage of the fibre reinforcement is the improved impact resistance of the whole volume of the concrete, reducing the sensitivity to defects that may occur during handling, in particular at the edges of the component.

Traditionally, steel fibres have been used for this purpose, for example Magnetti *et al.* [86]. In recent years, polymeric fibres have also been applied, and this was facilitated with the advent of structural polymeric fibres. Durability and reduced sensitivity to corrosion, especially in water-retaining components, have been cited as a driving force for the application of these fibres.

The content of the fibres for these applications, steel as well as polymers, is usually in the range of 0.3–0.7% by volume, depending on the application and the nature of the fibres. These contents are significantly higher than those used for plastic shrinkage crack control, which are usually 0.1% volume or less. The design of the fibre reinforced concrete, in particular the determination of the fibre content, is based on the principles discussed in Section 14.2. The required flexural performance is usually determined by one of the following alternatives: (i) specifications or standards (e.g. flexural loads required in bending tests of full-size components such as pipes); (ii) calculations of the loads imposed on a particular element (e.g. wind loads, loads during lifting) and (iii) flexural resistance which is calculated from a conventional design in which steel mesh is used. The content of the fibres is determined by considering the equivalent flexural strength or the residual flexural strength of the composite required for achieving the flexural resistance obtained by the mesh reinforcement.

Another type of application in the precast industry is the use of thin sheets of fibre reinforced cements as permanent formwork (known also as lost formwork and stay-in-place forms) for reinforced concrete components [87,88]. In this case the composite provides also finishing and protection to the reinforced concrete member, in particular with respect to corrosion of steel in concrete. The composite can also be taken into account as a tensile reinforcement, and thus lead to savings in steel.

The formwork is frequently used in applications where the span required is small, such as between girders in bridge decks (Figure 14.19) or in small and medium

Figure 14.19 Stay in place fromwork panels from fibre reinforced concrete, Photo by Finn Hubbard Wisconsin Department of Transporation.

size construction, such as family houses and apartment buildings. Glass fibre reinforced cements [87], and textile-reinforced cementitious matrices [88], have been applied and considered for these applications. Formwork using other fibres, such as carbon are also being developed [89]. Other fibres can also be potentially suitable, in particular systems for producing high performance fibre reinforced cement composites, such as those discussed in Chapters 12 and 13, made from discrete fibres or fabrics (textile-reinforced concrete).

The design of the formwork should be such that it will enable sufficient bonding to the concrete cast above it (Figure 14.20).

A variety of special precast components, such as utility poles [90], ducts, pipes have been produced from glass fibre reinforced cements, using technologies such as spray and filament winding.

14.7 Cladding

Fibre reinforced cements are extensively used in a variety of cladding applications, ranging from small shingles up to large facades units of several metres in size [91–96]. In these applications the composites used are usually thin section components, where the fibres are the primary reinforcement. Glass, polypropylene, carbon and cellulose fibres have been used for these applications. The mechanical and physical properties of these composites can be quite different, as demonstrated in Figure 14.21 and Table 14.3 [97].

Several types of fibre-reinforced thin sheets and components are available commercially, as outlined:

- Cellulose cement sheets: cement reinforced with cellulose fibres derived by pulping processes. These composites are produced by the Hatschek process and are applied in many instances as asbestos replacement or as the processing fibre in a hybrid composite (see Chapter 11);
- Wood particle cement sheets: cement reinforced with wood chips obtained by a variety of processes. The methods of production of the board can be based on simple casting techniques or it can involve pressing and heat treatments;

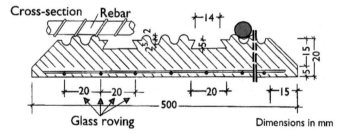

Figure 14.20 Stay in Place (SIP) form system approved in Germany (after Reinhardt [88]).

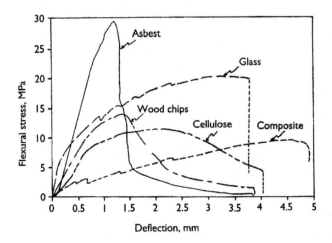

Figure 14.21   Load (calculated as flexural stress) vs. deflection curves of some commercial fibre-reinforced thin sheet cement composites, after Baum and Bentur [97].

Table 14.3 Properties of several commercial fibre-reinforced thin sheet cement composites, after Baum and Bentur [97]

Type of fibre	Density (kg/m³)	Flexural strength (MPa)	Swelling (%)
Asbestos	1600	32	0.09
Glass	1700	25	0.13
Cellulose	1100	10	0.30
Wood particle	1000	9	1.75
Composite[a]	1200	10	0.15

Note
a ~10–15 mm thick sandwich board, with lightweight mortar core and mesh reinforced skins.

- Glass fibre reinforced cement/concrete (GRC; GFRC): Cement mortar reinforced with glass fibre roving (see Chapter 8);
- Polypropylene fibre reinforced cement sheets: reinforcement with short fibres or continuous net film (see Chapter 10);
- Composite cement sheets: sheets of 10–15 mm thickness with a core of lightweight aggregate mortar and external skins of cement reinforced with glass fibre mat;
- Cement reinforced with a variety of modern man-made fibres: mainly carbon and PVA, or a hybrid of these fibres with cellulose as a processing fibre.

Since in these applications the composites are non-load-bearing, the major concerns in the design are usually the development of internal stresses which may lead to cracking and deformation (e.g. warping) as well as ageing effects which may result in changes in the mechanical properties which may aggravate the cracking sensitivity.

The driving forces for cracking are usually associated with environmental effects leading to length changes due to moisture movements and thermal effects. The moisture sensitivity can be quite large, depending on the type of fibre and the curing of the composite. Some values determined for commercial products are presented in Table 14.3. Differential movements can arise also from specific exposure conditions, where the external face of the cladding unit is exposed to radiation, temperature and moist environment which is substantially different from the inner face. These conditions are the kind that induces bowing and which eventually will result also in cracking [97]. The fibre reinforcement should provide the composite with tensile strength and ductility to enable it to resist such effects and perform properly over the whole life span of the component. However, this may not always be the case, especially when the driving forces are rather large and when the properties of the composite change with time (e.g. ageing), in particular loss of ductility (embrittlement). Ageing effects of this kind where discussed in general in Chapter 5, as well as for specific composites (e.g. Chapters 8 and 11, for glass and cellulose fibres, respectively).

In the design of the composite material special attention is given to minimize volume changes and ageing effects. Concepts for controlling such properties are covered in Chapters 5, 8 and 11. In the application of the composites for cladding, there is a need to take these characteristics into account, and come up with a design that can accommodate the volume changes and the variations in properties over time. These considerations will be briefly reviewed here. Detailed treatments of this topic can be found in various sources in the literature [92–96,98–100].

In the design for loads, such as dead weight and wind, the strength to be taken in the case of glass fibre reinforced cements is recommended to be the aged ultimate strength or the first crack stress in the unaged composite. A critical issue in the design is to minimize the restraint in the composite, in order not to give rise to stresses which result from differential movements. Such restraint is particularly large in sandwich elements, and the risk there for bowing and cracking can be quite high [98]. Therefore, there is preference in the construction of big units to have the cladding element placed over a steel-studded frame. The connection between the composite skin to the frame should be based on anchors which can transfer the required horizontal or vertical load, but at the same time be sufficiently flexible. Examples of such anchors are given in Figure 14.22. The stresses and stress distribution in systems of this kind has been modelled and calculated by various means [92,93]. Other types of connections which provide flexibility have been used in cellulose–cement boards as seen for example in Figure 14.23. Additional means to minimize such stresses include careful design of the spacing between joints and the joint themselves and the execution of joints.

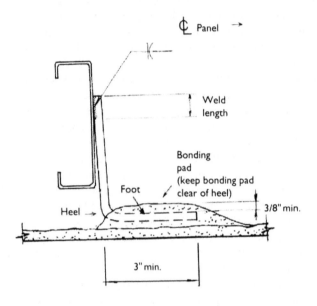

Figure 14.22 Flexible connections recommended for glass fibre-reinforced panels, PCI [95].

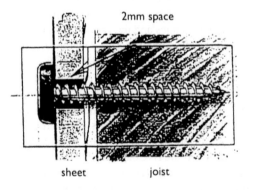

Figure 14.23 Example of a connections recommended for attaching cellulose-cement board to wooden studs which offers some freedom of movement.

Precautions of similar nature are taken even when using small components like shingles [91].

It should be noted that restraint can also be induced internally, when the skin geometry is not flat, but rather of a complex shape. A 3D geometry can therefore

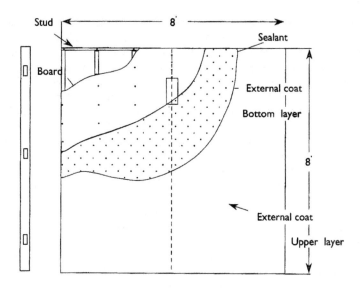

Figure 14.24 Schematic description of a system of thin sheet fibre reinforced cement including board mounted on a stud system, including the coating, joints and sealants.

result in cracking unless special precautions are taken in the design to analyse for stresses induced by such effects [92].

In cement composites which are especially susceptible to volume instability or fire resistance, the solution used can be based on a protecting layer which is applied on the FRC sheet. A schematic description of a system of this kind, including the coating layers, joints and sealant is shown in Figure 14.24.

Overall, the design of cladding units from FRC requires special attention to the details, making sure that the whole assembly is viewed as a system which takes into consideration the special property of the FRC cladding material. In view of the range of properties of these materials, the systems may be quite different, but equally effective in overall performance. In view of that, the evaluation of such systems is frequently based on performance tests, such as ISO 8336, in which a whole assembly is being tested by simulation of rain and heat cycles.

14.8   Concluding remarks

There are a host of other applications of FRC, some of them already used in practice, some still in the development stage. It is, unfortunately, beyond the scope of this book to describe them all. In any event, FRC is such a rapidly advancing field that, within a very few years, even more varied uses for this versatile material will emerge.

References

1. G. Fischer, 'Current U.S. guidelines on fiber reinforced concrete and implementation in structural design', in A. Ahmad, M. di Prisco, C. Meyer and S.P. Shah (eds) *Fiber Reinforced Concrete – From Theory to Practice*, Int. Workshop on Advances in Fiber Reinforced Concrete, Starrylink-Editrice, Bergamo, Italy, 2005, pp. 13–22.

2. B. Barr and M.K. Lee 'FRC guidelines in the UK, with emphasis on SFRC in floor slabs', in A. Ahmad, M. di Prisco, C. Meyer and S.P. Shah (eds) *Fiber Reinforced Concrete – From Theory to Practice*, Int. Workshop on Advances in Fiber Reinforced Concrete, Starrylink-Editrice, Bergamo, Italy, 2005, pp. 29–38.

3. C. Yan and S. Mindess, 'Bond between epoxy coated reinforcing bars and concrete under impact loading', *Can. J. Civ. Eng.* 21, 1994, 89–100.

4. C. Yan and S. Mindess, 'Effects of loading rate on bond behaviour under dynamic loading', in W. Burns (ed.) *Concrete and Blast Effects*, SP-175, American Concrete Institute, Farmington Hills, MI, 1998, pp. 179–197.

5. A.K. Panda, R.A. Spencer and S. Mindess, 'Bond of deformed bars in steel fibre reinforced concrete under cyclic loading', *Int. J. Cement Composites and Lightweight Concrete.* 8, 1986, 239–249.

6. R.J. Craig, J. Decker, L. Dombrowski, Jr R. Laurencelle and J. Federovich, 'Inelastic behaviour of reinforced fibrous concrete', *J. Struct. Eng. ASCE* 113, 1987, 802–817.

7. T.-Y. Lim, P. Paramasivam and S.-L. Lee, 'Behaviour of reinforced steel-fibre-concrete beams in flexure', *J. Struct. Eng. ASCE* 113, 1987, 2439–2458.

8. A. Bentur and S. Mindess, 'Concrete beams reinforced with conventional steel bars and steel fibres: Properties in static loading', *Int. J. Cement Composites and Lightweight Concrete.* 5, 1983, 199–202.

9. T. Kanazu, Y. Aoyagi and T. Nakano, 'Mechanical behaviour of concrete tension members reinforced with deformed bars and steel fibre', *Trans. Japan Concrete Institute.* 4, 1982, 395–402.

10. R.N. Swamy and H.M. Bahia, 'The effectiveness of steel fibres as shear reinforcement', *Concr. Int.: Design and Construction.* 7 (3), 1985, 35–40.

11. T.-Y. Lim, P. Paramasivam and S.-L. Lee, 'Shear and moment capacities of reinforced steel-fibre-concrete beams', *Magazine of Concrete Research.* 39, 1987, 148–160.

12. R. Narayanan and I.W.S. Darwish, 'Shear in prestressed concrete beams containing steel fibres', *Int. J. Cement Composites and Lightweight Concrete.* 9, 1987, 81–90.

13. A.K. Sharma, 'Shear strength of steel fiber reinforced concrete beams', *J. American Concrete Institute.* 83, 1986, 624–628.

14. P. Adebar, S. Mindess, D. St.-Pierre and B. Olund, 'Shear tests of fiber concrete beams without stirrups', *ACI Struct. J.* 94, 1997, 68–76.

15. S. Mindess, P. Adebar and J. Henley, 'Testing of fiber-reinforced structural concrete elements', in V.M. Malhotra (ed.) *High Performance Concrete*, SP-172, American Concrete Institute, Farmington Hills, MI, 1997, pp. 495–515.

16. T. Nagasaka, 'Effectiveness of steel fiber as web reinforcement in reinforced concrete columns', *Trans. Japan Concrete Institute.* 4, 1982, 493–500.

17. G. Campione, S. Mindess and G. Zingone, 'Compressive stress-strain behaviour of normal and high-strength carbon-fiber concrete reinforced with steel spirals', *ACI Mater. J.* 96, 1999, 27–34.

18. G. Campione, S. Mindess and G. Zingone, 'Behavior of fiber concrete reinforced with steel spirals under cyclic compressive loading', in A. Azizinamini, D. Darwin

and C. French (eds) *High Strength Concrete*, American Society of Civil Engineers, Reston, VA, 1999, pp. 136–148.

19. G. Campione and S. Mindess, 'Compressive toughness characterization of normal and high-strength fiber concrete reinforced with steel spirals', in N. Banthia, C. Mac-Donald and P. Tatnall (eds) *Structural Applications of Fiber Reinforced Concrete*, SP-182, American Concrete Institute, Farmington Hills, MI, 1999, pp. 141–161.

20. N. Hisabe, I. Yoshitake, H. Tanaka and S. Hamada, 'Mechanical behavior of fiber reinforced concrete elements subjected to pure shearing stress', in *International Workshop on High Perfromance Fiber Reinforced Cementitious Composites in Structural Applications*, Honolulu, 2005, CD-ROM.

21. RILEM TC162-TDF, 'Test and design methods for steel fibre reinforced concrete – σ-ε-design method', *Mater. Struct. (RILEM)*. 33, 2000, 75–81.

22. RILEM TC162-TDF, 'Test and design methods for steel fibre reinforced concrete–σ-ε-design method (final recommendation)', *Mater. Struct. (RILEM)*. 36, 2003, 560–567.

23. European pre-standard, ENV 1992–1991–1991: Eurocode 2: Design of concrete structures – Part 1: general rules and rules for buildings, 1991.

24. European pre-standard, prEN 1992–1991 (2nd draft): Eurocode 2: Design of concrete structures – part 1: general rules and rules for buildings, 2001.

25. UNI U73041440 (draft standard), 2004, 'Design, production and control of steel fibre reinforced structural elements' (reproduced in di Prisco *et al.* Ref. 33).

26. N. Lakshmanan and T.S. Krishnamoorthy, 'Guidelines for design of reinforced concrete structural elements with high strength steel fibres in concrete matrix', in J.J. Biernacki, S.P. Shah, N. Lakshmanan and S. Gopalakrishnan (eds) *High-Performance Cement-Based Concrete Composites*, The American Ceramic Society, Westerville, OH, 2005, pp. 83–92.

27. G.B. Batson and J. Kim 'Steel fibers for shear reinforcement in reinforced concrete beams', in A. Ahmad, M. di Prisco, C. Meyer and S.P. Shah (eds) *Fiber Reinforced Concrete – From Theory to Practice*, Int. Workshop on Advances in Fiber Reinforced Concrete, Starrylink-Editrice, Bergamo, Italy, 2005, pp. 181–192.

28. C.H. Henager and T.J. Doherty, 'Analysis of fibrous reinforced concrete beams', *J. Struct. Eng. ASCE*. 102, 1976, 177–188.

29. ACI Committee 544, *Design Considerations for Steel Fiber Reinforced Concrete*, ACI 544.4R-88, American Concrete Institute, Farmington Hills, MI, 1996.

30. R.N. Swamy and S.A. Al-Ta'an, 'Deformation and ultimate strength in flexure of reinforced concrete beams made with steel fibre concrete', *J. American Concrete Institute*. 78, 1981, 395–405.

31. RILEM TC162-TDF 'Test and design methods for steel fibre reinforced concrete – bending test (final recommendation)', *Mater. Struct. (RILEM)*. 35, 2002, 579–582.

32. M. Teutsch, 'German guideline "steelfibre concrete" ', in A. Ahmad, M. di Prisco, C. Meyer and S.P. Shah (eds) *Fiber Reinforced Concrete – From Theory to Practice*, Int. Workshop on Advances in Fiber Reinforced Concrete, Starrylink-Editrice, Bergamo, Italy, 2005, pp. 23–28.

33. M. di Prisco, G. Toniolo, G.A. Plizzari, S. Cangiano and C. Failla, 'Italian guidelines on SFRC', in A. Ahmad, M. di Prisco, C. Meyer and S.P. Shah (eds) *Fiber Reinforced Concrete – From Theory to Practice*, Int. Workshop on Advances in Fiber Reinforced Concrete, Starrylink-Editrice, Bergamo, Italy, 2005, pp. 39–72.

34. RILEM TC162-TDF, 'Test and design methods for steel fibre reinforced concrete. Design of steel fibre reinforced concrete using the σ-w method: principles and applications', *Mater. Struct. (RILEM).* 35, 2002, 262–278.

35. G. Parra-Montesinos, 'HPFRCC in earthquake-resistant structures: Current knowledge and future trends', in A.E. Naaman and H.W. Reinhardt (eds) *High Performance Fiber Reinforced Cement Composites (HPFRCC4)*, Proc. RILEM PRO 30, RILEM Publications, Bagneux, 2003, pp. 453–472.

36. B. Katzensteiner, S. Mindess, A. Filiatrault and N. Banthia, 'Dynamic tests of steel-fiber reinforced concrete frames', *Concr. Int.* 16 (9), 1994, 57–60.

37. K. Kosa and H. Goda, 'Improvement of the deformation capacity by the use of fibers', in *International Workshop on High Performance Fiber Reinforced Cementitious Composites in Structural Applications*, Honolulu, 2005, CD-ROM.

38. M.H. Harajli, 'Effect of fiber reinforcement on the response of R/Ccolumns under static and seismic loading', in M. di Prisco, R. Felicetti and G.A. Plizzari (eds) *Fiber-Reinforced Concretes, BEFIB 2004*, RILEM Proceedings PRO 39, RILEM Publications, Bagneux, 2004, Vol. 2, pp. 1207–1216.

39. G.J. Parra-Montesinos, B.A. Canbolat and K.Y. Kim, 'Fiber reinforced cement composites for seismic resistant elements with shear-dominated behavior', in M. di Prisco, R. Felicetti and G.A. Plizzari (eds) *Fiber-Reinforced Concretes, BEFIB 2004*, RILEM Proceedings PRO 39, RILEM Publications, Bagneux, 2004, Vol. 2, pp. 1237–1246.

40. D. Vachon and B. Massicotte, 'Seismic retrofitting of rectangular bridge piers with FRC jackets', in M. di Prisco, R. Felicetti and G.A. Plizzari (eds) *Fiber-Reinforced Concretes, BEFIB 2004*, RILEM Proceedings PRO 39, RILEM Publications, Bagneux, 2004, Vol. 2, pp. 1247–1256.

41. L. Vandewalle, 'Design recommendations – RILEM TC162-TDF: σ-ε-design method for steel fibre reinforced concrete', in A.E. Naaman and H.W. Reinhardt (eds) *High Performance Fiber Reinforced Cement Composites (HPFRCC4)*, Proc. RILEM PRO 30, RILEM Publications, Bagneux, 2003, pp. 531–541.

42. Vandewalle, L. (2004), Test and design method for steel fibre reinforced concrete based on the σ-ε-relation', in A.M. Brandt (ed.) *Some Aspects of Design and Application of High Performance Cement Based Materials*, AMAS Lecture Notes 18, Institute of Fundamental Technological Research, Warsaw, 2004, pp. 135–190.

43. G.L. Balazs and I. Kovacs, 'Effect of steel fibres on the cracking behaviour of RC members', in M. di Prisco, R. Felicetti and G.A. Plizzari (eds) *Fiber-Reinforced Concretes, BEFIB 2004*, RILEM Proceedings PRO 39, RILEM Publications, Bagneux, 2004, Vol. 2, pp. 1007–1016.

44. H.M. Westergaard, 'Stresses in concrete pavements computed by theoretical analysis', *Public Roads.* 7, 1926 (2), 25–35.

45. H.M. Westergaard, 'New formulas for stresses in concrete pavements of airfields', *ASCE Transactions.* 113, 1948, 425–439.

46. A. Losberg, *Structurally Reinforced Concrete Pavements*, PhD Thesis, Chalmers Tekniska Hogskola, Göteborg, Sweden, 1960.

47. G.G. Meyerhof, 'Load carrying capacity of concrete pavements', *ASCE J. Soil Mechanics and Foundation Division.* 88, SM3, 1962, 89–116.

48. R.A. Baumann and F.E. Weisgerber, 'Yield-line analysis of slabs-on-grade', *ASCE J. Struct. Eng.* 109, 1983, 1553–1568.

49. M. Hetenyi, *Beams on Elastic Foundations*, University of Michigan Press, Ann Arbor, MI, 1948.

50. A. Meda, G.A. Plizzari, L. Sorelli and B. Rossi, in *Concrete Structures: The Challenge of Creativity*, fib Symposium, Avignon, Association Francaise de Genie Civil. Bagneux, France, 2004.

51. Concrete Society, *Concrete Industrial Ground Floors: A Guide to Design and Construction*. Technical Report 34, Concrete Society, Berkshire, UK, 2003.

52. JCI, *Method of Tests for Flexural Strength and Flexural Toughness of Fiber Reinforced Concrete*, JCI Standards for Test Methods of Fiber Reinforced Concrete, JCI Standard SF-4, Japan Concrete Institute, 1984.

53. S. Altoubat, J.R. Roesler and K.-A. Rieder, 'Flexural capacity of synthetic fiber reinforced concrete slabs on ground based on beam toughness results', in M. di Prisco, R. Felicetti and G.A. Plizzari (eds) *Fiber-Reinforced Concretes, BEFIB 2004*, RILEM Proceedings PRO 39, RILEM Publications, Bagneux, 2004, Vol. 2, pp. 1063–1072.

54. Grace Construction Products, *Strux® 90/40 Slab-on- Ground Design and Specification*, Grace Construction Products, Cambridge, MA, 2004.

55. Bekaert, *Dramix® : Steel Fibre Reinforced Industrial Floor Design in Accordance with the Concrete Society TR34*, N.V. Bekaert, S.A. Zwevegem, Belgium, 1996.

56. R.F. Zollo and C.D. Hays, 'Fibers versus WWF as non-structural slab reinforcement', *Concr. Int.* 13 (11), 1991, 50–55.

57. J.-F. Trottier, M. Mahoney and D. Forgeron, 'Can synthetic fibers replace welded-wire fabric in slabs-on-ground?' *Concr. Int.* 24 (11), 2002, 59–68.

58. C.L. Roberts-Wollmann, M. Guirola and W.S. Easterling, 'Strength and performance of fiber-reinforced concrete composite slabs', *ASCE J. Struct. Eng.* 130, 2004, 520–528.

59. L. Sorelli, A. Meda, G.A. Plizzari and B. Rossi, 'Experimental investigation on slabs on grade : steel fibers vs. conventional reinforcement', in M. di Prisco, R. Felicetti and G.A. Plizzari (eds) *Fiber-Reinforced Concretes, BEFIB 2004*, RILEM Proceedings PRO 39, RILEM Publications, Bagneux, 2004, Vol. 2, pp. 1083–1092.

60. T. Voigt, V.K. Bui and S.P. Shah, 'Drying shrinkage of concrete reinforced with fibres and welded-wire fabric', *ACI Mater. J.* 101, 2004, 233–241.

61. H.S. Armelin, N. Banthia and S. Mindess, 'Kinematics of dry-mix shotcrete', *ACI Mater. J.* 3, 1999, 283–290.

62. ACI Committee 506, *Committee Report on Fiber Reinforced Shotcrete*, ACI 506.1R, American Concrete Institute, Farmington Hills, MI, 1998.

63. N.-O. Sandell, M. Dir and B. Westerdahl, 'System BESAB for high strength steel fibre reinforced shotcrete', in S.P. Shah and A. Skarendahl (eds) *Steel Fibre Concrete*, US–Sweden Joint Seminar, Swedish Cement and Concrete Research Institute, Elsevier Applied Science Publishers, Barking, Essex, Stockholm, 1985, pp. 25–39.

64. D.R. Morgan and D.N. Mowat, 'A comparative evaluation of plain mesh and steel fiber reinforced shotcrete', in G.C. Hoff (ed.) *Fiber Reinforced Concrete – International Symposium*, ACI SP-81, American Concrete Institute, Farmington Hills, MI, 1984, pp. 307–324.

65. N. Banthia 'Fibre reinforced shotcrete: Issues, challenges and opportunities', in H.W. Reinhardt and A.E. Naaman (eds) *High Performance Fiber Reinforced Composites (HPFRCC 3)*, RILEM Proceedings PRO 6, RILEM Publications, Bagneux, 1999, pp. 161–170.

66. EFNARC, *European Specification for Sprayed Concrete*, EFNARC, Farnham, Surrey, 1996.

67. F. Papworth, 'Design guidelines for the use of fiber-reinforced shotcrete in ground support', *Shotcrete.* 4 (2), 2002, 16–21.
68. V. Bindiganavile and N. Banthia, 'Rebound in dry-mix shotcrete: influence of type of mineral admixture', *ACI Mater. J.* 97, 2000, 115–119.
69. N. Banthia, J.-F. Trottier, D. Beaupre and D. Wood, 'Influence of fibre geometry in steel fiber reinforced wet-mix shotcrete', *Concr. Int.* 16 (6), 1994, 27–32.
70. H.S. Armelin and N. Banthia, 'Steel fiber rebound in shotcrete', *Concr. Int.* 20 (9), 1998, 74–79.
71. N. Banthia, P. Gupta, C. Yan and D.R. Morgan, 'How tough is fiber reinforced shotcrete? Part I, beam tests', *Concr. Int.* 21 (6), 1999, 59–62.
72. V. Wetzig and R. Weiss, 'Fibre reinforced shotcrete for long tunnel projects in Switzerland', in M. di Prisco, R. Felicetti and G.A. Plizzari (eds) *Fiber-Reinforced Concretes, BEFIB 2004*, RILEM Proceedings PRO 39, RILEM Publications, Bagneux, 2004, Vol. 1, pp. 545–552.
73. J. Sustersic, V. Jovicic, A. Zajc and R. Ercegovic, 'Evaluation of improvement in the bearing capacity of fibre reinforced shotcrete tunnel lining', in M. di Prisco, R. Felicetti and G.A. Plizzari (eds) *Fiber-Reinforced Concretes, BEFIB 2004*, RILEM Proceedings PRO 39, RILEM Publications, Bagneux, 2004, Vol. 2, pp. 985–994.
74. ACI Committee 506, *Guide to Certification of Shotcrete Nozzlemen*, ACI 506.3R, American Concrete Institute, Farmington Hills, MI, 1991.
75. P. Gilbride, D.R. Morgan and T.W. Bremner, 'Deterioration and rehabilitation of berth faces in tidal zones at the Port of Saint John', in V.M. Malhotra (ed.) *Concrete in Marine Environment*, ACI SP-109, American Concrete Institute, Farmington Hills, MI, 1988, pp. 199–227.
76. D.R. Morgan, L. Rich and A. Lobo, 'About face – repair at Port of Montreal', *Concr. Int.* 20 (9), 1998, 66–73.
77. Y.Y. Kim, H.J. Kong and V.C. Li, 'Development of sprayable engineered cementitious composites', in A.E. Naaman and H.W. Reinhardt (eds) *High Performance Fiber Reinforced Cement Composites (HPFRCC4)*, Proc. RILEM PRO 30, RILEM Publications, Bagneux, 2003, pp. 233–243.
78. F. Pasini, T. Garcia, R. Gettu and L. Agullp, 'Experimental study of the properties of flowable fibre reinforced concretes', in M. di Prisco, R. Felicetti and G.A. Plizzari (eds) *Fiber-Reinforced Concretes, BEFIB 2004*, RILEM Proceedings PRO 39, RILEM Publications, Bagneux, 2004, Vol. 1, pp. 279–288.
79. S. Grunewald, *Performance-Based Design of Self-Compacting Fibre Reinforced Concrete*. PhD Thesis, Technische Universiteit Delft, The Netherlands, 2004.
80. J. Ambroise, S. Rols and J. Pera, 'Properties of self-levelling concrete reinforced by steel fibres', in H.W. Reinhardt and A.E. Naaman (eds) *High Performance Fiber Reinforced Cement Composites (HPFRCC3)*, RILEM Proceedings PRO6, RILEM Publications, Bagneux, 1999, pp. 9–17.
81. E.N.B. Pereira, J.A.O. Barros, A.F. Ribeiro and A. Camoes, 'Post-cracking behaviour of self-compacting steel fibre reinforced concrete', in M. di Prisco, R. Felicetti and G.A. Plizzari (eds) *Fiber-Reinforced Concretes, BEFIB 2004*, RILEM Proceedings PRO 39, RILEM Publications, Bagneux, 2004, Vol. 2, pp. 1371–1380.
82. I. Markovic, J.C. Walraven and J.G.M. van Mier (2003), Development of high performance hybrid fibre concrete', in A.E. Naaman and H.W. Reinhardt (eds) *High Performance Fiber Reinforced Cement Composites (HPFRCC4)*, Proc. RILEM PRO 30, RILEM Publications, Bagneux, 2003, pp. 277–300.

83. J.-L. Granju, V. Sabathier, M. Alcantara, G. Pons and M. Mouret, 'Hybrid fibre rein-forcement of ordinary or self-compacting concrete', in M. di Prisco, R. Felicetti and G.A. Plizzari (eds) *Fiber-Reinforced Concretes, BEFIB 2004*, RILEM Proceedings PRO 39, RILEM Publications, Bagneux 2004, Vol. 2, pp. 1311–1320.

84. M. Borsa M. Molfetta, G.L. Guerrini, S. Lagomarsino and S. Podesta, 'First studies on a lightweight aggregate self compacting fibre reinforced concrete for structural applications', in M. di Prisco, R. Felicetti and G.A. Plizzari (eds) *Fiber-Reinforced Concretes, BEFIB 2004*, RILEM Proceedings PRO 39, RILEM Publications, Bagneux, 2004, Vol. 2, pp. 1291–1300.

85. R. Walter, H. Stang, N.J. Gimsing and J.F. Olesen, 'High performance composite bridge decks using SCSFRC', in A.E. Naaman and H.W. Reinhardt (eds) *High Performance Fiber Reinforced Cement Composites (HPFRCC4)*, Proc. RILEM PRO 30, RILEM Publications, Bagneux, 2003, pp. 495–504.

86. P. Magnetti, C. Failla and F. Pasini, 'Application of SFRC to precast buildings', in M. di Prisco, R. Felicetti and G.A. Plizzari (eds) *Fiber-Reinforced Concretes, BEFIB 2004*, RILEM Proceedings PRO 39, RILEM Publications, Bagneux, 2004, pp. 1153–1162.

87. G.F. True, 'Glass fibre reinforced cement permanent formwork', *Concrete*. 19(2), 1985, 31–33.

88. H.W. Reinhardt, 'Integral formwork panels made of GFRC', in A. Peled, S.P. Shah and N. Banthia (eds) *High Performance Fiber-Reinforced Concrete Thin Sheet Products*, SP-190, American Concrete Institute, Farmington Hills, MI, 2000, pp. 77–95.

89. N. Banthia, C. Yan, A.A. Mufti and B. Bakht, 'CFRC permanent formwork for steel free bridge deck', in A. Peled, S.P. Shah and N. Banthia (eds) *High Performance Fiber-Reinforced Concrete Thin Sheet Products*, SP-190, American Concrete Institute, Farmington Hills, MI, 2000, pp. 21–27.

90. G.T. Gibert and J.R. Mott, 'Filament winding of glassfibre reinforced cement utility poles and other products', in M. di Prisco, R. Felicetti and G.A. Plizzari (eds) *Fiber-Reinforced Concretes, BEFIB 2004*, RILEM Proceedings PRO 39, RILEM Publications, Bagneux, 2004, 1123–1132.

91. A. Bentur and S.A.S Akers, 'Thin sheet cementitious composites', in N. Banthia, A. Bentur and A. Mufti (eds) *Fiber Reinforced Concrete: Present and Future*, Canadian Society for Civil Engineers, Montreal, 1998, pp. 20–45.

92. R.G. Oesterle, D.M. Schultz and J.D. Gilkin 'Design considerations for GFRC facades', in J. Daniels and S.P. Shah (eds) *Thin Sheet Fiber Reinforced Concrete and Ferrocement*, SP-124, American Concrete Institute, Farmington Hills, MI, 1990, pp. 157–182.

93. N.W. Hansen, J.J. Roller and T.L. Weinerman, 'Manufacture and installation of GFRC facades', in J. Daniels and S.P. Shah (eds) *Thin Sheet Fiber Reinforced Concrete and Ferrocement*, SP-124, American Concrete Institute, Farmington Hills, MI, 1990, pp. 183–213.

94. M.W. Fordyce and R.G. Wodehouse, *GRC and Buildings*, Butterworths, England, 1983.

95. PCI, *Recommended Practice for Glass Fiber Reinforced Concrete Panels*, PCI Committee on Glass Fiber Reinforced Concrete Panels, Precast/Prestressed Concrete Institute, Chicago, IL, 1987.

96. Anon, *Design Guide: Glass Fibre Reinforced Cement*, Pilkington Brothers, L.P.O., UK, CemFIL GRC Technical Data Manual, 1984.

97. H. Baum and A. Bentur, 'Fiber reinforced cementitious materials for lightweight construction: development of criteria for evaluation of their long term performance', Research Report, National Building Research Institute, Technion, Haifa, Israel, 1994.

98. E. Haeussler, 'Considerations on warping and cracking in sandwich panels', *Betonwerk+Fertigteil-Technik*. 1984, 774–780.

99. K.N. Quinn, 'Design, performance and durability of GRC panels', in *Design Life of Buildings*, Thomas Telford, London, 1985, pp. 139–156.

100. G.R. Williamson, 'Evaluation of glass fibre reinforced concrete panels for use in military construction', in S. Diamond (ed.) Proc. Durability of Glass Fiber Reinforced Concrete Symp., Prestressed Concrete Institute, Chicago, IL, 1985, pp. 54–63.

101. PCI, *Manual for Quality Control for Plant and Production of Glass Fiber Reinforced Concrete Products*, Precast/Prestressed Concrete Institute, Chicago, IL, 1991.

Index

eBooks – at www.eBookstore.tandf.co.uk

A library at your fingertips!

eBooks are electronic versions of printed books. You can store them on your PC/laptop or browse them online.

They have advantages for anyone needing rapid access to a wide variety of published, copyright information.

eBooks can help your research by enabling you to bookmark chapters, annotate text and use instant searches to find specific words or phrases. Several eBook files would fit on even a small laptop or PDA.

NEW: Save money by eSubscribing: cheap, online access to any eBook for as long as you need it.

Annual subscription packages

We now offer special low-cost bulk subscriptions to packages of eBooks in certain subject areas. These are available to libraries or to individuals.

For more information please contact webmaster.ebooks@tandf.co.uk

We're continually developing the eBook concept, so keep up to date by visiting the website.

www.eBookstore.tandf.co.uk

9 780367 446239